INSTRUCTOR'S SOLUTIONS MANUAL

to accompany

LINEAR ALGEBRA with APPLICATIONS

Seventh Edition

by

GARETH WILLIAMS

Prepared by

W. J. Mourant and Gareth Williams

World Headquarters
Jones & Bartlett Learning
40 Tall Pine Drive
Sudbury, MA 01776
978-443-5000
info@jblearning.com
www.jblearning.com

Jones & Bartlett Learning Canada
6339 Ormindale Way
Mississauga, Ontario L5V 1J2
Canada

Jones & Bartlett Learning International
Barb House, Barb Mews
London W6 7PA
United Kingdom

Books and products by Jones & Bartlett Learning are available through most bookstores and online booksellers. To contact Jones & Bartlett Learning directly, call 800-832-0034, fax 978-443-8000, or visit our website, www.jblearning.com.

Substantial discounts on bulk quantities of Jones & Bartlett Learning publications are available to corporations, professional associations, and other qualified organizations. For details and specific discount information, contact the special sales department at Jones & Bartlett Learning via the above contact information or send an email to specialsales@jblearning.com.

ISBN-13: 9780763796167
ISBN-10: 0763796166

Production Credits
Publisher: David Pallai
Acquisitions Editor: Timothy Anderson
Editorial Assistant: Melissa Potter
Senior Production Editor: Katherine Crighton
Senior Marketing Manager: Andrea DeFronzo
Associate Marketing Manager: Lindsay Ruggiero
V.P., Manufacturing and Inventory Control: Therese Connell
Cover Design: Kristin E. Parker
Cover and Title Page Image: © 2009 Santiago Calatrava/Artists Rights Society (ARS), New York/VEGAP, Madrid. Photo provided © Antony Mcaulay/Dreamstime.com
Printing and Binding: Malloy, Inc.
Cover Printing: Malloy, Inc.

6048
Printed in the United States of America
14 13 12 11 10   10 9 8 7 6 5 4 3 2 1

CONTENTS

---

* Sections marked with an asterisk are optional. The instructor can use these sections to build around the core material to give the course the desired flavor.

Part 2 Vector Spaces

Part 3 Numerical Linear Algebra

Preface

This Instructor Solutions Manual contains worked-out solutions to all exercises in the text *Linear Algebra with Applications, 7th Edition,* by Gareth Williams. All references in this manual are to chapters, sections, and exercises in the text.

Chapter 1

Exercise Set 1.1

1. (a) 3×3 (b) 3×2 (c) 2×4 (d) 3×1 (e) 3×5 (f) 1×4

2. 1, 4, 9, −1, 3, 8

3. 4, 5, 6, 7, 2, 3

4. $\begin{bmatrix} 1 & 0 & 0 & 0 \\ 0 & 1 & 0 & 0 \\ 0 & 0 & 1 & 0 \\ 0 & 0 & 0 & 1 \end{bmatrix}$

5. (a) $\begin{bmatrix} 1 & 3 \\ 2 & -5 \end{bmatrix}$ and $\begin{bmatrix} 1 & 3 & 7 \\ 2 & -5 & -3 \end{bmatrix}$ (b) $\begin{bmatrix} 5 & 2 & -4 \\ 1 & 3 & 6 \\ 4 & 6 & -9 \end{bmatrix}$ and $\begin{bmatrix} 5 & 2 & -4 & 8 \\ 1 & 3 & 6 & 4 \\ 4 & 6 & -9 & 7 \end{bmatrix}$

   (c) $\begin{bmatrix} -1 & 3 & -5 \\ 2 & -2 & 4 \\ 1 & 3 & 0 \end{bmatrix}$ and $\begin{bmatrix} -1 & 3 & -5 & -3 \\ 2 & -2 & 4 & 8 \\ 1 & 3 & 0 & 6 \end{bmatrix}$ (d) $\begin{bmatrix} 5 & 4 \\ 2 & -8 \\ 1 & 2 \end{bmatrix}$ and $\begin{bmatrix} 5 & 4 & 9 \\ 2 & -8 & -4 \\ 1 & 2 & 3 \end{bmatrix}$

   (e) $\begin{bmatrix} 5 & 2 & -4 \\ 0 & 4 & 3 \\ 1 & 0 & -1 \end{bmatrix}$ and $\begin{bmatrix} 5 & 2 & -4 & 8 \\ 0 & 4 & 3 & 0 \\ 1 & 0 & -1 & 7 \end{bmatrix}$

   (f) $\begin{bmatrix} -1 & 3 & -9 \\ 1 & 0 & -4 \\ 1 & 8 & 0 \end{bmatrix}$ and $\begin{bmatrix} -1 & 3 & -9 & -4 \\ 1 & 0 & -4 & 11 \\ 1 & 8 & 0 & 1 \end{bmatrix}$ (g) $\begin{bmatrix} 1 & 0 & 0 \\ 0 & 1 & 0 \\ 0 & 0 & 1 \end{bmatrix}$ and $\begin{bmatrix} 1 & 0 & 0 & -3 \\ 0 & 1 & 0 & 12 \\ 0 & 0 & 1 & 8 \end{bmatrix}$

   (h) $\begin{bmatrix} -4 & 2 & -9 & 1 \\ 1 & 6 & -8 & -7 \\ 0 & -1 & 3 & -5 \end{bmatrix}$ and $\begin{bmatrix} -4 & 2 & -9 & 1 & -1 \\ 1 & 6 & -8 & -7 & 15 \\ 0 & -1 & 3 & -5 & 0 \end{bmatrix}$

6. (a) $\begin{aligned} x_1 + 2x_2 &= 3 \\ 4x_1 + 5x_2 &= 6 \end{aligned}$ (b) $\begin{aligned} 7x_1 + 9x_2 &= 8 \\ 6x_1 + 4x_2 &= -3 \end{aligned}$ (c) $\begin{aligned} x_1 + 9x_2 &= -3 \\ 5x_1 &= 2 \end{aligned}$

(d) $8x_1 + 7x_2 + 5x_3 = -1$
$4x_1 + 6x_2 + 2x_3 = 4$
$9x_1 + 3x_2 + 7x_3 = 6$

(e) $2x_1 - 3x_2 + 6x_3 = 4$
$7x_1 - 5x_2 - 2x_3 = 3$
$2x_2 + 4x_3 = 0$

(f) $-2x_2 = 4$
$5x_1 + 7x_2 = -3$
$6x_1 = 8$

(g) $x_1 = 3$
$x_2 = 8$
$x_3 = 4$

(h) $x_1 + 2x_2 - x_3 = 6$
$x_2 + 4x_3 = 5$
$x_3 = -2$

7. (a) $\begin{bmatrix} 1 & 3 & -2 & 0 \\ 1 & 2 & -3 & 6 \\ 8 & 3 & 2 & 5 \end{bmatrix}$ (b) $\begin{bmatrix} 2 & 7 & 5 & 1 \\ 0 & -8 & 4 & 3 \\ 3 & -5 & 8 & 9 \end{bmatrix}$ (c) $\begin{bmatrix} 1 & 2 & 3 & -1 \\ 0 & 3 & 10 & 0 \\ 0 & -8 & -1 & -1 \end{bmatrix}$

(d) $\begin{bmatrix} 1 & 0 & -1 & -6 \\ 0 & 1 & 2 & 1 \\ 0 & 0 & 11 & -1 \end{bmatrix}$ (e) $\begin{bmatrix} 1 & 0 & 0 & -23 \\ 0 & 1 & 0 & 17 \\ 0 & 0 & 1 & 5 \end{bmatrix}$ (f) $\begin{bmatrix} 1 & 0 & 2 & 7 \\ 0 & 1 & 5 & -3 \\ 0 & 0 & 1 & -4 \end{bmatrix}$

8. (a) Create zeros below the leading 1 in the first column.
x_1 is eliminated from all equations except the first.

(b) Normalize the (2,2) element, i.e., make the (2,2) element 1. This becomes a leading 1.
It is now possible to have x_2 in the second equation with coefficient 1.

(c) Need to have the leading 1 in row 2 to the left of leading 1 in row 3.
The second equation now contains an x_2 term.

(d) Create zeros above and below the leading 1 in row 2.
x_2 is eliminated from all equations except the second.

9. (a) Create zeros above the leading 1 in column 3.
x_3 is eliminated from all equations except the third.

(b) Need to have the leading 1 in row 1 to the left of leading 1s in other rows.
It is now possible to have x_1 in Equation 1 with leading coefficient 1.

(c) Normalize the (3,3) element, i.e., make the (3,3) element 1. This becomes a leading 1.
The coefficient of x_3 in the third equation becomes 1.

(d) Create zeros above the leading 1 in column 3.
x_3 is eliminated from all equations except the third.

10. (a) $\begin{bmatrix} 1 & -2 & -8 \\ 2 & -3 & -11 \end{bmatrix} \underset{R2+(-2)R1}{\approx} \begin{bmatrix} 1 & -2 & -8 \\ 0 & 1 & 5 \end{bmatrix} \underset{R1+(2)R2}{\approx} \begin{bmatrix} 1 & 0 & 2 \\ 0 & 1 & 5 \end{bmatrix}$,

so the solution is $x_1 = 2$ and $x_2 = 5$.

(b) $\begin{bmatrix} 2 & 2 & 4 \\ 3 & 2 & 3 \end{bmatrix} \underset{(1/2)R1}{\approx} \begin{bmatrix} 1 & 1 & 2 \\ 3 & 2 & 3 \end{bmatrix} \underset{R2+(-3)R1}{\approx} \begin{bmatrix} 1 & 1 & 2 \\ 0 & -1 & -3 \end{bmatrix} \underset{(-1)R2}{\approx} \begin{bmatrix} 1 & 1 & 2 \\ 0 & 1 & 3 \end{bmatrix}$

$\underset{R1+(-1)R2}{\approx} \begin{bmatrix} 1 & 0 & -1 \\ 0 & 1 & 3 \end{bmatrix}$, so the solution is $x_1 = -1$, $x_2 = 3$.

(c) $\begin{bmatrix} 1 & 0 & 1 & 3 \\ 0 & 2 & -2 & -4 \\ 0 & 1 & -2 & 5 \end{bmatrix} \underset{(1/2)R2}{\approx} \begin{bmatrix} 1 & 0 & 1 & 3 \\ 0 & 1 & -1 & -2 \\ 0 & 1 & -2 & 5 \end{bmatrix} \underset{R3+(-1)R2}{\approx} \begin{bmatrix} 1 & 0 & 1 & 3 \\ 0 & 1 & -1 & -2 \\ 0 & 0 & -1 & 7 \end{bmatrix}$

$\underset{(-1)R3}{\approx} \begin{bmatrix} 1 & 0 & 1 & 3 \\ 0 & 1 & -1 & -2 \\ 0 & 0 & 1 & -7 \end{bmatrix} \underset{\substack{R1+(-1)R3 \\ R2+R3}}{\approx} \begin{bmatrix} 1 & 0 & 0 & 10 \\ 0 & 1 & 0 & -9 \\ 0 & 0 & 1 & -7 \end{bmatrix}$,

so the solution is $x_1 = 10$, $x_2 = -9$, $x_3 = -7$.

(d) $\begin{bmatrix} 1 & 1 & 3 & 6 \\ 1 & 2 & 4 & 9 \\ 2 & 1 & 6 & 11 \end{bmatrix} \underset{\substack{R2+(-1)R1 \\ R3+(-2)R1}}{\approx} \begin{bmatrix} 1 & 1 & 3 & 6 \\ 0 & 1 & 1 & 3 \\ 0 & -1 & 0 & -1 \end{bmatrix} \underset{\substack{R1+(-1)R2 \\ R3+R2}}{\approx} \begin{bmatrix} 1 & 0 & 2 & 3 \\ 0 & 1 & 1 & 3 \\ 0 & 0 & 1 & 2 \end{bmatrix}$

$\underset{\substack{R1+(-2)R3 \\ R2+(-1)R3}}{} \begin{bmatrix} 1 & 0 & 0 & -1 \\ 0 & 1 & 0 & 1 \\ 0 & 0 & 1 & 2 \end{bmatrix}$, so the solution is $x_1 = -1$, $x_2 = 1$, $x_3 = 2$.

(e) $\begin{bmatrix} 1 & -1 & 3 & 3 \\ 2 & -1 & 2 & 2 \\ 3 & 1 & -2 & 3 \end{bmatrix} \underset{\substack{R2+(-2)R1 \\ R3+(-3)R1}}{\approx} \begin{bmatrix} 1 & -1 & 3 & 3 \\ 0 & 1 & -4 & -4 \\ 0 & 4 & -11 & -6 \end{bmatrix} \underset{\substack{R1+R2 \\ R3+(-4)R2}}{\approx} \begin{bmatrix} 1 & 0 & -1 & -1 \\ 0 & 1 & -4 & -4 \\ 0 & 0 & 5 & 10 \end{bmatrix}$

$\underset{(1/5)R3}{\approx} \begin{bmatrix} 1 & 0 & -1 & -1 \\ 0 & 1 & -4 & -4 \\ 0 & 0 & 1 & 2 \end{bmatrix} \underset{\substack{R1+R3 \\ R2+(4)R3}}{\approx} \begin{bmatrix} 1 & 0 & 0 & 1 \\ 0 & 1 & 0 & 4 \\ 0 & 0 & 1 & 2 \end{bmatrix}$,

so the solution is $x_1 = 1$, $x_2 = 4$, $x_3 = 2$.

(f) $\begin{bmatrix} -1 & 1 & -1 & -2 \\ 3 & 1 & 1 & 10 \\ 4 & 2 & 3 & 14 \end{bmatrix} \underset{(-1)R1}{\approx} \begin{bmatrix} 1 & -1 & 1 & 2 \\ 3 & 1 & 1 & 10 \\ 4 & 2 & 3 & 14 \end{bmatrix} \underset{\substack{R2+(-3)R1 \\ R3+(-4)R1}}{\approx} \begin{bmatrix} 1 & -1 & 1 & 2 \\ 0 & 4 & -2 & 4 \\ 0 & 6 & -1 & 6 \end{bmatrix}$

$\underset{(1/4)R2}{\approx} \begin{bmatrix} 1 & -1 & 1 & 2 \\ 0 & 1 & -1/2 & 1 \\ 0 & 6 & -1 & 6 \end{bmatrix} \underset{\substack{R1+R2 \\ R3+(-6)R2}}{\approx} \begin{bmatrix} 1 & 0 & 1/2 & 3 \\ 0 & 1 & -1/2 & 1 \\ 0 & 0 & 2 & 0 \end{bmatrix} \underset{(1/2)R3}{\approx} \begin{bmatrix} 1 & 0 & 1/2 & 3 \\ 0 & 1 & -1/2 & 1 \\ 0 & 0 & 1 & 0 \end{bmatrix}$

$\underset{\substack{R1+(-1/2)R3 \\ R2+(1/2)R3}}{\approx} \begin{bmatrix} 1 & 0 & 0 & 3 \\ 0 & 1 & 0 & 1 \\ 0 & 0 & 1 & 0 \end{bmatrix}$, so the solution is $x_1 = 3$, $x_2 = 1$, $x_3 = 0$.

11. (a) $\begin{bmatrix} 1 & 2 & 3 & 14 \\ 2 & 5 & 8 & 36 \\ 1 & -1 & 0 & -4 \end{bmatrix} \underset{\substack{R2+(-2)R1 \\ R3+(-1)R1}}{\approx} \begin{bmatrix} 1 & 2 & 3 & 14 \\ 0 & 1 & 2 & 8 \\ 0 & -3 & -3 & -18 \end{bmatrix} \underset{\substack{R1+(-2)R2 \\ R3+(3)R2}}{\approx} \begin{bmatrix} 1 & 0 & -1 & -2 \\ 0 & 1 & 2 & 8 \\ 0 & 0 & 3 & 6 \end{bmatrix}$

$\underset{(1/3)R3}{\approx} \begin{bmatrix} 1 & 0 & -1 & -2 \\ 0 & 1 & 2 & 8 \\ 0 & 0 & 1 & 2 \end{bmatrix} \underset{\substack{R1+R3 \\ R2+(-2)R3}}{\approx} \begin{bmatrix} 1 & 0 & 0 & 0 \\ 0 & 1 & 0 & 4 \\ 0 & 0 & 1 & 2 \end{bmatrix}$,

so the solution is $x_1 = 0$, $x_2 = 4$, $x_3 = 2$.

(b) $\begin{bmatrix} 1 & -1 & -1 & -1 \\ -2 & 6 & 10 & 14 \\ 2 & 1 & 6 & 9 \end{bmatrix} \underset{\substack{R2+(2)R1 \\ R3+(-2)R1}}{\approx} \begin{bmatrix} 1 & -1 & -1 & -1 \\ 0 & 4 & 8 & 12 \\ 0 & 3 & 8 & 11 \end{bmatrix} \underset{(1/4)R2}{\approx} \begin{bmatrix} 1 & -1 & -1 & -1 \\ 0 & 1 & 2 & 3 \\ 0 & 3 & 8 & 11 \end{bmatrix}$

$\underset{\substack{R1+R2 \\ R3+(-3)R2}}{\approx} \begin{bmatrix} 1 & 0 & 1 & 2 \\ 0 & 1 & 2 & 3 \\ 0 & 0 & 2 & 2 \end{bmatrix} \underset{(1/2)R3}{\approx} \begin{bmatrix} 1 & 0 & 1 & 2 \\ 0 & 1 & 2 & 3 \\ 0 & 0 & 1 & 1 \end{bmatrix} \underset{\substack{R1+(-1)R3 \\ R2+(-2)R3}}{\approx} \begin{bmatrix} 1 & 0 & 0 & 1 \\ 0 & 1 & 0 & 1 \\ 0 & 0 & 1 & 1 \end{bmatrix}$,

so the solution is $x_1 = 1$, $x_2 = 1$, $x_3 = 1$.

(c) $\begin{bmatrix} 2 & 2 & -4 & 14 \\ 3 & 1 & 1 & 8 \\ 2 & -1 & 2 & -1 \end{bmatrix} \underset{(1/2)R1}{\approx} \begin{bmatrix} 1 & 1 & -2 & 7 \\ 3 & 1 & 1 & 8 \\ 2 & -1 & 2 & -1 \end{bmatrix} \underset{\substack{R2+(-3)R1 \\ R3+(-2)R1}}{\approx} \begin{bmatrix} 1 & 1 & -2 & 7 \\ 0 & -2 & 7 & -13 \\ 0 & -3 & 6 & -15 \end{bmatrix}$

Let us swap rows 2 and rows 2 and 3 to avoid awkward fractions at the next step.

$$\underset{R2 \Leftrightarrow R3}{\approx}\begin{bmatrix}1 & 1 & -2 & 7\\0 & -3 & 6 & -15\\0 & -2 & 7 & -13\end{bmatrix}\underset{(-1/3)R2}{\approx}\begin{bmatrix}1 & 1 & -2 & 7\\0 & 1 & -2 & 5\\0 & -2 & 7 & -13\end{bmatrix}$$

$$\underset{\substack{R1+(-1)R2\\R3+(2)R2}}{\approx}\begin{bmatrix}1 & 0 & 0 & 2\\0 & 1 & -2 & 5\\0 & 0 & 3 & -3\end{bmatrix}\underset{(1/3)R3}{\approx}\begin{bmatrix}1 & 0 & 0 & 2\\0 & 1 & -2 & 5\\0 & 0 & 1 & -1\end{bmatrix}\underset{R2+(2)R3}{\approx}\begin{bmatrix}1 & 0 & 0 & 2\\0 & 1 & 0 & 3\\0 & 0 & 1 & -1\end{bmatrix},$$

so the solution is $x_1 = 2$, $x_2 = 3$, $x_3 = -1$.

(d) $$\begin{bmatrix}0 & 2 & 4 & 8\\2 & 2 & 0 & 6\\1 & 1 & 1 & 5\end{bmatrix}\underset{R1 \Leftrightarrow R2}{\approx}\begin{bmatrix}2 & 2 & 0 & 6\\0 & 2 & 4 & 8\\1 & 1 & 1 & 5\end{bmatrix}\underset{(1/2)R1}{\approx}\begin{bmatrix}1 & 1 & 0 & 3\\0 & 2 & 4 & 8\\1 & 1 & 1 & 5\end{bmatrix}$$

$$\underset{R3+(-1)R1}{\approx}\begin{bmatrix}1 & 1 & 0 & 3\\0 & 2 & 4 & 8\\0 & 0 & 1 & 2\end{bmatrix}\underset{(1/2)R2}{\approx}\begin{bmatrix}1 & 1 & 0 & 3\\0 & 1 & 2 & 4\\0 & 0 & 1 & 2\end{bmatrix}\underset{R1+(-1)R2}{\approx}\begin{bmatrix}1 & 0 & -2 & -1\\0 & 1 & 2 & 4\\0 & 0 & 1 & 2\end{bmatrix}$$

$$\underset{\substack{R1+(2)R3\\R2+(-2)R3}}{\approx}\begin{bmatrix}1 & 0 & 0 & 3\\0 & 1 & 0 & 0\\0 & 0 & 1 & 2\end{bmatrix},$$ so the solution is $x_1 = 3$, $x_2 = 0$, $x_3 = 2$.

(e) $$\begin{bmatrix}1 & 0 & -1 & 3\\-1 & 0 & 2 & -8\\3 & 1 & -1 & 0\end{bmatrix}\underset{\substack{R2+R1\\R2+(-3)R1}}{\approx}\begin{bmatrix}1 & 0 & -1 & 3\\0 & 0 & 1 & -5\\0 & 1 & 2 & -9\end{bmatrix}\underset{R2 \Leftrightarrow R3}{\approx}\begin{bmatrix}1 & 0 & -1 & 3\\0 & 1 & 2 & -9\\0 & 0 & 1 & -5\end{bmatrix}$$

$$\underset{\substack{R1+R3\\R2+(-2)R3}}{\approx}\begin{bmatrix}1 & 0 & 0 & -2\\0 & 1 & 0 & 1\\0 & 0 & 1 & -5\end{bmatrix},$$ so the solution is $x_1 = -2$, $x_2 = 1$, $x_3 = -5$.

12 **(a)** $$\begin{bmatrix}3/2 & 0 & 3 & 15\\-1 & 7 & -9 & -45\\2 & 0 & 5 & 22\end{bmatrix}\underset{(2/3)R1}{\approx}\begin{bmatrix}1 & 0 & 2 & 10\\-1 & 7 & -9 & -45\\2 & 0 & 5 & 22\end{bmatrix}\underset{\substack{R2+R1\\R3+(-2)R1}}{\approx}\begin{bmatrix}1 & 0 & 2 & 10\\0 & 7 & -7 & -35\\0 & 0 & 1 & 2\end{bmatrix}$$

$$\underset{(1/7)R2}{\approx}\begin{bmatrix}1 & 0 & 2 & 10\\0 & 1 & -1 & -5\\0 & 0 & 1 & 2\end{bmatrix}\underset{\substack{R1+(-2)R3\\R2+R3}}{\approx}\begin{bmatrix}1 & 0 & 0 & 6\\0 & 1 & 0 & -3\\0 & 0 & 1 & 2\end{bmatrix},$$

so the solution is $x_1 = 6$, $x_2 = -3$, $x_3 = 2$.

(b) $$\begin{bmatrix} -3 & -6 & -15 & -3 \\ 2 & 3 & 9 & 1 \\ -4 & -7 & -17 & -4 \end{bmatrix} \underset{(-1/3)R1}{\approx} \begin{bmatrix} 1 & 2 & 5 & 1 \\ 2 & 3 & 9 & 1 \\ -4 & -7 & -17 & -4 \end{bmatrix} \underset{\substack{R2+(-2)R1 \\ R3+(4)R1}}{\approx} \begin{bmatrix} 1 & 2 & 5 & 1 \\ 0 & -1 & -1 & -1 \\ 0 & 1 & 3 & 0 \end{bmatrix}$$

$$\underset{(-1)R2}{\approx} \begin{bmatrix} 1 & 2 & 5 & 1 \\ 0 & 1 & 1 & 1 \\ 0 & 1 & 3 & 0 \end{bmatrix} \underset{\substack{R1+(-2)R2 \\ R3+(-1)R2}}{\approx} \begin{bmatrix} 1 & 0 & 3 & -1 \\ 0 & 1 & 1 & 1 \\ 0 & 0 & 2 & -1 \end{bmatrix} \underset{(1/2)R3}{\approx} \begin{bmatrix} 1 & 0 & 3 & -1 \\ 0 & 1 & 1 & 1 \\ 0 & 0 & 1 & -1/2 \end{bmatrix}$$

$$\underset{\substack{R1+(-3)R3 \\ R2+(-1)R3}}{\approx} \begin{bmatrix} 1 & 0 & 0 & 1/2 \\ 0 & 1 & 0 & 3/2 \\ 0 & 0 & 1 & -1/2 \end{bmatrix}$$, so the solution is $x_1 = 1/2$, $x_2 = 3/2$, $x_3 = -1/2$.

(c) $$\begin{bmatrix} 3 & 6 & 0 & -3 & 3 \\ 1 & 3 & -1 & -4 & -12 \\ 1 & -1 & 1 & 2 & 8 \\ 2 & 3 & 0 & 0 & 8 \end{bmatrix} \underset{(1/3)R1}{\approx} \begin{bmatrix} 1 & 2 & 0 & -1 & 1 \\ 1 & 3 & -1 & -4 & -12 \\ 1 & -1 & 1 & 2 & 8 \\ 2 & 3 & 0 & 0 & 8 \end{bmatrix}$$

$$\underset{\substack{R2+(-1)R1 \\ R3+(-1)R1 \\ R4+(-2)R1}}{\approx} \begin{bmatrix} 1 & 2 & 0 & -1 & 1 \\ 0 & 1 & -1 & -3 & -13 \\ 0 & -3 & 1 & 3 & 7 \\ 0 & -1 & 0 & 2 & 6 \end{bmatrix} \underset{\substack{R1+(-2)R2 \\ R3+(3)R2 \\ R4+R2}}{\approx} \begin{bmatrix} 1 & 0 & 2 & 5 & 27 \\ 0 & 1 & -1 & -3 & -13 \\ 0 & 0 & -2 & -6 & -32 \\ 0 & 0 & -1 & -1 & -7 \end{bmatrix}$$

$$\underset{(-1/2)R3}{\approx} \begin{bmatrix} 1 & 0 & 2 & 5 & 27 \\ 0 & 1 & -1 & -3 & -13 \\ 0 & 0 & 1 & 3 & 16 \\ 0 & 0 & -1 & -1 & -7 \end{bmatrix} \underset{\substack{R1+(-2)R3 \\ R2+R3 \\ R4+R3}}{\approx} \begin{bmatrix} 1 & 0 & 0 & -1 & -5 \\ 0 & 1 & 0 & 0 & 3 \\ 0 & 0 & 1 & 3 & 16 \\ 0 & 0 & 0 & 2 & 9 \end{bmatrix}$$

$$\underset{(1/2)R4}{\approx} \begin{bmatrix} 1 & 0 & 0 & -1 & -5 \\ 0 & 1 & 0 & 0 & 3 \\ 0 & 0 & 1 & 3 & 16 \\ 0 & 0 & 0 & 1 & 9/2 \end{bmatrix} \underset{\substack{R1+R4 \\ R3+(-3)R4}}{\approx} \begin{bmatrix} 1 & 0 & 0 & 0 & -1/2 \\ 0 & 1 & 0 & 0 & 3 \\ 0 & 0 & 1 & 0 & 5/2 \\ 0 & 0 & 0 & 1 & 9/2 \end{bmatrix},$$

so the solution is $x_1 = -1/2$, $x_2 = 3$, $x_3 = 5/2$, $x_4 = 9/2$.

(d)
$$\begin{bmatrix} 1 & 2 & 2 & 5 & 11 \\ 2 & 4 & 2 & 8 & 14 \\ 1 & 3 & 4 & 8 & 19 \\ 1 & -1 & 1 & 0 & 2 \end{bmatrix} \begin{matrix} \approx \\ R2+(-2)R1 \\ R3+(-1)R1 \\ R4+(-1)R1 \end{matrix} \begin{bmatrix} 1 & 2 & 2 & 5 & 11 \\ 0 & 0 & -2 & -2 & -8 \\ 0 & 1 & 2 & 3 & 8 \\ 0 & -3 & -1 & -5 & -9 \end{bmatrix}$$

$$\begin{matrix} \approx \\ R2 \Leftrightarrow R3 \end{matrix} \begin{bmatrix} 1 & 2 & 2 & 5 & 11 \\ 0 & 1 & 2 & 3 & 8 \\ 0 & 0 & -2 & -2 & -8 \\ 0 & -3 & -1 & -5 & -9 \end{bmatrix} \begin{matrix} \approx \\ R1+(-2)R2 \\ R4+(3)R2 \end{matrix} \begin{bmatrix} 1 & 0 & -2 & -1 & -5 \\ 0 & 1 & 2 & 3 & 8 \\ 0 & 0 & -2 & -2 & -8 \\ 0 & 0 & 5 & 4 & 15 \end{bmatrix}$$

$$\begin{matrix} \approx \\ (-1/2)R3 \end{matrix} \begin{bmatrix} 1 & 0 & -2 & -1 & -5 \\ 0 & 1 & 2 & 3 & 8 \\ 0 & 0 & 1 & 1 & 4 \\ 0 & 0 & 5 & 4 & 15 \end{bmatrix} \begin{matrix} \approx \\ R1+(2)R3 \\ R2+(-2)R3 \\ R4+(-5)R3 \end{matrix} \begin{bmatrix} 1 & 0 & 0 & 1 & 3 \\ 0 & 1 & 0 & 1 & 0 \\ 0 & 0 & 1 & 1 & 4 \\ 0 & 0 & 0 & -1 & -5 \end{bmatrix}$$

$$\begin{matrix} \approx \\ (-1)R4 \end{matrix} \begin{bmatrix} 1 & 0 & 0 & 1 & 3 \\ 0 & 1 & 0 & 1 & 0 \\ 0 & 0 & 1 & 1 & 4 \\ 0 & 0 & 0 & 1 & 5 \end{bmatrix} \begin{matrix} \approx \\ R1+(-1)R4 \\ R2+(-1)R4 \\ R3+(-1)R4 \end{matrix} \begin{bmatrix} 1 & 0 & 0 & 0 & -2 \\ 0 & 1 & 0 & 0 & -5 \\ 0 & 0 & 1 & 0 & -1 \\ 0 & 0 & 0 & 1 & 5 \end{bmatrix},$$

so the solution is $x_1 = -2$, $x_2 = -5$, $x_3 = -1$, $x_4 = 5$.

(e)
$$\begin{bmatrix} 1 & 1 & 2 & 6 & 11 \\ 2 & 3 & 6 & 19 & 36 \\ 0 & 3 & 4 & 15 & 28 \\ 1 & -1 & -1 & -6 & -12 \end{bmatrix} \begin{matrix} \approx \\ R2+(-2)R1 \\ R4+(-1)R1 \end{matrix} \begin{bmatrix} 1 & 1 & 2 & 6 & 11 \\ 0 & 1 & 2 & 7 & 14 \\ 0 & 3 & 4 & 15 & 28 \\ 0 & -2 & -3 & -12 & -23 \end{bmatrix}$$

$$\begin{matrix} \approx \\ R1+(-1)R2 \\ R3+(-3)R2 \\ R4+(2)R2 \end{matrix} \begin{bmatrix} 1 & 0 & 0 & -1 & -3 \\ 0 & 1 & 2 & 7 & 14 \\ 0 & 0 & -2 & -6 & -14 \\ 0 & 0 & 1 & 2 & 5 \end{bmatrix} \begin{matrix} \approx \\ (-1/2)R3 \end{matrix} \begin{bmatrix} 1 & 0 & 0 & -1 & -3 \\ 0 & 1 & 2 & 7 & 14 \\ 0 & 0 & 1 & 3 & 7 \\ 0 & 0 & 1 & 2 & 5 \end{bmatrix}$$

$$\begin{matrix} \approx \\ R2+(-2)R3 \\ R4+(-1)R3 \end{matrix} \begin{bmatrix} 1 & 0 & 0 & -1 & -3 \\ 0 & 1 & 0 & 1 & 0 \\ 0 & 0 & 1 & 3 & 7 \\ 0 & 0 & 0 & -1 & -2 \end{bmatrix} \begin{matrix} \approx \\ (-1)R4 \end{matrix} \begin{bmatrix} 1 & 0 & 0 & -1 & -3 \\ 0 & 1 & 0 & 1 & 0 \\ 0 & 0 & 1 & 3 & 7 \\ 0 & 0 & 0 & 1 & 2 \end{bmatrix}$$

$$\underset{\substack{R1+R4\\R2+(-1)R4\\R3+(-3)R4}}{\approx}\begin{bmatrix}1&0&0&0&-1\\0&1&0&0&-2\\0&0&1&0&1\\0&0&0&1&2\end{bmatrix}$$, so the solution is $x_1 = -1,\ x_2 = -2,\ x_3 = 1,\ x_4 = 2$.

13. (a) $$\begin{bmatrix}1&2&3&4&3\\3&5&8&9&7\end{bmatrix}\underset{R2+(-3)R1}{\approx}\begin{bmatrix}1&2&3&4&3\\0&-1&-1&-3&-2\end{bmatrix}\underset{(-1)R2}{\approx}\begin{bmatrix}1&2&3&4&3\\0&1&1&3&2\end{bmatrix}\underset{R1-2R2}{\approx}$$

$$\begin{bmatrix}1&0&1&-2&-1\\0&1&1&3&2\end{bmatrix}$$, so $x_1 = 1,\ x_2 = 1$; $x_1 = -2,\ x_2 = 3$; and $x_1 = -1,\ x_2 = 2$.

(b) $$\begin{bmatrix}1&1&0&5&1\\2&3&1&13&2\end{bmatrix}\underset{R2+(-2)R1}{\approx}\begin{bmatrix}1&1&0&5&1\\0&1&1&3&0\end{bmatrix}\underset{R1+(-1)R2}{\approx}\begin{bmatrix}1&0&-1&2&1\\0&1&1&3&0\end{bmatrix},$$

so the solutions are in turn $x_1 = -1,\ x_2 = 1$; $x_1 = 2,\ x_2 = 3$; and $x_1 = 1,\ x_2 = 0$.

(c) $$\begin{bmatrix}1&-2&3&6&-5&4\\1&-1&2&5&-3&3\\2&-3&6&14&-8&9\end{bmatrix}\underset{\substack{R2+(-1)R1\\R3+(-2)R1}}{\approx}\begin{bmatrix}1&-2&3&6&-5&4\\0&1&-1&-1&2&-1\\0&1&0&2&2&1\end{bmatrix}$$

$$\underset{\substack{R1+(2)R2\\R3+(-1)R2}}{\approx}\begin{bmatrix}1&0&1&4&-1&2\\0&1&-1&-1&2&-1\\0&0&1&3&0&2\end{bmatrix}\underset{\substack{R1+(-1)R3\\R2+R3}}{\approx}\begin{bmatrix}1&0&0&1&-1&0\\0&1&0&2&2&1\\0&0&1&3&0&2\end{bmatrix},$$

so the solutions are in turn $x_1 = 1,\ x_2 = 2,\ x_3 = 3$; $x_1 = -1,\ x_2 = 2,\ x_3 = 0$; and $x_1 = 0,\ x_2 = 1,\ x_3 = 2$.

(d) $$\begin{bmatrix}1&2&-1&-1&6&0\\-1&-1&1&1&-4&-2\\3&7&-1&-1&18&-4\end{bmatrix}\underset{\substack{R2+R1\\R3+(-3)R1}}{\approx}\begin{bmatrix}1&2&-1&-1&6&0\\0&1&0&0&2&-2\\0&1&2&2&0&-4\end{bmatrix}$$

$$\underset{\substack{R1+(-2)R2\\R3+(-1)R2}}{\approx}\begin{bmatrix}1&0&-1&-1&2&4\\0&1&0&0&2&-2\\0&0&2&2&-2&-2\end{bmatrix}\underset{(1/2)R3}{\approx}\begin{bmatrix}1&0&-1&-1&2&4\\0&1&0&0&2&-2\\0&0&1&1&-1&-1\end{bmatrix}$$

$$\begin{bmatrix} 1 & 0 & 0 & 0 & 1 & 3 \\ 0 & 1 & 0 & 0 & 2 & -2 \\ 0 & 0 & 1 & 1 & -1 & -1 \end{bmatrix},$$

so the solutions are in turn $x_1 = 0$, $x_2 = 0$, $x_3 = 1$; $x_1 = 1$, $x_2 = 2$, $x_3 = -1$; and $x_1 = 3$, $x_2 = -2$, $x_3 = -1$.

Exercise Set 1.2

1. (a) Yes. (b) Yes.

   (c) No. The second column contains a leading 1, so other elements in that column should be zero.

   (d) No. The second row does not have 1 as the first nonzero number.

   (e) Yes. (f) Yes. (g) Yes.

   (h) No. The second row does not have 1 as the first nonzero number. (i) Yes.

2. (a) No. The leading 1 in row 3 is not to the right of the leading 1 in row 2.

   (b) Yes. (c) Yes.

   (d) No. The fourth and fifth columns contain leading 1s, so the other numbers in those columns should be zeros.

   (e) No. The row containing all zeros should be at the bottom of the matrix.

   (f) Yes.

   (g) No. The leading 1 in row 3 is not to the right of the leading 1 in row 2. Also, since column 3 contains a leading 1, all other numbers in that column should be zero.

   (h) No. The leading 1 in row 3 is not to the right of the leading 1s in rows 1 and 2.

   (i) Yes.

3. (a) $x_1 = 2$, $x_2 = 4$, $x_3 = -3$. (b) $x_1 = 3r + 4$, $x_2 = -2r + 8$, $x_3 = r$.

(c) $x_1 = -3r + 6, x_2 = r, x_3 = -2$. (d) There is no solution. The last row gives 0 = 1.

(e) $x_1 = -5r + 3,\ x_2 = -6r - 2,\ x_3 = -2r - 4,\ x_4 = r$.

(f) $x_1 = -3r + 2,\ x_2 = r,\ x_3 = 4,\ x_4 = 5$.

4. (a) $x_1 = -2r - 4s + 1,\ x_2 = 3r - 5s - 6,\ x_3 = r,\ x_4 = s$.

(b) $x_1 = 3r - 2s + 4,\ x_2 = r,\ x_3 = s,\ x_4 = -7$.

(c) $x_1 = 2r - 3s + 4,\ x_2 = r,\ x_3 = -2s + 9,\ x_4 = s,\ x_5 = 8$.

(d) $x_1 = -2r - 3s + 6,\ x_2 = -5r - 4s + 7,\ x_3 = r,\ x_4 = -9s - 3,\ x_5 = s$.

5. (a) $$\begin{bmatrix} 1 & 4 & 3 & 1 \\ 2 & 8 & 11 & 7 \\ 1 & 6 & 7 & 3 \end{bmatrix} \underset{\substack{R2+(-2)R1 \\ R3+(-1)R1}}{\approx} \begin{bmatrix} 1 & 4 & 3 & 1 \\ 0 & 0 & 5 & 5 \\ 0 & 2 & 4 & 2 \end{bmatrix} \underset{R2 \Leftrightarrow R3}{\approx} \begin{bmatrix} 1 & 4 & 3 & 1 \\ 0 & 2 & 4 & 2 \\ 0 & 0 & 5 & 5 \end{bmatrix}$$

$$\underset{(1/2)R2}{\approx} \begin{bmatrix} 1 & 4 & 3 & 1 \\ 0 & 1 & 2 & 1 \\ 0 & 0 & 5 & 5 \end{bmatrix} \underset{R1+(-4)R2}{\approx} \begin{bmatrix} 1 & 0 & -5 & -3 \\ 0 & 1 & 2 & 1 \\ 0 & 0 & 5 & 5 \end{bmatrix} \underset{(1/5)R3}{\approx} \begin{bmatrix} 1 & 0 & -5 & -3 \\ 0 & 1 & 2 & 1 \\ 0 & 0 & 1 & 1 \end{bmatrix}$$

$$\underset{\substack{R1+(5)R3 \\ R2+(-2)R3}}{\approx} \begin{bmatrix} 1 & 0 & 0 & 2 \\ 0 & 1 & 0 & -1 \\ 0 & 0 & 1 & 1 \end{bmatrix},$$ so the solution is $x_1 = 2,\ x_2 = -1,\ x_3 = 1$.

(b) $$\begin{bmatrix} 1 & 2 & 4 & 15 \\ 2 & 4 & 9 & 33 \\ 1 & 3 & 5 & 20 \end{bmatrix} \underset{\substack{R2+(-2)R1 \\ R3+(-1)R1}}{\approx} \begin{bmatrix} 1 & 2 & 4 & 15 \\ 0 & 0 & 1 & 3 \\ 0 & 1 & 1 & 5 \end{bmatrix} \underset{R2 \Leftrightarrow R3}{\approx} \begin{bmatrix} 1 & 2 & 4 & 15 \\ 0 & 1 & 1 & 5 \\ 0 & 0 & 1 & 3 \end{bmatrix}$$

$$\underset{R1+(-2)R2}{\approx} \begin{bmatrix} 1 & 0 & 2 & 5 \\ 0 & 1 & 1 & 5 \\ 0 & 0 & 1 & 3 \end{bmatrix} \underset{\substack{R1+(-2)R3 \\ R2+(-1)R3}}{\approx} \begin{bmatrix} 1 & 0 & 0 & -1 \\ 0 & 1 & 0 & 2 \\ 0 & 0 & 1 & 3 \end{bmatrix},$$

so the solution is $x_1 = -1,\ x_2 = 2,\ x_3 = 3$.

(c) $\begin{bmatrix} 1 & 1 & 1 & 7 \\ 2 & 3 & 1 & 18 \\ -1 & 1 & -3 & 1 \end{bmatrix} \underset{\substack{R2+(-2)R1 \\ R3+R1}}{\approx} \begin{bmatrix} 1 & 1 & 1 & 7 \\ 0 & 1 & -1 & 4 \\ 0 & 2 & -2 & 8 \end{bmatrix} \underset{\substack{R1+(-1)R2 \\ R3+(-2)R3}}{\approx} \begin{bmatrix} 1 & 0 & 2 & 3 \\ 0 & 1 & -1 & 4 \\ 0 & 0 & 0 & 0 \end{bmatrix}$,

so $x_1 + 2x_3 = 3$ and $x_2 - x_3 = 4$.
Thus the general solution is $x_1 = 3 - 2r$, $x_2 = 4 + r$, $x_3 = r$.

(d) $\begin{bmatrix} 1 & 4 & 1 & 2 \\ 1 & 2 & -1 & 0 \\ 2 & 6 & 0 & 3 \end{bmatrix} \underset{\substack{R2+(-1)R1 \\ R3+(-2)R1}}{\approx} \begin{bmatrix} 1 & 4 & 1 & 2 \\ 0 & -2 & -2 & -2 \\ 0 & -2 & -2 & -1 \end{bmatrix} \underset{(-1/2)R2}{\approx} \begin{bmatrix} 1 & 4 & 1 & 2 \\ 0 & 1 & 1 & 1 \\ 0 & -2 & -2 & -1 \end{bmatrix}$

$\underset{\substack{R1+(-4)R2 \\ R3+(2)R2}}{\approx} \begin{bmatrix} 1 & 0 & -3 & -2 \\ 0 & 1 & 1 & 1 \\ 0 & 0 & 0 & 1 \end{bmatrix}$, so there is no solution, since the last row of the matrix corresponds to the equation 0 = 1.

(e) $\begin{bmatrix} 1 & -1 & 1 & 3 \\ 2 & -1 & 4 & 7 \\ 3 & -5 & -1 & 7 \end{bmatrix} \underset{\substack{R2+(-2)R1 \\ R3+(-3)R1}}{\approx} \begin{bmatrix} 1 & -1 & 1 & 3 \\ 0 & 1 & 2 & 1 \\ 0 & -2 & -4 & -2 \end{bmatrix} \underset{\substack{R1+R2 \\ R3+(2)R2}}{\approx} \begin{bmatrix} 1 & 0 & 3 & 4 \\ 0 & 1 & 2 & 1 \\ 0 & 0 & 0 & 0 \end{bmatrix}$,

so $x_1 + 3x_3 = 4$ and $x_2 + 2x_3 = 1$.
Thus the general solution is $x_1 = 4 - 3r$, $x_2 = 1 - 2r$, $x_3 = r$.

(f) $\begin{bmatrix} 3 & -3 & 9 & 24 \\ 2 & -2 & 7 & 17 \\ -1 & 2 & -4 & -11 \end{bmatrix} \underset{(1/3)R1}{\approx} \begin{bmatrix} 1 & -1 & 3 & 8 \\ 2 & -2 & 7 & 17 \\ -1 & 2 & -4 & -11 \end{bmatrix} \underset{\substack{R2+(-2)R1 \\ R2+R1}}{\approx} \begin{bmatrix} 1 & -1 & 3 & 8 \\ 0 & 0 & 1 & 1 \\ 0 & 1 & -1 & -3 \end{bmatrix}$

$\underset{R2 \Leftrightarrow R3}{\approx} \begin{bmatrix} 1 & -1 & 3 & 8 \\ 0 & 1 & -1 & -3 \\ 0 & 0 & 1 & 1 \end{bmatrix} \underset{R1+R2}{\approx} \begin{bmatrix} 1 & 0 & 2 & 5 \\ 0 & 1 & -1 & -3 \\ 0 & 0 & 1 & 1 \end{bmatrix} \underset{\substack{R1+(-2)R3 \\ R2+R3}}{\approx} \begin{bmatrix} 1 & 0 & 0 & 3 \\ 0 & 1 & 0 & -2 \\ 0 & 0 & 1 & 1 \end{bmatrix}$,

so the solution is $x_1 = 3$, $x_2 = -2$, $x_3 = 1$.

6. (a) $\begin{bmatrix} 3 & 6 & -3 & 6 \\ -2 & -4 & -3 & -1 \\ 3 & 6 & -2 & 10 \end{bmatrix} \underset{(1/3)R1}{\approx} \begin{bmatrix} 1 & 2 & -1 & 2 \\ -2 & -4 & -3 & -1 \\ 3 & 6 & -2 & 10 \end{bmatrix} \underset{\substack{R2+(2)R1 \\ R3+(-3)R1}}{\approx} \begin{bmatrix} 1 & 2 & -1 & 2 \\ 0 & 0 & -5 & 3 \\ 0 & 0 & 1 & 4 \end{bmatrix}$

It is now clear that there is no solution. The last two rows give $-5x_3 = 3$ and $x_3 = 4$.

(b) $\begin{bmatrix} 1 & 2 & 1 & 7 \\ 1 & 2 & 2 & 11 \\ 2 & 4 & 3 & 18 \end{bmatrix} \underset{\substack{R2+(-1)R1 \\ R3+(-2)R1}}{\approx} \begin{bmatrix} 1 & 2 & 1 & 7 \\ 0 & 0 & 1 & 4 \\ 0 & 0 & 1 & 4 \end{bmatrix} \underset{\substack{R1+(-1)R2 \\ R3+(-1)R2}}{\approx} \begin{bmatrix} 1 & 2 & 0 & 3 \\ 0 & 0 & 1 & 4 \\ 0 & 0 & 0 & 0 \end{bmatrix}$

so $x_1 + 2x_2 = 3$ and $x_3 = 4$. Thus the general solution is
$x_1 = 3 - 2r$, $x_2 = r$, $x_3 = 4$.

(c) $\begin{bmatrix} 1 & 2 & -1 & 3 \\ 2 & 4 & -2 & 6 \\ 3 & 6 & 2 & -1 \end{bmatrix} \underset{\substack{R2+(-2)R1 \\ R3+(-3)R1}}{\approx} \begin{bmatrix} 1 & 2 & -1 & 3 \\ 0 & 0 & 0 & 0 \\ 0 & 0 & 5 & -10 \end{bmatrix} \underset{R2 \Leftrightarrow R3}{\approx} \begin{bmatrix} 1 & 2 & -1 & 3 \\ 0 & 0 & 5 & -10 \\ 0 & 0 & 0 & 0 \end{bmatrix}$

$\underset{(1/5)R2}{\approx} \begin{bmatrix} 1 & 2 & -1 & 3 \\ 0 & 0 & 1 & -2 \\ 0 & 0 & 0 & 0 \end{bmatrix} \underset{R1+R2}{\approx} \begin{bmatrix} 1 & 2 & 0 & 1 \\ 0 & 0 & 1 & -2 \\ 0 & 0 & 0 & 0 \end{bmatrix}$, so $x_1 + 2x_2 = 1$, $x_3 = -2$.

Thus the general solution is $x_1 = 1 - 2r$, $x_2 = r$, $x_3 = -2$.

(d) $\begin{bmatrix} 1 & 2 & 3 & 8 \\ 3 & 7 & 9 & 26 \\ 2 & 0 & 6 & 11 \end{bmatrix} \underset{\substack{R2+(-3)R1 \\ R3+(-2)R1}}{\approx} \begin{bmatrix} 1 & 2 & 3 & 8 \\ 0 & 1 & 0 & 2 \\ 0 & -4 & 0 & -5 \end{bmatrix}$, so there is no solution since the

last two rows give $x_2 = 2$ and $-4x_2 = -5$.

(e) $\begin{bmatrix} 0 & 1 & 2 & 5 \\ 1 & 2 & 5 & 13 \\ 1 & 0 & 2 & 4 \end{bmatrix} \underset{R1 \Leftrightarrow R2}{\approx} \begin{bmatrix} 1 & 2 & 5 & 13 \\ 0 & 1 & 2 & 5 \\ 1 & 0 & 2 & 4 \end{bmatrix} \underset{R3+(-1)R1}{\approx} \begin{bmatrix} 1 & 2 & 5 & 13 \\ 0 & 1 & 2 & 5 \\ 0 & -2 & -3 & -9 \end{bmatrix}$

$$\underset{\substack{R1+(-2)R2\\R3+(2)R2}}{\approx}\begin{bmatrix}1&0&1&3\\0&1&2&5\\0&0&1&1\end{bmatrix}\underset{\substack{R1+(-1)R3\\R2+(-2)R3}}{\approx}\begin{bmatrix}1&0&0&2\\0&1&0&3\\0&0&1&1\end{bmatrix},$$

so the solution is $x_1 = 2$, $x_2 = 3$, $x_3 = 1$.

(f) $$\begin{bmatrix}1&2&8&7\\2&4&16&14\\0&1&3&4\end{bmatrix}\underset{R2+(-2)R1}{\approx}\begin{bmatrix}1&2&8&7\\0&0&0&0\\0&1&3&4\end{bmatrix}\underset{R2\Leftrightarrow R3}{\approx}\begin{bmatrix}1&2&8&7\\0&1&3&4\\0&0&0&0\end{bmatrix}$$

$$\underset{R1+(-2)R2}{\approx}\begin{bmatrix}1&0&2&-1\\0&1&3&4\\0&0&0&0\end{bmatrix},$$ so $x_1 + 2x_3 = -1$, $x_2 + 3x_3 = 4$.

Thus the general solution is $x_1 = -1 - 2r$, $x_2 = 4 - 3r$, $x_3 = r$.

7. (a) $$\begin{bmatrix}1&1&-3&10\\-3&-2&4&-24\end{bmatrix}\underset{R2+(3)R1}{\approx}\begin{bmatrix}1&1&-3&10\\0&1&-5&6\end{bmatrix}\underset{R1+(-1)R2}{\approx}\begin{bmatrix}1&0&2&4\\0&1&-5&6\end{bmatrix},$$

so $x_1 + 2x_3 = 4$ and $x_2 - 5x_3 = 6$. Thus the general solution is $x_1 = 4 - 2r$, $x_2 = 6 + 5r$, $x_3 = r$.

(b) $$\begin{bmatrix}2&-6&-14&38\\-3&7&15&-37\end{bmatrix}\underset{(1/2)R1}{\approx}\begin{bmatrix}1&-3&-7&19\\-3&7&15&-37\end{bmatrix}\underset{R2+(3)R1}{\approx}\begin{bmatrix}1&-3&-7&19\\0&-2&-6&20\end{bmatrix}$$

$$\underset{(-1/2)R2}{\approx}\begin{bmatrix}1&-3&-7&19\\0&1&3&-10\end{bmatrix}\underset{R1+(3)R2}{\approx}\begin{bmatrix}1&0&2&-11\\0&1&3&-10\end{bmatrix},$$

so $x_1 + 2x_3 = -11$ and $x_2 + 3x_3 = -10$. Thus the general solution is $x_1 = -11 - 2r$, $x_2 = -10 - 3r$, $x_3 = r$.

(c) $$\begin{bmatrix}1&2&-1&-1&0\\1&2&0&1&4\\-1&-2&2&4&5\end{bmatrix}\underset{\substack{R2+(-1)R1\\R3+R1}}{\approx}\begin{bmatrix}1&2&-1&-1&0\\0&0&1&2&4\\0&0&1&3&5\end{bmatrix}\underset{\substack{R1+R2\\R3+(-1)R2}}{\approx}\begin{bmatrix}1&2&0&1&4\\0&0&1&2&4\\0&0&0&1&1\end{bmatrix}$$

$$\underset{\substack{R1+(-1)R3\\R2+(-2)R3}}{\approx}\begin{bmatrix}1&2&0&0&3\\0&0&1&0&2\\0&0&0&1&1\end{bmatrix}$$, so $x_1 + 2x_2 = 3$ and $x_3 = 2$, and $x_4 = 1$.

Thus the general solution is $x_1 = 3 - 2r$, $x_2 = r$, $x_3 = 2$, and $x_4 = 1$.

(d) $$\begin{bmatrix}1&2&0&4&0\\-2&-4&3&-2&0\end{bmatrix}\underset{R2+(2)R1}{\approx}\begin{bmatrix}1&2&0&4&0\\0&0&3&6&0\end{bmatrix}\underset{(1/3)R2}{\approx}\begin{bmatrix}1&2&0&4&0\\0&0&1&2&0\end{bmatrix},$$

so $x_1 + 2x_2 + 4x_4 = 0$ and $x_3 + 2x_4 = 0$. Thus the general solution is $x_1 = -2r - 4s$, $x_2 = r$, $x_3 = -2s$, $x_4 = s$.

(e) $$\begin{bmatrix}0&1&-3&1&0\\1&1&-1&4&0\\-2&-2&2&-8&0\end{bmatrix}\underset{R1\Leftrightarrow R2}{\approx}\begin{bmatrix}1&1&-1&4&0\\0&1&-3&1&0\\-2&-2&2&-8&0\end{bmatrix}$$

$$\underset{R3+(2)R1}{\approx}\begin{bmatrix}1&1&-1&4&0\\0&1&-3&1&0\\0&0&0&0&0\end{bmatrix}\underset{R1+(-1)R2}{\approx}\begin{bmatrix}1&0&2&3&0\\0&1&-3&1&0\\0&0&0&0&0\end{bmatrix},$$

so $x_1 + 2x_3 + 3x_4 = 0$ and $x_2 - 3x_3 + x_4 = 0$. Thus the general solution is $x_1 = -2r - 3s$, $x_2 = 3r - s$, $x_3 = r$, $x_4 = s$.

8. (a) $$\begin{bmatrix}1&1&1&-1&-3\\2&3&1&-5&-9\\1&3&-1&-6&-7\\-1&-1&-1&0&1\end{bmatrix}\underset{\substack{R2+(-2)R1\\R3+(-1)R1\\R4+R1}}{\approx}\begin{bmatrix}1&1&1&-1&-3\\0&1&-1&-3&-3\\0&2&-2&-5&-4\\0&0&0&-1&-2\end{bmatrix}$$

$$\underset{\substack{R1+(-1)R2\\R3+(-2)R2}}{\approx}\begin{bmatrix}1&0&2&2&0\\0&1&-1&-3&-3\\0&0&0&1&2\\0&0&0&-1&-2\end{bmatrix}\underset{\substack{R1+(2)R3\\R2+(-3)R3\\R4+R3}}{\approx}\begin{bmatrix}1&0&2&0&-4\\0&1&-1&0&3\\0&0&0&1&2\\0&0&0&0&0\end{bmatrix},$$

so $x_1 + 2x_3 = -4$, $x_2 - x_3 = 3$, $x_4 = 2$.
The general solution is $x_1 = -2r - 4$, $x_2 = r + 3$, $x_3 = r$, $x_4 = 2$.

(b)
$$\begin{bmatrix} 0 & 1 & 2 & 7 \\ 1 & -2 & -6 & -18 \\ -1 & -1 & -2 & -5 \\ 2 & -5 & -15 & -46 \end{bmatrix} \underset{R1 \Leftrightarrow R2}{\approx} \begin{bmatrix} 1 & -2 & -6 & -18 \\ 0 & 1 & 2 & 7 \\ -1 & -1 & -2 & -5 \\ 2 & -5 & -15 & -46 \end{bmatrix}$$

$$\underset{\substack{R3+R1 \\ R4+(-2)R1}}{\approx} \begin{bmatrix} 1 & -2 & -6 & -18 \\ 0 & 1 & 2 & 7 \\ 0 & -3 & -8 & -23 \\ 0 & -1 & -3 & -10 \end{bmatrix} \underset{\substack{R1+(2)R2 \\ R3+(3)R2 \\ R4+R2}}{\approx} \begin{bmatrix} 1 & 0 & -2 & 4 \\ 0 & 1 & 2 & 7 \\ 0 & 0 & -2 & -2 \\ 0 & 0 & -1 & -3 \end{bmatrix}$$, and there are no solutions because the last two rows of the matrix give, respectively, $x_3 = 1$ and $x_3 = 3$.

(c)
$$\begin{bmatrix} 2 & -4 & 16 & -14 & 10 \\ -1 & 5 & -17 & 19 & -2 \\ 1 & -3 & 11 & -11 & 4 \\ 3 & -4 & 18 & -13 & 17 \end{bmatrix} \underset{(1/2)R1}{\approx} \begin{bmatrix} 1 & -2 & 8 & -7 & 5 \\ -1 & 5 & -17 & 19 & -2 \\ 1 & -3 & 11 & -11 & 4 \\ 3 & -4 & 18 & -13 & 17 \end{bmatrix}$$

$$\underset{\substack{R2+R1 \\ R3+(-1)R1 \\ R4+(-3)R1}}{\approx} \begin{bmatrix} 1 & -2 & 8 & -7 & 5 \\ 0 & 3 & -9 & 12 & 3 \\ 0 & -1 & 3 & -4 & -1 \\ 0 & 2 & -6 & 8 & 2 \end{bmatrix} \underset{(1/3)R2}{\approx} \begin{bmatrix} 1 & -2 & 8 & -7 & 5 \\ 0 & 1 & -3 & 4 & 1 \\ 0 & -1 & 3 & -4 & -1 \\ 0 & 2 & -6 & 8 & 2 \end{bmatrix}$$

$$\underset{\substack{R1+(2)R2 \\ R3+R2 \\ R4+(-2)R2}}{\approx} \begin{bmatrix} 1 & 0 & 2 & 1 & 7 \\ 0 & 1 & -3 & 4 & 1 \\ 0 & 0 & 0 & 0 & 0 \\ 0 & 0 & 0 & 0 & 0 \end{bmatrix}$$, so $x_1+2x_3+x_4 = 7$ and $x_2-3x_3+4x_4=1$.

Thus the general solution is $x_1 = 7 - 2r - s$, $x_2 = 1 + 3r - 4s$, $x_3 = r$, $x_4 = s$.

(d)
$$\begin{bmatrix} 1 & -1 & 2 & 0 & 7 \\ 2 & -2 & 2 & -4 & 12 \\ -1 & 1 & -1 & 2 & -4 \\ -3 & 1 & -8 & -10 & -29 \end{bmatrix} \underset{\substack{R2+(-2)R1 \\ R3+R1 \\ R4+(3)R1}}{\approx} \begin{bmatrix} 1 & -1 & 2 & 0 & 7 \\ 0 & 0 & -2 & -4 & -2 \\ 0 & 0 & 1 & 2 & 3 \\ 0 & -2 & -2 & -10 & -8 \end{bmatrix}$$

$$\underset{R2 \Leftrightarrow R4}{\approx} \begin{bmatrix} 1 & -1 & 2 & 0 & 7 \\ 0 & -2 & -2 & -10 & -8 \\ 0 & 0 & 1 & 2 & 3 \\ 0 & 0 & -2 & -4 & -2 \end{bmatrix} \underset{(-1/2)R2}{\approx} \begin{bmatrix} 1 & -1 & 2 & 0 & 7 \\ 0 & 1 & 1 & 5 & 4 \\ 0 & 0 & 1 & 2 & 3 \\ 0 & 0 & -2 & -4 & -2 \end{bmatrix}$$

$$\underset{R1+R2}{\approx} \begin{bmatrix} 1 & 0 & 3 & 5 & 11 \\ 0 & 1 & 1 & 5 & 4 \\ 0 & 0 & 1 & 2 & 3 \\ 0 & 0 & -2 & -4 & -2 \end{bmatrix} \underset{\substack{R1+(-3)R3 \\ R2+(-1)R3 \\ R4+(2R3)}}{\approx} \begin{bmatrix} 1 & 0 & 0 & -1 & 2 \\ 0 & 1 & 0 & 3 & 1 \\ 0 & 0 & 1 & 2 & 3 \\ 0 & 0 & 0 & 0 & 4 \end{bmatrix}$$

The last row gives 0 = 4, so there is no solution.

(e) $$\begin{bmatrix} 1 & 6 & -1 & -4 & 0 \\ -2 & -12 & 5 & 17 & 0 \\ 3 & 18 & -1 & -6 & 0 \end{bmatrix} \underset{\substack{R2+(2)R1 \\ R3+(-3)R1}}{\approx} \begin{bmatrix} 1 & 6 & -1 & -4 & 0 \\ 0 & 0 & 3 & 9 & 0 \\ 0 & 0 & 2 & 6 & 0 \end{bmatrix}$$

$$\underset{(1/3)R2}{\approx} \begin{bmatrix} 1 & 6 & -1 & -4 & 0 \\ 0 & 0 & 1 & 3 & 0 \\ 0 & 0 & 2 & 6 & 0 \end{bmatrix} \underset{\substack{R1+R2 \\ R3+(-2)R2}}{\approx} \begin{bmatrix} 1 & 6 & 0 & -1 & 0 \\ 0 & 0 & 1 & 3 & 0 \\ 0 & 0 & 0 & 0 & 0 \end{bmatrix},$$

so $x_1 + 6x_2 - x_4 = 0$ and $x_3 + 3x_4 = 0$.
Thus the general solution is $x_1 = -6r + s$, $x_2 = r$, $x_3 = -3s$, $x_4 = s$.

(f) $$\begin{bmatrix} 4 & 8 & -12 & 28 \\ -1 & -2 & 3 & -7 \\ 2 & 4 & -8 & 16 \\ -3 & -6 & 9 & -21 \end{bmatrix} \underset{(1/4)R1}{\approx} \begin{bmatrix} 1 & 2 & -3 & 7 \\ -1 & -2 & 3 & -7 \\ 2 & 4 & -8 & 16 \\ -3 & -6 & 9 & -21 \end{bmatrix}$$

$$\underset{\substack{R2+R1 \\ R3+(-2)R1 \\ R4+(3)R1}}{\approx} \begin{bmatrix} 1 & 2 & -3 & 7 \\ 0 & 0 & 0 & 0 \\ 0 & 0 & -2 & 2 \\ 0 & 0 & 0 & 0 \end{bmatrix} \underset{R2 \Leftrightarrow R3}{\approx} \begin{bmatrix} 1 & 2 & -3 & 7 \\ 0 & 0 & -2 & 2 \\ 0 & 0 & 0 & 0 \\ 0 & 0 & 0 & 0 \end{bmatrix}$$

$$\underset{(-1/2)R2}{\approx} \begin{bmatrix} 1 & 2 & -3 & 7 \\ 0 & 0 & 1 & -1 \\ 0 & 0 & 0 & 0 \\ 0 & 0 & 0 & 0 \end{bmatrix} \underset{R1+(3)R2}{\approx} \begin{bmatrix} 1 & 2 & 0 & 4 \\ 0 & 0 & 1 & -1 \\ 0 & 0 & 0 & 0 \\ 0 & 0 & 0 & 0 \end{bmatrix},$$ so $x_1 + 2x_2 = 4$ and $x_3 = -1$.

Thus the general solution is $x_1 = 4 - 2r$, $x_2 = r$, $x_3 = -1$.

(g) $\begin{bmatrix} 1 & 1 & 2 \\ 2 & 3 & 3 \\ 1 & 3 & 0 \\ 1 & 2 & 1 \end{bmatrix} \underset{\substack{R2+(-2)R1 \\ R3+(-1)R1 \\ R4+(-1)R1}}{\approx} \begin{bmatrix} 1 & 1 & 2 \\ 0 & 1 & -1 \\ 0 & 2 & -2 \\ 0 & 1 & -1 \end{bmatrix} \underset{\substack{R1+(-1)R2 \\ R3+(-2)R2 \\ R4+(-1)R2}}{\approx} \begin{bmatrix} 1 & 0 & 3 \\ 0 & 1 & -1 \\ 0 & 0 & 0 \\ 0 & 0 & 0 \end{bmatrix}$,

so $x_1 = 3$, $x_2 = -1$.

9. (a) The system of equations

$$\begin{aligned} 3x_1 + 2x_2 - x_3 + x_4 &= 4 \\ 3x_1 + 2x_2 - x_3 + x_4 &= 1 \end{aligned}$$

clearly has no solution, since the equations are inconsistent. To make a system that is less obvious, add another equation to the system and perform an elementary transformation on this new system of three equations. For example, replace the second equation by the sum of the second equation and some multiple (2 in the example below) of the third equation:

$$\begin{aligned} 3x_1 + 2x_2 - x_3 + x_4 &= 4 \\ 5x_1 + 4x_2 - x_3 - x_4 &= 1 \\ x_1 + x_2 - x_4 &= 0 \end{aligned}$$

(b) Choose a solution, e.g., $x_1 = 1$, $x_2 = 2$. Now make up equations thinking of x_1 as 1 and x_2 as 2:

$$\begin{aligned} x_1 + x_2 &= 3 \\ x_1 + 2x_2 &= 5 \\ x_1 - 2x_2 &= -3 \end{aligned}$$

An easy way to ensure that there are no additional solutions is to include $x_1 = 1$ or $x_2 = 2$ as an equation in the system.

10. (a) $\begin{bmatrix} 1 & 0 & 0 \\ 0 & 1 & 0 \\ 0 & 0 & 1 \end{bmatrix}$ no solution, $\begin{bmatrix} 1 & 0 & * \\ 0 & 1 & * \\ 0 & 0 & 0 \end{bmatrix}$ unique solution, $\begin{bmatrix} 1 & * & 0 \\ 0 & 0 & 1 \\ 0 & 0 & 0 \end{bmatrix}$ no solution, $\begin{bmatrix} 1 & * & * \\ 0 & 0 & 0 \\ 0 & 0 & 0 \end{bmatrix}$ many solutions

(b) $\begin{bmatrix} 1 & 0 & 0 & * \\ 0 & 1 & 0 & * \\ 0 & 0 & 1 & * \end{bmatrix}$ unique solution $\begin{bmatrix} 1 & 0 & * & 0 \\ 0 & 1 & * & 0 \\ 0 & 0 & 0 & 1 \end{bmatrix}$ no solution $\begin{bmatrix} 1 & 0 & * & * \\ 0 & 1 & * & * \\ 0 & 0 & 0 & 0 \end{bmatrix}$ many solutions $\begin{bmatrix} 1 & * & 0 & 0 \\ 0 & 0 & 1 & 0 \\ 0 & 0 & 0 & 1 \end{bmatrix}$ no solution

$\begin{bmatrix} 1 & * & 0 & * \\ 0 & 0 & 1 & * \\ 0 & 0 & 0 & 0 \end{bmatrix}$ many solutions $\begin{bmatrix} 1 & * & * & 0 \\ 0 & 0 & 0 & 1 \\ 0 & 0 & 0 & 0 \end{bmatrix}$ no solution $\begin{bmatrix} 1 & * & * & * \\ 0 & 0 & 0 & 0 \\ 0 & 0 & 0 & 0 \end{bmatrix}$ many solutions

11. (a) If $ax_0 + by_0 = 0$ then $k(ax_0 + by_0) = 0$ so that $a(kx_0) + b(ky_0) = 0$. Thus $x = kx_0$, $y = ky_0$ is a solution. Likewise for the equation $cx + dy = 0$.

(b) If $ax_0 + by_0 = 0$ and $ax_1 + by_1 = 0$ then $ax_0 + by_0 + ax_1 + by_1 = 0 + 0 = 0$. But $ax_0 + by_0 + ax_1 + by_1 = ax_0 + ax_1 + by_0 + by_1 = a(x_0 + x_1) + b(y_0 + y_1)$ so that $a(x_0 + x_1) + b(y_0 + y_1) = 0$. Thus $x = x_0 + x_1$, $y = y_0 + y_1$ is a solution. Likewise for the equation $cx + dy = 0$.

12. $a(0) + b(0) = 0$ and $c(0) + d(0) = 0$, so $x = 0$, $y = 0$ is a solution.
Multiply 1st equation by c, 2nd by a to eliminate x. Get
$cax+cby=0$ and $acx+ady=0$. Subtract, $ady-cby=0$, $(ad-bc)y=0$. Similarly $(ad-bc)x=0$.
If $ad-bc\neq0$, $x=0,y=0$. If $ad-bc=0$ the x and y can be anything; thus many solutions.
Therefore $x=0,y=0$ is the only solution if and only if $ad-bc\neq0$.

13. (a) and (b), No. If the first system of equations has a unique solution, then the reduced echelon form of the matrix $[A:B_1]$ will be $[I_3:X]$. The reduced echelon form of $[A:B_2]$ must therefore be $[I_3:Y]$. So the second system must also have a unique solution.

(c) If the first system of equations has many solutions, then at least one row of the reduced echelon form of $[A:B_1]$ will consist entirely of zeros. Therefore the corresponding row(s) of the reduced echelon form of $[A:B_2]$ will have zeros in the first three columns. If any such row has a nonzero number in the fourth column, the system will have no solution.

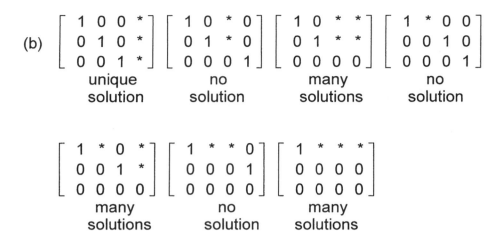

14. (a) $\begin{bmatrix} 1 & 1 & 5 & 2 & 3 \\ 1 & 2 & 8 & 5 & 2 \\ 2 & 4 & 16 & 10 & 4 \end{bmatrix} \underset{\substack{R2+(-1)R1 \\ R3+(-2)R1}}{\approx} \begin{bmatrix} 1 & 1 & 5 & 2 & 3 \\ 0 & 1 & 3 & 3 & -1 \\ 0 & 2 & 6 & 6 & -2 \end{bmatrix} \underset{\substack{R1+(-1)R2 \\ R3+(-2)R2}}{\approx} \begin{bmatrix} 1 & 0 & 2 & -1 & 4 \\ 0 & 1 & 3 & 3 & -1 \\ 0 & 0 & 0 & 0 & 0 \end{bmatrix}$,

so the general solution to the first system is $x_1 = -1 - 2r$, $x_2 = 3 - 3r$, $x_3 = r$, and the general solution to the second system is $x_1 = 4 - 2r$, $x_2 = -1 - 3r$, $x_3 = r$.

(b) $\begin{bmatrix} 1 & 2 & 4 & 8 & 5 \\ 1 & 1 & 2 & 5 & 3 \\ 2 & 3 & 6 & 13 & 11 \end{bmatrix} \underset{\substack{R2+(-1)R1 \\ R3+(-2)R1}}{\approx} \begin{bmatrix} 1 & 2 & 4 & 8 & 5 \\ 0 & -1 & -2 & -3 & -2 \\ 0 & -1 & -2 & -3 & 1 \end{bmatrix} \underset{(-1)R2}{\approx} \begin{bmatrix} 1 & 2 & 4 & 8 & 5 \\ 0 & 1 & 2 & 3 & 2 \\ 0 & -1 & -2 & -3 & 1 \end{bmatrix}$,

$\underset{\substack{R1+(-2)R2 \\ R3+R2}}{\approx} \begin{bmatrix} 1 & 0 & 0 & 2 & 1 \\ 0 & 1 & 2 & 3 & 2 \\ 0 & 0 & 0 & 0 & 3 \end{bmatrix}$, so the general solution to the first system is

$x_1 = 2$, $x_2 = 3 - 2r$, $x_3 = r$, and the second system has no solution.

15. A 3x3 matrix represents the equations of three lines in a plane. In order for there to be a unique solution, the three lines would have to meet in a point. For there to be many solutions, the three lines would all have to be the same. It is far more likely that the lines will meet in pairs (or that one pair will be parallel), i.e., that there will be no solution, the situation represented by the reduced echelon form I_3 .

16. A 3x4 matrix represents the equations of three planes. In order for there to be many solutions, the three planes must have at least one line in common. For there to be no solutions, either at least two of the three planes must be parallel or the line of intersection of two of the planes must lie in a plane that is parallel to the third plane. It is more likely that the three planes will meet in a single point, i.e., that there will be a unique solution. The reduced echelon form therefore will be $[I_3 :X]$.

17. The difference between no solution and at least one solution is the presence of a nonzero number in the last position of a row that otherwise consists entirely of zeros. Round-off error is more likely to produce a nonzero number when there should be a zero than the reverse. Thus the answer is (b). Thinking geometrically, a small move by one or more of the linear surfaces (round-off error) may destroy a solution if there is one, but probably won't produce a solution if there is none.

<u>Exercise Set 1.3</u>

1.

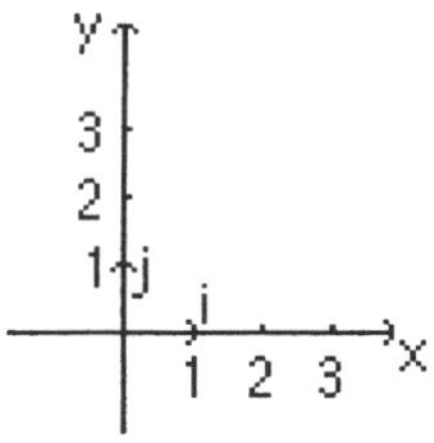
y
3
2
1
j
i
1 2 3
x

2.

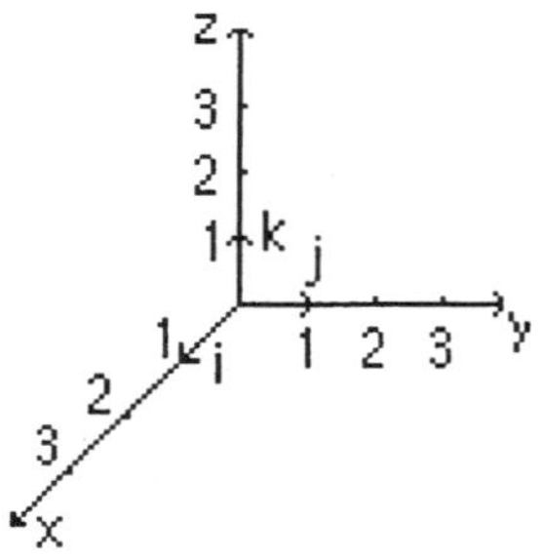
z
3
2
1
k
j
1 2 3
y
1
i
2
3
x

3. (a)

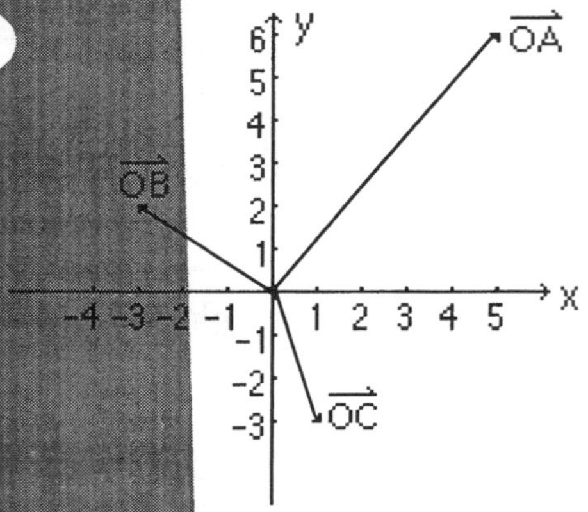
y
6
5
4
3
2
1
$\overrightarrow{OA}$
$\overrightarrow{OB}$
$\overrightarrow{OC}$
-4 -3 -2 -1 1 2 3 4 5
x
-1
-2
-3

(b)

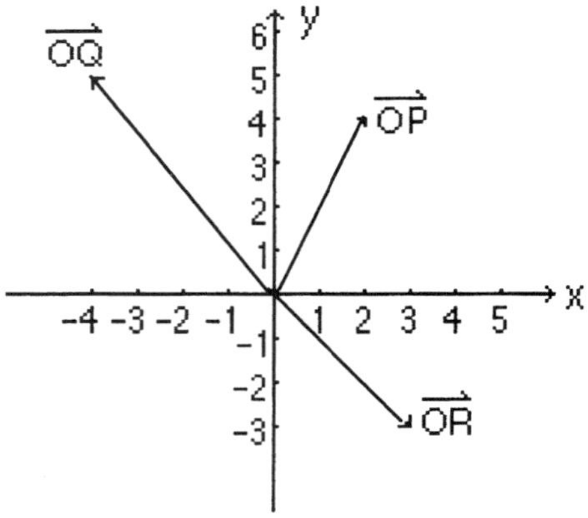
y
6
5
4
3
2
1
$\overrightarrow{OQ}$
$\overrightarrow{OP}$
$\overrightarrow{OR}$
-4 -3 -2 -1 1 2 3 4 5
x
-1
-2
-3

(c)

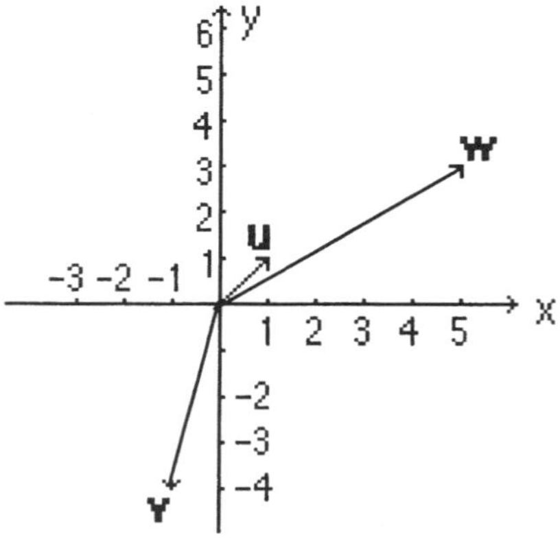
y
6
5
4
3
2
1
w
u
-3 -2 -1
1 2 3 4 5
x
-2
-3
-4
v

4. (a)

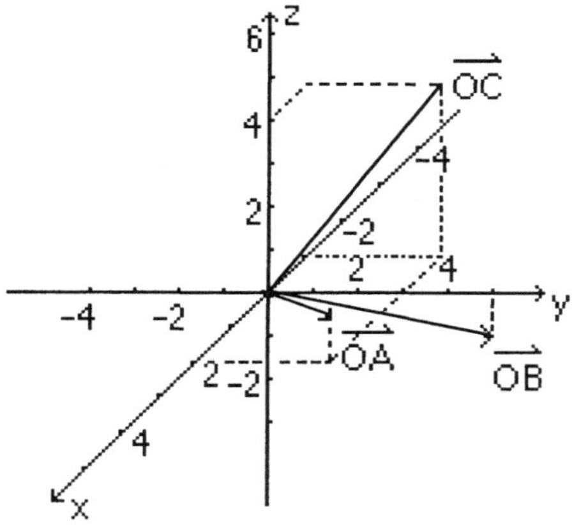

(b)

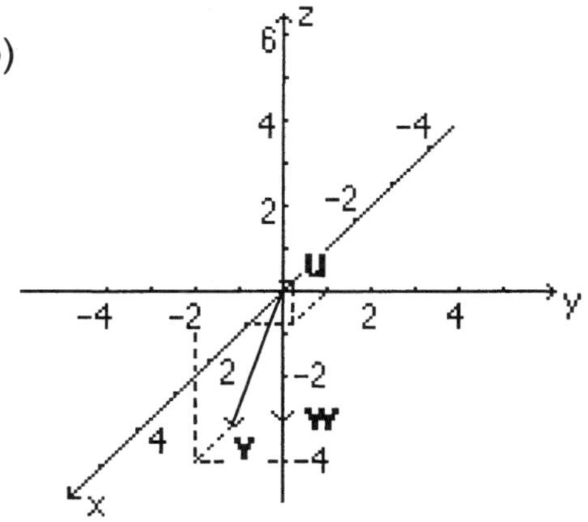

5. (a) 3(1,4) = (3,12). (b) −2(−1,3) = (2,−6). (c) (1/2)(2,6) = (1,3).

(d) (−1/2)(2,4,2) = (−1,−2,−1). (e) 3(−1,2,3) = (−3,6,9).

(f) 4(−1,2,3,−2) = (−4,8,12,−8). (g) −5(1,−4,3,−2,5) = (−5,20,−15,10,−25).

(h) 3(3,0,4,2,−1) = (9,0,12,6,−3).

6. (a) **u** + **w** = (1,2) + (−3,5) = (−2,7), (b) **u** + 3**v** = (1,2) + 3(4,−1) = (13,−1).

(c) **v** + **w** = (4,−1) + (−3,5) = (1,4).

(d) 2**u** + 3**v** − **w** = 2(1,2) + 3(4,−1) − (−3,5) = (17,−4).

(e) −3**u** + 4**v** − 2**w** = −3(1,2) + 4(4,−1) − 2(−3,5) = (19,−20).

7. (a) **u** + **w** = (2,1,3) + (2,4,−2) = (4,5,1). (b) 2**u** + **v** = 2(2,1,3) + (−1,3,2) = (3,5,8).

(c) **u** + 3**w** = (2,1,3) + 3(2,4,−2) = (8,13,−3).

(d) 5**u** − 2**v** + 6**w** = 5(2,1,3) − 2(−1,3,2) + 6(2,4,−2) = (24,23,−1).

(e) 2**u** − 3**v** − 4**w** = 2(2,1,3) − 3(−1,3,2) − 4(2,4,−2) = (−1,−23,8).

8. (a) $\mathbf{u} + \mathbf{v} = \begin{bmatrix} 2 \\ 3 \end{bmatrix} + \begin{bmatrix} -1 \\ -4 \end{bmatrix} = \begin{bmatrix} 1 \\ -1 \end{bmatrix}$. (b) $2\mathbf{v} - 3\mathbf{w} = 2\begin{bmatrix} -1 \\ -4 \end{bmatrix} - 3\begin{bmatrix} 4 \\ -6 \end{bmatrix} = \begin{bmatrix} -14 \\ 10 \end{bmatrix}$.

(c) $2\mathbf{u} + 4\mathbf{v} - \mathbf{w} = 2\begin{bmatrix} 2 \\ 3 \end{bmatrix} + 4\begin{bmatrix} -1 \\ -4 \end{bmatrix} - \begin{bmatrix} 4 \\ -6 \end{bmatrix} = \begin{bmatrix} -4 \\ -4 \end{bmatrix}$.

(d) $-3\mathbf{u} - 2\mathbf{v} + 4\mathbf{w} = -3\begin{bmatrix} 2 \\ 3 \end{bmatrix} - 2\begin{bmatrix} -1 \\ -4 \end{bmatrix} + 4\begin{bmatrix} 4 \\ -6 \end{bmatrix} = \begin{bmatrix} 12 \\ -25 \end{bmatrix}$.

9. (a) $\mathbf{u} + 2\mathbf{v} = \begin{bmatrix} 1 \\ 2 \\ -1 \end{bmatrix} + 2\begin{bmatrix} 3 \\ 0 \\ 1 \end{bmatrix} = \begin{bmatrix} 7 \\ 2 \\ 1 \end{bmatrix}$. (b) $-4\mathbf{v} + 3\mathbf{w} = -4\begin{bmatrix} 3 \\ 0 \\ 1 \end{bmatrix} + 3\begin{bmatrix} -1 \\ 0 \\ 5 \end{bmatrix} = \begin{bmatrix} -15 \\ 0 \\ 11 \end{bmatrix}$.

(c) $3\mathbf{u} - 2\mathbf{v} + 4\mathbf{w} = 3\begin{bmatrix} 1 \\ 2 \\ -1 \end{bmatrix} - 2\begin{bmatrix} 3 \\ 0 \\ 1 \end{bmatrix} + 4\begin{bmatrix} -1 \\ 0 \\ 5 \end{bmatrix} = \begin{bmatrix} -7 \\ 6 \\ 15 \end{bmatrix}$.

$$2\mathbf{u} + 3\mathbf{v} - 8\mathbf{w} = 2\begin{bmatrix} 1 \\ 2 \\ -1 \end{bmatrix} + 3\begin{bmatrix} 3 \\ 0 \\ 1 \end{bmatrix} - 8\begin{bmatrix} -1 \\ 0 \\ 5 \end{bmatrix} = \begin{bmatrix} 19 \\ 4 \\ -39 \end{bmatrix}.$$

10. (a) a(1,2) + b(-1,3) = (1,7). a-b=1, 2a+3b=7. Unique solution a=2, b=1. (1,7)=2(1,2) + (-1,3).
(1,7) is a linear combination of (1,2) and (-1,3).

(b) a(1,1) + b(3,2) = (1,2). a+3b=1, a+2b=2. Unique solution a=4,b=-1. (1,2)=4(1,1) - (3,2).
(1,2) is a linear combination of (1,1) and (3,2).

(c) a(1,-3) + b(-2,6) = (3,5). a-2b=3, -3a+6b=5. No solution.
(3,5) is not a linear combination of (1,-3) and (-2,6).

(d) a(2,4) + b(-4,-8) = (6,2). 2a-4b=6, 4a-8b=2. No solution.
(6,2) is not a linear combination of (2,4) and (-4,-8).

11. (a) a(1,1,2) + b(1,2,1) + c(2,3,4) = (7,9,15). a+b+2c = 7, a+2b+3c = 9, 2a+b+4c = 15.
Unique solution, a=2, b=-1, c=3. Is a linear combination.

(b) a(1,1,1) + b(1,2,4) + c(0,1,-3) = (6,13,9). a+b = 6, a+2b+c = 13, a+4b-3c = 9.
Unique solution, a=2, b=4, c=3. Is a linear combination.

(c) a(1,2,0) + b(-1,-1,2) + c(1,3,2) = (1,2,-1). a-b+c = 1, 2a-b+3c = 2, 2b+2c = -1.
No solution . Not a linear combination.

(d) a(1,2,3) + b(2,5,7) + c(0,0,1) = (-1,-1,2). a+2b = -1, 2a+5b = -1, 3a+7b+c = 2.
Unique solution, a=-3, b=1, c=4. Is a linear combination.

(e) a(1,1,1) + b(1,2,-1) + c(5,7,1) = (5,8,1). a+b+5c = 5, a+2b+7c = 8, a-b+c = 1.
No solution . Not a linear combination.

(f) $a(1,1,2) + b(2,2,4) + c(1,-1,1) = (5,-1,7)$. $a+2b+c = 5$, $a+2b-c = -1$, $2a+4b+c = 7$.
Many solutions, $a=-2r+2$, $b=r$, $c=3$. $(5,-1,7) = (-2r+2)(1,1,2) + r(2,2,4) + 3(1,-1,1)$.
There are many linear combinations.
For example, when $r=1$ we get $(5,-1,7) = 0(1,1,2) + (2,2,4) + 3(1,-1,1)$.
When $r=2$, $(5,-1,7) = -2(1,1,2) + 2(2,2,4) + 3(1,-1,1)$.

12. (a) $\mathbf{u} + (\mathbf{v} + \mathbf{w}) = (u_1, u_2, \ldots, u_n) + ((v_1, v_2, \ldots, v_n) + (w_1, w_2, \ldots, w_n))$
$= (u_1, u_2, \ldots, u_n) + (v_1 + w_1, v_2 + w_2, \ldots, v_n + w_n)$
$= (u_1 + (v_1 + w_1), u_2 + (v_2 + w_2), \ldots, u_n + (v_n + w_n))$
$= ((u_1 + v_1) + w_1, (u_2 + v_2) + w_2, \ldots, (u_n + v_n) + w_n)$
$= ((u_1 + v_1), (u_2 + v_2), \ldots, (u_n + v_n)) + (w_1, w_2, \ldots, w_n)$
$= ((u_1, u_2, \ldots, u_n) + (v_1, v_2, \ldots, v_n)) + (w_1, w_2, \ldots, w_n) = (\mathbf{u} + \mathbf{v}) + \mathbf{w}$.

(b) $\mathbf{u} + (-\mathbf{u}) = (u_1, u_2, \ldots, u_n) + (-1)(u_1, u_2, \ldots, u_n)$
$= (u_1, u_2, \ldots, u_n) + (-u_1, -u_2, \ldots, -u_n) = (u_1 - u_1, u_2 - u_2, \ldots, u_n - u_n)$
$= (0,0, \ldots, 0) = \mathbf{0}$.

(c) $(c+d)\mathbf{u} = (c+d)(u_1, u_2, \ldots, u_n) = ((c+d)u_1, (c+d)u_2, \ldots, (c+d)u_n)$
$= (cu_1 + du_1, cu_2 + du_2, \ldots, cu_n + du_n)$
$= (cu_1, cu_2, \ldots, cu_n) + (du_1, du_2, \ldots, du_n)$
$= c(u_1, u_2, \ldots, u_n) + d(u_1, u_2, \ldots, u_n) = c\mathbf{u} + d\mathbf{u}$.

(d) $1\mathbf{u} = 1(u_1, u_2, \ldots, u_n) = (1xu_1, 1xu_2, \ldots, 1xu_n) = (u_1, u_2, \ldots, u_n) = \mathbf{u}$.

<u>Exercise Set 1.4</u>

1. (a) Let W be the subset of vectors of the form (a,3a). Let **u**=(a,3a), **v**=(b,3b) and k be a scalar. Then **u**+**v**=((a+b), 3(a+b)) and k**u**=(ka,3ka).
The second component of both **u**+**v** and k**u** is 3 times the first component. Thus W is closed under addition and scalar multiplication - it is a subspace.

(b) Let W be the subset of vectors of the form (a,-a). Let **u**=(a,-a), **v**=(b,-b) and k be a scalar. Then **u**+**v**=((a+b), -(a+b)) and k**u**=(ka,-ka).
The second component of both **u**+**v** and k**u** is minus the first component. Thus W is closed under addition and scalar multiplication - it is a subspace.

(c) Let W be the subset of vectors of the form (a,0). Let **u**=(a,0), **v**=(b,0) and k be a scalar. Then **u**+**v**=((a+b), 0) and k**u**=(ka,0).
The second component of both **u**+**v** and k**u** is zero. Thus W is closed under addition and scalar multiplication - it is a subspace.

(d) Let W be the subset of vectors of the form (2a,3a). Let **u**=(2a,3a), **v**=(2b,3b) and k be a scalar. Then **u**+**v**=(2(a+b), 3(a+b)) and k**u**=(2ka,3ka).
The second component of both **u**+**v** and k**u** is 3/2 times the first component. Thus W is closed under addition and scalar multiplication - it is a subspace.

2. (a) Let W be the subset of vectors of the form (a,b,b). Let **u**=(a,b,b), **v**=(c,d,d) and k be a scalar. Then **u**+**v**=(a+c, b+d,b+d) and k**u**=(ka,kb,kb).
The second and third components of **u**+**v** are the same; so are those of k**u**. Thus W is closed under addition and scalar multiplication - it is a subspace.

(b) Let W be the subset of vectors of the form (a,-a,b). Let **u**=(a,-a,b), **v**=(c,-c,d) and k be a scalar. Then **u**+**v**=(a+c,-(a+c),b+d) and k**u**=(ka,-ka,kb).
The second component of **u**+**v** equals minus the first component; and same for k**u**. Thus
W is closed under addition and scalar multiplication - it is a subspace.

(c) Let W be the subset of vectors of the form (a,2a,-a). Let **u**=(a,2a,-a), **v**=(b,2b,-b) and k be a scalar. Then **u**+**v**=(a+b,2(a+b),-(a+b)) and k**u**=(ka,2ka,-ka).
The second component of **u**+**v** is twice the first, and the third component is minus the first; and same for k**u**. Thus W is closed under addition and scalar multiplication - it is a subspace.

(d) Let W be the subset of vectors of the form (a,a,b,b). Let **u**=(a,a,b,b), **v**=(c,c,d,d) and k be a scalar. Then **u**+**v**=(a+c,a+c,b+d,b+d) and k**u**=(ka,ka,kb,kb).
The 1st and 2nd components of **u**+**v** are the same, so are 3rd and 4th; same for k**u**. Thus W is closed under addition and scalar multiplication - it is a subspace.

3. (a) Let W be the subset of vectors of the form (a,b,2a+3b). Let **u**=(a,b,2a+3b), **v**=(c,d,2c+3d) and k be a scalar. Then **u**+**v**=(a+c, b+d,2(a+c)+3(b+d)) and k**u**=(ka,kb,2ka+3kb).The third component of **u**+**v** is twice the first plus three times the second; same for k**u**. Thus W is closed under addition and scalar multiplication - it is a subspace.

(b) Let W be the subset of vectors of the form (a,b,3). Let **u**=(a,b,3), **v**=(c,d,3) and k be a scalar. Then **u**+**v**=(a+c,b+d,6). The third component is not 6. Thus **u**+**v** is not in W. W is not a subspace. Let us check closure under scalar multiplication. k**u**=(ka,kb,3k). Thus unless k=1, k**u** is not in W. W is not closed under scalar multiplication either.

(c) Let W be the subset of vectors of the form (a,a+2,b). Let **u**=(a,a+2,b), **v**=(c,c+2,d) and k be a scalar. Then **u**+**v**=(a+c,a+c+4,b+d). The second component is not the first plus 2. Thus **u**+**v** is not in W. W is not a subspace. Let us check closure under scalar multiplication. k**u**=(ka,ka+2k,kb). Thus unless k=1 k**u** is not in W. W is not closed under scalar multiplication either.

(d) Let W be the subset of vectors of the form (a,-a,0). Let **u**=(a,-a,0), **v**=(b,-b,0) and k be a scalar. Then **u**+**v**=(a+b,-(a+b),0) and

ku=(ka,-ka,0).The second component of **u**+**v** is minus the first and the last component is zero; same for **ku**. Thus W is closed under addition and scalar multiplication - it is a subspace.

4. $\begin{bmatrix} 1 & 3 & 1 & 0 \\ 0 & 1 & 1 & 0 \\ 2 & 7 & 3 & 0 \end{bmatrix} \approx \begin{bmatrix} 1 & 3 & 1 & 0 \\ 0 & 1 & 1 & 0 \\ 0 & 1 & 1 & 0 \end{bmatrix} \approx \begin{bmatrix} 1 & 0 & -2 & 0 \\ 0 & 1 & 1 & 0 \\ 0 & 0 & 0 & 0 \end{bmatrix}$. General solution (2r,-r,r).

Let W be the subset of vectors of the form (2r,-r,r). Let **u**=(2r,-r,r), **v**=(2s,-s,s) and k be a scalar. Then **u**+**v**=(2(r+s),-(r+s),r+s) and k**u**=(2kr, -kr, kr).The first component of **u**+**v** is twice the last component, and the second component is minus the last component; same for k**u**. Thus W is closed under addition and scalar multiplication - it is a subspace. Line defined by vector (2, -1, 1).

5. $\begin{bmatrix} 1 & 1 & -7 & 0 \\ 0 & 1 & -4 & 0 \\ 1 & 0 & -3 & 0 \end{bmatrix} \approx \begin{bmatrix} 1 & 1 & -7 & 0 \\ 0 & 1 & -4 & 0 \\ 0 & -1 & 4 & 0 \end{bmatrix} \approx \begin{bmatrix} 1 & 0 & -3 & 0 \\ 0 & 1 & -4 & 0 \\ 0 & 0 & 0 & 0 \end{bmatrix}$. General solution (3r,4r,r).

Let W be the subset of vectors of the form (3r,4r,r). Let **u**=(3r,4r,r), **v**=(3s,4s,s) and k be a
scalar. Then **u**+**v**=(3(r+s),4(r+s),r+s) and k**u**=(3kr, 4kr, kr).The first component of **u**+**v** is three times the last component, and the second component is four times the last component; same for k**u**. Thus W is closed under addition and scalar multiplication - it is a subspace. It is a line defined by vector (3,4,1).

6. $\begin{bmatrix} 1 & -2 & 3 & 0 \\ -1 & 2 & 1 & 0 \\ 1 & -2 & 4 & 0 \end{bmatrix} \approx \begin{bmatrix} 1 & -2 & 3 & 0 \\ 0 & 0 & 4 & 0 \\ 0 & 0 & 1 & 0 \end{bmatrix} \approx \begin{bmatrix} 1 & -2 & 0 & 0 \\ 0 & 0 & 1 & 0 \\ 0 & 0 & 0 & 0 \end{bmatrix}$. General solution (2r, r, 0).

Let W be the subset of vectors of the form (2r,r,0). Let **u**=(2r,r,0), **v**=(2s,s,0) and k be a scalar. Then **u**+**v**=(2(r+s),r+s,0) and k**u**=(2kr, kr, 0).The first component of **u**+**v** is twice the second component, and the last component is zero; same for k**u**. Thus W is closed under addition and scalar multiplication - it is a subspace. Line defined by vector (2, 1, 0).

7. $\begin{bmatrix} 1 & 2 & -1 & 0 \\ 1 & 3 & 1 & 0 \\ 3 & 7 & -1 & 0 \end{bmatrix} \approx \begin{bmatrix} 1 & 2 & -1 & 0 \\ 0 & 1 & 2 & 0 \\ 0 & 1 & 2 & 0 \end{bmatrix} \approx \begin{bmatrix} 1 & 0 & -5 & 0 \\ 0 & 1 & 2 & 0 \\ 0 & 0 & 0 & 0 \end{bmatrix}$. General solution (5r,-2r,r).

Let W be the subset of vectors of the form (5r,-2r,r). Let **u**=(5r,-2r,r), **v**=(5s,-2s,s) and k be a scalar. Then **u**+**v**=(5(r+s),-2(r+s),r+s) and k**u**=(5kr, -2kr, kr).The first component of **u**+**v** is five times the last component, and the second component is minus two times the last component; same for k**u**. Thus W is closed under addition and scalar multiplication - it is a subspace. Line defined by vector (5, -2, 1).

8. General solution is (2r-2s, r -3s, r, s). Let **u**=(2r-2s, r -3s, r, s) and **v**=(2p-2q, p-3q, p, q). Then **u**+**v**=(2r-2s+2p-2q, r-3s+p-3q, r+p, s+q) = (2(r+p)-2(s+q), (r+p)-3(s+q), r+p, s+q). The first component is twice the third minus twice the fourth. The second component is the third minus three times the fourth - the required form. Thus space is closed under addition. Let k be a scalar. Then k**u**= (2kr -2ks, kr -3ks, kr, ks) – the required form. Thus the set of solutions is closed under addition and scalar multiplication; it is a subspace.

9. General solution is (3r + s, -r - 4s, r, s). (a) Two specific solutions: r=1, s=1, **u**(4, -5, 1, 1); r=-1, s=2, **v**(-1, -7, -1, 2). (b) Other solutions: **u**+**v**=(3, -12, 0, 3) and say, -2**u**=(-8, 10, -2, -2), 4**v**=(-4, -28, -4, 8). (-2**u**)+(4**v**)=(-12, -18, -6, 6). (c) Are solutions for r=0,s=3; r=-2,s=-2; r=-4,s=8; r=-6,s=6 respectively.

10. General solution is (2r, 3r, r). (a) Two specific solutions: r=1 gives **u**(2, 3, 1); r=2 gives **v**(4, 6, 2). (b) Other solutions: **u**+**v** = (6, 9, 3); 4u = (8, 12, 4); -**v** = (-4, -6, - 2); **u** – **v** = (-2, -3, -1). Are solutions for r = 3. r = 4, r = -2, r = -1 respectively.

11. General solution is (2r - s, -3r - 2s r, s). (a) Two specific solutions: r=1, s=1, **u**(1, -5, 1, 1); r=2, s=-1, **v**(5, -4, 2, -1). (b) Other solutions: **u**+**v**=(6, -9, 3, 0). and say 2**u**=(2, -10, 2, 2), 3**v**=(15, -12, 6, -3). (2**u**)+(3**v**)=(17, -22, 8, -1). (c) Are solutions for r=3,s=0; r=2,s=2; r=6,s=-3; r=8,s=-1 respectively.

Exercise Set 1.5

1. Standard basis for $\mathbf{R}^2$: {(1, 0), (0, 1)}. (a) Let (a, b) be an arbitrary vector in $\mathbf{R}^2$. We can write
(a, b) = a(1, 0) + b(0, 1). Thus vectors (1, 0) and (0, 1) span $\mathbf{R}^2$. (b) Let us examine the identity p(1, 0) + q(0, 1) = (0, 0). This gives (p, 0) + (0, q) = 0, (p, q) = (0, 0). Thus p=0 and q=0. The vectors are linearly independent.

2. Standard basis for $\mathbf{R}^4$: {(1,0,0,0), (0,1,0,0), (0,0,1,0), (0,0,0,1)}.
(a) Let (a,b,c,d) be an arbitrary vector in $\mathbf{R}^4$. We can write
(a,b,c,d) = a(1,0,0,0) +b(0,1,0,0) +c(0,0,1,0) +d(0,0,0,1). Thus vectors in basis span $\mathbf{R}^4$.
(b) Let us examine the identity p(1,0,0,0)+q(0,1,0,0)+r(0,0,1,0)+s(0,0,0,1) = (0,0,0,0). This gives (p,q,r,s) = 0. Thus p=0, q=0, r=0, s=0. The vectors are linearly independent.

3. (a) (a, a, b) + (c, c, d) = (a+c, a+c, b+d). 1st components same. Closed under addition. k(a, a, b) = (ka, ka, kb). 1st components same. Closed under scalar multiplication. Subspace. (a, a, b) = a(1, 1, 0) + b(0, 0, 1). Vectors (1, 1, 0) and (0, 0, 1) span space and are linearly independent. {(1, 1, 0), (0, 0, 1)} is a basis. Dimension is 2.

(b) (a, 2a, b) + (c, 2c, d) = (a+c, 2(a+c), b+d). 2nd component is twice first. Closed under addition. k(a, 2a, b) = (ka, 2ka, kb). 2nd component is twice 1st. Closed under scalar multiplication. Subspace. (a, 2a, b) = a(1, 2, 0) + b(0, 0, 1). Vectors (1, 2, 0) and (0, 0, 1) span the space and are linearly independent. {(1, 2, 0, (0, 0, 1)} is a basis. Dimension is 2.

(c) (a, 2a, 4a) + (b, 2b, 4b) = (a+b, 2a+2b, 4a+4b) =(a+b, 2(a+b), 4(a+b)). 2nd component is twice 1st, 3rd component four times 1st. Closed under addition. k(a, 2a, 4a) = (ka, 2ka, 4ka).
2nd component is twice 1st, 3rd component four times 1st. Closed under scalar multiplication. Subspace. (a, 2a, 4a) = a(1, 2, 4). {(1, 2, 4)} is a basis. Dimension is 1. Space is a line defined by the vector (1, 2, 4).

(d) (a, -a, 0) + (b, -b, 0) = (a+b, -a-b, 0) = (a+b, -(a+b), 0). 2nd component is negative of 1st. Closed under addition. k(a, -a, 0) = (ka, -ka, 0). 2nd component is negative first. Closed under scalar multiplication. Subspace. (a, -a, 0) = a(1, -1, 0). {(1, -1, 0)} is a basis. Dimension is 1. Space is line defined by the vector (1, -1, 0).

4. (a) (a, b, a) + (c, d, c) = (a+c, b+d, a+c). 3rd component same as 1st. Closed under addition. k(a, b, a) = (ka, kb, ka). 3rd component same as 1st. Closed under scalar multiplication. Subspace. (a, b, a) = a(1, 0, 1) + b(0, 1, 0). {(1, 0, 1), (0, 1, 0)} is a basis. Dimension is 2.

(b) (a, b, 0) + (c, d, 0) = (a+c, b+d, 0). Last component is zero. Closed under addition. k(a, b, 0) = (ka, kb, 0). Last component is zero. Closed under scalar multiplication. Subspace. (a, b, 0) = a(1, 0, 0) + b(0, 1, 0). {(1, 0, 0), (0, 1, 0)} is a basis. Dimension is 2. It is xy-plane.

(c) (a, b, 2) + (c, d, 2) = (a+c, b+d, 4). Last component not 2. Not closed under addition. Not a subspace.

(d) (a, a, a+3) + (b, b, b+3) = (a+b, a+b, a+b+6). Last component is not 1st plus 3. Not closed under addition. Not a subspace.

5. (a) a(1, 2, 3) + b(1, 2, 3) = (a+b)(1, 2, 3). Sum is a scalar multiple of (1, 2, 3). Closed under addition. ka(1, 2, 3) = (ka)(1, 2, 3). It is a scalar multiple of (1, 2, 3). Closed under scalar multiplication. Subspace of $\mathbf{R}^3$. Basis {(1, 2, 3)}. Dimension is 1. Space is line defined by vector (1, 2, 3).

(b) (a, 0, 0) + (b, 0, 0) = (a+b, 0, 0). Last two components zero. Closed under addition. k(a, 0, 0) = (ka, 0, 0). Last two components zero. Closed under scalar multiplication. Subspace of $\mathbf{R}^3$. Basis {(1, 0, 0)}. Dimension is 1. Space is the x-axis.

(c) (a, 2a) + (b, 2b) = (a+b, 2a+2b) = (a+b, 2(a+b)). 2nd component is twice 1st. Closed under addition. k(a, 2a) = (ka, 2ka). 2nd component is twice 1st. Closed under scalar multiplication. Subspace. Basis {(1, 2)}. Dimension is 1. Space is line defined by vector (1, 2) in $\mathbf{R}^2$.

(d) (a, b, c, 1) + (d, e, f, 1) = (a+d, b+e, c+f, 2). Last component is not 1. Not closed under addition. Not a subspace of $\mathbf{R}^4$.

6. (a) True: Arbitrary vector can be expressed as a linear combination of (1, 0) and (0, 1). (a, b) = a(1, 0) + b(0, 1).

(b) True: (a, b) = a(1, 0) + b(0, 1) + 0(1, 1). (Some of the scalars can be zero).

(c) True: p(1, 0) + q(0, 1) = (0, 0) has the unique solution p=0, q=0.

(d) False: Consider the identity p(1, 0) + q(0, 1) + r(0, 2) = (0, 0). Does this have the unique solution p=0, q=0, r=0? No, can have 0(1, 0) + 2(0, 1) - 1(0, 2) = (0, 0). Vectors are not linearly independent - we say that they are linearly dependent.

(e) True: (x, y) = x(1, 0) - y(0, -1). Thus (1, 0) and (0, -1) span $\mathbf{R}^2$. Further, p(1, 0)+q(0, -1)=(0, 0) has the unique solution p=0, q=0. Vectors are linearly independent.

(f) True: (x, y) = $\frac{x}{2}$(2, 0) + $\frac{y}{3}$(0, 3). Thus (2, 0) and (0, 3) span $\mathbf{R}^2$. Further, p(2, 0) + q(0, 3) = (0, 0) ⇔ (2p, 3q) = (0, 0), has the unique solution p=0, q=0. Vectors linearly independent.

7. (a) True: (1, 0, 0) and (0, 1, 0) span the subset of vectors of the form (a, b, 0). Further, p(1, 0, 0) + q(0, 1, 0) = (0, 0, 0) has the unique solution p=0, q=0. Vectors are linearly independent. Subspace is 2D since 2 base vectors. The subspace is the xy plane.

(b) True: The vector (1, 0, 0) spans the subset of vectors of the form a(1, 0, 0). Further p(1, 0, 0)=(0,0,0) has unique solution p=0. Subspace is line defined by vector (1, 0, 0). One vector in basis, thus 1D.

(c) True: Can write (a, 2a, b) in the form (a, 2a, b) = a(1, 2, 0) + b(0, 0, 1).

(d) True: Can write (a, b, 2a-b) in the form (a, b, 2a-b) = a(1, 0, 2) + b(0, 1, -1).

(e) False: 1(1, 0, 0) + 1(0, 1, 0) -1(1, 1, 0) = (0, 0, 0). Thus vectors not linearly independent.

(f) False: $\mathbf{R}^2$ is not a subset of $\mathbf{R}^3$. e.g., (1, 2) is an element of $\mathbf{R}^2$, but not of $\mathbf{R}^3$.

8. (a) Let (x, y) be an arbitrary vector in $\mathbf{R}^2$. Then (x, y) = x(1, 0) + y(0, 1). Thus (1, 0), (0, 1) span $\mathbf{R}^2$. Notice that both vectors are needed to span $\mathbf{R}^2$ - we cannot just use one of them. Further, a(1, 0) + b(0, 1) = (0, 0) => (a, b) = (0, 0) => a=0, and b=0. Thus (1, 0) and (0, 1) are linearly independent. They form a basis for $\mathbf{R}^2$.

(b) (x, y) can be expressed (x, y) = x(1, 0) + 3y(0, 1) - y(0, 2), or (x, y) = x(1, 0) + 5y(0, 1) - 2y(0, 2); there are many ways. Thus (1, 0), (0, 1), (0, 2) spans $\mathbf{R}^2$. But it is not an efficient spanning set. The vector (0, 2) is not really needed.

(c) 0(1, 0) + 2(0, 1) - (0, 2) = (0, 0). Thus {(1, 0), (0, 1), (0, 2)} is linearly dependent. It is not a basis.

9. (a) Let (x, y) be an arbitrary vector in $\mathbf{R}^2$. Then (x, y) = x(1, 0) + y(0, 1). Thus (1, 0), (0, 1) span $\mathbf{R}^2$. Notice that both vectors are needed to span $\mathbf{R}^2$ - we cannot just use one of them. Further, a(1, 0) + b(0, 1) = (0, 0) => (a, b) = (0, 0) => a=0, and b=0. Thus (1, 0) and (0, 1) are linearly independent. They form a basis for $\mathbf{R}^2$.

(b) (x, y) can be expressed (x, y) = x(1, 0) + y(0, 1) + 0(1, 1); or (x, y) = (x+y)(1, 0) + 2y(0, 1) - y(1,1) - there are many ways. Thus (1, 0), (0, 1), (1, 1) span $\mathbf{R}^2$. But it is not an efficient spanning set. The vector (1, 1) is not really needed.

(c) 1(1, 0) + 1(0, 1) - 1(1, 1) = (0, 0). Thus {(1, 0), (0, 1), (1, 1)} is linearly dependent. It is not a basis.

10. (a) Let (x, y, z) be an arbitrary vector in $\mathbf{R}^3$. Then (x, y, z) = x(1, 0, 0) + y(0, 1, 0) + z(0, 0, 1). Thus (1, 0, 0), (0, 1, 0), (0, 0, 1) span $\mathbf{R}^3$. Notice that all three vectors are needed to span $\mathbf{R}^3$ - we cannot just use two of them. Further, a(1, 0, 0) + b(0, 1, 0) + c(0, 0, 1) = (0, 0, 0) => (a, b, c) = (0, 0, 0) => a=0, b=0, and c=0. Thus (1, 0, 0) and (0, 1, 0, (0, 0, 1) are linearly independent. They form a basis for $\mathbf{R}^3$.

(b) (x, y, z) can be expressed (x, y, z) = x(1, 0, 0) + y(0, 1, 0) + z(0, 0, 1) + 0(0, 1, 1); or

$$(x, y, z) = x(1, 0, 0) + \frac{(y-z)}{2}(0, 1, 0) + \frac{(-y+z)}{2}(0, 0, 1) + \frac{(y+z)}{2}(0, 1, 1).$$

- there are many ways. Thus (1, 0, 0), (0, 1, 0), (0, 0, 1), (0, 1, 1) span $\mathbf{R}^2$. But it is not an efficient spanning set. The vector (0, 1, 1) is not really needed.

(c) 1(1, 0, 0) + 1(0, 1, 0) + 1(0, 0, 1) - 0(0, 1, 1) = (0, 0, 0). Thus (1, 0, 0), (0, 1, 0), (0, 0, 1), (0, 1, 1) are linearly dependent. Do not form a basis.

11. (a) Vectors (1, 0, 0), (0, 1, 0), (0, 0, 1), and (1, 1, 1) span $\mathbf{R}^3$ but are not linearly independent.

(b) Vectors (1, 0, 0), (0, 1, 0) are linearly independent but do not span $\mathbf{R}^3$.

12. Separate the variables in the general solution, (2r - 2s, r - 3s, r, s)= r(2, 1, 1, 0) + s(-2, -3, 0, 1).

Vectors (2, 1, 1, 0) and (-2, 3, 0, 1) thus span W.

Also, identity p(2, 1, 1, 0) + q(-2, 3, 0, 1) = (0, 0, 0, 0) leads to p=0, q=0. The two vectors are

thus linearly independent. Set {(2, 1, 1, 0), (-2, 3, 0, 1)} is therefore a basis for W.

Dimension of W is 2. Solutions form a plane through the origin in $\mathbf{R}^4$.

13. Separate the variables in the general solution, (3r + s, -r - 4s, r, s)= r(3, -1, 1, 0)+s(1, -4, 0, 1).
Vectors (3, -1, 1, 0) and (1, -4, 0, 1) thus span W.
Also, identity p(3, -1, 1, 0) + q(1, -4, 0, 1) = (0, 0, 0, 0) leads to p=0, q=0. The two vectors are thus linearly independent. Set {(3, -1, 1, 0), (1, -4, 0, 1)} is therefore a basis for W.
Dimension of W is 2. Solutions form a plane through the origin in $\mathbf{R}^4$.

14. (2r, 3r, r) = r(2, 3, 1). The set of solutions form the line through the origin in $\mathbf{R}^3$ defined by the vector (2, 3, 1). Set {(2, 3, 1)} is a basis. Dimension of W is 1.

15. Separate the variables in the solution, (2r - s, -3r - 2s, r, s)= r(2, -3, 1, 0) + s(-1, -2, 0, 1).
Vectors (2, -3, 1, 0) and (-1, -2, 0, 1) thus span W.
Also, identity p(2, -3, 1, 0) + q(-1, -2, 0, 1) = (0, 0, 0, 0) leads to p=0, q=0. The two vectors are thus linearly independent. The set {(2, -3, 1, 0), (-1, -2, 0, 1)} is therefore a basis for W.
Dimension of W is 2. Solutions form a plane through the origin in $\mathbf{R}^4$.

16. (2r, -r, 4r, r) = r(2, -1, 4, 1). The set of solutions form the line through the origin in $\mathbf{R}^4$ defined by the vector (2,-1,4,1). Set {(2, -1, 4)} is a basis. Dimension of W is 1.

17. Separate the variables in the solution, (3r-s, r, s)= r(3, 1, 0) + s(-1, 0, 1).
Vectors (3, 1, 0) and (-1, 0, 1) thus span W.
Also, identity p(3, 1, 0) + q(-1, 0, 1) = (0, 0, 0) leads to p=0, q=0. The two vectors are thus linearly independent. The set {(3, 1, 0), (-1, 0, 1)} is therefore a basis for W.
Dimension of W is 2. Solutions form a plane through the origin in $\mathbf{R}^3$.

18. Separate the variables in the solution, (2r + t, 3r – 2s, r, s, t) = r(2, 3, 1, 0, 0) + s(0, -2, 0, 1, 0) + t(1, 0, 0, 0, 1). Vectors (2, 3, 1, 0, 0), (0, -2, 0, 1, 0), (1, 0, 0, 0, 1) thus span W. The identity p(2, 3, 1, 0, 0) + q(0, -2, 0, 1, 0) + h(1, 0, 0, 0, 1) = (0, 0, 0, 0, 0) leads to p = 0, q = 0, h = 0. The vectors are thus linearly independent. The set

{(2, 3, 1, 0, 0), (0, -2, 0, 1, 0), (1, 0, 0, 0, 1)} is a basis for W. The dimension of W is 3. Solutions form a three-dimensional subspace of the five-dimensional space $\mathbf{R}^5$.

19. (a) a(1, 0, 2) + b(1, 1, 0) + c(5, 3, 6)=**0** gives a+b+5c=0, b+3c=0, 2a+6c=0.
Unique solution a=0, b=0, c=0. Thus linearly independent.

(b) a(1,1,1) + b(2, -1, 1) + c(3, -3, 0)=**0** gives a + 2b + 3c = 0, a – b -3c = 0, a + b = 0.
Unique solution a=0, b=0, c=0. Thus linearly independent.

(c) a(1, -1, 1) +b(2, 1, 0)+c(4, -1, 2)=**0** gives a+2b+4c=0, -a+b-c=0, a+2c=0.
Many solutions, a=-2r, b=-r, c=r, where r is a real number. Thus linearly dependent.

(d) a(1, 2, 1)+b(-2, 1, 3)+c(-1, 8, 9)=**0** gives a-2b-c=0, 2a+b+8c=0, a+3b+9c=0.
Many solutions, a=-3r, b=-2r, c=r. Thus linearly dependent.

(e) a(-2,0,3) + b(5,2,1) + c(10,6,9)=0 gives –2a+5b+10c=0, 2b+6c=0, 3a+b+9c=0.
Unique solution a=0, b=0, c=0. Thus linearly independent.

(f) a(3,4,1)+b(2,1,0)+c(9,7,1)=**0** gives 3a+2b+9c=0, 4a+b+7c=0, a+c=0.
Many solutions, a=-r, b=-3r, c=r. Thus linearly dependent.

Exercise Set 1.6

1. (a) (2,1)·(3,4) = 2x3 + 1x4 = 6 + 4 = 10

(b) (1,−4)·(3,0) = 1x3 + −4x0 = 3

(c) (2,0)·(0,−1) = 2x0 + 0x−1= 0

(d) (5,−2)·(−3,−4) = 5x−3 + −2x−4 = −15 + 8 = −7

2. (a) (1,2,3)·(4,1,0) = 1x4 + 2x1 + 3x0 = 4 + 2 + 0 = 6

(b) (3,4,−2)·(5,1,−1) = 3x5 + 4x1 + −2x−1 = 15 + 4 + 2 = 21

(c) (7,1,−2)·(3,−5,8) = 7x3 + 1x−5 + −2x8 = 21 − 5 − 16 = 0

(d) (3, 2, 0).(5, -2, 8) = 3x5 + 2x-2 + 0x8 = 15 - 4 + 0 = 11

3. (a) (5,1)·(2,−3) = 5x2 + 1x−3 = 10 − 3 = 7

(b) $(-3,1,5)\cdot(2,0,4) = -3x2 + 1x0 + 5x4 = -6 + 0 + 20 = 14$

(c) $(7,1,2,-4)\cdot(3,0,-1,5) = 7x3 + 1x0 + 2x-1 + -4x5 = 21 + 0 - 2 -20 = -1$

(d) $(2,3,-4,1,6)\cdot(-3,1,-4,5,-1) = 2x-3 + 3x1 + -4x-4 + 1x5 + 6x-1$
$= -6 + 3 + 16 + 5 - 6 = 12$

(e) $(1,2,3,0,0,0)\cdot(0,0,0,-2,-4,9) = 1x0 + 2x0 + 3x0 + 0x-2 + 0x-4 + 0x9 = 0$

4. (a) $\begin{bmatrix} 1 \\ 3 \end{bmatrix} \cdot \begin{bmatrix} -2 \\ 5 \end{bmatrix} = 1x\text{-}2 + 3x5 = -2 + 15 = 13$

(b) $\begin{bmatrix} 5 \\ 0 \end{bmatrix} \cdot \begin{bmatrix} 4 \\ -6 \end{bmatrix} = 5x4 + 0x\text{-}6 = 20 + 0 = 20$

(c) $\begin{bmatrix} 2 \\ 0 \\ -5 \end{bmatrix} \cdot \begin{bmatrix} 3 \\ 6 \\ -4 \end{bmatrix} = 2x3 + 0x6 + \text{-}5x\text{-}4 = 6 + 0 + 20 = 26$

(d) $\begin{bmatrix} 1 \\ 3 \\ -7 \end{bmatrix} \cdot \begin{bmatrix} -2 \\ 8 \\ -3 \end{bmatrix} = 1x\text{-}2 + 3x8 + \text{-}7x\text{-}3 = -2 + 24 + 21 = 43$

5. (a) $\|(1, 2)\| = \sqrt{1^2 + 2^2} = \sqrt{5}$

(b) $\|(3, -4)\| = \sqrt{3^2 + (-4)^2} = \sqrt{25} = 5$

(c) $\|(4, 0)\| = \sqrt{4^2 + 0^2} = \sqrt{16} = 4$

(d) $\|(-3, 1)\| = \sqrt{(-3)^2 + 1^2} = \sqrt{10}$

(e) $\|(0, 27)\| = \sqrt{0^2 + 27^2} = 27$

6. (a) $\|(1,3,-1)\| = \sqrt{1^2 + 3^2 + (-1)^2} = \sqrt{11}$

(b) $\|(3,0,4)\| = \sqrt{3^2 + 0^2 + 4^2} = \sqrt{25} = 5$

(c) $\|(5,1,1)\| = \sqrt{5^2 + 1^2 + 1^2} = \sqrt{27} = 3\sqrt{3}$

(d) $\|(0,5,0)\| = \sqrt{0^2 + 5^2 + 0^2} = \sqrt{25} = 5$

(e) $\|(7,-2,-3)\| = \sqrt{7^2 + (-2)^2 + (-3)^2} = \sqrt{62}$

7. (a) $||(5,2)|| = \sqrt{5^2 + 2^2} = \sqrt{29}$

(b) $||(-4,2,3)|| = \sqrt{(-4)^2 + 2^2 + 3^2} = \sqrt{29}$

(c) $||(1,2,3,4)|| = \sqrt{1^2 + 2^2 + 3^2 + 4^2} = \sqrt{30}$

(d) $||(4,-2,1,3)|| = \sqrt{4^2 + (-2)^2 + 1^2 + 3^2} = \sqrt{30}$

(e) $||(-3,0,1,4,2)|| = \sqrt{(-3)^2 + 0^2 + 1^2 + 4^2 + 2^2} = \sqrt{30}$

(f) $||(0,0,0,7,0,0)|| = \sqrt{0^2 + 0^2 + 0^2 + 7^2 + 0^2 + 0^2} = \sqrt{49} = 7$

8. (a) $\left|\left|\begin{bmatrix}3\\4\end{bmatrix}\right|\right| = \sqrt{3^2 + 4^2} = \sqrt{25} = 5$

(b) $\left|\left|\begin{bmatrix}2\\-7\end{bmatrix}\right|\right| = \sqrt{2^2 + (-7)^2} = \sqrt{53}$

(c) $\left|\left|\begin{bmatrix}1\\2\\3\end{bmatrix}\right|\right| = \sqrt{1^2 + 2^2 + 3^2} = \sqrt{14}$

(d) $\left|\left|\begin{bmatrix}-2\\0\\5\end{bmatrix}\right|\right| = \sqrt{(-2)^2 + 0^2 + 5^2} =$ $\sqrt{29}$

(e) $\left|\left|\begin{bmatrix}2\\3\\5\\9\end{bmatrix}\right|\right| = \sqrt{2^2 + 3^2 + 5^2 + 9^2} = \sqrt{119}$

9. (a) $\frac{(1,3)}{||(1,3)||} = \left(\frac{1}{\sqrt{10}}, \frac{3}{\sqrt{10}}\right)$

(b) $\frac{(2,-4)}{||(2,-4)||} = \left(\frac{2}{2\sqrt{5}}, \frac{-4}{2\sqrt{5}}\right) = \left(\frac{1}{\sqrt{5}}, \frac{-2}{\sqrt{5}}\right)$

(c) $\frac{(1,2,3)}{||(1,2,3)||} = \left(\frac{1}{\sqrt{14}}, \frac{2}{\sqrt{14}}, \frac{3}{\sqrt{14}}\right)$

(d) $\frac{(-2,4,0)}{||(-2,4,0)||} = \left(\frac{-2}{\sqrt{20}}, \frac{4}{\sqrt{20}}, 0\right) = \left(\frac{-1}{\sqrt{5}}, \frac{2}{\sqrt{5}}, 0\right)$

(e) $\frac{(0,5,0)}{||(0,5,0)||} = (0,1,0)$

10. (a) $\frac{(4,2)}{||(4,2)||} = \left(\frac{4}{2\sqrt{5}}, \frac{2}{2\sqrt{5}}\right) = \left(\frac{2}{\sqrt{5}}, \frac{1}{\sqrt{5}}\right)$

(b) $\dfrac{(4,1,1)}{\|(4,1,1)\|} = \left(\dfrac{4}{3\sqrt{2}}, \dfrac{1}{3\sqrt{2}}, \dfrac{1}{3\sqrt{2}}\right)$

(c) $\dfrac{(7,2,0,1)}{\|(7,2,0,1)\|} = \left(\dfrac{7}{3\sqrt{6}}, \dfrac{2}{3\sqrt{6}}, 0, \dfrac{1}{3\sqrt{6}}\right)$

(d) $\dfrac{(3,-1,1,2)}{\|(3,-1,1,2)\|} = \left(\dfrac{3}{\sqrt{15}}, \dfrac{-1}{\sqrt{15}}, \dfrac{1}{\sqrt{15}}, \dfrac{2}{\sqrt{15}}\right)$

(e) $\dfrac{(0,0,0,7,0,0)}{\|(0,0,0,7,0,0)\|} = (0,0,0,1,0,0)$

11. (a) $\begin{bmatrix} 4 \\ 3 \end{bmatrix} / \left\| \begin{bmatrix} 4 \\ 3 \end{bmatrix} \right\| = \begin{bmatrix} 4/5 \\ 3/5 \end{bmatrix}$

(b) $\begin{bmatrix} 1 \\ -3 \end{bmatrix} / \left\| \begin{bmatrix} 1 \\ -3 \end{bmatrix} \right\| = \begin{bmatrix} 1/\sqrt{10} \\ -3/\sqrt{10} \end{bmatrix}$

(c) $\begin{bmatrix} 3 \\ 4 \\ 0 \end{bmatrix} / \left\| \begin{bmatrix} 3 \\ 4 \\ 0 \end{bmatrix} \right\| = \begin{bmatrix} 3/5 \\ 4/5 \\ 0 \end{bmatrix}$

(d) $\begin{bmatrix} -1 \\ 2 \\ -5 \end{bmatrix} / \left\| \begin{bmatrix} -1 \\ 2 \\ -5 \end{bmatrix} \right\| = \begin{bmatrix} -1/\sqrt{30} \\ 2/\sqrt{30} \\ -5/\sqrt{30} \end{bmatrix}$

(e) $\begin{bmatrix} 3 \\ 0 \\ 1 \\ 8 \end{bmatrix} / \left\| \begin{bmatrix} 3 \\ 0 \\ 1 \\ 8 \end{bmatrix} \right\| = \begin{bmatrix} 3/\sqrt{74} \\ 0 \\ 1/\sqrt{74} \\ 8/\sqrt{74} \end{bmatrix}$

12. (a) $\cos\theta = \dfrac{(-1,1)\supseteq(0,1)}{\|(-1,1)\|\,\|(0,1)\|} = \dfrac{1}{\sqrt{2}}$. $\theta = \dfrac{\pi}{4} = 45°$

(b) $\cos\theta = \dfrac{(2,0)\cdot(1,\sqrt{3})}{\|(2,0)\|\,\|(1,\sqrt{3})\|} = \dfrac{2}{4} = \dfrac{1}{2}$. $\theta = \dfrac{\pi}{3} = 60°$

(c) $\cos\theta = \dfrac{(2,3)\cdot(3,-2)}{\|(2,3)\|\,\|(3,-2)\|} = 0$. $\theta = \dfrac{\pi}{2} = 90°$

(d) $\cos\theta = \dfrac{(5,2)\cdot(-5,-2)}{\|(5,2)\|\,\|(-5,-2)\|} = \dfrac{-29}{29} = -1$. $\theta = \pi = 180°$

13. (a) $\cos\theta = \dfrac{(4,-1)\cdot(2,3)}{\|(4,-1)\|\,\|(2,3)\|} = \dfrac{5}{\sqrt{17}\sqrt{13}}$ ($\theta = 70.3462^0$)

(b) $\cos\theta = \dfrac{(3,-1,2)\cdot(4,1,1)}{\|(3,-1,2)\|\,\|(4,1,1)\|} = \dfrac{13}{\sqrt{14}\sqrt{18}} = \dfrac{13}{6\sqrt{7}}$ ($\theta = 35.0229^0$)

(c) $\cos\theta = \dfrac{(2,-1,0)\cdot(5,3,1)}{\|(2,-1,0)\|\,\|(5,3,1)\|} = \dfrac{7}{\sqrt{5}\sqrt{35}} = \dfrac{7}{5\sqrt{7}} = \dfrac{\sqrt{7}}{5}$ ($\theta = 58.0519^0$)

(d) $\cos\theta = \dfrac{(7,1,0,0)\cdot(3,2,1,0)}{\|(7,1,0,0)\|\,\|(3,2,1,0)\|} = \dfrac{23}{\sqrt{50}\sqrt{14}} = \dfrac{23}{10\sqrt{7}}$ ($\theta = 29.6205^0$)

(e) $\cos\theta = \dfrac{(1,2,-1,3,1)\cdot(2,0,1,0,4)}{\|(1,2,-1,3,1)\|\,\|(2,0,1,0,4)\|} = \dfrac{5}{4\sqrt{21}}$ ($\theta = 74.1707^0$)

14. (a) $\cos\theta = \dfrac{\begin{bmatrix}1\\2\end{bmatrix}\cdot\begin{bmatrix}-1\\4\end{bmatrix}}{\left\|\begin{bmatrix}1\\2\end{bmatrix}\right\|\,\left\|\begin{bmatrix}-1\\4\end{bmatrix}\right\|} = \dfrac{7}{\sqrt{5}\sqrt{17}}$ ($\theta = 40.6013^0$)

(b) $\cos\theta = \dfrac{\begin{bmatrix}5\\1\end{bmatrix}\cdot\begin{bmatrix}0\\-3\end{bmatrix}}{\left\|\begin{bmatrix}5\\1\end{bmatrix}\right\|\,\left\|\begin{bmatrix}0\\-3\end{bmatrix}\right\|} = \dfrac{-3}{\sqrt{26}\sqrt{9}} = \dfrac{-1}{\sqrt{26}}$ ($\theta = 101.3099^0$)

(c) $\cos\theta = \dfrac{\begin{bmatrix}1\\-3\\0\end{bmatrix}\cdot\begin{bmatrix}2\\5\\-1\end{bmatrix}}{\left\|\begin{bmatrix}1\\-3\\0\end{bmatrix}\right\|\,\left\|\begin{bmatrix}2\\5\\-1\end{bmatrix}\right\|} = \dfrac{-13}{\sqrt{10}\sqrt{30}} = \dfrac{-13}{10\sqrt{3}}$ ($\theta = 138.6385^0$)

(d) $\cos\theta = \dfrac{\begin{bmatrix}-2\\3\\-4\end{bmatrix}\cdot\begin{bmatrix}2\\5\\-1\end{bmatrix}}{\left\|\begin{bmatrix}-2\\3\\-4\end{bmatrix}\right\|\,\left\|\begin{bmatrix}2\\5\\-1\end{bmatrix}\right\|} = \dfrac{15}{\sqrt{29}\sqrt{30}} = \dfrac{\sqrt{15}}{\sqrt{29}\sqrt{2}}$ ($\theta = 59.4329^0$)

15. (a) $(1,3)\cdot(3,-1) = 1\text{x}3 + 3\text{x}-1 = 0$, thus the vectors are orthogonal.

(b) $(-2,4)\cdot(4,2) = -2\text{x}4 + 4\text{x}2 = 0$, thus the vectors are orthogonal.

(c) $(3,0)\cdot(0,-2) = 3x0 + 0x{-2} = 0$, thus the vectors are orthogonal.

(d) $(7,-1)\cdot(1,7) = 7x1 + -1x7 = 0$, thus the vectors are orthogonal.

16. (a) $(3,-5)\cdot(5,3) = 3x5 + -5x3 = 0$, thus the vectors are orthogonal.

(b) $(1,2,-3)\cdot(4,1,2) = 1x4 + 2x1 + -3x2 = 0$, thus the vectors are orthogonal.

(c) $(7,1,0)\cdot(2,-14,3) = 7x2 + 1x{-14} + 0x3 = 0$, thus the vectors are orthogonal.

(d) $(5,1,0,2)\cdot(-3,7,9,4) = 5x{-3} + 1x7 + 0x9 + 2x4 = 0$, thus the vectors are orthogonal.

(e) $(1,-1,2,-5,9)\cdot(4,7,4,1,0) = 1x4 + -1x7 + 2x4 + -5x1 + 9x0 = 0$, thus the vectors are orthogonal.

17. (a) $\begin{bmatrix} 1 \\ 2 \end{bmatrix} \cdot \begin{bmatrix} -6 \\ 3 \end{bmatrix}$ = 1x-6 + 2x3 = 0, thus the vectors are orthogonal.

(b) $\begin{bmatrix} 5 \\ -2 \end{bmatrix} \cdot \begin{bmatrix} 4 \\ 10 \end{bmatrix}$ = 5x4 + -2x10 = 0, thus the vectors are orthogonal.

(c) $\begin{bmatrix} 4 \\ -1 \\ 0 \end{bmatrix} \cdot \begin{bmatrix} 2 \\ 8 \\ -1 \end{bmatrix}$ = 4x2 + -1x8 + 0x-1 = 0, thus the vectors are orthogonal.

(d) $\begin{bmatrix} -2 \\ 3 \\ 2 \end{bmatrix} \cdot \begin{bmatrix} 2 \\ 6 \\ -7 \end{bmatrix}$ = -2x2 + 3x6 + 2x-7 = 0, thus the vectors are orthogonal.

18. (a) If (a,b) is orthogonal to $(1,3)$, then $(a,b)\cdot(1,3) = a + 3b = 0$, so $a = -3b$. Thus any vector of the form $(-3b,b)$ is orthogonal to $(1,3)$.

(b) If (a,b) is orthogonal to $(7,-1)$, then $(a,b)\cdot(7,-1) = 7a - b = 0$, so $b = 7a$. Thus any vector of the form $(a,7a)$ is orthogonal to $(7,-1)$.

(c) If (a,b) is orthogonal to $(-4,-1)$, then $(a,b)\cdot(-4,-1) = -4a - b = 0$, so $b = -4a$. Thus any vector of the form $(a,-4a)$ is orthogonal to $(-4,-1)$.

(d) If (a,b) is orthogonal to $(-3,0)$, then $(a,b)\cdot(-3,0) = -3a = 0$, so $a = 0$. Thus any vector of the form $(0,b)$ is orthogonal to $(-3,0)$.

19. (a) If (a,b) is orthogonal to (5,−1), then (a,b)·(5,−1) = 5a − b = 0, so b = 5a. Thus any vector of the form (a,5a) is orthogonal to (5,−1).

(b) If (a,b,c) is orthogonal to (1,−2,3), then (a,b,c)·(1,−2,3) = a − 2b +3c = 0, so a = 2b−3c. Thus any vector of the form (2b−3c,b,c) is orthogonal to (1,−2,3).

(c) If (a,b,c) is orthogonal to (5,1,−1), then (a,b,c)·(5,1,−1) = 5a + b − c = 0, so c = 5a+b. Thus any vector of the form (a,b,5a+b) is orthogonal to (5,1,−1).

(d) If (a,b,c,d) is orthogonal to (5,0,1,1), then (a,b,c,d)·(5,0,1,1) = 5a + c + d = 0, so d = −5a−c. Thus any vector of the form (a,b,c,−5a−c) is orthogonal to (5,0,1,1).

(e) If (a,b,c,d) is orthogonal to (6,−1,2,3), then (a,b,c,d)·(6,−1,2,3) = 6a − b + 2c + 3d = 0, so b = 6a+2c+3d. Thus any vector of the form (a,6a+2c+3d, c,d) is orthogonal to (6,−1,2,3).

(f) If (a,b,c,d,e) is orthogonal to (0,−2,3,1,5), then (a,b,c,d,e)·(0,−2,3,1,5) = −2b +3c + d + 5e = 0, so d = 2b−3c−5e. Thus any vector of the form (a,b,c,2b−3c−5e,e) is orthogonal to (0,−2,3,1,5).

20. If (a,b,c) is orthogonal to both (1,2,−1) and (3,1,0), then (a,b,c)·(1,2,−1) = a + 2b − c = 0
and (a,b,c)·(3,1,0) = 3a + b = 0. These equations yield the solution b = −3a and c = −5a, so any vector of the form (a,−3a,−5a) is orthogonal to both (1,2,−1) and (3,1,0).

21. Let (a,b,c) be in W. Then (a,b,c) is orthogonal to (-1,1,1). (a,b,c)·(-1,1,1)=0, -a+b+c=0, c=a-b. W consists of vectors of the form (a,b,a-b). Separate the variables. (a,b,a-b)=a(1,0,1)+b(0,1,-1). (1,0,1), (0,1,-1) span W. Vectors are also linearly independent. {(1,0,1),(0,1,-1)} is a basis for W. The dimension of W is 2. It is a plane spanned by (1,0,1) and (0,1,-1).

22. Let (a,b,c) be in W. Then (a,b,c) is orthogonal to (-3,4,1). (a,b,c)·(-3,4,1)=0, -3a+4b+c=0, c=3a-4b. W consists of vectors of the form (a,b,3a-4b). Separate the variables. (a,b,3a-4b)=a(1,0,3)+b(0,1,-4). {(1,0,3), ((0,1,-4)} is a basis for W. The dimension of W is 2. It is a plane spanned by (1,0,3) and (0,1,-4).

23. Let (a,b,c) be in W. Then (a,b,c) is orthogonal to (1,-2,5). (a,b,c)·(1,-2,5)=0, a-2b+5c=0,
a=2b-5c. W consists of vectors of the form (2b-5c,b,c). Separate the variables. (2b-5c,b,c)=b(2,1,0)+c(-5,0,1). {(2,1,0), ((-5,0,1)} is a basis for W. The dimension of W is 2. It is a plane spanned by (2,1,0) and (-5,0,1).

24. Let (a,b,c,d) be in W. Then (a,b,c,d) is orthogonal to (1,-3,7,4).(a,b,c,d)·(1,-3,7,4)=0,

a-3b+7c+4d=0, a=3b-7c-4d. W consists of vectors of the form (3b-7c-d,b,c,d). Separate the variables. (3b-7c-4d,b,c,d)=b(3,1,0,0)+c(-7,0,1,0) +d(-4,0,0,1). {(3,1,0,0), (-7,0,1,0), (-4,0,0,1)} is a basis for W. The dimension of W is 3.

25. (a) $d = \sqrt{(6-2)^2+(5-2)^2} = 5.$ (b) $d = \sqrt{(3+4)^2+(1-0)^2} = \sqrt{50} = 5\sqrt{2}$.

(c) $d = \sqrt{(7-2)^2+(-3-2)^2} = \sqrt{50} = 5\sqrt{2}$. (d) $d = \sqrt{(1-5)^2+(-3-1)^2} = 4\sqrt{2}$.

26. (a) $d = \sqrt{(4-2)^2+(1+3)^2} = \sqrt{20} = 2\sqrt{5}$. (b) $d = \sqrt{(1-2)^2+(2-1)^2+(3-0)^2} = \sqrt{11}$.

(c) $d = \sqrt{(-3-4)^2+(1+1)^2+(2-1)^2} = \sqrt{54} = 3\sqrt{6}$.

(d) $d^2 = (5-2)^2 + (1-0)^2 + (0-1)^2 + (0-3)^2 = 20$, so $d = \sqrt{20} = 2\sqrt{5}$.

(e) $d^2 = (-3-2)^2 + (1-1)^2 + (1-4)^2 + (0-1)^2 + (2+1)^2 = 44$, so $d = \sqrt{44} = 2\sqrt{11}$.

27. (a) $(\mathbf{u} + \mathbf{v})\cdot\mathbf{w} = (u_1 + v_1)w_1 + (u_2 + v_2)w_2 + \dots + (u_n + v_n)w_n$
$= u_1 w_1 + v_1 w_1 + u_2 w_2 + v_2 w_2 + \dots + u_n w_n + v_n w_n$
$= u_1 w_1 + u_2 w_2 + \dots + u_n w_n + v_1 w_1 + v_2 w_2 + \dots + v_n w_n = \mathbf{u}\cdot\mathbf{w} + \mathbf{v}\cdot\mathbf{w}$.

(b) $c\mathbf{u}\cdot\mathbf{v} = cu_1 v_1 + cu_2 v_2 + \dots + cu_n v_n = c(u_1 v_1 + u_2 v_2 + \dots + u_n v_n) = c(\mathbf{u}\cdot\mathbf{v})$, and
$cu_1 v_1 + cu_2 v_2 + \dots + cu_n v_n = u_1 cv_1 + u_2 cv_2 + \dots + u_n cv_n = \mathbf{u}\cdot c\mathbf{v}$.

28. **u** is a positive scalar multiple of **v** so it has the same direction as **v**. The magnitude of **u** is

$$||\mathbf{u}|| = \frac{1}{||\mathbf{v}||}\sqrt{(v_1)^2+(v_2)^2+\dots+(v_n)^2} = \frac{||\mathbf{v}||}{||\mathbf{v}||} = 1,$$ so **u** is a unit vector.

29. **u** and **v** are orthogonal if and only if the cosine of the angle θ between them is zero.

$$\cos\theta = \frac{\mathbf{u}\cdot\mathbf{v}}{||\mathbf{u}||\,||\mathbf{v}||} = 0$$ if and only if $\mathbf{u}\cdot\mathbf{v} = 0$.

30. If $\mathbf{u}\cdot\mathbf{v} = \mathbf{u}\cdot\mathbf{w}$ then $\mathbf{u}\cdot(\mathbf{v}-\mathbf{w}) = 0$ for all vectors **u** in U. Since $\mathbf{v}-\mathbf{w}$ is a vector in U this means that $(\mathbf{v}-\mathbf{w})\cdot(\mathbf{v}-\mathbf{w}) = 0$. Therefore $\mathbf{v}-\mathbf{w} = \mathbf{0}$, so $\mathbf{v} = \mathbf{w}$.

31. $\mathbf{u}\cdot(a_1\mathbf{v}_1 + a_2\mathbf{v}_2 + \ldots + a_n\mathbf{v}_n) = \mathbf{u}\cdot(a_1\mathbf{v}_1) + \mathbf{u}\cdot(a_2\mathbf{v}_2 + \ldots + a_n\mathbf{v}_n)$
$= \mathbf{u}\cdot(a_1\mathbf{v}_1) + \mathbf{u}\cdot(a_2\mathbf{v}_2) + \mathbf{u}\cdot(a_3\mathbf{v}_3 + \ldots + a_n\mathbf{v}_n) = \ldots = \mathbf{u}\cdot(a_1\mathbf{v}_1) + \mathbf{u}\cdot(a_2\mathbf{v}_2) + \ldots$
$+ \mathbf{u}\cdot(a_n\mathbf{v}_n)$
$= a_1(\mathbf{u}\cdot\mathbf{v}_1) + a_2(\mathbf{u}\cdot\mathbf{v}_2) + \ldots + a_n(\mathbf{u}\cdot\mathbf{v}_n) = a_1\,\mathbf{u}\cdot\mathbf{v}_1 + a_2\,\mathbf{u}\cdot\mathbf{v}_2 + \ldots + a_n\,\mathbf{u}\cdot\mathbf{v}_n$.

32. (a) vector (b) not valid (c) not valid (d) scalar

(e) not valid (f) scalar (g) not valid

(h) not valid

33. $||c(3,0,4)|| = \sqrt{3c\times3c+4c\times4c} = |c|\sqrt{9+16} = 5|c| = 15$, so $|c| = 3$ and $c = \pm3$.

34. $||\mathbf{u}+\mathbf{v}||^2 = (u_1+v_1)^2 + (u_2+v_2)^2 + \ldots + (u_n+v_n)^2$

$= u_1^2 + 2u_1v_1 + v_1^2 + u_2^2 + 2u_2v_2 + v_2^2 + \ldots + u_n^2 + 2u_nv_n + v_n^2$

$= u_1^2 + u_2^2 + \ldots + u_n^2 + 2u_1v_1 + 2u_2v_2 + \ldots + 2u_nv_n + v_1^2 + v_2^2 + \ldots +$
v_n^2

$= u_1^2 + u_2^2 + \ldots + u_n^2 + 2(u_1v_1 + u_2v_2 + \ldots + u_nv_n) + v_1^2 + v_2^2 + \ldots + v_n^2$

$= ||\mathbf{u}||^2 + 2(\mathbf{u}\cdot\mathbf{v}) + ||\mathbf{v}||^2 = ||\mathbf{u}||^2 + ||\mathbf{v}||^2$ if and only if $\mathbf{u}\cdot\mathbf{v} = 0$, i.e., if and only if **u** and **v** are orthogonal.

35. $(a,b)\cdot(-b,a) = a\times -b + b\times a = 0$, so $(-b,a)$ is orthogonal to (a,b).

36. $(\mathbf{u}+\mathbf{v})\cdot(\mathbf{u}-\mathbf{v}) = (u_1+v_1)(u_1-v_1) + (u_2+v_2)(u_2-v_2) + \ldots + (u_n+v_n)(u_n-v_n)$
$= u_1^2 - v_1^2 + u_2^2 - v_2^2 + \ldots + u_n^2 - v_n^2$
$= u_1^2 + u_2^2 + \ldots + u_n^2 - v_1^2 - v_2^2 - \ldots - v_n^2 = ||\mathbf{u}|| - ||\mathbf{v}||$.

Thus $||\mathbf{u}|| - ||\mathbf{v}||=0$ if and only if $(\mathbf{u}+\mathbf{v})\cdot(\mathbf{u}-\mathbf{v}) = 0$. That is $||\mathbf{u}|| = ||\mathbf{v}||$ if and only if **u** + **v** and **u** - **v** are orthogonal.

37. (a) $||\mathbf{u}||^2 = u_1^2 + u_2^2 + \ldots + u_n^2 \geq 0$, so $||\mathbf{u}|| \geq 0$.

(b) $||\mathbf{u}|| = 0$ if and only if $u_1^2 + u_2^2 + \ldots + u_n^2 = 0$ if and only if $u_1 = u_2 = \ldots = u_n = 0$.

(c) $||c\mathbf{u}||^2 = (cu_1)^2 + (cu_2)^2 + \ldots + (cu_n)^2 = c^2(u_1^2 + u_2^2 + \ldots + u_n^2) = c^2\ ||\mathbf{u}||^2$, so $||c\mathbf{u}|| = |c|\ ||\mathbf{u}||$.

38. (a) $||\mathbf{u}|| = |u_1| + |u_2| + \ldots + |u_n| \geq 0$ since each term is equal to or greater than zero.

$|u_1| + |u_2| + \ldots + |u_n| = 0$ if and only if each term is zero.

$||c\mathbf{u}|| = |cu_1| + |cu_2| + \ldots + |cu_n| = |c||u_1| + |c||u_2| + \ldots + |c||u_n|$

$= |c|(|u_1| + |u_2| + \ldots + |u_n|) = |c|\ ||\mathbf{u}||$.

$||(1,2)|| = |1| + |2| = 3$, $||(-3,4)|| = |-3| + |4| = 7$, $||(1,2,-5)|| = |1| + |2| + |-5| = 8$,

and $||(0,-2,7)|| = |0| + |-2| + |7| = 9$.

(b) $||\mathbf{u}|| = \max\limits_{i=1,\ldots,n} |u_i| \geq 0$ since the absolute value of any number is equal to or greater than zero.

$\max\limits_{i=1,\ldots,n} |u_i| = 0$ if and only if all $|u_i| = 0$.

$||c\mathbf{u}|| = \max\limits_{i=1,\ldots,n} |cu_i| = |c| \max\limits_{i=1,\ldots,n} |u_i| = |c|\ ||\mathbf{u}||$.

$||(1,2)|| = |2| = 2$, $||(-3,4)|| = |4| = 4$, $||(1,2,-5)|| = |-5| = 5$, and $||(0,-2,7)|| = |7| = 7$.

39. (a) $d(\mathbf{x},\mathbf{y}) = ||\mathbf{x}-\mathbf{y}|| \geq 0$.

(b) $d(\mathbf{x},\mathbf{y}) = ||\mathbf{x}-\mathbf{y}|| = 0$ if and only if $\mathbf{x}-\mathbf{y} = \mathbf{0}$ if and only if $\mathbf{x} = \mathbf{y}$.

(c) $d(\mathbf{x},\mathbf{z}) = ||\mathbf{x}-\mathbf{y}+\mathbf{y}-\mathbf{z}|| \leq ||\mathbf{x}-\mathbf{y}|| + ||\mathbf{y}-\mathbf{z}|| = d(\mathbf{x},\mathbf{y}) + d(\mathbf{y},\mathbf{z})$, from the triangle inequality.

Exercise Set 1.7

Exercises 1, 2, and 3 can be solved simultaneously since the coefficient matrices are the same for all three.

$$\begin{aligned} a_0 + a_1 + a_2 &= b_1 \\ a_0 + 2a_1 + 4a_2 &= b_2 \\ a_0 + 3a_1 + 9a_2 &= b_3 \end{aligned}$$, where b_1 , b_2 , b_3 are the y values 2, 2, 4 in Exercise 1;

14, 22, 32 in Exercise 2; and 5, 7, 9 in Exercise 3.

$$\begin{bmatrix} 1 & 1 & 1 & 2 & 14 & 5 \\ 1 & 2 & 4 & 2 & 22 & 7 \\ 1 & 3 & 9 & 4 & 32 & 9 \end{bmatrix} \underset{\substack{R2+(-1)R1 \\ R3+(-1)R1}}{\approx} \begin{bmatrix} 1 & 1 & 1 & 2 & 14 & 5 \\ 0 & 1 & 3 & 0 & 8 & 2 \\ 0 & 2 & 8 & 2 & 18 & 4 \end{bmatrix} \underset{\substack{R1+(-1)R2 \\ R3+(-2)R2}}{\approx} \begin{bmatrix} 1 & 0 & -2 & 2 & 6 & 3 \\ 0 & 1 & 3 & 0 & 8 & 2 \\ 0 & 0 & 2 & 2 & 2 & 0 \end{bmatrix}$$

$$\underset{(1/2)R3}{\approx} \begin{bmatrix} 1 & 0 & -2 & 2 & 6 & 3 \\ 0 & 1 & 3 & 0 & 8 & 2 \\ 0 & 0 & 1 & 1 & 1 & 0 \end{bmatrix} \underset{\substack{R1+(2)R3 \\ R2+(-3)R3}}{\approx} \begin{bmatrix} 1 & 0 & 0 & 4 & 8 & 3 \\ 0 & 1 & 0 & -3 & 5 & 2 \\ 0 & 0 & 1 & 1 & 1 & 0 \end{bmatrix}$$, so the values of a_0, a_1, a_2

are 4, −3, 1 for Exercise 1; 8, 5, 1 for Exercise 2; and 3, 2, 0 for Exercise 3. Thus the equations of the polynomials are:

1. $4 - 3x + x^2 = y$ 2. $8 + 5x + x^2 = y$ 3. $3 + 2x = y$

4. $$\begin{aligned} a_0 + a_1 + a_2 &= 8 \\ a_0 + 3a_1 + 9a_2 &= 26 \\ a_0 + 5a_1 + 25a_2 &= 60 \end{aligned}$$ $$\begin{bmatrix} 1 & 1 & 1 & 8 \\ 1 & 3 & 9 & 26 \\ 1 & 5 & 25 & 60 \end{bmatrix} \underset{\substack{R2+(-1)R1 \\ R3+(-1)R1}}{\approx} \begin{bmatrix} 1 & 1 & 1 & 8 \\ 0 & 2 & 8 & 18 \\ 0 & 4 & 24 & 52 \end{bmatrix} (1/2)R2$$

$$\begin{bmatrix} 1 & 1 & 1 & 8 \\ 0 & 1 & 4 & 9 \\ 0 & 4 & 24 & 52 \end{bmatrix} \underset{\substack{R1+(-1)R2 \\ R3+(-4)R2}}{\approx} \begin{bmatrix} 1 & 0 & -3 & -1 \\ 0 & 1 & 4 & 9 \\ 0 & 0 & 8 & 16 \end{bmatrix} \underset{(1/8)R3}{\approx} \begin{bmatrix} 1 & 0 & -3 & -1 \\ 0 & 1 & 4 & 9 \\ 0 & 0 & 1 & 2 \end{bmatrix} \underset{\substack{R1+(3)R3 \\ R2+(-4)R3}}{\approx} \begin{bmatrix} 1 & 0 & 0 & 5 \\ 0 & 1 & 0 & 1 \\ 0 & 0 & 1 & 2 \end{bmatrix},$$

so $a_0 = 5$, $a_1 = 1$, $a_2 = 2$, and the equation is $5 + x + 2x^2 = y$.
When $x = 2$, $y = 5 + 2 + 8 = 15$.

5. $$\begin{aligned} a_0 - a_1 + a_2 &= -1 \\ a_0 \phantom{{}+a_1+a_2} &= 1 \\ a_0 + a_1 + a_2 &= -3 \end{aligned}$$ $$\begin{bmatrix} 1 & -1 & 1 & -1 \\ 1 & 0 & 0 & 1 \\ 1 & 1 & 1 & -3 \end{bmatrix} \underset{\substack{R2+(-1)R1 \\ R3+(-1)R1}}{\approx} \begin{bmatrix} 1 & -1 & 1 & -1 \\ 0 & 1 & -1 & 2 \\ 0 & 2 & 0 & -2 \end{bmatrix}$$

$$\underset{\substack{R1+R2 \\ R3+(-2)R2}}{\approx} \begin{bmatrix} 1 & 0 & 0 & 1 \\ 0 & 1 & -1 & 2 \\ 0 & 0 & 2 & -6 \end{bmatrix} \underset{(1/2)R3}{\approx} \begin{bmatrix} 1 & 0 & 0 & 1 \\ 0 & 1 & -1 & 2 \\ 0 & 0 & 1 & -3 \end{bmatrix} \underset{R2+R3}{\approx} \begin{bmatrix} 1 & 0 & 0 & 1 \\ 0 & 1 & 0 & -1 \\ 0 & 0 & 1 & -3 \end{bmatrix}$$

so $a_0 = 1$, $a_1 = -1$, $a_2 = -3$, and the equation is $1 - x - 3x^2 = y$.
When $x = 3$, $y = 1 - 3 - 27 = -29$.

6. $$\begin{aligned} a_0 + a_1 + a_2 + a_3 &= -3 \\ a_0 + 2a_1 + 4a_2 + 8a_3 &= -1 \\ a_0 + 3a_1 + 9a_2 + 27a_3 &= 9 \\ a_0 + 4a_1 + 16a_2 + 64a_3 &= 33 \end{aligned}$$

$$\begin{bmatrix} 1 & 1 & 1 & 1 & -3 \\ 1 & 2 & 4 & 8 & -1 \\ 1 & 3 & 9 & 27 & 9 \\ 1 & 4 & 16 & 64 & 33 \end{bmatrix} \begin{matrix} \approx \\ R2+(-1)R1 \\ R3+(-1)R1 \\ R4+(-1)R1 \end{matrix} \begin{bmatrix} 1 & 1 & 1 & 1 & -3 \\ 0 & 1 & 3 & 7 & 2 \\ 0 & 2 & 8 & 26 & 12 \\ 0 & 3 & 15 & 63 & 36 \end{bmatrix}$$

$$\begin{matrix} \approx \\ R1+(-1)R2 \\ R3+(-2)R2 \\ R4+(-3)R2 \end{matrix} \begin{bmatrix} 1 & 0 & -2 & -6 & -5 \\ 0 & 1 & 3 & 7 & 2 \\ 0 & 0 & 2 & 12 & 8 \\ 0 & 0 & 6 & 42 & 30 \end{bmatrix} \begin{matrix} \approx \\ (1/2)R3 \\ (1/6)R4 \end{matrix} \begin{bmatrix} 1 & 0 & -2 & -6 & -5 \\ 0 & 1 & 3 & 7 & 2 \\ 0 & 0 & 1 & 6 & 4 \\ 0 & 0 & 1 & 7 & 5 \end{bmatrix} \begin{matrix} \approx \\ R1+(2)R3 \\ R2+(-3)R3 \\ R4+(-1)R3 \end{matrix} \begin{bmatrix} 1 & 0 & 0 & 6 & 3 \\ 0 & 1 & 0 & -11 & -10 \\ 0 & 0 & 1 & 6 & 4 \\ 0 & 0 & 0 & 1 & 1 \end{bmatrix}$$

$$\begin{matrix} \approx \\ R1+(-6)R4 \\ R2+(11)R4 \\ R3+(-6)R4 \end{matrix} \begin{bmatrix} 1 & 0 & 0 & 0 & -3 \\ 0 & 1 & 0 & 0 & 1 \\ 0 & 0 & 1 & 0 & -2 \\ 0 & 0 & 0 & 1 & 1 \end{bmatrix}$$, so $a_0 = -3$, $a_1 = 1$, $a_2 = -2$, $a_3 = 1$ and the equation

is $-3 + x - 2x^2 + x^3 = y$.

7. $$\begin{aligned} I_1 + I_2 - I_3 &= 0 \\ 2I_1 + 4I_3 &= 34 \\ 4I_2 + 4I_3 &= 28 \end{aligned}$$
so that $I_1 = 5$, $I_2 = 1$, $I_3 = 6$.

8. $$\begin{aligned} I_1 + I_2 - I_3 &= 0 \\ I_1 + 2I_3 &= 9 \\ 3I_2 + 2I_3 &= 17 \end{aligned}$$
so that $I_1 = 1$, $I_2 = 3$, $I_3 = 4$.

9. $$\begin{aligned} I_1 + I_2 - I_3 &= 0 \\ 3I_3 &= 9 \\ 4I_2 + 3I_3 &= 13 \end{aligned}$$
so that $I_1 = 2$, $I_2 = 1$, $I_3 = 3$.

10. $$\begin{aligned} I_1 + I_2 - I_3 &= 0 \\ 2I_1 + 2I_3 &= 4 \\ 4I_2 + 2I_3 &= 2 \end{aligned}$$
so that $I_1 = 1$, $I_2 = 0$, $I_3 = 1$.

11. $$\begin{aligned} I_1 - I_2 - I_3 &= 0 \\ I_1 + 3I_2 &= 31 \\ I_1 + 7I_3 &= 31 \end{aligned}$$
so that $I_1 = 10$, $I_2 = 7$, $I_3 = 3$.

12.
$$\begin{aligned} I_1 - I_2 - I_3 \qquad\qquad &= 0 \\ I_3 - I_4 - I_5 &= 0 \\ I_2 \qquad\qquad &= 4 \\ I_4 \qquad &= 4 \\ I_5 &= 4 \end{aligned}$$

so that $I_1 = 12$, $I_2 = 4$, $I_3 = 8$, $I_4 = 4$, $I_5 = 4$.

13.
$$\begin{aligned} I_1 - I_2 - I_3 \qquad\qquad &= 0 \\ I_3 - I_4 + I_5 &= 0 \\ I_1 + I_2 \qquad\qquad &= 4 \\ I_1 \qquad + 2I_4 \qquad &= 4 \\ 2I_4 + 2I_5 &= 2 \end{aligned}$$

so that $I_1 = 7/3$, $I_2 = 5/3$, $I_3 = 2/3$, $I_4 = 5/6$, $I_5 = 1/6$.

14. Assume I_3 flows from A to B.
$$\begin{aligned} I_1 - I_2 - I_3 &= 0 \\ I_1 \qquad + 2I_3 &= 4 \\ I_2 - 2I_3 &= 9 \end{aligned}$$

gives $I_1 = 6$, $I_2 = 7$, $I_3 = -1$, so the current in AB is 1 amp flowing from B to A.

15. Let I_1 be the current in the direction from the 16volt battery to A, let I_2 be the current from A to B, and let I_3 be the current in the direction from C to B.

$$\begin{aligned} I_1 - I_2 - I_3 &= 0 : 0 \\ 5I_1 + I_2 \qquad &= 16 : 16 \\ -I_2 + 5I_3 &= 9 : 23 \end{aligned}$$

We solve the two systems simultaneously:

$$\begin{bmatrix} 1 & -1 & -1 & 0 & 0 \\ 5 & 1 & 0 & 16 & 16 \\ 0 & -1 & 5 & 9 & 23 \end{bmatrix} \underset{R2+(-5)R1}{} \begin{bmatrix} 1 & -1 & -1 & 0 & 0 \\ 0 & 6 & 5 & 16 & 16 \\ 0 & -1 & 5 & 9 & 23 \end{bmatrix} \underset{R2 \Leftrightarrow R3}{} \begin{bmatrix} 1 & -1 & -1 & 0 & 0 \\ 0 & -1 & 5 & 9 & 23 \\ 0 & 6 & 5 & 16 & 16 \end{bmatrix}$$

$$\underset{(-1)R2}{\approx} \begin{bmatrix} 1 & -1 & -1 & 0 & 0 \\ 0 & 1 & -5 & -9 & -23 \\ 0 & 6 & 5 & 16 & 16 \end{bmatrix} \underset{\substack{R1+R2 \\ R3+(-6)R2}}{\approx} \begin{bmatrix} 1 & 0 & -6 & -9 & -23 \\ 0 & 1 & -5 & -9 & -23 \\ 0 & 0 & 35 & 70 & 154 \end{bmatrix} \underset{(1/35)R3}{\approx} \begin{bmatrix} 1 & 0 & -6 & -9 & -23 \\ 0 & 1 & -5 & -9 & -23 \\ 0 & 0 & 1 & 2 & 22/5 \end{bmatrix}$$

$$\underset{\substack{R1+(6)R3 \\ R2+(5)R3}}{\approx} \begin{bmatrix} 1 & 0 & 0 & 3 & 17/5 \\ 0 & 1 & 0 & 1 & -1 \\ 0 & 0 & 1 & 2 & 22/5 \end{bmatrix}$$

(a) $I_1 = 3,\ I_2 = 1,\ I_3 = 2.$ (b) $I_1 = 17/5,\ I_2 = -1,\ I_3 = 22/5.$

If $I_2 = 0$, let the voltage at C be denoted by V.

$$\begin{aligned} I_1 - I_3 &= 0 \\ 5I_1 &= 16 \\ 5I_3 &= V \end{aligned}$$

Thus $I_1 = I_3 = 16/5$, so $V = 16$.

16. A: $x_1 + x_4 = 300$ B: $x_1 + x_2 = 250$
C: $x_2 + x_3 = 100$ D: $x_3 + x_4 = 150$

$$\begin{bmatrix} 1 & 0 & 0 & 1 & 300 \\ 1 & 1 & 0 & 0 & 250 \\ 0 & 1 & 1 & 0 & 100 \\ 0 & 0 & 1 & 1 & 150 \end{bmatrix} \approx \cdots \approx \begin{bmatrix} 1 & 0 & 0 & 1 & 300 \\ 0 & 1 & 0 & -1 & -50 \\ 0 & 0 & 1 & 1 & 150 \\ 0 & 0 & 0 & 0 & 0 \end{bmatrix}.$$

$x_1 = -x_4 + 300,\ x_2 = x_4 - 50,\ x_3 = -x_4 + 150.$

Two solutions are $x_1 = 250,\ x_2 = 0,\ x_3 = 100,\ x_4 = 50$
and $x_1 = 150,\ x_2 = 100,\ x_3 = 0,\ x_4 = 150.$

The minimum value of x_1 comes from taking the maximum value of x_4, which is 150. Thus, the minimum value of x_1 is 150.

17. A: $x_1 - x_4 = 100$ B: $x_1 - x_2 = 200$

C: $-x_2 + x_3 = 150$ D: $x_3 - x_4 = 50$

$$\begin{bmatrix} 1 & 0 & 0 & -1 & 100 \\ 1 & -1 & 0 & 0 & 200 \\ 0 & -1 & 1 & 0 & 150 \\ 0 & 0 & 1 & -1 & 50 \end{bmatrix} \approx \cdots \approx \begin{bmatrix} 1 & 0 & 0 & -1 & 100 \\ 0 & 1 & 0 & -1 & -100 \\ 0 & 0 & 1 & -1 & 50 \\ 0 & 0 & 0 & 0 & 0 \end{bmatrix}.$$

$x_1 = x_4 + 100$, $x_2 = x_4 - 100$, $x_3 = x_4 + 50$.

$x_2 = 0$ is min flow along BC; i.e., BC, closed to traffic.
In that case $x_4 = 100$, $x_1 = 200$, $x_3 = 150$. Then alternative route would have to be provided to get from B to towns in direction of C, and D.

18. $x_1 = x_2+100$, $x_3=x_2+90$, $x_3=x_4+130$, $x_5=x_4+110$, $x_5=x_6+80$, $x_7=x_6+75$, $x_7=x_8+120$, $x_1=x_8+155$. Since a flow cannot be negative these equs give:
$x_1 \geq 100$, $x_3 \geq 90$, $x_3 \geq 130$, $x_5 \geq 110$, $x_5 \geq 80$, $x_7 \geq 75$, $x_7 \geq 120$, $x_1 \geq 155$.
Thus, must have $x_1 \geq 155$, $x_3 \geq 130$, $x_5 \geq 110$, $x_7 \geq 120$.
Is $x_1=155$ possible, i.e., does it result in nonneg flows? Yes, gives
$x_2=55$, $x_3=145$, $x_4=15$, $x_5=125$, $x_6=45$, $x_7=120$, $x_8=0$.
Minimum flow allowable along x_1 is 155. Note that this is attained by closing x_8 to traffic. Alternative routes (clearly labeled diversions!) will then have to be provided for the x_7 traffic wanting to get to some of the towns that are accessed from other exits of the roundabout.

19. A: $x_1 + x_2 = 200$ B: $x_1 - x_3 - x_4 = 0$

E: $x_2 + x_3 - x_5 = 0$ D: $x_4 + x_5 = 200$

$$\begin{bmatrix} 1 & 1 & 0 & 0 & 0 & 200 \\ 1 & 0 & -1 & -1 & 0 & 0 \\ 0 & 1 & 1 & 0 & -1 & 0 \\ 0 & 0 & 0 & 1 & 1 & 200 \end{bmatrix} \approx \cdots \approx \begin{bmatrix} 1 & 0 & -1 & 0 & 1 & 200 \\ 0 & 1 & 1 & 0 & -1 & 0 \\ 0 & 0 & 0 & 1 & 1 & 200 \\ 0 & 0 & 0 & 0 & 0 & 0 \end{bmatrix}.$$

$x_1 = x_3 - x_5 + 200$, $x_2 = -x_3 + x_5$, $x_4 = -x_5 + 200$.

Total time $= k(x_1 + 2x_2 + x_3 + 2x_4 + x_5)$
$= 4(x_3 - x_5 + 200 + 2(-x_3 + x_5) + x_3 + 2(-x_5 + 200) + x_5) = 4(600) = 2400$ minutes.

This averages to 12 minutes per car. It is interesting to note that the time is independent of the actual distribution of traffic for this model. This will not be true if k differs on different stretches of road due to different road conditions. Students can examine these situations.

20. Let $y = a_0 + a_1x + a_2x^2$. These polynomials must pass through (1, 2) and (3, 4). Thus

$$\begin{aligned} a_0 + a_1 + a_2 &= 2 \\ a_0 + 3a_1 + 9a_2 &= 4 \end{aligned}. \quad \begin{bmatrix} 1 & 1 & 1 & 2 \\ 1 & 3 & 9 & 4 \end{bmatrix} \approx \begin{bmatrix} 1 & 0 & -3 & 1 \\ 0 & 1 & 4 & 1 \end{bmatrix}. \quad a_0 = 3a_2+1,\ a_1 = -4a_2+1.$$

Let $a_2 = r$. The family of polynomials is $y = (3r+1) + (-4r+1)x + rx^2$. $r=0$ gives the line $y=1+x$ that passes through these points. When $r>0$ the polynomials open up and when $r<0$ the polynomials open down.

21. Let $y = a_0 + a_1x + a_2x^2 + a_3x^3$. Polynomials must pass through (1, 2), (3, 4) (4,8). Thus

$$\begin{aligned} a_0 + a_1 + a_2 + a_3 &= 2 \\ a_0 + 3a_1 + 9a_2 + 27a_3 &= 4 \\ a_0 + 4a_1 + 16a_2 + 64a_3 &= 8 \end{aligned}. \quad \begin{bmatrix} 1 & 1 & 1 & 1 & 2 \\ 1 & 3 & 9 & 27 & 4 \\ 1 & 4 & 16 & 64 & 8 \end{bmatrix} \approx \begin{bmatrix} 1 & 0 & 0 & 12 & 4 \\ 0 & 1 & 0 & -19 & -3 \\ 0 & 0 & 1 & 8 & 1 \end{bmatrix}.$$

$a_0 = -12a_3+4$, $a_1 = 19a_3-3$, $a_2 = -8a_3+1$.

Let $a_2 = r$. The family of polynomials is $y = (-12r+4) + (19r-3)x + (-8r+1)x^2 + rx^3$. When $r=1$, $y = -8 + 16x - 7x^2 + x^3$.

Chapter 1 Review Exercises

1. (a) 2×3 (b) 2×2 (c) 1×4 (d) 3×1 (e) 4×6

2. 0, 6, 5, 1, 9

3. $I_5 = \begin{bmatrix} 1 & 0 & 0 & 0 & 0 \\ 0 & 1 & 0 & 0 & 0 \\ 0 & 0 & 1 & 0 & 0 \\ 0 & 0 & 0 & 1 & 0 \\ 0 & 0 & 0 & 0 & 1 \end{bmatrix}$

4. (a) $\begin{bmatrix} 1 & 2 \\ 4 & -3 \end{bmatrix}, \begin{bmatrix} 1 & 2 & 6 \\ 4 & -3 & -1 \end{bmatrix}$ (b) $\begin{bmatrix} 2 & 1 & -4 \\ 1 & -2 & 8 \\ 3 & 5 & -7 \end{bmatrix}, \begin{bmatrix} 2 & 1 & -4 & 1 \\ 1 & -2 & 8 & 0 \\ 3 & 5 & -7 & -3 \end{bmatrix}$

(c) $\begin{bmatrix} -1 & 2 & -7 \\ 3 & -1 & 5 \\ 4 & 3 & 0 \end{bmatrix}, \begin{bmatrix} -1 & 2 & -7 & -2 \\ 3 & -1 & 5 & 3 \\ 4 & 3 & 0 & 5 \end{bmatrix}$ (d) $\begin{bmatrix} 1 & 0 & 0 \\ 0 & 1 & 0 \\ 0 & 0 & 1 \end{bmatrix}, \begin{bmatrix} 1 & 0 & 0 & 1 \\ 0 & 1 & 0 & 5 \\ 0 & 0 & 1 & -3 \end{bmatrix}$

(e) $\begin{bmatrix} -2 & 3 & -8 & 5 \\ 1 & 5 & 0 & -6 \\ 0 & -1 & 2 & 3 \end{bmatrix}, \begin{bmatrix} -2 & 3 & -8 & 5 & -2 \\ 1 & 5 & 0 & -6 & 0 \\ 0 & -1 & 2 & 3 & 5 \end{bmatrix}$

5. (a) $\begin{aligned} 4x_1 + 2x_2 &= 0 \\ -3x_1 + 7x_2 &= 8 \end{aligned}$ (b) $\begin{aligned} x_1 + 9x_2 &= -3 \\ 3x_2 &= 2 \end{aligned}$ (c) $\begin{aligned} x_1 + 2x_2 + 3x_3 &= 4 \\ 5x_1 \qquad\quad -3x_3 &= 6 \end{aligned}$

(d) $\begin{aligned} x_1 \qquad\qquad &= 5 \\ x_2 \qquad &= -8 \\ x_3 &= 2 \end{aligned}$ (e) $\begin{aligned} x_1 + 4x_2 - x_3 &= 7 \\ x_2 + 3x_3 &= 8 \\ x_3 &= -5 \end{aligned}$

6. (a) Yes. (b) Yes.

(c) No. There is a 2 (a non zero element) above the leading 1 of row 2.

(d) Yes.

(e) No. The leading 1 in row 3 is not positioned to the right of the leading 1 in row 2.

7. (a) $\begin{bmatrix} 2 & 4 & 2 \\ 3 & 7 & 2 \end{bmatrix} \underset{(1/2)R1}{} \begin{bmatrix} 1 & 2 & 1 \\ 3 & 7 & 2 \end{bmatrix} \underset{R2+(-3)R1}{} \begin{bmatrix} 1 & 2 & 1 \\ 0 & 1 & -1 \end{bmatrix} \underset{R1+(-2)R2}{} \begin{bmatrix} 1 & 0 & 3 \\ 0 & 1 & -1 \end{bmatrix},$

so the solution is $x_1 = 3$ and $x_2 = -1$.

(b) $\begin{bmatrix} 1 & -2 & -6 & -17 \\ 2 & -6 & -16 & -46 \\ 1 & 2 & -1 & -5 \end{bmatrix} \underset{\substack{R2+(-2)R1 \\ R3+(-1)R1}}{\approx} \begin{bmatrix} 1 & -2 & -6 & -17 \\ 0 & -2 & -4 & -12 \\ 0 & 4 & 5 & 12 \end{bmatrix} \underset{(-1/2)R2}{\approx} \begin{bmatrix} 1 & -2 & -6 & -17 \\ 0 & 1 & 2 & 6 \\ 0 & 4 & 5 & 12 \end{bmatrix}$

$\underset{\substack{R1+(2)R2 \\ R3+(-4)R2}}{\approx} \begin{bmatrix} 1 & 0 & -2 & -5 \\ 0 & 1 & 2 & 6 \\ 0 & 0 & -3 & -12 \end{bmatrix} \underset{(-1/3)R3}{\approx} \begin{bmatrix} 1 & 0 & -2 & -5 \\ 0 & 1 & 2 & 6 \\ 0 & 0 & 1 & 4 \end{bmatrix} \underset{\substack{R1+(2)R3 \\ R2+(-2)R3}}{\approx} \begin{bmatrix} 1 & 0 & 0 & 3 \\ 0 & 1 & 0 & -2 \\ 0 & 0 & 1 & 4 \end{bmatrix},$

so that $x_1 = 3$, $x_2 = -2$, $x_3 = 4$.

(c) $\begin{bmatrix} 0 & 1 & 2 & 6 & 21 \\ 1 & -1 & 1 & 5 & 12 \\ 1 & -1 & -1 & -4 & -9 \\ 3 & -2 & 0 & -6 & -4 \end{bmatrix} \underset{R1 \Leftrightarrow R2}{\approx} \begin{bmatrix} 1 & -1 & 1 & 5 & 12 \\ 0 & 1 & 2 & 6 & 21 \\ 1 & -1 & -1 & -4 & -9 \\ 3 & -2 & 0 & -6 & -4 \end{bmatrix} \underset{\substack{R3+(-1)R1 \\ R4+(-3)R1}}{\approx} \begin{bmatrix} 1 & -1 & 1 & 5 & 12 \\ 0 & 1 & 2 & 6 & 21 \\ 0 & 0 & -2 & -9 & -21 \\ 0 & 1 & -3 & -21 & -40 \end{bmatrix}$

$$\underset{\substack{R1+R2\\R4+(-1)R2}}{\approx}\begin{bmatrix}1&0&3&11&33\\0&1&2&6&21\\0&0&-2&-9&-21\\0&0&-5&-27&-61\end{bmatrix}\underset{(-1/2)R3}{\approx}\begin{bmatrix}1&0&3&11&33\\0&1&2&6&21\\0&0&1&9/2&21/2\\0&0&-5&-27&-61\end{bmatrix}$$

$$\underset{\substack{R1+(-3)R3\\R2+(-2)R3\\R4+(5)R3}}{\approx}\begin{bmatrix}1&0&0&-5/2&3/2\\0&1&0&-3&0\\0&0&1&9/2&21/2\\0&0&0&-9/2&-17/2\end{bmatrix}\underset{(-2/9)R4}{\approx}\begin{bmatrix}1&0&0&-5/2&3/2\\0&1&0&-3&0\\0&0&1&9/2&21/2\\0&0&0&1&17/9\end{bmatrix}$$

$$\underset{\substack{R1+(5/2)R4\\R2+(3)R4\\R3+(-9/2)R4}}{\approx}\begin{bmatrix}1&0&0&0&56/9\\0&1&0&0&17/3\\0&0&1&0&2\\0&0&0&1&17/9\end{bmatrix}$$, so that $x_1 = 56/9$, $x_2 = 17/3$, $x_3 = 2$, $x_4 = 17/9$.

8. (a) $$\begin{bmatrix}1&-1&1&3\\-2&3&1&-8\\4&-2&10&10\end{bmatrix}\underset{\substack{R2+(2)R1\\R3+(-4)R1}}{\approx}\begin{bmatrix}1&-1&1&3\\0&1&3&-2\\0&2&6&-2\end{bmatrix}\underset{\substack{R1+R2\\R3+(-2)R2}}{\approx}\begin{bmatrix}1&0&4&1\\0&1&3&-2\\0&0&0&2\end{bmatrix}$$

There is no need to continue. The last row gives 0 = 2 so there is no solution.

(b) $$\begin{bmatrix}1&3&6&-2&-7\\-2&-5&-10&3&10\\1&2&4&0&0\\0&1&2&-3&-10\end{bmatrix}\underset{\substack{R2+(2)R1\\R3+(-1)R1}}{\approx}\begin{bmatrix}1&3&6&-2&-7\\0&1&2&-1&-4\\0&-1&-2&2&7\\0&1&2&-3&-10\end{bmatrix}$$

$$\underset{\substack{R1+(-3)R2\\R3+R2\\R4+(-1)R2}}{\approx}\begin{bmatrix}1&0&0&1&5\\0&1&2&-1&-4\\0&0&0&1&3\\0&0&0&-2&-6\end{bmatrix}\underset{\substack{R1+(-1)R3\\R2+R3\\R4+(2)R3}}{\approx}\begin{bmatrix}1&0&0&0&2\\0&1&2&0&-1\\0&0&0&1&3\\0&0&0&0&0\end{bmatrix},$$

so there are many solutions, and the general solution is

$x_1 = 2$, $x_2 = -1 - 2r$, $x_3 = r$, $x_4 = 3$.

9. If a matrix A is in reduced echelon form, it is clear from the definition that the leading 1 in any row cannot be to the left of the diagonal element in that row. Therefore if $A \neq I_n$,

there must be some row that has its leading 1 to the right of the diagonal element in that row. Suppose row j is such a row and the leading 1 is in position (j, k) where $j < k \le n$. Then, if rows $j + 1, j + 2, \ldots, j + (n - k) < n$ all contain nonzero terms, the leading 1 in these rows must be at least as far to the right as columns $k + 1, k + 2, \ldots, k + (n - k) = n$, respectively. The leading 1 in row $j + (n - k) + 1$ must then be to the right of column n. But there is no column to the right of column n, so row $j + (n - k) + 1$ must consist of all zeros.

10. Let E be the reduced echelon form of A. Since B is row equivalent to A, B is also row equivalent to E. But since E is in reduced echelon form, it must be the reduced echelon form of B.

11.

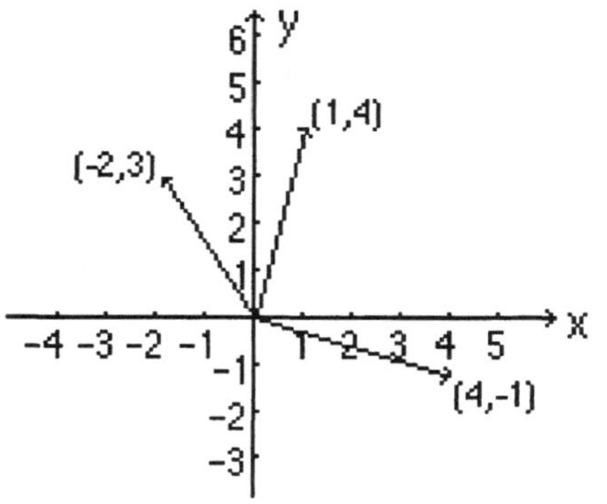

12.

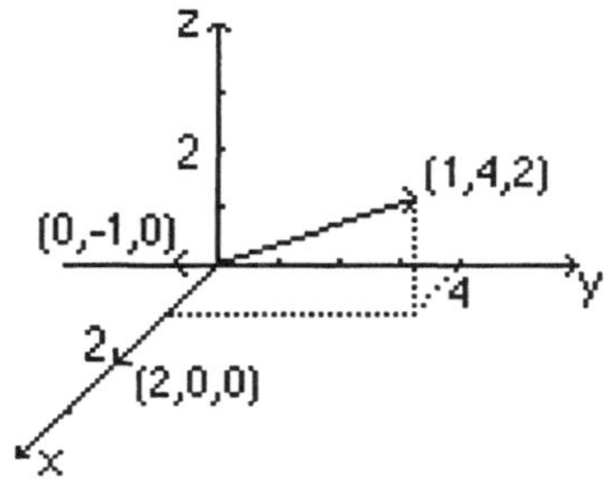

13. (a) $\mathbf{u} + \mathbf{w} = (3,-1,5) + (0,1,-3) = (3,0,2)$.

(b) $3\mathbf{u} + \mathbf{v} = 3(3,-1,5) + (2,3,7) = (11,0,22)$.

(c) $\mathbf{u} - 2\mathbf{w} = (3,-1,5) - 2(0,1,-3) = (3,-3,11)$.

(d) $4\mathbf{u} - 2\mathbf{v} + 3\mathbf{w} = 4(3,-1,5) - 2(2,3,7) + 3(0,1,-3) = (8,-7,-3)$.

(e) $2\mathbf{u} - 5\mathbf{v} - \mathbf{w} = 2(3,-1,5) - 5(2,3,7) - (0,1,-3) = (-4,-18,-22)$.

14. $a(1, 2, 0) + b(0, 1, 4) + c(-3, 2, 7) = (-1, 3, -5)$. $a-3c = -1$, $2a+b+2c=3$, $4b+7c=-5$. Unique solution, $a=2$, $b=-3$, $c=1$. Is a linear combination.

15. (a) Let W be the subset of vectors of the form $(a,b,a+2b)$. Let $\mathbf{u}=(a,b,a+2b)$, $\mathbf{v}=(c,d,c+2d)$, and k be a scalar. Then $\mathbf{u}+\mathbf{v}=(a+c, b+d,(a+c)+2(b+d))$ and $k\mathbf{u}=(ka,kb,ka+2kb)$.The third component of $\mathbf{u}+\mathbf{v}$ is the first plus two times the second; same for $k\mathbf{u}$. Thus W is closed under addition and scalar multiplication - it is a subspace.

(b) Let W be the subset of vectors of the form $(a,b,a+4)$. Let $\mathbf{u}=(a,b,a+4)$, $\mathbf{v}=(c,d,c+4)$ and k be a scalar. Then $\mathbf{u}+\mathbf{v}=(a+c,b+d,(a+c)+8)$. Third component is not $(a+c)+4$. Thus

u+**v** is not in W. W is not a subspace. Let us check closure under scalar multiplication. k**u**=(ka,kb,ka+4k)). Unless k=1 the 4th component is not the first component plus 4. **ku** is not in general in W. W Is not closed under scalar multiplication either.

16. Separate the variables in the general solution, (2r + s, 3s, r, s)= r(2, 0, 1, 0) + s(1, 3, 0, 1). Vectors (2, 0, 1, 0) and (1, 3, 0, 1) thus span W.
Also, identity p(2, 0, 1, 0) + q(1, 3, 0, 1) = (0, 0, 0, 0) leads to p=0, q=0. The two vectors are thus linearly independent. The set {(2, 0, 1, 0), (1, 3, 0, 1)} is a basis for W; dimension is 2.

17. a(1, -1, 2) + b(1, 0, 1) + c(6, -2, 10)=**0** gives a+b+6c=0, -a-2c=0, 2a+b+10c=0.
Unique solution a=0, b=0, c=0. Thus linearly independent.

18. (a) $(1,2)\cdot(3,-4) = 1\text{x}3 + 2\text{x}{-4} = 3 - 8 = -5$.

(b) $(1,-2,3)\cdot(4,2,-7) = 1\text{x}4 + -2\text{x}2 + 3\text{x}{-7} = 4 - 4 - 21 = -21$.

(c) $(2,2,-5)\cdot(3,2,-1) = 2\text{x}3 + 2\text{x}2 + -5\text{x}{-1} = 6 + 4 + 5 = 15$.

19. (a) $||(1,-4)|| = \sqrt{(1,-4)\cdot(1,-4)} = \sqrt{1\text{x}1+-4\text{x}-4} = \sqrt{17}$.

(b) $||(-2,1,3)|| = \sqrt{(-2,1,3)\cdot(-2,1,3)} = \sqrt{-2\text{x}-2+1\text{x}1+3\text{x}3} = \sqrt{14}$.

(c) $||(1,-2,3,4)|| = \sqrt{(1,-2,3,4)\cdot(1,-2,3,4)} = \sqrt{1\text{x}1+-2\text{x}-2+3\text{x}3+4\text{x}4} = \sqrt{30}$.

20. (a) $\cos\theta = \dfrac{(-1,1)\cdot(2,3)}{||(-1,1)||\,||(2,3)||} = \dfrac{1}{\sqrt{2}\sqrt{13}} = \dfrac{1}{\sqrt{26}}$.

(b) $\cos\theta = \dfrac{(1,2,-3)\cdot(4,1,2)}{||(1,2,-3)||\,||(4,1,2)||} = 0$.

21. The vector (1,2,0) is orthogonal to (−2,1,5).

22. (a) $d = \sqrt{(1-5)^2+(-2-3)^2} = \sqrt{16+25} = \sqrt{41}$.

(b) $d = \sqrt{(3-7)^2+(2-1)^2+(1-2)^2} = \sqrt{18} = 3\sqrt{2}$.

(c) $d^2 = (3-4)^2 + (1-1)^2 + (-1-6)^2 + (2-2)^2 = 50$, so $d = \sqrt{50} = 5\sqrt{2}$.

23. $||c(1,2,3)|| = |c|\ ||(1,2,3)|| = |c|\sqrt{(1,2,3)\cdot(1,2,3)} = |c|\sqrt{1x1+2x2+3x3} = |c|\sqrt{14}$, so if

$||c(1,2,3)|| = 196$ then $c = \pm\dfrac{196}{\sqrt{14}} = \pm 14\sqrt{14}$.

24. Let (a,b,c) be in W. Then (a,b,c) is orthogonal to (-3,5,1). (a,b,c)·(-3,5,1)=0, -3a+5b+c=0, c=3a-5b. W consists of vectors of the form (a,b,3a-5b). Separate the variables. (a,b,3a-5b)=a(1,0,3)+b(0,1,-5). {(1,0,3), {(0,1,-5)} is a basis for W. The dimension of W is 2. It is the plane defined by (1,0,3) and (0,1,-5).

25. The equation is of the form $a_0 + a_1 x + a_2 x^2 = y$, so the system of equations to be solved is

$$\begin{aligned} a_0 + a_1 + a_2 &= 3 \\ a_0 + 2a_1 + 4a_2 &= 6 \\ a_0 + 3a_1 + 9a_2 &= 13 \end{aligned}$$

$$\begin{bmatrix} 1 & 1 & 1 & 3 \\ 1 & 2 & 4 & 6 \\ 1 & 3 & 9 & 13 \end{bmatrix} \underset{\substack{R2+(-1)R1 \\ R3+(-1)R1}}{\approx} \begin{bmatrix} 1 & 1 & 1 & 3 \\ 0 & 1 & 3 & 3 \\ 0 & 2 & 8 & 10 \end{bmatrix} \underset{\substack{R1+(-1)R2 \\ R3+(-2)R2}}{\approx} \begin{bmatrix} 1 & 0 & -2 & 0 \\ 0 & 1 & 3 & 3 \\ 0 & 0 & 2 & 4 \end{bmatrix}$$

$$\underset{(1/2)R3}{\approx} \begin{bmatrix} 1 & 0 & -2 & 0 \\ 0 & 1 & 3 & 3 \\ 0 & 0 & 1 & 2 \end{bmatrix} \underset{\substack{R1+(2)R3 \\ R2+(-3)R3}}{\approx} \begin{bmatrix} 1 & 0 & 0 & 4 \\ 0 & 1 & 0 & -3 \\ 0 & 0 & 1 & 2 \end{bmatrix},$$ so $a_0=4$, $a_1=-3$, $a_2=2$, and the

equation is $4 - 3x + 2x^2 = y$.

26. $$\begin{aligned} I_1 - I_2 - I_3 &= 0 \\ 2I_1 \qquad\quad + I_3 &= 7 \\ 3I_2 - I_3 &= 5 \end{aligned}$$ gives $I_1 = 3$, $I_2 = 2$, $I_3 = 1$.

27. $x_1 = x_8+160$, $x_1=x_2+100$, $x_3=x_2+80$, $x_3=x_4+120$, $x_5=x_4+90$, $x_5=x_6+130$, $x_7=x_6+100$, $x_7=x_8+80$. Since a flow cannot be negative these equs give:
$x_1\geq160$, $x_1\geq100$, $x_3\geq80$, $x_3\geq120$, $x_5\geq90$, $x_5\geq130$, $x_7\geq100$, $x_7\geq80$.
Thus, must have $x_1\geq160$, $x_3\geq120$, $x_5\geq130$, $x_7\geq100$.
Let us look at the equs involving x_8, in the light of these restrictions;
$x_1 = x_8+160$ and $x_7=x_8+80$.
For $x_7\geq100$, must have $x_8\geq20$. Is $x_8=20$ possible, i.e., does it result in nonneg flows?
Yes, $x_8=20$ implies that $x_1=180$, $x_2=80$, $x_3=160$, $x_4=40$, $x_5=130$, $x_6=0$, $x_7=100$.
Minimum flow allowable along x_8 is 20. Note that this is attained by closing x_6 to traffic.
Alternative routes (clearly labeled diversions!) will then have to be provided for the x_5

traffic wanting to get to some of the towns that are accessed from other exits of the roundabout.

Chapter 2

Exercise Set 2.1

1. (a) $A + B = \begin{bmatrix} 5-3 & 4+0 \\ -1+4 & 7+2 \\ 9+5 & -3-7 \end{bmatrix} = \begin{bmatrix} 2 & 4 \\ 3 & 9 \\ 14 & -10 \end{bmatrix}$.

(b) $2B = \begin{bmatrix} -6 & 0 \\ 8 & 4 \\ 10 & -14 \end{bmatrix}$.

(c) $-D = \begin{bmatrix} -9 & 5 \\ -3 & 0 \end{bmatrix}$.

(d) $C + D = \begin{bmatrix} 10 & -3 \\ 6 & 4 \end{bmatrix}$.

(e) A + D does not exist.

(f) $2A + B = \begin{bmatrix} 10-3 & 8+0 \\ -2+4 & 14+2 \\ 18+5 & -6-7 \end{bmatrix} = \begin{bmatrix} 7 & 8 \\ 2 & 16 \\ 23 & -13 \end{bmatrix}$.

(g) $A - B = \begin{bmatrix} 8 & 4 \\ -5 & 5 \\ 4 & 4 \end{bmatrix}$.

2. (a) A + B does not exist.

(b) $4B = \begin{bmatrix} 0 & -4 & 16 \\ 24 & -32 & 8 \\ -16 & 20 & 36 \end{bmatrix}$.

(c) $-3D = \begin{bmatrix} 9 \\ 0 \\ -6 \end{bmatrix}$.

(d) $B - 3C = \begin{bmatrix} 0-3 & -1-6 & 4+15 \\ 6+21 & -8-27 & 2-9 \\ -4-15 & 5+12 & 9-0 \end{bmatrix} = \begin{bmatrix} -3 & -7 & 19 \\ 27 & -35 & -7 \\ -19 & 17 & 9 \end{bmatrix}$.

(e) $-A = \begin{bmatrix} -9 \\ -2 \\ 1 \end{bmatrix}$.

(f) $3A + 2D = \begin{bmatrix} 27-6 \\ 6+0 \\ -3+4 \end{bmatrix} = \begin{bmatrix} 21 \\ 6 \\ 1 \end{bmatrix}$.

(g) $A + D = \begin{bmatrix} 6 \\ 2 \\ 1 \end{bmatrix}$.

3. (a) $AB = \begin{bmatrix} (1x0)+(0x-2) & (1x1)+(0x5) \\ (0x0)+(1x-2) & (0x1)+(1x5) \end{bmatrix} = \begin{bmatrix} 0 & 1 \\ -2 & 5 \end{bmatrix} = B$.

(b) BA = B.

(c) AC = C.

(d) CA does not exist.

(e) AD = D.

(f) DC does not exist.

(g) $BD = \begin{bmatrix} (0x-1)+(1x5) & (0x0)+(1x7) & (0x3)+(1x2) \\ (-2x-1)+(5x5) & (-2x0)+(5x7) & (-2x3)+(5x2) \end{bmatrix} = \begin{bmatrix} 5 & 7 & 2 \\ 27 & 35 & 4 \end{bmatrix}$.

(h) $A^2 = \begin{bmatrix} 1 & 0 \\ 0 & 1 \end{bmatrix}$.

4. (a) $BA = \begin{bmatrix} 27 \\ 23 \\ 9 \end{bmatrix}$. (b) AB does not exist. (c) CB = [10 13 -5].

(d) CA = [27]. (e) DA does not exist. (f) DB does not exist.

(g) $AC = \begin{bmatrix} (-1x-2) & (-1x0) & (-1x5) \\ (2x-2) & (2x0) & (2x5) \\ (5x-2) & (5x0) & (5x5) \end{bmatrix} = \begin{bmatrix} 2 & 0 & -5 \\ -4 & 0 & 10 \\ -10 & 0 & 25 \end{bmatrix}$.

(h) $B^2 = \begin{bmatrix} (0x0)+(1x3)+(5x2) & (0x1)+(1x-7)+(5x3) & (0x5)+(1x8)+(5x1) \\ (3x0)+(-7x3)+(8x2) & (3x1)+(-7x-7)+(8x3) & (3x5)+(-7x8)+(8x1) \\ (2x0)+(3x3)+(1x2) & (2x1)+(3x-7)+(1x3) & (2x5)+(3x8)+(1x1) \end{bmatrix}$

$= \begin{bmatrix} 13 & 8 & 13 \\ -5 & 76 & -33 \\ 11 & -16 & 35 \end{bmatrix}$.

5. (a) $2A - 3(BC) = \begin{bmatrix} -12 & 2 \\ -9 & -24 \\ -20 & -24 \end{bmatrix}$. (b) AB does not exist. (c) $AC - BD = \begin{bmatrix} 8 & 2 \\ 4 & 1 \\ 0 & 6 \end{bmatrix}$.

(d) $CD - 2D = \begin{bmatrix} -30 & 0 \\ 15 & 0 \end{bmatrix}$. (e) BA does not exist.

(f) AD + 2(DC) does not exist. (AD is a 3x2 matrix and DC is 2x2)

(g) $C^3 + 2(D^2) = \begin{bmatrix} -14 & 0 \\ 12 & 10 \end{bmatrix}$.

6. $A + O_3 = \begin{bmatrix} 1+0 & -8+0 & 4+0 \\ 5+0 & -6+0 & 3+0 \\ 2+0 & 0+0 & -1+0 \end{bmatrix} = \begin{bmatrix} 1 & -8 & 4 \\ 5 & -6 & 3 \\ 2 & 0 & -1 \end{bmatrix} = A$. In the same manner $O_3 + A = A$.

$BO_3 = \begin{bmatrix} 0x0+2x0-3x0 & 0x0+2x0-3x0 & 0x0+2x0-3x0 \\ 5x0+6x0+7x0 & 5x0+6x0+7x0 & 5x0+6x0+7x0 \\ -1x0+0x0+4x0 & -1x0+0x0+4x0 & -1x0+0x0+4x0 \end{bmatrix} = \begin{bmatrix} 0 & 0 & 0 \\ 0 & 0 & 0 \\ 0 & 0 & 0 \end{bmatrix} = O_3$.

$$BI_3 = \begin{bmatrix} 0x1+2x0-3x0 & 0x0+2x1-3x0 & 0x0+2x0-3x1 \\ 5x1+6x0+7x0 & 5x0+6x1+7x0 & 5x0+6x0+7x1 \\ -1x1+0x0+4x0 & -1x0+0x1+4x0 & -1x0+0x0+4x1 \end{bmatrix} = \begin{bmatrix} 0 & 2 & -3 \\ 5 & 6 & 7 \\ -1 & 0 & 4 \end{bmatrix} = B.$$

In the same manner $O_3 B = O_3$ and $I_3 B = B$.

7. (a) AX = X. $A\begin{bmatrix} 1 \\ \vdots \\ 1 \end{bmatrix} = \begin{bmatrix} 1 \\ \vdots \\ 1 \end{bmatrix}$. Thus $\begin{bmatrix} \text{sum of elements in row 1} \\ \vdots \\ \text{sum of elements in row n} \end{bmatrix} = \begin{bmatrix} 1 \\ \vdots \\ 1 \end{bmatrix}$

The sum of the elements in each row of A is 1.

(b) XA = X. [1 . . . 1]A = [1 . . . 1].

[sum of column 1 . . . sum of column n] = [1 . . . 1].

The sum of the elements in each column of A is 1.

8. (a) 3x2 (b) 4x2 (c) does not exist (d) 3x2 (e) 4x2

(f) 3x2 (g) does not exist

9. (a) 2x2 (b) 2x3 (c) 2x2 (d) does not exist (e) 3x2

(f) does not exist (g) does not exist

10. (a) c_{31} = (1x-1)+(0x5)+(-2x0) = -1. (b) c_{23} = (2x-3)+(6x2)+(3x6) = 24.

(c) d_{12} = (-1x3)+(2x6)+(-3x0) = 9. (d) d_{22} = (5x3)+(7x6)+(2x0) = 57.

11. (a) r_{21} = (4x0)+(6x0) = 0. (b) r_{33} = (-1x3)+(3x4) = 9.

(c) s_{11} = (0x1)+(1x4)+(3x-1) = 1. (d) s_{23} does not exist. S is a 2x2 matrix.

12. (a) d_{12} = (1x2)+(-3x0) + 2x- 4 = - 6. (b) d_{23} = (0x-3)+(4x-1)+ 2x0 = -4.

13. (a) d_{11} = 2[(1x1)+(-3x3)+(0x-1)] + (2x2)+(0x4)+(-2x1) = -14.

(b) $d_{21} = 2[(4x1)+(5x3)+(1x-1)] + (4x2)+(7x4)+(-5x1) = 67.$

(c) $d_{32} = 2[(3x1)+(8x0)+(0x3)] + (1x0)+(0x7)+(-1x0) = 6.$

14. (a) $AB_1=\begin{bmatrix} 1 & 2 \\ 3 & 0 \end{bmatrix}\begin{bmatrix} -2 \\ 4 \end{bmatrix} = \begin{bmatrix} 6 \\ -6 \end{bmatrix}$, $AB_2=\begin{bmatrix} 1 & 2 \\ 3 & 0 \end{bmatrix}\begin{bmatrix} 3 \\ 1 \end{bmatrix} = \begin{bmatrix} 5 \\ 9 \end{bmatrix}$. $AB=\begin{bmatrix} 6 & 5 \\ -6 & 9 \end{bmatrix}$.

(b) $AC_1=\begin{bmatrix} 1 & 2 \\ 3 & 0 \end{bmatrix}\begin{bmatrix} 1 \\ 4 \end{bmatrix} = \begin{bmatrix} 9 \\ 3 \end{bmatrix}$, $AC_2=\begin{bmatrix} 1 & 2 \\ 3 & 0 \end{bmatrix}\begin{bmatrix} -2 \\ 0 \end{bmatrix} = \begin{bmatrix} -2 \\ -6 \end{bmatrix}$,

$AC_3=\begin{bmatrix} 1 & 2 \\ 3 & 0 \end{bmatrix}\begin{bmatrix} 3 \\ 5 \end{bmatrix} = \begin{bmatrix} 13 \\ 9 \end{bmatrix}$. $AC = \begin{bmatrix} 9 & -2 & 13 \\ 3 & -6 & 9 \end{bmatrix}$.

(c) $BC_1=\begin{bmatrix} -2 & 3 \\ 4 & 1 \end{bmatrix}\begin{bmatrix} 1 \\ 4 \end{bmatrix} = \begin{bmatrix} 10 \\ 8 \end{bmatrix}$, $BC_2=\begin{bmatrix} -2 & 3 \\ 4 & 1 \end{bmatrix}\begin{bmatrix} -2 \\ 0 \end{bmatrix} = \begin{bmatrix} 4 \\ -8 \end{bmatrix}$,

$BC_3=\begin{bmatrix} -2 & 3 \\ 4 & 1 \end{bmatrix}\begin{bmatrix} 3 \\ 5 \end{bmatrix} = \begin{bmatrix} 9 \\ 17 \end{bmatrix}$. $BC = \begin{bmatrix} 10 & 4 & 9 \\ 8 & -8 & 17 \end{bmatrix}$.

15. (a) $AB = 4\begin{bmatrix} 3 \\ 4 \\ 8 \end{bmatrix} + 3\begin{bmatrix} -2 \\ 2 \\ -5 \end{bmatrix} - 5\begin{bmatrix} 0 \\ 7 \\ 6 \end{bmatrix}$.

(b) $PQ = -3\begin{bmatrix} 3 \\ 5 \end{bmatrix} + 2\begin{bmatrix} 0 \\ 6 \end{bmatrix} + 1\begin{bmatrix} 2 \\ 7 \end{bmatrix} + 5\begin{bmatrix} 1 \\ 3 \end{bmatrix}$.

16. row2 of AB = (row2 of A)xB = $\begin{bmatrix} 4 & 0 & 3 \end{bmatrix}\begin{bmatrix} 8 & 1 & 3 \\ 2 & 1 & 0 \\ 4 & 6 & 3 \end{bmatrix} = \begin{bmatrix} 44 & 22 & 21 \end{bmatrix}$.

17. The third row of AB is the third row of A times each of the columns of B in turn. Since the third row of A is all zeros, each of the products is zero.

18. The second column of CD is the rows of C in turn multiplied by the second column of D. Since the second column of D is all zeros, each of the products is zero.

19. $C = AB$. $B = [B_1\ B_2\ \dots\ B_n]$, and $C = [C_1\ C_2\ \dots\ C_n]$.

Thus $[C_1\ C_2\ \dots\ C_n] = A\,[B_1\ B_2\ \dots\ B_n]$
$= [AB_1\ AB_2\ \dots\ AB_n]$.

The columns of the matrix on the left must be equal to the columns of the matrix on the right. Thus
$C_1 = AB_1$, $C_2 = AB_2$, $C_n = AB_n$. In general, $C_j = AB_j$,

20. $AB_3 = \begin{bmatrix} 1 & 2 & 3 \\ 0 & 4 & 1 \\ 2 & 5 & 0 \end{bmatrix}\begin{bmatrix} 4 \\ 1 \\ 5 \end{bmatrix} = \begin{bmatrix} 21 \\ 9 \\ 13 \end{bmatrix}$.

21. (a) Submatrix products: $\begin{bmatrix} 2 \\ -1 \end{bmatrix}[3 \quad 0] + \begin{bmatrix} 1 \\ 0 \end{bmatrix}[2 \quad 1] = \begin{bmatrix} 6 & 0 \\ -3 & 0 \end{bmatrix} + \begin{bmatrix} 2 & 1 \\ 0 & 0 \end{bmatrix} = \begin{bmatrix} 8 & 1 \\ -3 & 0 \end{bmatrix}$,

$[3]\,[3 \quad 0] + [1]\,[2 \quad 1] = [9 \quad 0] + [2 \quad 1] = [11 \quad 1]$. $AB = \begin{bmatrix} 8 & 1 \\ -3 & 0 \\ 11 & 1 \end{bmatrix}$.

(b) Submatrix products: $\begin{bmatrix} 1 & 2 \\ 3 & 0 \end{bmatrix}\begin{bmatrix} 2 & 4 \\ 0 & -1 \end{bmatrix} + \begin{bmatrix} -1 \\ 1 \end{bmatrix}[1 \quad 3] = \begin{bmatrix} 2 & 2 \\ 6 & 12 \end{bmatrix} + \begin{bmatrix} -1 & -3 \\ 1 & 3 \end{bmatrix}$.

$AB = \begin{bmatrix} 1 & -1 \\ 7 & 15 \end{bmatrix}$.

(c) Submatrix products: $\begin{bmatrix} 1 & 2 \\ 3 & -1 \end{bmatrix}\begin{bmatrix} 3 & -1 \\ 2 & 5 \end{bmatrix} + \begin{bmatrix} 0 \\ 1 \end{bmatrix}[0 \quad 1] = \begin{bmatrix} 7 & 9 \\ 7 & -8 \end{bmatrix} + \begin{bmatrix} 0 & 0 \\ 0 & 1 \end{bmatrix} =$
$\begin{bmatrix} 7 & 9 \\ 7 & -8 \end{bmatrix}$

$[4 \quad -2]\begin{bmatrix} 3 & -1 \\ 2 & 5 \end{bmatrix} + [0]\,[0 \quad 1] = [8 \quad -14] + [0 \quad 0] = [8 \quad -14]$. $AB = \begin{bmatrix} 7 & 9 \\ 7 & -8 \\ 8 & -14 \end{bmatrix}$.

22.(a) $B = \left[\begin{array}{c|c} -1 & -2 \\ \hline 0 & 3 \\ 4 & 1 \end{array}\right]$ or $\left[\begin{array}{cc} -1 & -2 \\ \hline 0 & 3 \\ 4 & 1 \end{array}\right]$ (b) $B = \left[\begin{array}{c|c} -1 & -2 \\ \hline 0 & 3 \\ 4 & 1 \end{array}\right]$ or $\left[\begin{array}{cc} -1 & -2 \\ \hline 0 & 3 \\ 4 & 1 \end{array}\right]$

(c) $B = \left[\begin{array}{c|c} -1 & -2 \\ \hline 0 & 3 \\ \hline 4 & 1 \end{array}\right]$ or $\left[\begin{array}{cc} -1 & -2 \\ \hline 0 & 3 \\ \hline 4 & 1 \end{array}\right]$

23.(a) $A = \left[\begin{array}{cc|cc} 2 & 0 & 3 & -1 \\ \hline 6 & 2 & -5 & 9 \\ \hline 1 & 1 & 1 & 1 \end{array}\right]$, or $\left[\begin{array}{cc|cc} 2 & 0 & 3 & -1 \\ \hline 6 & 2 & -5 & 9 \\ 1 & 1 & 1 & 1 \end{array}\right]$, $\left[\begin{array}{cc|cc} 2 & 0 & 3 & -1 \\ 6 & 2 & -5 & 9 \\ \hline 1 & 1 & 1 & 1 \end{array}\right]$, $\left[\begin{array}{cc|cc} 2 & 0 & 3 & -1 \\ 6 & 2 & -5 & 9 \\ 1 & 1 & 1 & 1 \end{array}\right]$

(b) $A = \left[\begin{array}{c|ccc} 2 & 0 & 3 & -1 \\ \hline 6 & 2 & -5 & 9 \\ \hline 1 & 1 & 1 & 1 \end{array}\right]$, or $\left[\begin{array}{c|ccc} 2 & 0 & 3 & -1 \\ \hline 6 & 2 & -5 & 9 \\ 1 & 1 & 1 & 1 \end{array}\right]$, $\left[\begin{array}{c|ccc} 2 & 0 & 3 & -1 \\ 6 & 2 & -5 & 9 \\ \hline 1 & 1 & 1 & 1 \end{array}\right]$, $\left[\begin{array}{c|ccc} 2 & 0 & 3 & -1 \\ 6 & 2 & -5 & 9 \\ 1 & 1 & 1 & 1 \end{array}\right]$

(c) $A = \left[\begin{array}{ccc|c} 2 & 0 & 3 & -1 \\ \hline 6 & 2 & -5 & 9 \\ \hline 1 & 1 & 1 & 1 \end{array}\right]$, or $\left[\begin{array}{ccc|c} 2 & 0 & 3 & -1 \\ \hline 6 & 2 & -5 & 9 \\ 1 & 1 & 1 & 1 \end{array}\right]$, $\left[\begin{array}{ccc|c} 2 & 0 & 3 & -1 \\ 6 & 2 & -5 & 9 \\ \hline 1 & 1 & 1 & 1 \end{array}\right]$, $\left[\begin{array}{ccc|c} 2 & 0 & 3 & -1 \\ 6 & 2 & -5 & 9 \\ 1 & 1 & 1 & 1 \end{array}\right]$

24.(a) Partition: $\left[\begin{array}{c|ccc} 2 & 3 & 2 & 1 \\ \hline 4 & 0 & 0 & 0 \\ 1 & 0 & 0 & 0 \\ 5 & 0 & 0 & 0 \end{array}\right]\left[\begin{array}{cc} 1 & 2 \\ \hline -1 & 3 \\ 4 & 0 \\ 2 & 5 \end{array}\right]$. $[2][1 \;\; 2] + [3 \;\; 2 \;\; 1]\begin{bmatrix} -1 & 3 \\ 4 & 0 \\ 2 & 5 \end{bmatrix} = [2 \;\; 4] + [7 \;\; 14] = [\,9 \;\; 18]$.

$\begin{bmatrix} 4 \\ 1 \\ 5 \end{bmatrix}[1 \;\; 2] + \begin{bmatrix} 0 & 0 & 0 \\ 0 & 0 & 0 \\ 0 & 0 & 0 \end{bmatrix}\begin{bmatrix} -1 & 3 \\ 4 & 0 \\ 2 & 5 \end{bmatrix} = \begin{bmatrix} 4 & 8 \\ 1 & 2 \\ 5 & 10 \end{bmatrix} + \begin{bmatrix} 0 & 0 \\ 0 & 0 \\ 0 & 0 \end{bmatrix} = \begin{bmatrix} 4 & 8 \\ 1 & 2 \\ 5 & 10 \end{bmatrix}$. Product is $\begin{bmatrix} 9 & 18 \\ 4 & 8 \\ 1 & 2 \\ 5 & 10 \end{bmatrix}$

(b) Partition: $\left[\begin{array}{ccc|c} 1 & 0 & 0 & 2 \\ 0 & 1 & 0 & 3 \\ 0 & 0 & 1 & 4 \\ \hline 1 & 1 & 1 & -2 \end{array}\right]\left[\begin{array}{cc} 1 & 2 \\ 3 & 4 \\ 5 & 6 \\ \hline -1 & 3 \end{array}\right]$. $\begin{bmatrix} 1 & 0 & 0 \\ 0 & 1 & 0 \\ 0 & 0 & 1 \end{bmatrix}\begin{bmatrix} 1 & 2 \\ 3 & 4 \\ 5 & 6 \end{bmatrix} + \begin{bmatrix} 2 \\ 3 \\ 4 \end{bmatrix}[-1 \;\; 3] = \begin{bmatrix} 1 & 2 \\ 3 & 4 \\ 5 & 6 \end{bmatrix} + \begin{bmatrix} -2 & 6 \\ -3 & 9 \\ -4 & 12 \end{bmatrix} = \begin{bmatrix} -1 & 8 \\ 0 & 13 \\ 1 & 18 \end{bmatrix}$.

$[1 \;\; 1 \;\; 1]\begin{bmatrix} 1 & 2 \\ 3 & 4 \\ 5 & 6 \end{bmatrix} + [-2][-1 \;\; 3] = [9 \;\; 12] + [2 \;\; -6] = [11 \;\; 6]$. Product is $\begin{bmatrix} -1 & 8 \\ 0 & 13 \\ 1 & 18 \\ 11 & 6 \end{bmatrix}$

(c) Partition: $\left[\begin{array}{ccc|c} 1 & 0 & 0 & 1 \\ 0 & 1 & 0 & 2 \\ 0 & 0 & 1 & 3 \\ \hline 0 & 0 & 0 & 4 \\ 0 & 0 & 0 & 5 \end{array}\right]\left[\begin{array}{c} -1 \\ 2 \\ 6 \\ \hline 3 \end{array}\right]$. $\begin{bmatrix} 1 & 0 & 0 \\ 0 & 1 & 0 \\ 0 & 0 & 1 \end{bmatrix}\begin{bmatrix} -1 \\ 2 \\ 6 \end{bmatrix} + \begin{bmatrix} 1 \\ 2 \\ 3 \end{bmatrix}[3] = \begin{bmatrix} -1 \\ 2 \\ 6 \end{bmatrix} + \begin{bmatrix} 3 \\ 6 \\ 9 \end{bmatrix} = \begin{bmatrix} 2 \\ 8 \\ 15 \end{bmatrix}$

$\begin{bmatrix} 0 & 0 & 0 \\ 0 & 0 & 0 \end{bmatrix}\begin{bmatrix} -1 \\ 2 \\ 6 \end{bmatrix} + \begin{bmatrix} 4 \\ 5 \end{bmatrix}[3] = \begin{bmatrix} 0 \\ 0 \end{bmatrix} + \begin{bmatrix} 12 \\ 15 \end{bmatrix} = \begin{bmatrix} 12 \\ 15 \end{bmatrix}$. Product is $\begin{bmatrix} 2 \\ 8 \\ 15 \\ 12 \\ 15 \end{bmatrix}$

25. (a) True: If A+B exists then A and B are same size. If B+C exists B and C are same size. Thus A and C are the same size. A+C exists.

(b) False: Let A be mxn, B nxr, C rxq, all indices different. Then AB and BC exist. But A has n columns, C has r rows; AC does not exist.

(c) False: For example, let A and B be square with A being the identity matrix I. Then AB=B and BA=B. This is called a counter example.

(d) True: Let A be mx1, B be 1xm. Then AB is mxm, square.

(e) True: For example, consider elements in the third column; a_{33} is on main diagonal, a_{43} is below it, a_{23} is above it. If i>j then a_{ij} is below ajj in the jth column, it is below the main diagonal.

Exercise Set 2.2

1. (a) $AB = \begin{bmatrix} 4 & 7 & 10 \\ 0 & -5 & -4 \end{bmatrix}$. BA does not exist.

(b) $AC = \begin{bmatrix} 14 & 5 \\ -2 & -3 \end{bmatrix}$. $CA = \begin{bmatrix} -1 & 4 \\ 5 & 12 \end{bmatrix}$.

(c) $AD = \begin{bmatrix} 4 & 4 \\ -2 & 2 \end{bmatrix}$. $DA = \begin{bmatrix} 4 & 4 \\ -2 & 2 \end{bmatrix}$.

2. $A(BC) = \begin{bmatrix} 1 & 2 \\ -1 & 0 \\ 1 & 1 \end{bmatrix}\begin{bmatrix} 10 \\ 4 \end{bmatrix} = \begin{bmatrix} 18 \\ -10 \\ 14 \end{bmatrix}$. $(AB)C = \begin{bmatrix} -2 & 10 \\ -2 & -4 \\ 0 & 7 \end{bmatrix}\begin{bmatrix} 1 \\ 2 \end{bmatrix} = \begin{bmatrix} 18 \\ -10 \\ 14 \end{bmatrix}$.

3. $A(BC) = \begin{bmatrix} 1 & 2 \\ -1 & 3 \end{bmatrix}\begin{bmatrix} 8 & 10 \\ 14 & 20 \end{bmatrix} = \begin{bmatrix} 36 & 50 \\ 34 & 50 \end{bmatrix}$.

$(AB)C = \begin{bmatrix} 11 & 7 & 4 \\ 9 & 8 & 1 \end{bmatrix}\begin{bmatrix} 1 & 2 \\ 3 & 4 \\ 1 & 0 \end{bmatrix} = \begin{bmatrix} 36 & 50 \\ 34 & 50 \end{bmatrix}$.

4. (a) $2A + 3B = 2\begin{bmatrix} 1 & 2 \\ 3 & 4 \end{bmatrix} + 3\begin{bmatrix} 2 & -3 \\ 0 & 1 \end{bmatrix} = \begin{bmatrix} 2 & 4 \\ 6 & 8 \end{bmatrix} + \begin{bmatrix} 6 & -9 \\ 0 & 3 \end{bmatrix} = \begin{bmatrix} 8 & -5 \\ 6 & 11 \end{bmatrix}$.

(b) $A + 2B + 4C = \begin{bmatrix} 1 & 2 \\ 3 & 4 \end{bmatrix} + 2\begin{bmatrix} 2 & -3 \\ 0 & 1 \end{bmatrix} + 4\begin{bmatrix} -2 & 0 \\ 3 & 4 \end{bmatrix} = \begin{bmatrix} 1 & 2 \\ 3 & 4 \end{bmatrix} + \begin{bmatrix} 4 & -6 \\ 0 & 2 \end{bmatrix} + \begin{bmatrix} -8 & 0 \\ 12 & 16 \end{bmatrix}$

$= \begin{bmatrix} -3 & -4 \\ 15 & 22 \end{bmatrix}.$

(c) $3A + B - 2C = 3\begin{bmatrix} 1 & 2 \\ 3 & 4 \end{bmatrix} + \begin{bmatrix} 2 & -3 \\ 0 & 1 \end{bmatrix} - 2\begin{bmatrix} -2 & 0 \\ 3 & 4 \end{bmatrix} = \begin{bmatrix} 3 & 6 \\ 9 & 12 \end{bmatrix} + \begin{bmatrix} 2 & -3 \\ 0 & 1 \end{bmatrix} - \begin{bmatrix} -4 & 0 \\ 6 & 8 \end{bmatrix}$

$= \begin{bmatrix} 9 & 3 \\ 3 & 5 \end{bmatrix}.$

5. (a) $(AB)^2 = \begin{bmatrix} -2 & 2 \\ 11 & 19 \end{bmatrix}^2 = \begin{bmatrix} 26 & 34 \\ 187 & 383 \end{bmatrix}.$

(b) $A - 3B^2 = \begin{bmatrix} 2 & 0 \\ -1 & 5 \end{bmatrix} - 3\begin{bmatrix} 3 & 3 \\ 6 & 18 \end{bmatrix} = \begin{bmatrix} -7 & -9 \\ -19 & -49 \end{bmatrix}.$

(c) $A^2 B + 2C^3 = \begin{bmatrix} 4 & 0 \\ -7 & 25 \end{bmatrix}\begin{bmatrix} -1 & 1 \\ 2 & 4 \end{bmatrix} + 2\begin{bmatrix} 27 & 76 \\ 0 & 8 \end{bmatrix}$

$= \begin{bmatrix} -4 & 4 \\ 57 & 93 \end{bmatrix} + \begin{bmatrix} 54 & 152 \\ 0 & 16 \end{bmatrix} = \begin{bmatrix} 50 & 156 \\ 57 & 109 \end{bmatrix}.$

(d) $2A^2 - 2A + 3I_2 = \begin{bmatrix} 8 & 0 \\ -14 & 50 \end{bmatrix} - \begin{bmatrix} 4 & 0 \\ -2 & 10 \end{bmatrix} + \begin{bmatrix} 3 & 0 \\ 0 & 3 \end{bmatrix} = \begin{bmatrix} 7 & 0 \\ -12 & 43 \end{bmatrix}.$

6. (a) does not exist (b) 4x3 (c) 3x6 (d) 6x6

(e) does not exist, since A^2 does not exist

7. (a) 3x3 (b) 3x1 (c) does not exist, since PR does not exist

(d) does not exist, since SP does not exist (e) 2x3

8. AB can exist only if the number of rows in B is n, and BA can exist only if the number of columns in B is m. So for both AB and BA to exist, B must be nxm.

9. (a) The (i,j)th element of A + (B + C) is $a_{ij} + (b_{ij} + c_{ij})$. The (i,j)th element of (A + B) + C is $(a_{ij} + b_{ij}) + c_{ij} = a_{ij} + (b_{ij} + c_{ij})$. Since their elements are the same, A + (B + C) = (A + B) + C.

(b) The (i,j)th element of c(A + B) is $c(a_{ij} + b_{ij})$. The (i,j)th element of cA + cB is $ca_{ij} + cb_{ij} = c(a_{ij} + b_{ij})$. Since their elements are the same, c(A + B) = cA + cB.

(c) The only nonzero element in the jth column of I_n is the 1 in the jth row. Thus the (i,j)th element of AI_n is $a_{ij}(1) = a_{ij}$. Likewise, the only nonzero element in the ith row of I_n is the 1 in column i. Thus the (i,j)th element in $I_n A$ is $(1)a_{ij} = a_{ij}$. Since their elements are the same, $AI_n = I_n A = A$.

10. The only nonzero element in column j of I_n is the 1 in row j. Thus the (i,j)th element of AI_n is $a_{ij}(1) = a_{ij}$. Since their elements are the same, $AI_n = A$.

11. The (i,j)th element of cA is ca_{ij}. If $cA = O_{mn}$, then $ca_{ij} = 0$ for all i and j. So either c = 0 or all $a_{ij} = 0$, in which case $A = O_{mn}$.

12. (a) $A(A - 4B) + 2B(A + B) - A^2 + 7B^2 + 3AB$
$= A^2 - 4AB + 2BA + 2B^2 - A^2 + 7B^2 + 3AB = 9B^2 - AB + 2BA.$

(b) $B(2I_n - BA) + B(4I_n + 5A)B - 3BAB + 7B^2 A$
$= 2B - B^2 A + 4B + 5BAB - 3BAB + 7B^2 A = 6B + 2BAB + 6B^2 A.$

(c) $(A - B)(A + B) - (A + B)^2 = A^2 - BA + AB - B^2 - (A^2 + AB + BA + B^2)$
$= -2BA - 2B^2.$

13. (a) $A(A + B) - B(A + B) = A^2 + AB - BA - B^2.$

(b) $A(A - B)B + B2AB - 3A^2 = A^2 B - AB^2 + 2BAB - 3A^2.$

(c) $(A + B)^3 - 2A^3 - 3ABA - A3B^2 - B^3$
$= A^3 + A^2B + BA^2 + BAB + ABA + AB^2 + B^2A + B^3 - 2A^3 - 3ABA - 3AB^2 - B^3$
$= -A^3 + A^2B + BA^2 + BAB - 2ABA - 2AB^2 + B^2A.$

14. (a) $\begin{bmatrix} 1 & 0 \\ -1 & 0 \end{bmatrix}\begin{bmatrix} a & b \\ c & d \end{bmatrix} = \begin{bmatrix} a & b \\ -a & -b \end{bmatrix}$ and $\begin{bmatrix} a & b \\ c & d \end{bmatrix}\begin{bmatrix} 1 & 0 \\ -1 & 0 \end{bmatrix} = \begin{bmatrix} a-b & 0 \\ c-d & 0 \end{bmatrix}$.

For equality, it is necessary to have a = a-b, b = 0, and -a = c-d. Thus those matrices that commute with the given matrix are all matrices of the form

$$\begin{bmatrix} a & 0 \\ c & c+a \end{bmatrix}$$

where a and c can take any real values.

(b) $\begin{bmatrix} 1 & 0 \\ 0 & 2 \end{bmatrix}\begin{bmatrix} a & b \\ c & d \end{bmatrix} = \begin{bmatrix} a & b \\ 2c & 2d \end{bmatrix}$ and $\begin{bmatrix} a & b \\ c & d \end{bmatrix}\begin{bmatrix} 1 & 0 \\ 0 & 2 \end{bmatrix} = \begin{bmatrix} a & 2b \\ c & 2d \end{bmatrix}$.

For equality, it is necessary to have b = 2b and c = 2c, so that b = c = 0. Thus those matrices that commute with the given matrix are all matrices of the form

$$\begin{bmatrix} a & 0 \\ 0 & d \end{bmatrix}$$

where a and d can take any real values.

(c) $\begin{bmatrix} 0 & 1 & 0 \\ 0 & 0 & 1 \\ 0 & 0 & 0 \end{bmatrix}\begin{bmatrix} a & b & c \\ d & e & f \\ g & h & i \end{bmatrix} = \begin{bmatrix} d & e & f \\ g & h & i \\ 0 & 0 & 0 \end{bmatrix}$ and $\begin{bmatrix} a & b & c \\ d & e & f \\ g & h & i \end{bmatrix}\begin{bmatrix} 0 & 1 & 0 \\ 0 & 0 & 1 \\ 0 & 0 & 0 \end{bmatrix} = \begin{bmatrix} 0 & a & b \\ 0 & d & e \\ 0 & g & h \end{bmatrix}$.

For equality, it is necessary to have d = g = h = 0, a = e = i, and b = f. Thus those matrices that commute with the given matrix are all matrices of the form

$$\begin{bmatrix} a & b & c \\ 0 & a & b \\ 0 & 0 & a \end{bmatrix}$$

where a, b, and c can take any real values.

15. $AX_1 = AX_2$ does not imply that $X_1 = X_2$.

16. (a) $A^2 = AA$. Both matrices are nxn, so the number of columns in the first is the same as the number of rows in the second and they can be multiplied. The product matrix will have n rows because the first matrix has n rows, and it will have n columns because the second matrix has n columns.

(b) $A^2 = AA$. The first matrix is mxn and the second matrix is mxn, so the number of columns in the first matrix is not equal to the number of rows in the second matrix. Therefore they cannot be multiplied.

17. $(A + B)^2 = (A + B)(A + B) = A(A + B) + B(A + B)$ (distributive law)
$= A^2 + AB + BA + B^2$.

If $AB = BA$ then $(A + B)^2 = A^2 + 2AB + B^2$.

18. $(AB)^2 = (AB)(AB) = ABAB$.
If $AB = BA$ then $ABAB = A(BA)B = A(AB)B = AABB = A^2 B^2$.

Example: Let $A = \begin{bmatrix} 1 & -1 \\ 0 & 2 \end{bmatrix}$, $B = \begin{bmatrix} 1 & 1 \\ 1 & 0 \end{bmatrix}$. $AB = \begin{bmatrix} 0 & 1 \\ 2 & 0 \end{bmatrix}$ and $(AB)^2 = \begin{bmatrix} 2 & 0 \\ 0 & 2 \end{bmatrix}$.

$A^2 = \begin{bmatrix} 1 & -3 \\ 0 & 4 \end{bmatrix}$, $B^2 = \begin{bmatrix} 2 & 1 \\ 1 & 1 \end{bmatrix}$, and $A^2 B^2 = \begin{bmatrix} -1 & -2 \\ 4 & 4 \end{bmatrix} \neq (AB)^2$.

19. $(AB)^n = ABABAB \ldots ABAB = AABABA \ldots BABB = AAABAB \ldots ABBB$
$= AAAABA \ldots BBBB = \ldots = AAA \ldots AABBB \ldots BB = A^n B^n$.

The example from Exercise 18 will serve here also.

20. $A^r A^s = A^{r+s} = A^{s+r} = A^s A^r$.

21. (a) The (i,j)th element of A + B is $a_{ij} + b_{ij} = 0 + 0$ if $i \neq j$, so A + B is a diagonal matrix.

(b) The (i,j)th element of cA is $ca_{ij} = 0$ if $i \neq j$, so cA is a diagonal matrix.

(c) The (i,j)th element of AB is $a_{i1} b_{1j} + a_{i2} b_{2j} + \ldots + a_{in} b_{nj}$, which is zero unless i = j, since only a_{ii} and b_{jj} can be nonzero. Thus only the diagonal elements of AB can be nonzero, so AB is a diagonal matrix.

22. The nondiagonal elements of AB and of BA are zero. (See Exercise 21(c).)
The diagonal elements of AB are the elements $a_{i1}b_{1i} + a_{i2}b_{2i} + \ldots + a_{in}b_{ni} = a_{ii}b_{ii}$
and
the diagonal elements of BA are the elements $b_{i1}a_{1i} + b_{i2}a_{2i} + \ldots + b_{in}a_{ni} = b_{ii}a_{ii} = a_{ii}b_{ii}$. Thus AB = BA.

23. Call the diagonal matrix B. The (i,j)th element of AB is $a_{i1}b_{1j} + a_{i2}b_{2j} + \ldots + a_{in}b_{nj} = a_{ij}b_{jj}$ and the (i,j)th element of BA is $b_{i1}a_{1j} + b_{i2}a_{2j} + \ldots + b_{in}a_{nj} = b_{ii}a_{ij}$. If $b_{ii} \neq b_{jj}$, then a_{ij} must be zero for $i \neq j$. Thus A is a diagonal matrix.

24. (a) Yes, $\begin{bmatrix} 1 & 0 \\ 0 & 1 \end{bmatrix}^2 = \begin{bmatrix} 1 & 0 \\ 0 & 1 \end{bmatrix}$. (b) Yes, $\begin{bmatrix} 1 & 0 \\ 0 & 0 \end{bmatrix}^2 = \begin{bmatrix} 1 & 0 \\ 0 & 0 \end{bmatrix}$.

(c) No, $\begin{bmatrix} 0 & 1 \\ 1 & 0 \end{bmatrix}^2 = \begin{bmatrix} 1 & 0 \\ 0 & 1 \end{bmatrix}$. (d) Yes, $\begin{bmatrix} 3 & -6 \\ 1 & -2 \end{bmatrix}^2 = \begin{bmatrix} 3 & -6 \\ 1 & -2 \end{bmatrix}$.

(e) Yes, $\begin{bmatrix} 1 & 2 & 2 \\ 0 & 0 & -1 \\ 0 & 0 & 1 \end{bmatrix}^2 = \begin{bmatrix} 1 & 2 & 2 \\ 0 & 0 & -1 \\ 0 & 0 & 1 \end{bmatrix}$ (f) No, $\begin{bmatrix} 1 & 3 & 0 \\ 0 & 0 & 1 \\ 0 & 0 & 0 \end{bmatrix}^2 = \begin{bmatrix} 1 & 3 & 3 \\ 0 & 0 & 0 \\ 0 & 0 & 0 \end{bmatrix}$.

25. $\begin{bmatrix} 1 & b \\ c & d \end{bmatrix}^2 = \begin{bmatrix} 1+bc & b+bd \\ c+cd & bc+d^2 \end{bmatrix}$, so the matrix will be idempotent if

$$1 = 1 + bc,\ \ b = b + bd,\ \ c = c + cd,\ \text{ and }\ d = bc + d^2 .$$

The first equation implies that bc = 0 and substituting bc = 0 in the last equation gives $d = d^2$. Thus d must be either zero or 1. If one of b and c is nonzero, the second and third equations give d = 0. If both b and c are zero, then d can be either zero or 1. Thus, the idempotent matrices of the given form are

$$\begin{bmatrix} 1 & 0 \\ 0 & 0 \end{bmatrix}, \begin{bmatrix} 1 & 0 \\ 0 & 1 \end{bmatrix}, \begin{bmatrix} 1 & b \\ 0 & 0 \end{bmatrix}, \text{ and } \begin{bmatrix} 1 & 0 \\ c & 0 \end{bmatrix}.$$

26. $\begin{bmatrix} a & 0 \\ c & d \end{bmatrix}^2 = \begin{bmatrix} a^2 & 0 \\ ac+cd & d^2 \end{bmatrix}$, so the matrix will be idempotent if

$$a = a^2,\ \ d = d^2,\ \text{ and }\ c = ac + cd = c(a + d).$$

Thus a is zero or 1, d is zero or 1, and if c is nonzero, a + d = 1. If c = 0, the sum of a and
d is unimportant. The idempotent matrices of the given form are

$$\begin{bmatrix} 0 & 0 \\ 0 & 0 \end{bmatrix}, \begin{bmatrix} 1 & 0 \\ 0 & 0 \end{bmatrix}, \begin{bmatrix} 1 & 0 \\ 0 & 1 \end{bmatrix}, \begin{bmatrix} 0 & 0 \\ 0 & 1 \end{bmatrix}, \begin{bmatrix} 1 & 0 \\ c & 0 \end{bmatrix}, \text{ and } \begin{bmatrix} 0 & 0 \\ c & 1 \end{bmatrix}.$$

27. $(AB)^2 = (AB)(AB) = A(BA)B$. If $AB = BA$, then $(AB)^2 = A(AB)B = A^2B^2 = AB$. Thus AB is idempotent.

28. $A^2 = A$, $A^3 = A^2A = AA = A$, $A^4 = A^3A = AA = A$, . . . , and $A^n = A^{n-1}A = AA = A$.

29. (a) $\begin{bmatrix} 1 & 1 \\ -1 & -1 \end{bmatrix}^2 = \begin{bmatrix} 0 & 0 \\ 0 & 0 \end{bmatrix}$. (b) $\begin{bmatrix} -4 & 8 \\ -2 & 4 \end{bmatrix}^2 = \begin{bmatrix} 0 & 0 \\ 0 & 0 \end{bmatrix}$.

(c) $\begin{bmatrix} 3 & -9 \\ 1 & -3 \end{bmatrix}^2 = \begin{bmatrix} 0 & 0 \\ 0 & 0 \end{bmatrix}$.

30. $$\begin{bmatrix} 0 & 1 & 0 \\ 0 & 0 & 1 \\ 0 & 0 & 0 \end{bmatrix}^3 = \begin{bmatrix} 0 & 1 & 0 \\ 0 & 0 & 1 \\ 0 & 0 & 0 \end{bmatrix}^2 \begin{bmatrix} 0 & 1 & 0 \\ 0 & 0 & 1 \\ 0 & 0 & 0 \end{bmatrix} = \begin{bmatrix} 0 & 0 & 1 \\ 0 & 0 & 0 \\ 0 & 0 & 0 \end{bmatrix} \begin{bmatrix} 0 & 1 & 0 \\ 0 & 0 & 1 \\ 0 & 0 & 0 \end{bmatrix} = \begin{bmatrix} 0 & 0 & 0 \\ 0 & 0 & 0 \\ 0 & 0 & 0 \end{bmatrix}.$$

31. (a) $\begin{bmatrix} 2 & 3 \\ 3 & -8 \end{bmatrix} \begin{bmatrix} x_1 \\ x_2 \end{bmatrix} = \begin{bmatrix} 4 \\ -1 \end{bmatrix}$. (b) $\begin{bmatrix} 4 & 7 \\ -2 & 3 \end{bmatrix} \begin{bmatrix} x_1 \\ x_2 \end{bmatrix} = \begin{bmatrix} -2 \\ -4 \end{bmatrix}$.

(c) $\begin{bmatrix} -9 & -3 \\ 6 & -2 \end{bmatrix} \begin{bmatrix} x_1 \\ x_2 \end{bmatrix} = \begin{bmatrix} -4 \\ 7 \end{bmatrix}$.

32. (a) $\begin{bmatrix} 1 & 8 & -2 \\ 4 & -7 & 1 \\ -2 & -5 & -2 \end{bmatrix} \begin{bmatrix} x_1 \\ x_2 \\ x_3 \end{bmatrix} = \begin{bmatrix} 3 \\ -3 \\ 1 \end{bmatrix}$. (b) $\begin{bmatrix} 5 & 2 \\ 4 & -3 \\ 3 & 1 \end{bmatrix} \begin{bmatrix} x_1 \\ x_2 \end{bmatrix} = \begin{bmatrix} 6 \\ -2 \\ 9 \end{bmatrix}$.

(c) $\begin{bmatrix} 1 & -3 & 6 \\ 7 & 5 & 1 \end{bmatrix}\begin{bmatrix} x_1 \\ x_2 \\ x_3 \end{bmatrix} = \begin{bmatrix} 2 \\ -9 \end{bmatrix}$. (d) $\begin{bmatrix} 2 & 5 & -3 & 4 \\ 1 & 0 & 9 & 5 \\ 3 & -3 & -8 & 5 \end{bmatrix}\begin{bmatrix} x_1 \\ x_2 \\ x_3 \\ x_4 \end{bmatrix} = \begin{bmatrix} 4 \\ 12 \\ -2 \end{bmatrix}$.

33. Since X_1 is a solution then the scalar multiple of X_1 by a is a solution. aX_1 is a solution. Similarly bX_2 is a solution. Since the sum of two solutions is a solution $aX_1 + bX_2$ is a solution.

34. e.g., $X_3 = \begin{bmatrix} 5 \\ -2 \\ 1 \\ 0 \end{bmatrix} + \begin{bmatrix} 11 \\ -4 \\ 1 \\ 1 \end{bmatrix} = \begin{bmatrix} 16 \\ -6 \\ 2 \\ 1 \end{bmatrix}$. $X_4 = 2\begin{bmatrix} 5 \\ -2 \\ 1 \\ 0 \end{bmatrix} = \begin{bmatrix} 10 \\ -4 \\ 2 \\ 0 \end{bmatrix}$. $X_5 = 3\begin{bmatrix} 11 \\ -4 \\ 1 \\ 1 \end{bmatrix} = \begin{bmatrix} 33 \\ -12 \\ 3 \\ 3 \end{bmatrix}$. $X_6 = X_4 + X_5 = \begin{bmatrix} 10 \\ -4 \\ 2 \\ 0 \end{bmatrix} + \begin{bmatrix} 33 \\ -12 \\ 3 \\ 3 \end{bmatrix} = \begin{bmatrix} 43 \\ -16 \\ 5 \\ 3 \end{bmatrix}$.

By Exercise 33 the linear combinations $a\begin{bmatrix} 5 \\ -2 \\ 1 \\ 0 \end{bmatrix} + b\begin{bmatrix} 11 \\ -4 \\ 1 \\ 1 \end{bmatrix}$ are solutions for all values of a and b.

Want $x_1=1, x_2=0$ in solution. Thus $5a+11b=1$. $-2a-4b=0$. Gives $a=-2$, $b=1$.

The solution is $-2\begin{bmatrix} 5 \\ -2 \\ 1 \\ 0 \end{bmatrix} + 1\begin{bmatrix} 11 \\ -4 \\ 1 \\ 1 \end{bmatrix} = \begin{bmatrix} 1 \\ 0 \\ -1 \\ 1 \end{bmatrix}$.

35. e.g., $X_3 = \begin{bmatrix} 0 \\ -1 \\ 1 \\ -1 \end{bmatrix} + \begin{bmatrix} 3 \\ 7 \\ 2 \\ 1 \end{bmatrix} = \begin{bmatrix} 3 \\ 6 \\ 3 \\ 0 \end{bmatrix}$. $X_4 = 4\begin{bmatrix} 0 \\ -1 \\ 1 \\ -1 \end{bmatrix} = \begin{bmatrix} 0 \\ -4 \\ 4 \\ -4 \end{bmatrix}$. $X_5 = -2\begin{bmatrix} 3 \\ 7 \\ 2 \\ 1 \end{bmatrix} = \begin{bmatrix} -6 \\ -14 \\ -4 \\ -2 \end{bmatrix}$. $X_6 = X_4 + X_5 = \begin{bmatrix} 0 \\ -4 \\ 4 \\ -4 \end{bmatrix} + \begin{bmatrix} -6 \\ -14 \\ -4 \\ -2 \end{bmatrix} = \begin{bmatrix} -6 \\ -18 \\ 0 \\ -6 \end{bmatrix}$.

By Exercise 33 the linear combinations $a\begin{bmatrix} 0 \\ -1 \\ 1 \\ -1 \end{bmatrix} + b\begin{bmatrix} 3 \\ 7 \\ 2 \\ 1 \end{bmatrix}$ are solutions for all values of a and b.

Want $x_1 = 6$, $x_2 = 9$ in solution. Thus 0a+3b=6, b=2. –a+7b=9. Thus a=5, b=2.

The solution is $5\begin{bmatrix}0\\-1\\1\\-1\end{bmatrix}+2\begin{bmatrix}3\\7\\2\\1\end{bmatrix}=\begin{bmatrix}6\\9\\9\\-3\end{bmatrix}$.

36. (a) Corresponding system of homogeneous linear equations:

$$\begin{aligned} x_1 \qquad + x_3 &= 0\\ x_1 + x_2 - 2x_3 &= 0\\ 2x_1 + x_2 - x_3 &= 0 \end{aligned}$$

General solution of nonhomogeneous system is (-r+2, 3r+3, r) = (-r, 3r, r) + (2, 3, 0). Thus general solution of homogeneous system is (-r, 3r, r). (Check that it satisfies equations.)

(b) Homog system: Write general solution as r(-1, 3, 1). Basis {(-1, 3, 1)}. Subspace of solutions is a line through the origin defined by the vector (-1, 3, 1).

(c) Nonhomog system: (-r+2, 3r+3, r) = r(-1, 3, 1) + (2, 3, 0). Solution is the line defined by the vector (-1, 3, 1) slid in a manner defined by the vector (2, 3, 0). i.e. it is the line through the point (2, 3, 0) parallel to the line defined by the vector (-1, 3, 1).

37. (a) Corresponding system of homogeneous linear equations:

$$\begin{aligned} x_1 + x_2 + 5x_3 &= 0\\ x_2 + 3x_3 &= 0\\ x_1 + 2x_2 + 8x_3 &= 0 \end{aligned}$$

General solution of nonhomogeneous system is (-2r+1, -3r+4, r) = (-2r, -3r, r) + (1, 4, 0). Thus general solution of homogeneous system is (-2r, -3r, r).

(b) Homog system: Write general solution as r(-2, -3, 1). Basis {(-2, -3, 1)}. Subspace of solutions is a line through the origin defined by the vector (-2, -3, 1).

(c) Nonhomog system: (-2r+1, -3r+4, r) = r(-2, -3, 1) + (1, 4, 0). Solution is the line defined by the vector (-2, -3, 1) slid in a manner defined by the vector (1, 4, 0). i.e. it is the line through the point (1, 4, 0) parallel to the line defined by the vector (-2, -3, 1).

38. (a) Corresponding system of homogeneous linear equations:

$$\begin{aligned} x_1 + x_2 + x_3 &= 0\\ 2x_1 - x_2 - 4x_3 &= 0\\ x_1 - x_2 - 3x_3 &= 0 \end{aligned}$$

General solution of nonhomogeneous system is (r-1, -2r+3, r) = (r, -2r, r) + (-1, 3, 0). Thus general solution of homogeneous system is (r, -2r, r).

(b) Homog system: Write general solution as r(1, -2, 1). Basis {(1, -2, 1)}. Subspace of solutions is a line through the origin defined by the vector (1, -2, 1).

(c) Nonhomog system: (r-1, -2r+3, r) = r(1, -2, 1) + (-1, 3, 0). Solution is the line defined by the vector (1, -2, 1) slid in a manner defined by the vector (-1, 3, 0). i.e. it is the line through the point (-1, 3, 0) parallel to the line defined by the vector (1, -2,1).

39. (a) Corresponding system of homogeneous linear equations: $x_1 + 2x_2 - x_3 = 0$

$$2x_1 + 4x_2 - 2x_3 = 0$$
$$-3x_1 - 6x_2 + 3x_3 = 0$$

General solution of nonhomogeneous system is (-2r+s+3, r, s) = (-2r+s, r, s) + (3, 0, 0).

Thus general solution of homogeneous system is (-2r+s, r, s).

(b) Homog system: Write general solution as r(-2, 1, 0) + s(1, 0, 1). Basis {(-2, 1, 0), (1, 0, 1)}. Subspace of solutions is a plane through the origin defined by the vectors (-2, 1, 0) and (1, 0, 1).

(c) Nonhomog system: (-2r+s+3, r, s) = r(-2, 1, 0) + s(1, 0, 1)+ (3, 0, 0). Solution is the plane defined by the vectors (-2, 1, 0 and (1, 0, 1) slid in a manner defined by the vector (3, 0, 0). i.e. it is the plane through the point (3, 0, 0) parallel to the plane defined by the vectors (-2, 1, 0) and (1, 0, 1).

40. $$\begin{bmatrix} 1 & 2 & -3 & 5 \\ 0 & 1 & -1 & 2 \\ 2 & 5 & -7 & 12 \end{bmatrix} \underset{R3-2R1}{\approx} \begin{bmatrix} 1 & 2 & -3 & 5 \\ 0 & 1 & -1 & 2 \\ 0 & 1 & -1 & 2 \end{bmatrix} \underset{\substack{R1-2R2 \\ R3-R2}}{\approx} \begin{bmatrix} 1 & 0 & -1 & 1 \\ 0 & 1 & -1 & 2 \\ 0 & 0 & 0 & 0 \end{bmatrix},$$

so $x_1 - x_3 = 1$ and $x_2 - x_3 = 2$. General solution is x_1=r+1, x_2=r+2, x_3=r.
Write as (r+1, r+2, r). Consider two solutions: r=1, gives (2,3,1) and r=2, gives (3, 4, 2).
Check addition: (2,3,1)+(3,4,2) = (5,7,3). Not of form (r+1, r+2, r). Thus not closed under addition, not a subspace. Check scalar multiplication: e.g., 2(2,3,1) = (4,6,2).
Not of form (r+1, r+2, r). Thus not closed under scalar multiplication either.
Geometry: (r+1, r+2, r) = r(1,1,1)+(1,2,0). Set of solutions is line through origin in $\mathbf{R}^3$ defined by r(1,1,1) slid in manner described by (1,2,0). (Line is set of solutions to corresponding homogeneous system.)

41. Let AX = B be a system of nonhomogeneous linear equations. Thus B≠ 0. Let X_1 be a solution. Then AX_1 = B. Let c be an arbitrary scalar. Then AX_1= B gives cAX_1= cB. By the properties of matrices, this gives $A(cX_1) = cB \neq B$. The set is not closed under scalar multiplication.

42. Consider the system AX = B. Let $A_1, \ldots, A_n$ be the columns of A. Suppose B is a linear combination of $A_1, \ldots, A_n$. Let B= $x_1A_1 + \ldots + x_nA_n$. This can be written B = AX where X = $\begin{bmatrix} x_1 \\ \vdots \\ x_n \end{bmatrix}$ (see Section 2.1). This value of X is a solution to the system.

43. (a) False: $(A+B)(A-B) = A^2 + BA - AB - B^2 = A^2 - B^2$ only if BA - AB = 0; that is AB = BA.

(b) True: A(B+C+D) = A((B+C)+D) = A(B+C)+AD = AB+AC+AD.

(c) False: ABC is mq matrix; m rows, q columns. Thus mq elements.

(d) True: Given A^2=A. Thus $A^3=A^2A=AA=A^2=A$; $A^4=A^3A=AA=A^2=A$, and so on.

(e) True: $AX_1 = B_1$, $AX_2 = B_2$. Add, $AX_1+AX_2 = B_1+B_2$, $A(X_1+X_2) = B_1+B_2$.

44. (a) Since the terminals are connected directly, $V_1=V_2$ (1st equ). Voltage drop across the resistance is V_1. The current through the resistance is I_1-I_2. Thus $V_1= (I_1-I_2)R$ (2nd equ). Have two equations, $V_2 = V_1 + 0I_1$ and $I_2 = -V_1/R+I_1$. Combine into matrix form
$\begin{bmatrix} V_2 \\ I_2 \end{bmatrix} = \begin{bmatrix} 1 & 0 \\ -1/R & 1 \end{bmatrix}\begin{bmatrix} V_1 \\ I_1 \end{bmatrix}$. Transmission matrix is $\begin{bmatrix} 1 & 0 \\ -1/R & 1 \end{bmatrix}$.

(b) Current through resistance R_1 is $I_1 - I_2$. Drop in voltage across R_1 is V_1. Ohm's Law gives $V_1 = (I_1 - I_2)R_1$, $I_2 = I_1 - V_1/R_1$.
Current thro' resistance R_2 is I_2. Drop in voltage across R_2 is V_1-V_2. Thus $V_1-V_2=I_2R_2$,
$V_2 = V_1-I_2R_2 = V_1-(I_1 - V_1/R_1)R_2 = V_1(1+R_2/R_1) - I_1$.
Combine $V_2 = V1(1+R_2/R_1) - I_1$ and $I_2 = I_1 - V_1/R_1$.

$$\begin{bmatrix} V_2 \\ I_2 \end{bmatrix} = \begin{bmatrix} 1+R_2/R_1 & -1 \\ -1/R_1 & 1 \end{bmatrix}\begin{bmatrix} V_1 \\ I_1 \end{bmatrix}. \text{ Transmission matrix is } \begin{bmatrix} 1+R_2/R_1 & -1 \\ -1/R_1 & 1 \end{bmatrix}.$$

(c) Current through resistance R_1 is I_1. Drop in voltage across R_1 is V_1-V_2. Thus $V_1-V_2 = I_1R_1$. Gives $V_2 = V_1 - I_1 R_1$.
Current through resistance R_2 is $I_1 - I_2$. Drop in voltage across R_2 is V_2. Thus $V_2=(I_1 - I_2)R_2$.
$I_2 = I_1-V_2/R_2 = I_1- (V_1 - I_1 R_1)/R_2 = -V_1/R_2 + I_1(1 + R_1/R_2)$.
Combine $V_2 = V_1 - I_1 R_1$ and $I_2 = -V_1/R_2 + I_1(1 + R_1/R_2)$.

$$\begin{bmatrix} V_2 \\ I_2 \end{bmatrix} = \begin{bmatrix} 1 & -R_1 \\ -1/R_2 & 1+R_1/R_2 \end{bmatrix}\begin{bmatrix} V_1 \\ I_1 \end{bmatrix}. \text{ Transmission matrix is } \begin{bmatrix} 1 & -R_1 \\ -1/R_2 & 1+R_1/R_2 \end{bmatrix}.$$

45. (a) Transmission matrix = $\begin{bmatrix} 3 & -1 \\ -1 & 1 \end{bmatrix}\begin{bmatrix} 1 & 0 \\ 1 & 1 \end{bmatrix}\begin{bmatrix} 1 & -1 \\ 0 & 1 \end{bmatrix} = \begin{bmatrix} 2 & -3 \\ 0 & 1 \end{bmatrix}$

(b) Outputs $\begin{bmatrix} 2 & -3 \\ 0 & 1 \end{bmatrix}\begin{bmatrix} 5 \\ 2 \end{bmatrix} = \begin{bmatrix} 4 \\ 2 \end{bmatrix}$ V=4 volts, I=2 amps.

<u>Exercise Set 2.3</u>

1. (a) $A^t = \begin{bmatrix} -1 & 2 \\ 2 & -3 \end{bmatrix}$. symmetric (b) $B^t = \begin{bmatrix} 1 & 0 \\ 2 & 3 \end{bmatrix}$. not symmetric

(c) $C^t = \begin{bmatrix} 3 & 2 \\ -1 & 4 \end{bmatrix}$. not symmetric (d) $D^t = \begin{bmatrix} 4 & -2 & 7 \\ 5 & 3 & 0 \end{bmatrix}$. not symmetric

(e) $E^t = \begin{bmatrix} 4 & -1 & 0 \\ 5 & 2 & 1 \\ 6 & 3 & 2 \end{bmatrix}$. not symmetric (f) $F^t = \begin{bmatrix} 1 & -1 & 3 \\ -1 & 2 & 0 \\ 3 & 0 & 4 \end{bmatrix}$. symmetric

(g) $G^t = \begin{bmatrix} -2 & 1 \\ 4 & 0 \\ 5 & 3 \\ 7 & -7 \end{bmatrix}$. not symmetric (h) $H^t = \begin{bmatrix} 1 & 4 & -2 \\ -2 & 5 & 6 \\ 3 & 6 & 7 \end{bmatrix}$. not symmetric

(i) $K^t = \begin{bmatrix} 7 & 0 & 0 \\ 0 & -3 & 0 \\ 0 & 0 & 9 \end{bmatrix}$. symmetric

2. (a) $\begin{bmatrix} 1 & 2 & 4 \\ 2 & 6 & 5 \\ 4 & 5 & 2 \end{bmatrix}$ (b) $\begin{bmatrix} 3 & 5 & -3 \\ 5 & 8 & 4 \\ -3 & 4 & 3 \end{bmatrix}$ (c) $\begin{bmatrix} -3 & -4 & 8 & 9 \\ -4 & 7 & 2 & 7 \\ 8 & 2 & 6 & 4 \\ 9 & 7 & 4 & 9 \end{bmatrix}$

3. (a) 4x2 (b) 4x3 (c) Does not exist. (d) 4x4 (e) 4x3

4. (a) The (i,j)th element of $(A + B)^t$ is the (j,i)th element of A + B, which is $a_{ji} + b_{ji}$.
The (i,j)th element of A^t is the (j,i)th element of A, which is a_{ji}.
The (i,j)th element of B^t is the (j,i)th element of B, which is b_{ji}.
So the (i,j)th element of $A^t + B^t$ is $a_{ji} + b_{ji}$. Hence $(A + B)^t = A^t + B^t$.

(b) The (i,j)th element of $(cA)^t$ is the (j,i)th element of cA, which is ca_{ji}.
The (i,j)th element of cA^t is c times the (j,i)th element of A, i.e., ca_{ji}. Hence $(cA)^t = cA^t$.

(c) The (i,j)th element of $(A^t)^t$ is the (j,i)th element of A^t, which is the (i,j)th element of A. Thus $(A^t)^t = A$.

5. (a) $(A + B + C)^t = (A + (B + C))^t = A^t + (B + C)^t = A^t + B^t + C^t$.

(b) $(ABC)^t = (A(BC))^t = (BC)^t A^t = C^t B^t A^t$.

6. The (i,j)th element of A^t is the (j,i)th element of A. However, since A is a diagonal matrix, $a_{ji} = a_{ij} = 0$ if $i \neq j$. If $i = j$ then $a_{ij} = a_{ii} = a_{ji}$. Hence $A = A^t$.

7. $(A^n)^t = (\underbrace{AA\ldots A}_{n \text{ terms}})^t = \underbrace{A^t A^t \ldots A^t}_{n \text{ terms}}$ (by extension of result of 5(b)) $= (A^t)^n$.

8. (a) Suppose A is symmetric. Then $A = A^t$. (I,j)th element of A = (I,j)th element of A^t = (j,I)th element of A. Thus $a_{ij} = a_{ji}$.

   (b) Suppose $a_{ij} = a_{ji}$. Then (I,j)th element of A = (j,I)th element of A = (I,j)th element of A^t. Thus $A = A^t$. A is symmetric.

9. If A is symmetric then $A = A^t$. By Theorem 2.4, $A^t = A = (A^t)^t$. So A^t is symmetric.

10. $A = A^t$ and $B = B^t$. Thus $A + B = A^t + B^t = (A + B)^t$ by Theorem 2.4. Thus A + B is symmetric. Closed under addition.
    Further, $A = A^t$, so $cA = cA^t = (cA)^t$ by Theorem 2.4. So cA is symmetric. Closed under scalar multiplication.

11. (a) $\begin{bmatrix} 0 & -1 \\ 1 & 0 \end{bmatrix}$

    (b) If $A = -A^t$, then diagonal elements are such that $a_{ii} = -a_{ii}$. Thus $a_{ii} = 0$.

    (c) If A and B are antisymmetric, then $A + B = (-A^t) + (-B^t) = -(A^t + B^t) = -(A + B)^t$, so A + B is antisymmetric.

    (d) If A is antisymmetric, then $A = -A^t$ and
    $cA = c(-A^t) = c((-1)A^t) = c(-1)A^t = -cA^t = -(cA)^t$, so cA is antisymmetric.

12. (a) $(A + A^t)^t = A^t + (A^t)^t = A^t + A = A + A^t$, so $A + A^t$ is symmetric.

(b) $(A - A^t)^t = A^t - (A^t)^t = A^t - A = -(A - A^t)$, so $A - A^t$ is antisymmetric.

13. $B = (1/2)(A + A^t)$ is symmetric and $C = (1/2)(A - A^t)$ is antisymmetric (Based on exercise 12). $A = B + C$.

14. (a) If A is idempotent, then $A^2 = A$. $(A^t)^2 = (A^2)^t$ (ex 7) $= A^t$, so A^t is idempotent.

(b) If A^t is idempotent, then $(A^t)^2 = A^t$. Therefore $(A^2)^t = (A^t)^2 = A^t$, so that $A^2 = ((A^2)^t)^t = (A^t)^t = A$, and A is idempotent.

15. (a) $2 + (-4) = -2$. (b) $5 + (-3) + 8 = 10$. (c) $0 + 5 - 7 + 1 = -1$.

16. (a) $tr(cA) = ca_{11} + ca_{22} + \ldots + ca_{nn} = c(a_{11} + a_{22} + \ldots + a_{nn}) = ctr(A)$.

(b) The diagonal elements of AB are

$$a_{11}b_{11} + a_{12}b_{21} + \ldots + a_{1n}b_{n1} = b_{11}a_{11} + b_{21}a_{12} + \ldots + b_{n1}a_{1n},$$

$$a_{21}b_{12} + a_{22}b_{22} + \ldots + a_{2n}b_{n2} = b_{12}a_{21} + a_{22}b_{22} + \ldots + b_{n2}a_{2n},$$

$$\vdots$$

$$a_{m1}b_{1m} + a_{m2}b_{2m} + \ldots + a_{mn}b_{nm} = b_{1m}a_{m1} + b_{2m}a_{m2} + \ldots + b_{nm}a_{mn}$$

.

The diagonal elements of BA are

$$b_{11}a_{11} + b_{12}a_{21} + \ldots + b_{1m}a_{m1},$$

$$b_{21}a_{12} + b_{22}a_{22} + \ldots + b_{2m}a_{m2},$$

$$\vdots$$

$b_{n1}a_{1n} + b_{n2}a_{2n} + \ldots + b_{nm}a_{mn}$. Since A and B are square n=m.

Notice that the terms that appear in the diagonal elements of AB are the same terms that appear in the diagonal elements of BA, although not necessarily in the same locations. Thus the sum of the diagonal elements of AB will be the same as the sum of the diagonal elements of BA. So tr(AB) = tr(BA).

(c) $tr(A) = a_{11} + a_{22} + \ldots + a_{nn} = tr(A^t)$.

17. $tr(A + B + C) = tr(A + (B + C)) = tr(A) + tr(B + C) = tr(A) + tr(B) + tr(C)$.

18. If $A^t = B^t$, the (i,j)th element of A^t equals the (i,j)th element of B^t. The (i,j)th element of A^t is the (j,i)th element of A and the (i,j)th element of B^t is the (j,i)th element of B, so the (j,i)th element of A equals the (j,i)th element of B and A = B. To show that if A = B then $A^t = B^t$, simply interchange A and A^t and B and B^t everywhere in the above argument.

19. If the number of columns in A is equal to the number of rows in B, then from the definition of matrix multiplication AB exists and its (i,j)th element is obtained by multiplying the corresponding elements of row i of A and column j of B. If AB exists, then from the definition of matrix multiplication the number of columns in A equals the number of rows in B.

20. If $A = O_n$, then for any nxn matrix B the (i,j)th element of AB is $0 \times b_{1j} + 0 \times b_{2j} + \ldots + 0 \times b_{nj} = 0$, so $AB = O_n$. Suppose the (i,j)th element of A is not zero. Let B be the matrix with all elements zero except the (j,k)th element. Then the (i,k)th element of AB is $a_{i1} \times 0 + \ldots + a_{ij-1} \times 0 + a_{ij} \times b_{jk} + a_{ij+1} \times 0 + \ldots + a_{in} \times 0 = a_{ij} \times b_{jk} \neq 0$, so $AB \neq O_n$. Thus the only matrix for which $AB = O_n$ for all nxn matrices B is $A = O_n$.

21. $A + B = \begin{bmatrix} 3+i & 8+i \\ 5+2i & 4-2i \end{bmatrix}$.

$$AB = \begin{bmatrix} 5(-2+i)+(3-i)(3-i) & 5(5+2i)+(3-i)(4+3i) \\ (2+3i)(-2+i)+(-5i)(3-i) & (2+3i)(5+2i)+(-5i)(4+3i) \end{bmatrix} = \begin{bmatrix} -2-i & 40+15i \\ -12-19i & 19-i \end{bmatrix}.$$

$$BA = \begin{bmatrix} (-2+i)5+(5+2i)(2+3i) & (-2+i)(3-i)+(5+2i)(-5i) \\ (3-i)5+(4+3i)(2+3i) & (3-i)(3-i)+(4+3i)(-5i) \end{bmatrix} = \begin{bmatrix} -6+24i & 5-20i \\ 14+13i & 23-26i \end{bmatrix}.$$

22. $A + B = \begin{bmatrix} 6+2i & -1-3i \\ 8+2i & 5-6i \end{bmatrix}$.

$AB = \begin{bmatrix} 11 & -19-25i \\ 12+8i & -19-15i \end{bmatrix}$. $\quad BA = \begin{bmatrix} -11 & 4-i \\ 42-20i & 3-15i \end{bmatrix}$

23. $\bar{A} = \begin{bmatrix} 2+3i & -5i \\ 2 & 5+4i \end{bmatrix}$ $A^* = \bar{A}^t = \begin{bmatrix} 2+3i & 2 \\ -5i & 5+4i \end{bmatrix}$ not hermitian

$\bar{B} = \begin{bmatrix} 4 & 5+i \\ 5-i & 6 \end{bmatrix}$ $B^* = \bar{B}^t = \begin{bmatrix} 4 & 5-i \\ 5+i & 6 \end{bmatrix}$ hermitian

$\bar{C} = \begin{bmatrix} -7i & 4+3i \\ 6-8i & -9 \end{bmatrix}$ $C^* = \bar{C}^t = \begin{bmatrix} -7i & 6-8i \\ 4+3i & -9 \end{bmatrix}$ not hermitian

$\bar{D} = \begin{bmatrix} -2 & 3+5i \\ 3-5i & 9 \end{bmatrix}$ $D^* = \bar{D}^t = \begin{bmatrix} -2 & 3-5i \\ 3+5i & 9 \end{bmatrix}$ hermitian

24. $\bar{A} = \begin{bmatrix} 3 & 7-2i \\ 7+2i & 5 \end{bmatrix}$ $A^* = \bar{A}^t = \begin{bmatrix} 3 & 7+2i \\ 7-2i & 5 \end{bmatrix}$ hermitian

$\bar{B} = \begin{bmatrix} 3-5i & 1+2i \\ 1-2i & 5-6i \end{bmatrix}$ $B^* = \bar{B}^t = \begin{bmatrix} 3-5i & 1-2i \\ 1+2i & 5-6i \end{bmatrix}$ not hermitian

$\bar{C} = \begin{bmatrix} 1 & 2 \\ 2 & 4 \end{bmatrix}$ $C^* = \bar{C}^t = \begin{bmatrix} 1 & 2 \\ 2 & 4 \end{bmatrix}$ hermitian

$\bar{D} = \begin{bmatrix} 9 & 3i \\ -3i & 8 \end{bmatrix}$ $D^* = \bar{D}^t = \begin{bmatrix} 9 & -3i \\ 3i & 8 \end{bmatrix}$ hermitian

25. (a) The (i,j)th element of $A^* + B^*$ is $\overline{a_{ji}} + \overline{b_{ji}} = \overline{a_{ji} + b_{ji}}$, the (i,j)th element of $(A+B)^*$.

(b) The (i,j)th element of $(zA)^*$ is $\overline{z a_{ji}} = \bar{z}\,\overline{a_{ji}}$, the (i,j)th element of $\bar{z} A^*$.

(c) The (i,j)th element of $(AB)^*$ is the (j,i)th element of $\overline{AB}$:

$$\overline{a_{j1}}\,\overline{b_{1i}} + \overline{a_{j2}}\,\overline{b_{2i}} + \dots + \overline{a_{jn}}\,\overline{b_{ni}} = [\,\overline{a_{j1}} \;\; \overline{a_{j2}} \;\; \dots \;\; \overline{a_{jn}}\,] \begin{bmatrix} \overline{b_{1i}} \\ \overline{b_{2i}} \\ \vdots \\ \overline{b_{ni}} \end{bmatrix}$$

Row i of B^* is $[\overline{b_{1i}}\ \overline{b_{2i}}\ \dots\ \overline{b_{ni}}]$, and column j of A^* is $\begin{bmatrix} \overline{a_{j1}} \\ \overline{a_{j2}} \\ \vdots \\ \overline{a_{jn}} \end{bmatrix}$.

Thus the (i,j)th element of $B^* A^*$ is also $\overline{a_{j1}}\,\overline{b_{1i}} + \overline{a_{j2}}\,\overline{b_{2i}} + \dots + \overline{a_{jn}}\,\overline{b_{ni}}$.

(d) The (i,j)th element of $(A^*)^*$ is the conjugate of the (j,i)th element of A^*, which is the conjugate of the (i,j)th element of A, so the (i,j)th element of $(A^*)^*$ is the conjugate of the conjugate of the (i,j)th element of A. That is, it is the (i,j)th element of A.

26. If $A = A^*$, then for all the diagonal elements $a_{ii} = \overline{a_{ii}}$. Thus all the diagonal elements are real.

27. (a) $G = \begin{bmatrix} 1 & 0 & 1 \\ 0 & 1 & 1 \\ 1 & 1 & 2 \end{bmatrix}$.

$P = \begin{bmatrix} 2 & 1 \\ 1 & 2 \end{bmatrix}$.

$g_{12} = 0$, $g_{13} = 1$, $g_{23} = 1$,
so 1→3→2 or 2→3→1.

$p_{12} = 1$, so 1→2 or 2→1,
gives no information.

(b) $G = \begin{bmatrix} 1 & 0 & 1 & 0 \\ 0 & 2 & 1 & 1 \\ 1 & 1 & 2 & 0 \\ 0 & 1 & 0 & 1 \end{bmatrix}$.

$P = \begin{bmatrix} 2 & 1 & 1 \\ 1 & 2 & 0 \\ 1 & 0 & 2 \end{bmatrix}$.

$g_{12} = 0$, $g_{13} = 1$, $g_{14} = 0$,
$g_{23} = 1$, $g_{24} = 1$, $g_{34} = 0$,
so 1→3→2→4 or 4→2→3→1.

$p_{12} = 1$, $p_{13} = 1$, $p_{23} = 0$,
so 2→1→3 or 3→1→2.

(c) $G = \begin{bmatrix} 2 & 1 & 2 \\ 1 & 3 & 3 \\ 2 & 3 & 4 \end{bmatrix}$.

$P = \begin{bmatrix} 2 & 1 & 2 & 1 \\ 1 & 2 & 2 & 2 \\ 2 & 2 & 3 & 2 \\ 1 & 2 & 2 & 2 \end{bmatrix}$.

$g_{12} = 1$, $g_{13} = 2$, $g_{23} = 3$,

$p_{12} = 1$, $p_{13} = 2$, $p_{14} = 1$,

so 1→3→2 or 2→3→1.

$p_{23} = 2$, $p_{24} = 2$, $p_{34} = 2$,
so 1→3→ {2,4} or {2,4}→3→1.

(d) $G = \begin{bmatrix} 1 & 0 & 1 & 0 \\ 0 & 2 & 0 & 1 \\ 1 & 0 & 2 & 1 \\ 0 & 1 & 1 & 2 \end{bmatrix}$.

$P = \begin{bmatrix} 2 & 1 & 1 & 0 \\ 1 & 1 & 0 & 0 \\ 1 & 0 & 2 & 1 \\ 0 & 0 & 1 & 2 \end{bmatrix}$.

$g_{12} = 0$, $g_{13} = 1$, $g_{14} = 0$,
$g_{23} = 0$, $g_{24} = 1$, $g_{34} = 1$,
so 1→3→4→2 or 2→4→3→1.

$p_{12} = 1$, $p_{13} = 1$, $p_{14} = 0$,
$p_{23} = 0$, $p_{24} = 0$, $p_{34} = 1$,
so 2→1→3→4 or 4→3→1→2.

(e) $G = \begin{bmatrix} 2 & 1 & 0 & 1 \\ 1 & 1 & 0 & 0 \\ 0 & 0 & 2 & 1 \\ 1 & 0 & 1 & 2 \end{bmatrix}$.

$P = \begin{bmatrix} 2 & 0 & 1 & 0 \\ 0 & 2 & 1 & 1 \\ 1 & 1 & 2 & 0 \\ 0 & 1 & 0 & 1 \end{bmatrix}$.

$g_{12} = 1$, $g_{13} = 0$, $g_{14} = 1$,
$g_{23} = 0$, $g_{24} = 0$, $g_{34} = 1$,
so 2→1→4→3 or 3→4→1→2.

$p_{12} = 0$, $p_{13} = 1$, $p_{14} = 0$,
$p_{23} = 1$, $p_{24} = 1$, $p_{34} = 0$,
so 1→3→2→4 or 4→2→3→1.

(f) $G = \begin{bmatrix} 1 & 0 & 0 & 1 & 1 \\ 0 & 3 & 1 & 0 & 2 \\ 0 & 1 & 1 & 0 & 0 \\ 1 & 0 & 0 & 1 & 1 \\ 1 & 2 & 0 & 1 & 3 \end{bmatrix}$.

$P = \begin{bmatrix} 2 & 1 & 1 & 2 \\ 1 & 3 & 0 & 1 \\ 1 & 0 & 2 & 1 \\ 2 & 1 & 1 & 2 \end{bmatrix}$.

$g_{12} = 0$, $g_{13} = 0$, $g_{14} = 1$, $g_{15} = 1$,
$g_{23} = 1$, $g_{24} = 0$, $g_{25} = 2$,
$g_{34} = 0$, $g_{35} = 0$, $g_{45} = 1$,
so 3→2→5→{1,4}
or {1´4}→5→2→3.

$p_{12} = 1$, $p_{13} = 1$, $p_{14} = 2$,
$p_{23} = 0$, $p_{24} = 1$, $p_{34} = 1$,
so 2→{1´4}→3 or 3→{1´4}→2.

28. (a) $G^t = (AA^t)^t = (A^t)^t A^t = AA^t = G$, and $P^t = (A^t A)^t = A^t (A^t)^t = A^t A = P$.

(b) The number of types of pottery common to graves i and j is the number of types of pottery common to graves j and i, so the (i,j)th element of G equals the (j,i)th element of G which is the (i,j)th element of G^t, and so $G = G^t$. Likewise, the number of graves in which pottery types i and j both appear is the same as the number of graves in which pottery types j and i both appear, so the (i,j)th element of P equals the (j,i)th element of P, which is the (i,j)th element of P^t , and so $P = P^t$.

29. $g_{ii} = (a_{i1})^2 + \ldots (a_{in})^2$, sum of squares of elements in ith row of AA^t. Every a_{ik} will be 1 or 0 depending on whether the pottery type k is in grave i or not. Thus g_{ii} = number of pottery types in grave i.

In general ith diagonal element of AA^t is $(a_{i1})^2 + \ldots (a_{in})^2$, but elements need not be 1 or 0. Thus ith diagonal gives sum of squares of elements in row i of AA^t.

30. $p_{ij} = a_{1i}\,a_{1j} + a_{2i}\,a_{2j} + \ldots + a_{mi}\,a_{mj}$, where m is the number of graves. $a_{ki}\,a_{kj}$ is 1 if grave k contains both the ith and jth types of pottery and is zero otherwise. So the sum of these terms is the number of graves that contain both types of pottery.

31. (a) $f_{ij} = a_{i1}\,a_{j1} + a_{i2}\,a_{j2} + \ldots + a_{in}a_{jn}$ as in the graves model. $a_{ik}a_{jk}$ is 1 if both person i and person j are friends of person k and is zero otherwise. So the sum of these terms is the number of friends person i and person j have in common.

(b) The matrix A is symmetric when friendships are mutual and is not symmetric when friendships are not mutual. Define a_{ij} to be 1 if person i considers person j to be his friend and zero otherwise. f_{ij} will then be the number of people both person i and person j consider to be their friends.

Exercise Set 2.4

1. (a) $AB = BA = I_2$, so B is the inverse of A.

(b) $AB = BA = I_2$, so B is the inverse of A.

(c) $AB = \begin{bmatrix} 2 & -4 \\ -5 & 3 \end{bmatrix}\begin{bmatrix} 3 & 1 \\ 5 & 2 \end{bmatrix} = \begin{bmatrix} -14 & -6 \\ 0 & 1 \end{bmatrix}$, so B is not the inverse of A.

(d) $AB = BA = I_2$, so B is the inverse of A.

2. (a) $AB = BA = I_3$, so B is the inverse of A.

(b) $AB = BA = I_3$, so B is the inverse of A.

(c) $AB = \begin{bmatrix} 0 & 1 & -1 \\ 2 & -2 & -1 \\ -1 & 1 & 1 \end{bmatrix}\begin{bmatrix} 1 & 2 & 3 \\ 1 & 1 & 2 \\ 0 & 1 & 1 \end{bmatrix} = \begin{bmatrix} 1 & 0 & 1 \\ 0 & 1 & 1 \\ 0 & 0 & 0 \end{bmatrix}$, so B is not the inverse of A.

3. (a) $\begin{bmatrix} 1 & 0 & 1 & 0 \\ 2 & 1 & 0 & 1 \end{bmatrix} \underset{R2+(-2)R1}{\approx} \begin{bmatrix} 1 & 0 & 1 & 0 \\ 0 & 1 & -2 & 1 \end{bmatrix}$, and the inverse matrix is $\begin{bmatrix} 1 & 0 \\ -2 & 1 \end{bmatrix}$.

(b) $\begin{bmatrix} 1 & 2 & 1 & 0 \\ 9 & 4 & 0 & 1 \end{bmatrix} \underset{R2+(-9)R1}{\approx} \begin{bmatrix} 1 & 2 & 1 & 0 \\ 0 & -14 & -9 & 1 \end{bmatrix} \underset{(-1/14)R2}{\approx} \begin{bmatrix} 1 & 2 & 1 & 0 \\ 0 & 1 & 9/14 & -1/14 \end{bmatrix}$

$\underset{R1-(2)R2}{\approx} \begin{bmatrix} 1 & 0 & -2/7 & 1/7 \\ 0 & 1 & 9/14 & -1/14 \end{bmatrix}$, and the inverse matrix is $\begin{bmatrix} -2/7 & 1/7 \\ 9/14 & -1/14 \end{bmatrix}$.

(c) $\begin{bmatrix} 2 & 1 & 1 & 0 \\ 4 & 3 & 0 & 1 \end{bmatrix} \underset{(1/2)R1}{\approx} \begin{bmatrix} 1 & 1/2 & 1/2 & 0 \\ 4 & 3 & 0 & 1 \end{bmatrix} \underset{R2+(-4)R1}{\approx} \begin{bmatrix} 1 & 1/2 & 1/2 & 0 \\ 0 & 1 & -2 & 1 \end{bmatrix}$

$\underset{R1+(-1/2)R2}{\approx} \begin{bmatrix} 1 & 0 & 3/2 & -1/2 \\ 0 & 1 & -2 & 1 \end{bmatrix}$, and the inverse is $\begin{bmatrix} 3/2 & -1/2 \\ -2 & 1 \end{bmatrix}$.

(d) $\begin{bmatrix} 0 & 1 & 1 & 0 \\ 1 & 3 & 0 & 1 \end{bmatrix} \underset{R1 \Leftrightarrow R2}{\approx} \begin{bmatrix} 1 & 3 & 0 & 1 \\ 0 & 1 & 1 & 0 \end{bmatrix} \underset{R1+(-3)R2}{\approx} \begin{bmatrix} 1 & 0 & -3 & 1 \\ 0 & 1 & 1 & 0 \end{bmatrix}$,

and the inverse is $\begin{bmatrix} -3 & 1 \\ 1 & 0 \end{bmatrix}$.

(e) $\begin{bmatrix} 1 & 2 & 1 & 0 \\ 3 & 6 & 0 & 1 \end{bmatrix} \underset{R2+(-3)R1}{\approx} \begin{bmatrix} 1 & 2 & 1 & 0 \\ 0 & 0 & -3 & 1 \end{bmatrix}$, and the inverse does not exist.

(f) $\begin{bmatrix} 2 & -3 & 1 & 0 \\ 6 & -7 & 0 & 1 \end{bmatrix} \underset{(1/2)R1}{\approx} \begin{bmatrix} 1 & -3/2 & 1/2 & 0 \\ 6 & -7 & 0 & 1 \end{bmatrix} \underset{R2+(-6)R1}{\approx} \begin{bmatrix} 1 & -3/2 & 1/2 & 0 \\ 0 & 2 & -3 & 1 \end{bmatrix}$

$\underset{(1/2)R2}{\approx} \begin{bmatrix} 1 & -3/2 & 1/2 & 0 \\ 0 & 1 & -3/2 & 1/2 \end{bmatrix} \underset{R1+(3/2)R2}{\approx} \begin{bmatrix} 1 & 0 & -7/4 & 3/4 \\ 0 & 1 & -3/2 & 1/2 \end{bmatrix}$, and the inverse is $\begin{bmatrix} -7/4 & 3/4 \\ -3/2 & 1/2 \end{bmatrix}$.

4. (a) $$\left[\begin{array}{ccc|ccc} 1 & 2 & 3 & 1 & 0 & 0 \\ 0 & 1 & 2 & 0 & 1 & 0 \\ 4 & 5 & 3 & 0 & 0 & 1 \end{array}\right] \underset{R2+(-4)R1}{\approx} \left[\begin{array}{ccc|ccc} 1 & 2 & 3 & 1 & 0 & 0 \\ 0 & 1 & 2 & 0 & 1 & 0 \\ 0 & -3 & -9 & -4 & 0 & 1 \end{array}\right]$$

$$\underset{\substack{R1+(-2)R2 \\ R3+(3)R2}}{\approx} \left[\begin{array}{ccc|ccc} 1 & 0 & -1 & 1 & -2 & 0 \\ 0 & 1 & 2 & 0 & 1 & 0 \\ 0 & 0 & -3 & -4 & 3 & 1 \end{array}\right] \underset{(-1/3)R3}{\approx} \left[\begin{array}{ccc|ccc} 1 & 0 & -1 & 1 & -2 & 0 \\ 0 & 1 & 2 & 0 & 1 & 0 \\ 0 & 0 & 1 & 4/3 & -1 & -1/3 \end{array}\right]$$

$$\underset{\substack{R1+R3 \\ R2+(-2)R3}}{\approx} \left[\begin{array}{ccc|ccc} 1 & 0 & 0 & 7/3 & -3 & -1/3 \\ 0 & 1 & 0 & -8/3 & 3 & 2/3 \\ 0 & 0 & 1 & 4/3 & -1 & -1/3 \end{array}\right]$$, and the inverse is $\begin{bmatrix} 7/3 & -3 & -1/3 \\ -8/3 & 3 & 2/3 \\ 4/3 & -1 & -1/3 \end{bmatrix}$.

(b) $$\left[\begin{array}{ccc|ccc} 2 & 0 & 4 & 1 & 0 & 0 \\ -1 & 3 & 1 & 0 & 1 & 0 \\ 0 & 1 & 2 & 0 & 0 & 1 \end{array}\right] \underset{(1/2)R1}{\approx} \left[\begin{array}{ccc|ccc} 1 & 0 & 2 & 1/2 & 0 & 0 \\ -1 & 3 & 1 & 0 & 1 & 0 \\ 0 & 1 & 2 & 0 & 0 & 1 \end{array}\right]$$

$$\underset{R2+R1}{\approx} \left[\begin{array}{ccc|ccc} 1 & 0 & 2 & 1/2 & 0 & 0 \\ 0 & 3 & 3 & 1/2 & 1 & 0 \\ 0 & 1 & 2 & 0 & 0 & 1 \end{array}\right] \underset{(1/3)R2}{\approx} \left[\begin{array}{ccc|ccc} 1 & 0 & 2 & 1/2 & 0 & 0 \\ 0 & 1 & 1 & 1/6 & 1/3 & 0 \\ 0 & 1 & 2 & 0 & 0 & 1 \end{array}\right]$$

$$\underset{R3+(-1)R2}{\approx} \left[\begin{array}{ccc|ccc} 1 & 0 & 2 & 1/2 & 0 & 0 \\ 0 & 1 & 1 & 1/6 & 1/3 & 0 \\ 0 & 0 & 1 & -1/6 & -1/3 & 1 \end{array}\right] \underset{\substack{R1+(-2)R3 \\ R2+(-1)R3}}{\approx} \left[\begin{array}{ccc|ccc} 1 & 0 & 0 & 5/6 & 2/3 & -2 \\ 0 & 1 & 0 & 1/3 & 2/3 & -1 \\ 0 & 0 & 1 & -1/6 & -1/3 & 1 \end{array}\right],$$

and the inverse is $\begin{bmatrix} 5/6 & 2/3 & -2 \\ 1/3 & 2/3 & -1 \\ -1/6 & -1/3 & 1 \end{bmatrix}$.

(c) $$\left[\begin{array}{ccc|ccc} 1 & 2 & -3 & 1 & 0 & 0 \\ 1 & -2 & 1 & 0 & 1 & 0 \\ 5 & -2 & -3 & 0 & 0 & 1 \end{array}\right] \underset{\substack{R2+(-1)R1 \\ R3+(-5)R1}}{\approx} \left[\begin{array}{ccc|ccc} 1 & 2 & -3 & 1 & 0 & 0 \\ 0 & -4 & 4 & -1 & 1 & 0 \\ 0 & -12 & 12 & -5 & 0 & 1 \end{array}\right]$$

$$\underset{(-1/4)R2}{\approx}\left[\begin{array}{ccc|ccc}1&2&-3&1&0&0\\0&1&-1&1/4&-1/4&0\\0&-12&12&-5&0&1\end{array}\right]\underset{\substack{R1+(-2)R2\\R3+(12)R2}}{\approx}\left[\begin{array}{ccc|ccc}1&0&-1&1/2&1/2&0\\0&1&-1&1/4&-1/4&0\\0&0&0&-2&-3&1\end{array}\right],$$

so the inverse does not exist.

(d) $$\left[\begin{array}{ccc|ccc}1&2&-1&1&0&0\\3&-1&0&0&1&0\\2&-3&1&0&0&1\end{array}\right]\underset{\substack{R2+(-3)R1\\R3+(-2)R1}}{\approx}\left[\begin{array}{ccc|ccc}1&2&-1&1&0&0\\0&-7&3&-3&1&0\\0&-7&3&-2&0&1\end{array}\right]$$

$$\underset{(-1/7)R2}{\approx}\left[\begin{array}{ccc|ccc}1&2&-1&1&0&0\\0&1&-3/7&3/7&-1/7&0\\0&-7&3&-2&0&1\end{array}\right]\underset{\substack{R1+(-2)R2\\R3+(7)R2}}{\approx}\left[\begin{array}{ccc|ccc}1&0&-1/7&1/7&2/7&0\\0&1&-3/7&3/7&-1/7&0\\0&0&0&1&-1&1\end{array}\right],$$

and the inverse does not exist.

5. (a) $$\left[\begin{array}{ccc|ccc}1&2&3&1&0&0\\2&-1&4&0&1&0\\0&-1&1&0&0&1\end{array}\right]\underset{R2+(-2)R1}{\approx}\left[\begin{array}{ccc|ccc}1&2&3&1&0&0\\0&-5&-2&-2&1&0\\0&-1&1&0&0&1\end{array}\right]$$

$$\underset{R2\Leftrightarrow R3}{\approx}\left[\begin{array}{ccc|ccc}1&2&3&1&0&0\\0&-1&1&0&0&1\\0&-5&-2&-2&1&0\end{array}\right]\underset{(-1)R2}{\approx}\left[\begin{array}{ccc|ccc}1&2&3&1&0&0\\0&1&-1&0&0&-1\\0&-5&-2&-2&1&0\end{array}\right]$$

$$\underset{\substack{R1+(-2)R2\\R3+(5)R2}}{\approx}\left[\begin{array}{ccc|ccc}1&0&5&1&0&2\\0&1&-1&0&0&-1\\0&0&-7&-2&1&-5\end{array}\right]\underset{(-1/7)R3}{\approx}\left[\begin{array}{ccc|ccc}1&0&5&1&0&2\\0&1&-1&0&0&-1\\0&0&1&2/7&-1/7&5/7\end{array}\right]$$

$$\underset{\substack{R1+(-5)R3\\R2+R3}}{\approx}\left[\begin{array}{ccc|ccc}1&0&0&-3/7&5/7&-11/7\\0&1&0&2/7&-1/7&-2/7\\0&0&1&2/7&-1/7&5/7\end{array}\right],$$ and the inverse is $$\begin{bmatrix}-3/7&5/7&-11/7\\2/7&-1/7&-2/7\\2/7&-1/7&5/7\end{bmatrix}.$$

(b) $$\left[\begin{array}{ccc|ccc}1&2&-1&1&0&0\\2&4&-3&0&1&0\\1&-2&0&0&0&1\end{array}\right]\underset{\substack{R2+(-2)R1\\R3+(-1)R1}}{\approx}\left[\begin{array}{ccc|ccc}1&2&-1&1&0&0\\0&0&-1&-2&1&0\\0&-4&1&-1&0&1\end{array}\right]$$

$$\underset{R2 \Leftrightarrow R3}{\approx}\begin{bmatrix} 1 & 2 & -1 & 1 & 0 & 0 \\ 0 & -4 & 1 & -1 & 0 & 1 \\ 0 & 0 & -1 & -2 & 1 & 0 \end{bmatrix} \underset{(-1/4)R2}{\approx}\begin{bmatrix} 1 & 2 & -1 & 1 & 0 & 0 \\ 0 & 1 & -1/4 & 1/4 & 0 & -1/4 \\ 0 & 0 & -1 & -2 & 1 & 0 \end{bmatrix}$$

$$\underset{R1+(-2)R2}{\approx}\begin{bmatrix} 1 & 0 & -1/2 & 1/2 & 0 & 1/2 \\ 0 & 1 & -1/4 & 1/4 & 0 & -1/4 \\ 0 & 0 & -1 & -2 & 1 & 0 \end{bmatrix} \underset{(-1)R3}{\approx}\begin{bmatrix} 1 & 0 & -1/2 & 1/2 & 0 & 1/2 \\ 0 & 1 & -1/4 & 1/4 & 0 & -1/4 \\ 0 & 0 & 1 & 2 & -1 & 0 \end{bmatrix}$$

$$\underset{\substack{R1+(1/2)R3 \\ R2+(1/4)R3}}{\approx}\begin{bmatrix} 1 & 0 & 0 & 3/2 & -1/2 & 1/2 \\ 0 & 1 & 0 & 3/4 & -1/4 & -1/4 \\ 0 & 0 & 1 & 2 & -1 & 0 \end{bmatrix}$$, and the inverse is $\begin{bmatrix} 3/2 & -1/2 & 1/2 \\ 3/4 & -1/4 & -1/4 \\ 2 & -1 & 0 \end{bmatrix}$.

(c) $\begin{bmatrix} 1 & -2 & -1 & 1 & 0 & 0 \\ -2 & 4 & 6 & 0 & 1 & 0 \\ 0 & 0 & 5 & 0 & 0 & 1 \end{bmatrix} \underset{R2+(2)R1}{\approx}\begin{bmatrix} 1 & -2 & -1 & 1 & 0 & 0 \\ 0 & 0 & 4 & 2 & 1 & 0 \\ 0 & 0 & 5 & 0 & 0 & 1 \end{bmatrix}$,

and the inverse does not exist.

(d) $\begin{bmatrix} 7 & 0 & 0 & 1 & 0 & 0 \\ 0 & -3 & 0 & 0 & 1 & 0 \\ 0 & 0 & 1/5 & 0 & 0 & 1 \end{bmatrix} \underset{\substack{(1/7)R1 \\ (-1/3)R2 \\ 5R3}}{\approx}\begin{bmatrix} 1 & 0 & 0 & 1/7 & 0 & 0 \\ 0 & 1 & 0 & 0 & -1/3 & 0 \\ 0 & 0 & 1 & 0 & 0 & 5 \end{bmatrix}$, the inverse is

$\begin{bmatrix} 1/7 & 0 & 0 \\ 0 & -1/3 & 0 \\ 0 & 0 & 5 \end{bmatrix}$.

6. (a) $\begin{bmatrix} -3 & -1 & 1 & -2 & 1 & 0 & 0 & 0 \\ -1 & 3 & 2 & 1 & 0 & 1 & 0 & 0 \\ 1 & 2 & 3 & -1 & 0 & 0 & 1 & 0 \\ -2 & 1 & -1 & -3 & 0 & 0 & 0 & 1 \end{bmatrix} \underset{R1 \Leftrightarrow R3}{\approx}\begin{bmatrix} 1 & 2 & 3 & -1 & 0 & 0 & 1 & 0 \\ -1 & 3 & 2 & 1 & 0 & 1 & 0 & 0 \\ -3 & -1 & 1 & -2 & 1 & 0 & 0 & 0 \\ -2 & 1 & -1 & -3 & 0 & 0 & 0 & 1 \end{bmatrix}$

$$\underset{\substack{R2+R1 \\ R3+(3)R1 \\ R4+(2)R1}}{\approx}\begin{bmatrix} 1 & 2 & 3 & -1 & 0 & 0 & 1 & 0 \\ 0 & 5 & 5 & 0 & 0 & 1 & 1 & 0 \\ 0 & 5 & 10 & -5 & 1 & 0 & 3 & 0 \\ 0 & 5 & 5 & -5 & 0 & 0 & 2 & 1 \end{bmatrix} \underset{(1/5)R2}{\approx}\begin{bmatrix} 1 & 2 & 3 & -1 & 0 & 0 & 1 & 0 \\ 0 & 1 & 1 & 0 & 0 & 1/5 & 1/5 & 0 \\ 0 & 5 & 10 & -5 & 1 & 0 & 3 & 0 \\ 0 & 5 & 5 & -5 & 0 & 0 & 2 & 1 \end{bmatrix}$$

$$\underset{\substack{R1+(-2)R2\\R3+(-5)R2\\R4+(-5)R2}}{\approx}\begin{bmatrix}1&0&1&-1&0&-2/5&3/5&0\\0&1&1&0&0&1/5&1/5&0\\0&0&5&-5&1&-1&2&0\\0&0&0&-5&0&-1&1&1\end{bmatrix}$$

$$\underset{(1/5)R3}{\approx}\begin{bmatrix}1&0&1&-1&0&-2/5&3/5&0\\0&1&1&0&0&1/5&1/5&0\\0&0&1&-1&1/5&-1/5&2/5&0\\0&0&0&-5&0&-1&1&1\end{bmatrix}$$

$$\underset{\substack{R1+(-1)R3\\R2+(-1)R3}}{\approx}\begin{bmatrix}1&0&0&0&-1/5&-1/5&1/5&0\\0&1&0&1&-1/5&2/5&-1/5&0\\0&0&1&-1&1/5&-1/5&2/5&0\\0&0&0&-5&0&-1&1&1\end{bmatrix}\underset{(-1/5)R3}{\approx}\begin{bmatrix}1&0&0&0&-1/5&-1/5&1/5&0\\0&1&0&1&-1/5&2/5&-1/5&0\\0&0&1&-1&1/5&-1/5&2/5&0\\0&0&0&1&0&1/5&-1/5&-1/5\end{bmatrix}$$

$$\underset{\substack{R2+(-1)R4\\R3+R4}}{\approx}\begin{bmatrix}1&0&0&0&-1/5&-1/5&1/5&0\\0&1&0&0&-1/5&1/5&0&1/5\\0&0&1&0&1/5&0&1/5&-1/5\\0&0&0&1&0&1/5&-1/5&-1/5\end{bmatrix},$$ so the inverse is

$$\begin{bmatrix}-1/5&-1/5&1/5&0\\-1/5&1/5&0&1/5\\1/5&0&1/5&-1/5\\0&1/5&-1/5&-1/5\end{bmatrix}.$$

(b) $$\begin{bmatrix}1&1&0&0&1&0&0&0\\0&1&1&0&0&1&0&0\\1&0&0&1&0&0&1&0\\0&0&1&1&0&0&0&1\end{bmatrix}\underset{R3+(-1)R1}{\approx}\begin{bmatrix}1&1&0&0&1&0&0&0\\0&1&1&0&0&1&0&0\\0&-1&0&1&-1&0&1&0\\0&0&1&1&0&0&0&1\end{bmatrix}$$

$$\underset{\substack{R1+(-1)R2\\R3+R2}}{\approx}\begin{bmatrix}1&0&-1&0&1&-1&0&0\\0&1&1&0&0&1&0&0\\0&0&1&1&-1&1&1&0\\0&0&1&1&0&0&0&1\end{bmatrix}$$

$$\approx \begin{matrix} \\ R1+R3 \\ R2+(-1)R3 \\ R4+(-1)R3 \end{matrix} \left[\begin{array}{cccccccc} 1 & 0 & 0 & 1 & 0 & 0 & 1 & 0 \\ 0 & 1 & 0 & -1 & 1 & 0 & -1 & 0 \\ 0 & 0 & 1 & 1 & -1 & 1 & 1 & 0 \\ 0 & 0 & 0 & 0 & 1 & -1 & -1 & 1 \end{array}\right]$$, so the inverse does not exist.

(c) $$\left[\begin{array}{cccccccc} -1 & 0 & -1 & -1 & 1 & 0 & 0 & 0 \\ -3 & -1 & 0 & -1 & 0 & 1 & 0 & 0 \\ 5 & 0 & 4 & 3 & 0 & 0 & 1 & 0 \\ 3 & 0 & 3 & 2 & 0 & 0 & 0 & 1 \end{array}\right] \begin{matrix} \\ \approx \\ (-1)R1 \\ \\ \end{matrix} \left[\begin{array}{cccccccc} 1 & 0 & 1 & 1 & -1 & 0 & 0 & 0 \\ -3 & -1 & 0 & -1 & 0 & 1 & 0 & 0 \\ 5 & 0 & 4 & 3 & 0 & 0 & 1 & 0 \\ 3 & 0 & 3 & 2 & 0 & 0 & 0 & 1 \end{array}\right]$$

$$\approx \begin{matrix} \\ R2+(3)R1 \\ R3+(-5)R1 \\ R4+(-3)R1 \end{matrix} \left[\begin{array}{cccccccc} 1 & 0 & 1 & 1 & -1 & 0 & 0 & 0 \\ 0 & -1 & 3 & 2 & -3 & 1 & 0 & 0 \\ 0 & 0 & -1 & -2 & 5 & 0 & 1 & 0 \\ 0 & 0 & 0 & -1 & 3 & 0 & 0 & 1 \end{array}\right] \approx \begin{matrix} \\ (-1)R2 \\ (-1)R3 \\ (-1)R4 \end{matrix} \left[\begin{array}{cccccccc} 1 & 0 & 1 & 1 & -1 & 0 & 0 & 0 \\ 0 & 1 & -3 & -2 & 3 & -1 & 0 & 0 \\ 0 & 0 & 1 & 2 & -5 & 0 & -1 & 0 \\ 0 & 0 & 0 & 1 & -3 & 0 & 0 & -1 \end{array}\right]$$

$$\approx \begin{matrix} \\ R1+(-1)R3 \\ R2+(3)R3 \\ \\ \end{matrix} \left[\begin{array}{cccccccc} 1 & 0 & 0 & -1 & 4 & 0 & 1 & 0 \\ 0 & 1 & 0 & 4 & -12 & -1 & -3 & 0 \\ 0 & 0 & 1 & 2 & -5 & 0 & -1 & 0 \\ 0 & 0 & 0 & 1 & -3 & 0 & 0 & -1 \end{array}\right]$$

$$\approx \begin{matrix} \\ R1+R4 \\ R2+(-4)R4 \\ R3+(-2)R4 \end{matrix} \left[\begin{array}{cccccccc} 1 & 0 & 0 & 0 & 1 & 0 & 1 & -1 \\ 0 & 1 & 0 & 0 & 0 & -1 & -3 & 4 \\ 0 & 0 & 1 & 0 & 1 & 0 & -1 & 2 \\ 0 & 0 & 0 & 1 & -3 & 0 & 0 & -1 \end{array}\right]$$

$$\approx \begin{matrix} \\ R1+R4 \\ R2+(-4)R4 \\ R3+(-2)R4 \end{matrix} \left[\begin{array}{cccccccc} 1 & 0 & 0 & 0 & 1 & 0 & 1 & -1 \\ 0 & 1 & 0 & 0 & 0 & -1 & -3 & 4 \\ 0 & 0 & 1 & 0 & 1 & 0 & -1 & 2 \\ 0 & 0 & 0 & 1 & -3 & 0 & 0 & -1 \end{array}\right]$$, so the inverse is $\begin{bmatrix} 1 & 0 & 1 & -1 \\ 0 & -1 & -3 & 4 \\ 1 & 0 & -1 & 2 \\ -3 & 0 & 0 & -1 \end{bmatrix}$.

7. $\begin{bmatrix} a & b \\ c & d \end{bmatrix} \frac{1}{ad-bc} \begin{bmatrix} d & -b \\ -c & a \end{bmatrix} = \frac{1}{ad-bc} \begin{bmatrix} a & b \\ c & d \end{bmatrix} \begin{bmatrix} d & -b \\ -c & a \end{bmatrix}$

$= \frac{1}{ad-bc} \begin{bmatrix} ad-bc & 0 \\ 0 & ad-bc \end{bmatrix} = \begin{bmatrix} 1 & 0 \\ 0 & 1 \end{bmatrix}$.

Likewise $\frac{1}{ad-bc} \begin{bmatrix} d & -b \\ -c & a \end{bmatrix} \begin{bmatrix} a & b \\ c & d \end{bmatrix} = \begin{bmatrix} 1 & 0 \\ 0 & 1 \end{bmatrix}$. Thus $\frac{1}{ad-bc} \begin{bmatrix} d & -b \\ -c & a \end{bmatrix} = A^{-1}$.

The inverses of the given matrices are:

(a) $\begin{bmatrix} 3 & -8 \\ -1 & 3 \end{bmatrix}$, (b) $\frac{1}{2}\begin{bmatrix} 4 & -5 \\ -2 & 3 \end{bmatrix} = \begin{bmatrix} 2 & -5/2 \\ -1 & 3/2 \end{bmatrix}$,

(c) $\frac{1}{2}\begin{bmatrix} 4 & -6 \\ -3 & 5 \end{bmatrix} = \begin{bmatrix} 2 & -3 \\ -3/2 & 5/2 \end{bmatrix}$, (d) $\frac{1}{4}\begin{bmatrix} -2 & 6 \\ -2 & 4 \end{bmatrix} = \begin{bmatrix} -1/2 & 3/2 \\ -1/2 & 1 \end{bmatrix}$.

8. (a) The inverse of the coefficient matrix is $\begin{bmatrix} -5 & 2 \\ 3 & -1 \end{bmatrix}$.

$$\begin{bmatrix} x_1 \\ x_2 \end{bmatrix} = \begin{bmatrix} -5 & 2 \\ 3 & -1 \end{bmatrix}\begin{bmatrix} 2 \\ 4 \end{bmatrix} = \begin{bmatrix} -2 \\ 2 \end{bmatrix}.$$

(b) $$\begin{bmatrix} x_1 \\ x_2 \end{bmatrix} = \begin{bmatrix} -9 & 5 \\ 2 & -1 \end{bmatrix}\begin{bmatrix} -1 \\ 3 \end{bmatrix} = \begin{bmatrix} 24 \\ -5 \end{bmatrix}.$$

(c) $$\begin{bmatrix} x_1 \\ x_2 \end{bmatrix} = \begin{bmatrix} -1/5 & 3/5 \\ 2/5 & -1/5 \end{bmatrix}\begin{bmatrix} 5 \\ 10 \end{bmatrix} = \begin{bmatrix} 5 \\ 0 \end{bmatrix}.$$

(d) $$\begin{bmatrix} x_1 \\ x_2 \end{bmatrix} = \begin{bmatrix} 3/2 & -1/2 \\ -2 & 1 \end{bmatrix}\begin{bmatrix} 4 \\ 6 \end{bmatrix} = \begin{bmatrix} 3 \\ -2 \end{bmatrix}.$$

(e) $$\begin{bmatrix} x_1 \\ x_2 \end{bmatrix} = \begin{bmatrix} 2 & -1 \\ -3/4 & 1/2 \end{bmatrix}\begin{bmatrix} 6 \\ 1 \end{bmatrix} = \begin{bmatrix} 11 \\ -4 \end{bmatrix}.$$

(f) $$\begin{bmatrix} x_1 \\ x_2 \end{bmatrix} = \begin{bmatrix} 7/3 & -3 \\ -2/3 & 1 \end{bmatrix}\begin{bmatrix} 9 \\ 4 \end{bmatrix} = \begin{bmatrix} 9 \\ -2 \end{bmatrix}.$$

9. (a) $$\begin{bmatrix} x_1 \\ x_2 \\ x_3 \end{bmatrix} = \begin{bmatrix} 1/9 & 3/9 & 5/9 \\ 3/9 & 0 & -3/9 \\ -2/9 & 3/9 & -1/9 \end{bmatrix}\begin{bmatrix} 2 \\ 0 \\ 1 \end{bmatrix} = \begin{bmatrix} 7/9 \\ 3/9 \\ -5/9 \end{bmatrix}.$$

(b) $\begin{bmatrix} x_1 \\ x_2 \\ x_3 \end{bmatrix} = \begin{bmatrix} 3/4 & -1/4 & 1/2 \\ -1/4 & -1/4 & 1/2 \\ -1/4 & 3/4 & -1/2 \end{bmatrix} \begin{bmatrix} 1 \\ 2 \\ 0 \end{bmatrix} = \begin{bmatrix} 1/4 \\ -3/4 \\ 5/4 \end{bmatrix}.$

(c) $\begin{bmatrix} x_1 \\ x_2 \\ x_3 \end{bmatrix} = \begin{bmatrix} -40 & 16 & 9 \\ 13 & -5 & -3 \\ 5 & -2 & -1 \end{bmatrix} \begin{bmatrix} 1 \\ 3 \\ 15 \end{bmatrix} = \begin{bmatrix} 143 \\ -47 \\ -16 \end{bmatrix}.$

(d) $\begin{bmatrix} x_1 \\ x_2 \\ x_3 \end{bmatrix} = \begin{bmatrix} -1 & -10 & -8 \\ -1 & -6 & -5 \\ 0 & -1 & -1 \end{bmatrix} \begin{bmatrix} 3 \\ 2 \\ -1 \end{bmatrix} = \begin{bmatrix} -15 \\ -10 \\ -1 \end{bmatrix}.$

(e) $\begin{bmatrix} x_1 \\ x_2 \\ x_3 \end{bmatrix} = \begin{bmatrix} 2 & 3 & 1 \\ 3 & 3 & 1 \\ 2 & 4 & 1 \end{bmatrix} \begin{bmatrix} 5 \\ -2 \\ 1 \end{bmatrix} = \begin{bmatrix} 5 \\ 10 \\ 3 \end{bmatrix}.$

10. $\begin{bmatrix} x_1 \\ x_2 \\ x_3 \\ x_4 \end{bmatrix} = \begin{bmatrix} -14/17 & 8/17 & 4/17 & 5/17 \\ 3/17 & -9/17 & 4/17 & 5/17 \\ 5/17 & 2/17 & 1/17 & -3/17 \\ 18/17 & -3/17 & -10/17 & -4/17 \end{bmatrix} \begin{bmatrix} 5 \\ 6 \\ 1 \\ 7 \end{bmatrix} = \begin{bmatrix} 1 \\ 0 \\ 1 \\ 2 \end{bmatrix}.$

11. We solve the three sets of equations simultaneously:

$$\begin{bmatrix} 1/9 & 3/9 & 5/9 \\ 3/9 & 0 & -3/9 \\ -2/9 & 3/9 & -1/9 \end{bmatrix} \begin{bmatrix} 1 & 0 & 5 \\ 2 & 1 & 2 \\ 3 & 4 & -3 \end{bmatrix} = \begin{bmatrix} 22/9 & 23/9 & -4/9 \\ -6/9 & -12/9 & 24/9 \\ 1/9 & -1/9 & -1/9 \end{bmatrix}.$$

Thus the solutions are, in turn,

$$\begin{bmatrix} x_1 \\ x_2 \\ x_3 \end{bmatrix} = \begin{bmatrix} 22/9 \\ -6/9 \\ 1/9 \end{bmatrix}, \begin{bmatrix} x_1 \\ x_2 \\ x_3 \end{bmatrix} = \begin{bmatrix} 23/9 \\ -12/9 \\ -1/9 \end{bmatrix}, \begin{bmatrix} x_1 \\ x_2 \\ x_3 \end{bmatrix} = \begin{bmatrix} -4/9 \\ 24/9 \\ -1/9 \end{bmatrix}.$$

12. Use properties of matrices, and the definition of matrix inverse.

(a) $(cA)(\frac{1}{c}A^{-1}) = (c\frac{1}{c})(AA^{-1}) = I$, and $(\frac{1}{c}A^{-1})(cA) = (\frac{1}{c}c)(A^{-1}A) = I$. Thus $(cA)^{-1} = \frac{1}{c}A^{-1}$.

(b) $(A^n)(A^{-1})^n = A \ldots AAA^{-1}A^{-1} \ldots A^{-1} = A \ldots AIA^{-1} \ldots A^{-1} = \ldots = I$, and

$A^{-1} \ldots A^{-1}A^{-1}AA \ldots A = A^{-1} \ldots A^{-1}IA \ldots A = \ldots = I$. Thus $(A^n)^{-1} = (A^{-1})^n$.

(c) $(A^t)(A^{-1})^t = (A^{-1}A)^t = I^t = I$, and $(A^{-1})^t(A^t) = (AA^{-1})^t = I^t = I$. Thus $(A^t)^{-1} = (A^{-1})^t$.

13. $A = \begin{bmatrix} 2 & 1 \\ 4 & 3 \end{bmatrix}^{-1} = \begin{bmatrix} 3/2 & -1/2 \\ -2 & 1 \end{bmatrix}$.

14. $A = 2\begin{bmatrix} -3 & 2 \\ -10 & 6 \end{bmatrix}^{-1} = 2\begin{bmatrix} 3 & -1 \\ 5 & -3/2 \end{bmatrix} = \begin{bmatrix} 6 & -2 \\ 10 & -3 \end{bmatrix}$.

15. (a) $(3A)^{-1} = \frac{1}{3}\begin{bmatrix} 2 & -1 \\ -5 & 3 \end{bmatrix} = \begin{bmatrix} 2/3 & -1/3 \\ -5/3 & 3/3 \end{bmatrix}$.   (b) $(A^2)^{-1} = (A^{-1})^2 = \begin{bmatrix} 9 & -5 \\ -25 & 14 \end{bmatrix}$

(c) $A^{-2} = (A^{-1})^2 = \begin{bmatrix} 9 & -5 \\ -25 & 14 \end{bmatrix}$.   (d) $(A^t)^{-1} = (A^{-1})^t = \begin{bmatrix} 2 & -5 \\ -1 & 3 \end{bmatrix}$.

16. (a) $(2A^t)^{-1} = \frac{1}{2}(A^t)^{-1} = \frac{1}{2}(A^{-1})^t = \frac{1}{2}\begin{bmatrix} 2 & -9 \\ -1 & 5 \end{bmatrix}$. (b) $A^{-3} = (A^{-1})^3 = \begin{bmatrix} 89 & -48 \\ -432 & 233 \end{bmatrix}$.

(c) $(AA^t)^{-1} = (A^t)^{-1}A^{-1} = (A^{-1})^t A^{-1} = \begin{bmatrix} 2 & -9 \\ -1 & 5 \end{bmatrix}\begin{bmatrix} 2 & -1 \\ -9 & 5 \end{bmatrix} = \begin{bmatrix} 85 & -47 \\ -47 & 26 \end{bmatrix}$.

In exercises 17 – 19 it is convenient to use the result of exercise 7 to compute matrix inverse.

17. $\begin{bmatrix} 2x & 7 \\ 1 & 2 \end{bmatrix}^{-1} = \begin{bmatrix} 2 & -7 \\ -1 & 4 \end{bmatrix}$, $\begin{bmatrix} 2x & 7 \\ 1 & 2 \end{bmatrix} = \begin{bmatrix} 2 & -7 \\ -1 & 4 \end{bmatrix}^{-1} = \frac{1}{8-7}\begin{bmatrix} 4 & 7 \\ 1 & 2 \end{bmatrix} = \begin{bmatrix} 4 & 7 \\ 1 & 2 \end{bmatrix}$. Thus 2x=4, x=2.

18. $2\begin{bmatrix} 2x & x \\ 5 & 3 \end{bmatrix}^{-1} = \begin{bmatrix} 3 & -2 \\ -5 & 4 \end{bmatrix}$, $\begin{bmatrix} 2x & x \\ 5 & 3 \end{bmatrix}^{-1} = \frac{1}{2}\begin{bmatrix} 3 & -2 \\ -5 & 4 \end{bmatrix}$, $\begin{bmatrix} 2x & x \\ 5 & 3 \end{bmatrix} = 2\begin{bmatrix} 3 & -2 \\ -5 & 4 \end{bmatrix}^{-1} = \frac{2}{12-10}\begin{bmatrix} 4 & 2 \\ 5 & 3 \end{bmatrix} = \begin{bmatrix} 4 & 2 \\ 5 & 3 \end{bmatrix}$.

x=2.

19. $4A^t = \begin{bmatrix} 2 & 3 \\ -4 & -4 \end{bmatrix}^{-1} = \frac{1}{4}\begin{bmatrix} -4 & -3 \\ 4 & 2 \end{bmatrix}$, so $A^t = \frac{1}{16}\begin{bmatrix} -4 & -3 \\ 4 & 2 \end{bmatrix} = \begin{bmatrix} -1/4 & -3/16 \\ 1/4 & 1/8 \end{bmatrix}$ and

$A = \begin{bmatrix} -1/4 & 1/4 \\ -3/16 & 1/8 \end{bmatrix}$.

20. $(ABC)^{-1} = C^{-1}(AB)^{-1} = C^{-1}B^{-1}A^{-1}$.

21. $(A^tB^t)^{-1} = (B^t)^{-1}(A^t)^{-1} = (B^{-1})^t(A^{-1})^t = (A^{-1}B^{-1})^t$.

22. e.g., Let $A = \begin{bmatrix} 3 & 1 \\ 5 & 2 \end{bmatrix}$ and $B = \begin{bmatrix} 2 & -1 \\ -9 & 5 \end{bmatrix}$. Then $(A + B)^{-1} = \begin{bmatrix} 5 & 0 \\ -4 & 7 \end{bmatrix}^{-1} = \frac{1}{35}\begin{bmatrix} 7 & 0 \\ 4 & 5 \end{bmatrix}$

and $A^{-1} + B^{-1} = \begin{bmatrix} 2 & -1 \\ -5 & 3 \end{bmatrix} + \begin{bmatrix} 5 & 1 \\ 9 & 2 \end{bmatrix} = \begin{bmatrix} 7 & 0 \\ 4 & 5 \end{bmatrix}$.

23. Suppose A has no inverse but A^t has an inverse $(A^t)^{-1}$. Then $(A^t)^{-1}A^t = A^t(A^t)^{-1} = I$.

Take the transpose. $A((A^t)^{-1})^t = ((A^t)^{-1})^tA = I^t = I$. i.e., $((A^t)^{-1})^t$ is the inverse of A. So if A has no inverse, then A^t has no inverse.

24. (a) AB = AC. Multiply by A^{-1} on the left. $A^{-1}AB = A^{-1}AC$. i.e. B = C.

(b) AB = O = AO. Thus from part (a) B = O.

25. (a) If A is an nxn matrix with row i equal to row j, then A is not row equivalent to I_n and so cannot have an inverse.

(b) The columns of A are the rows of A^t so from part (a) A^t has no inverse. From Exercise 23, that means that $(A^t)^t$ = A also has no inverse.

(c) A has a column of zeros. Thus $[A:I_n]$ cannot become $[I_n:B]$; we cannot create a zero in the diagonal location when we reach the column of zeros. Thus A has no inverse.

26. Let A be a diagonal matrix. (a) Assume A has all non-zero diagonal elements, $a_{ii} \neq 0$. Multiplying each ith row of A by $1/a_{ii}$ gives $[A|I] \approx \ldots \approx [I|B]$. B is the inverse of A.

(b) Assume A has an inverse. Then $A \approx \ldots \approx I$. If A has a diagonal element $a_{ii}=0$ then A will have a row of zeros. The echelon form of A cannot be I. Thus A has all non-zero diagonal elements.
Suppose A is diagonal and invertible with elements a_{ii}. Let B be the diagonal matrix with elements $1/a_{ii}$. Then ABxBA = I and BAxAB = I. B is the inverse of A.

27. Let $A=\begin{bmatrix} 1 & 0 \\ 0 & 1 \end{bmatrix}$ and $B = \begin{bmatrix} -1 & 0 \\ 0 & -1 \end{bmatrix}$. Then $A^{-1}=\begin{bmatrix} 1 & 0 \\ 0 & 1 \end{bmatrix}$ and $B^{-1}=\begin{bmatrix} -1 & 0 \\ 0 & -1 \end{bmatrix}$. A and B are invertible.
But $A+B = \begin{bmatrix} 0 & 0 \\ 0 & 0 \end{bmatrix}$, not invertible. $0A=\begin{bmatrix} 0 & 0 \\ 0 & 0 \end{bmatrix}$, not invertible.

28. (a) True: A is invertible. Thus there exists a matrix denoted A^{-1}, such that $AA^{-1} = A^{-1}A = I$. This also shows that the inverse of A^{-1} is A.

(b) True: $(A^2)^{-1} = (AA)^{-1} = A^{-1}A^{-1}$ (using $(AB) = B^{-1}A^{-1}$ with A = B).

(c) False: e.g., Let $A = \begin{bmatrix} 1 & 0 & 0 \\ 0 & 0 & 1 \\ 0 & 1 & 0 \end{bmatrix}$. This is identity matrix with 2nd and 3rd rows int.
We can swap last two rows to get identity matrix. Expect it to be invertible.

Then $\begin{bmatrix} 1 & 0 & 0 & 1 & 0 & 0 \\ 0 & 0 & 1 & 0 & 1 & 0 \\ 0 & 1 & 0 & 0 & 0 & 1 \end{bmatrix} \approx \begin{bmatrix} 1 & 0 & 0 & 1 & 0 & 0 \\ 0 & 1 & 0 & 0 & 0 & 1 \\ 0 & 0 & 1 & 0 & 1 & 0 \end{bmatrix}$. $A^{-1} = \begin{bmatrix} 1 & 0 & 0 \\ 0 & 0 & 1 \\ 0 & 1 & 0 \end{bmatrix}$.

(d) True: $(AB)^{-1} = B^{-1}A^{-1}$. Thus if A^{-1} does not exist, $(AB)^{-1}$ does not exist.

(e) True: In finding A^{-1}, $[A:I_n]$ is transformed to $[I_n: A^{-1}]$. Thus $A^{-1} \approx \ldots \approx I_n$.

29. (a) multiplications: $\frac{25^3}{2} + \frac{25^2}{2} = 8{,}125$; additions: $\frac{25^3}{2} - \frac{25}{2} = 7{,}800$.

(b) multiplications: $25^3 + 25^2 = 16{,}250$; additions: $25^3 - 25^2 = 15{,}000$.

30. To find the inverse of A one finds the reduced echelon form of the matrix

$\begin{bmatrix} a_{11} & a_{12} & 1 & 0 \\ a_{21} & a_{22} & 0 & 1 \end{bmatrix}$. To find the first column of the inverse only, one would ignore the

last column of the matrix. This is the same amount of computation as finding the reduced echelon form of the matrix $\begin{bmatrix} a_{11} & a_{12} & b_1 \\ a_{21} & a_{22} & b_2 \end{bmatrix}$ for any $Y = \begin{bmatrix} b_1 \\ b_2 \end{bmatrix}$. That is, it is the same as finding the solution to the system of equations AX = Y using Gauss-Jordan elimination.

31. $\begin{bmatrix} 0 & 1 & 0 \\ 1 & 0 & 0 \\ 0 & 0 & 1 \end{bmatrix}\begin{bmatrix} a & b & c \\ d & e & f \\ g & h & i \end{bmatrix} = \begin{bmatrix} d & e & f \\ a & b & c \\ g & h & i \end{bmatrix}$. row 1 <-> row 2.

$\begin{bmatrix} 1 & 0 & 0 \\ 0 & 1 & 0 \\ 0 & 0 & -2 \end{bmatrix}\begin{bmatrix} a & b & c \\ d & e & f \\ g & h & i \end{bmatrix} = \begin{bmatrix} a & b & c \\ d & e & f \\ -2g & -2h & -2i \end{bmatrix}$. -2 row 3.

$\begin{bmatrix} 1 & 0 & 0 \\ 0 & 1 & 0 \\ 4 & 0 & 1 \end{bmatrix}\begin{bmatrix} a & b & c \\ d & e & f \\ g & h & i \end{bmatrix} = \begin{bmatrix} a & b & c \\ d & e & f \\ g+4a & h+4b & i+4c \end{bmatrix}$. row 3 + 4 row 1.

32. T_1, -3row1:

$E_1 = \begin{bmatrix} -3 & 0 & 0 \\ 0 & 1 & 0 \\ 0 & 0 & 1 \end{bmatrix}$. Check: $\begin{bmatrix} -3 & 0 & 0 \\ 0 & 1 & 0 \\ 0 & 0 & 1 \end{bmatrix}\begin{bmatrix} a & b & c \\ d & e & f \\ g & h & i \end{bmatrix} = \begin{bmatrix} -3a & -3b & -3c \\ d & e & f \\ g & h & i \end{bmatrix}$.

T_2, row1 + 3row2:

$E_2 = \begin{bmatrix} 1 & 3 & 0 \\ 0 & 1 & 0 \\ 0 & 0 & 1 \end{bmatrix}$. Check: $\begin{bmatrix} 1 & 3 & 0 \\ 0 & 1 & 0 \\ 0 & 0 & 1 \end{bmatrix}\begin{bmatrix} a & b & c \\ d & e & f \\ g & h & i \end{bmatrix} = \begin{bmatrix} a+3d & b+3e & c+3f \\ d & e & f \\ g & h & i \end{bmatrix}$.

33. (a) $\begin{bmatrix} 0 & 0 & 1 \\ 0 & 1 & 0 \\ 1 & 0 & 0 \end{bmatrix}$ interchanges rows 1 and 3. The inverse row operation will still interchange rows 1 and 3. $\begin{bmatrix} 0 & 0 & 1 \\ 0 & 1 & 0 \\ 1 & 0 & 0 \end{bmatrix}^{-1} = \begin{bmatrix} 0 & 0 & 1 \\ 0 & 1 & 0 \\ 1 & 0 & 0 \end{bmatrix}$.

(b) $\begin{bmatrix} 0 & 1 & 0 \\ 1 & 0 & 0 \\ 0 & 0 & 1 \end{bmatrix}$ interchanges rows 1 and 2. $\begin{bmatrix} 0 & 1 & 0 \\ 1 & 0 & 0 \\ 0 & 0 & 1 \end{bmatrix}^{-1} = \begin{bmatrix} 0 & 1 & 0 \\ 1 & 0 & 0 \\ 0 & 0 & 1 \end{bmatrix}$.

(c) $\begin{bmatrix} 1 & 0 & 0 \\ 0 & 1 & 0 \\ 0 & 0 & 4 \end{bmatrix}$ multiplies row 3 by 4. The inverse row operation divides row 3 by 4.

$$\begin{bmatrix} 1 & 0 & 0 \\ 0 & 1 & 0 \\ 0 & 0 & 4 \end{bmatrix}^{-1} = \begin{bmatrix} 0 & 1 & 0 \\ 1 & 0 & 0 \\ 0 & 0 & 1/4 \end{bmatrix}.$$

(d) $\begin{bmatrix} 1/3 & 0 & 0 \\ 0 & 1 & 0 \\ 0 & 0 & 1 \end{bmatrix}$ multiplies row 1 by 1/3. $\begin{bmatrix} 1/3 & 0 & 0 \\ 0 & 1 & 0 \\ 0 & 0 & 1 \end{bmatrix}^{-1} = \begin{bmatrix} 3 & 0 & 0 \\ 0 & 1 & 0 \\ 0 & 0 & 1 \end{bmatrix}.$

(e) $\begin{bmatrix} 1 & 0 & 0 \\ 0 & 1 & 0 \\ -4 & 0 & 1 \end{bmatrix}$ adds −4 times row 1 to row 3.

The inverse adds 4 times row 1 to row 3. $\begin{bmatrix} 1 & 0 & 0 \\ 0 & 1 & 0 \\ -4 & 0 & 1 \end{bmatrix}^{-1} = \begin{bmatrix} 1 & 0 & 0 \\ 0 & 1 & 0 \\ 4 & 0 & 1 \end{bmatrix}.$

(f) $\begin{bmatrix} 1 & 0 & 0 \\ 0 & 1 & 0 \\ 0 & 3 & 1 \end{bmatrix}$ adds 3 times row 2 to row 3.

The inverse adds -3 times row 2 to row 3. $\begin{bmatrix} 1 & 0 & 0 \\ 0 & 1 & 0 \\ 0 & 3 & 1 \end{bmatrix}^{-1} = \begin{bmatrix} 1 & 0 & 0 \\ 0 & 1 & 0 \\ 0 & -3 & 1 \end{bmatrix}.$

In general, to find the inverse of an elementary matrix E:
Swapping rows: $E^{-1} = E$.
Multiply row k by c: Elementary matrix E has $e_{kk} = c$. E^{-1} has $e_{kk} = 1/c$.
Add c times row i to row k: Element e_{ki} of E is c. Element e_{ki} of E^{-1} is -c.
(Other elements of E and E^{-1} are the same.)

34. $$\begin{bmatrix} Y \\ I \\ Q \end{bmatrix} = \begin{bmatrix} .299 & .587 & .114 \\ .596 & -.275 & -.321 \\ .212 & -.523 & .311 \end{bmatrix} \begin{bmatrix} R \\ G \\ B \end{bmatrix} = \begin{bmatrix} .299R+.587G+.114B \\ .596R-.275G-.321B \\ .212R-.523G+.311B \end{bmatrix}$$

Thus Y = .299R + .587G + .114B, I = .596R - .275G -.321B, Q = .212R -.523G +.311B.

$0 \le R \le 255$, $0 \le G \le 255$, $0 \le B \le 255$. Using these intervals and allowing for negatives in the expressions for Y, I, and Q:
Get Y_{max} when R=G=B=255. $Y_{max}=255$.
Y_{min} when R=G=B=0. $Y_{min}=0$.

Get I_{max} when R=255, G=B=0. I_{max}=.596x255=151.98.
I_{min} when R=0, G=B=255. I_{min}=-.275x255-.321x255=-151.98.

Get Q_{max} when R=255,G=0,B=255. Q_{max}=.212x255+.311x255=133.365.
Q_{min} when R=B=0, G=255. Q_{min}=-.523x255=-133.365.

Thus 0≤Y≤255, -151.98≤I≤151.89, -133.365≤Q≤133.365.

35. Use $\begin{bmatrix} R \\ G \\ B \end{bmatrix} = \begin{bmatrix} 1 & .956 & .620 \\ 1 & -.272 & -.647 \\ 1 & -1.108 & 1.705 \end{bmatrix} \begin{bmatrix} Y \\ I \\ Q \end{bmatrix}$. Get

YIQ	RGB	
(176, -111, -33) ~	(49.424, 227.543, 242.723)	Aqua
(184, 62, -18) ~	(232.112, 178.782, 84.614)	Orange
(171, 5, -19) ~	(164.000, 181.933, 133.065)	Mossy green
(165, -103, -23)		Sky blue

36. (a) Use $\begin{bmatrix} R \\ G \\ B \end{bmatrix} = \begin{bmatrix} 1 & .956 & .620 \\ 1 & -.272 & -.647 \\ 1 & -1.108 & 1.705 \end{bmatrix} \begin{bmatrix} Y \\ I \\ Q \end{bmatrix}$. Get $\begin{bmatrix} 1 & .956 & .620 \\ 1 & -.272 & -.647 \\ 1 & -1.108 & 1.705 \end{bmatrix} \begin{bmatrix} s \\ 0 \\ 0 \end{bmatrix} = \begin{bmatrix} s \\ s \\ s \end{bmatrix}$.

Thus (s,0,0) ~ (s,s,s). Happens since the first column of the matrix is all 1s and the last two elements of the column matrix are zeros.

(b) (255,0,0)~(255,255,255); (200,0,0)~(200,200,200); (150,0,0)~(150,150,150); (100,0,0)~(100,100,100); (0,0,0)~(0,0,0). (255,0,0) is white. Effect of decreasing a in (a,0,0) to zero is to gradually darken, getting shades of gray, until we finally get black.

37.

R	E	T	R	E	A	T
18	5	20	18	5	1	20

The vectors are $\begin{bmatrix} 18 \\ 5 \end{bmatrix}, \begin{bmatrix} 20 \\ 18 \end{bmatrix}, \begin{bmatrix} 5 \\ 1 \end{bmatrix}$, and $\begin{bmatrix} 20 \\ 27 \end{bmatrix}$. Get

$$\begin{bmatrix} 4 & -3 \\ 3 & -2 \end{bmatrix} \begin{bmatrix} 18 & 20 & 5 & 20 \\ 5 & 18 & 1 & 27 \end{bmatrix} = \begin{bmatrix} 57 & 26 & 17 & -1 \\ 44 & 24 & 13 & 6 \end{bmatrix}$$

Coded message is 57, 44, 26, 24, 17, 13, -1, 6.

38. T H E - B R I T I S H - A R E - C O M I N G

20 8 5 27 2 18 9 20 9 19 8 27 1 18 5 27 3 15 13 9 14 7

The vectors are $\begin{bmatrix} 20 \\ 8 \\ 5 \end{bmatrix}, \begin{bmatrix} 27 \\ 2 \\ 18 \end{bmatrix}, \begin{bmatrix} 9 \\ 20 \\ 9 \end{bmatrix}, \begin{bmatrix} 19 \\ 8 \\ 27 \end{bmatrix}, \begin{bmatrix} 1 \\ 18 \\ 5 \end{bmatrix}, \begin{bmatrix} 27 \\ 3 \\ 15 \end{bmatrix}, \begin{bmatrix} 13 \\ 9 \\ 14 \end{bmatrix}$, and $\begin{bmatrix} 7 \\ 27 \\ 27 \end{bmatrix}$. Get

$$\begin{bmatrix} 1 & 2 & 1 \\ 2 & 3 & 1 \\ -2 & 0 & 1 \end{bmatrix}\begin{bmatrix} 20 & 27 & 9 & 19 & 1 & 27 & 13 & 7 \\ 8 & 2 & 20 & 8 & 18 & 3 & 9 & 27 \\ 5 & 18 & 9 & 27 & 5 & 15 & 14 & 27 \end{bmatrix} = \begin{bmatrix} 41 & 49 & 58 & 62 & 42 & 48 & 45 & 88 \\ 69 & 78 & 87 & 89 & 61 & 78 & 67 & 122 \\ -35 & -36 & -9 & -11 & 3 & -39 & -12 & 13 \end{bmatrix}$$

The coded message is 41, 69, -35, 49, 78, -36, 58, 87, -9, 62, 89, -11, 42, 61, 3, 48, 78, -39, 45, 67, -12, 88, 122, 13.

39. The decoding matrix is $\begin{bmatrix} 4 & -3 \\ 3 & -2 \end{bmatrix}^{-1} = \begin{bmatrix} -2 & 3 \\ -3 & 4 \end{bmatrix}$.

$\begin{bmatrix} -2 & 3 \\ -3 & 4 \end{bmatrix}\begin{bmatrix} 49 & -5 & -61 \\ 38 & -3 & -39 \end{bmatrix} = \begin{bmatrix} 16 & 1 & 5 \\ 5 & 3 & 27 \end{bmatrix}$, and the message is PEACE.

40. The decoding matrix is $\begin{bmatrix} 1 & 2 & 1 \\ 2 & 3 & 1 \\ -2 & 0 & 1 \end{bmatrix}^{-1} = \begin{bmatrix} 3 & -2 & -1 \\ -4 & 3 & 1 \\ 6 & -4 & -1 \end{bmatrix}$.

$$\begin{bmatrix} 3 & -2 & -1 \\ -4 & 3 & 1 \\ 6 & -4 & -1 \end{bmatrix}\begin{bmatrix} 71 & 28 & 84 & 63 & 69 & 88 \\ 100 & 43 & 122 & 98 & 102 & 126 \\ -1 & -5 & -11 & -27 & -12 & -3 \end{bmatrix} = \begin{bmatrix} 14 & 3 & 19 & 20 & 15 & 15 \\ 15 & 12 & 19 & 15 & 18 & 23 \\ 27 & 1 & 27 & 13 & 18 & 27 \end{bmatrix},$$

and the message is NO CLASS TOMORROW.

41. (a)

B	O	S	T	O	N	-	C	A	F	E	-	A	T	-	T	W	O
2	15	19	20	15	14	27	3	1	6	5	27	1	20	27	20	23	15

Write as matrix $\begin{bmatrix} 2 & 19 & 15 & 27 & 1 & 5 & 1 & 27 & 23 \\ 15 & 20 & 14 & 3 & 6 & 27 & 20 & 20 & 15 \end{bmatrix}$.

Let encoding matrix be $A = \begin{bmatrix} a & b \\ c & d \end{bmatrix}$.

Encoded message in matrix form is $\begin{bmatrix} 32 & 59 & 43 & 33 & 13 & 59 & 41 & 67 & 53 \\ 47 & 79 & 57 & 36 & 19 & 86 & 61 & 87 & 68 \end{bmatrix}$. Thus,

$$\begin{bmatrix} a & b \\ c & d \end{bmatrix}\begin{bmatrix} 2 & 19 & 15 & 27 & 1 & 5 & 1 & 27 & 23 \\ 15 & 20 & 14 & 3 & 6 & 27 & 20 & 20 & 15 \end{bmatrix} = \begin{bmatrix} 32 & 59 & 43 & 33 & 13 & 59 & 41 & 67 & 53 \\ 47 & 79 & 57 & 36 & 19 & 86 & 61 & 87 & 68 \end{bmatrix}$$

Compare (1,1) elements and (1,2) elements of each side: 2a+15b=32, 19a+20b=59.
Get a=1,b=2.
Compare (2,1) elements and (2,2) elements of each side: 2c+15d=47, 19c+20d=79.
Get c=1, d=3.
Other comparisons of corresponding elements just confirm these results.

Thus encoding matrix is $A = \begin{bmatrix} 1 & 2 \\ 1 & 3 \end{bmatrix}$.

(b) $A^{-1}=\begin{bmatrix} 3 & -2 \\ -1 & 1 \end{bmatrix}$. $\begin{bmatrix} 3 & -2 \\ -1 & 1 \end{bmatrix}\begin{bmatrix} 43 & 49 & 59 & 39 & 45 & 59 \\ 64 & 70 & 79 & 45 & 63 & 79 \end{bmatrix} = \begin{bmatrix} 1 & 7 & 19 & 27 & 9 & 19 \\ 21 & 21 & 20 & 6 & 18 & 20 \end{bmatrix}$.

1	21	7	21	19	20	27	6	9	18	19	20
A	U	G	U	S	T	-	F	I	R	S	T

42. Matrix A is used to send a message from base to agent, matrix B from agent to informer.
Base station $\xrightarrow{A}$ Agent $\xrightarrow{B}$ Informer. Let original message, in matrix form be X.
Thus X -> AX -> B(AX) = (BA)X. The informer receives the digital form (BA)X.
Encoding matrix for base station to send message directly to informer is BA.

$A=\begin{bmatrix} 4 & -3 \\ 3 & -2 \end{bmatrix}$, $B=\begin{bmatrix} 3 & 8 \\ 4 & 11 \end{bmatrix}$. Encoding matrix $=\begin{bmatrix} 4 & -3 \\ 3 & -2 \end{bmatrix}\begin{bmatrix} 3 & 8 \\ 4 & 11 \end{bmatrix} = \begin{bmatrix} 36 & -25 \\ 49 & -34 \end{bmatrix}$.

<u>Exercise Set 2.5</u>

1. $A\mathbf{x} = \begin{bmatrix} 8 \\ 1 \end{bmatrix}$, $A\mathbf{y} = \begin{bmatrix} 0 \\ 4 \end{bmatrix}$, and $A\mathbf{z} = \begin{bmatrix} -8 \\ 7 \end{bmatrix}$.

2. $A\mathbf{x} = \begin{bmatrix} -1 \\ 26 \end{bmatrix}$, $A\mathbf{y} = \begin{bmatrix} 0 \\ 14 \end{bmatrix}$, and $A\mathbf{z} = \begin{bmatrix} -14 \\ 4 \end{bmatrix}$.

3. $A\mathbf{x} = \begin{bmatrix} 1 \\ 4 \\ 1 \end{bmatrix}$, $A\mathbf{y} = \begin{bmatrix} 8 \\ 7 \\ 8 \end{bmatrix}$, and $A\mathbf{z} = \begin{bmatrix} 9 \\ 1 \\ 9 \end{bmatrix}$.

4. $T(\begin{bmatrix} x \\ y \end{bmatrix}) = \begin{bmatrix} -x \\ y \end{bmatrix}$ gives $T(\begin{bmatrix} x \\ y \end{bmatrix}) = \begin{bmatrix} -1 & 0 \\ 0 & 1 \end{bmatrix}\begin{bmatrix} x \\ y \end{bmatrix}$. $T(\begin{bmatrix} 3 \\ 2 \end{bmatrix}) = \begin{bmatrix} -1 & 0 \\ 0 & 1 \end{bmatrix}\begin{bmatrix} 3 \\ 2 \end{bmatrix} = \begin{bmatrix} -3 \\ 2 \end{bmatrix}$.

5. (a) $A = \begin{bmatrix} 0 & -1 \\ 1 & 0 \end{bmatrix}$ $\quad A\begin{bmatrix} 2 \\ 1 \end{bmatrix} = \begin{bmatrix} -1 \\ 2 \end{bmatrix}$

(b) $A = \begin{bmatrix} 0 & 1 \\ -1 & 0 \end{bmatrix}$ $\quad A\begin{bmatrix} 2 \\ 1 \end{bmatrix} = \begin{bmatrix} 1 \\ -2 \end{bmatrix}$

(c) $A = \begin{bmatrix} \frac{1}{\sqrt{2}} & \frac{-1}{\sqrt{2}} \\ \frac{1}{\sqrt{2}} & \frac{1}{\sqrt{2}} \end{bmatrix}$ $\quad A\begin{bmatrix} 2 \\ 1 \end{bmatrix} = \begin{bmatrix} \frac{1}{\sqrt{2}} \\ \frac{3}{\sqrt{2}} \end{bmatrix}$

(d) $A = \begin{bmatrix} -1 & 0 \\ 0 & -1 \end{bmatrix}$ $\quad A\begin{bmatrix} 2 \\ 1 \end{bmatrix} = \begin{bmatrix} -2 \\ -1 \end{bmatrix}$

(e) $A = \begin{bmatrix} 0 & -1 \\ 1 & 0 \end{bmatrix}$ $\quad A\begin{bmatrix} 2 \\ 1 \end{bmatrix} = \begin{bmatrix} -1 \\ 2 \end{bmatrix}$

(f) $A = \begin{bmatrix} \frac{\sqrt{3}}{2} & \frac{-1}{2} \\ \frac{1}{2} & \frac{\sqrt{3}}{2} \end{bmatrix}$ $\quad A\begin{bmatrix} 2 \\ 1 \end{bmatrix} = \begin{bmatrix} \sqrt{3} - \frac{1}{2} \\ 1 + \frac{\sqrt{3}}{2} \end{bmatrix}$

(g) $A = \begin{bmatrix} \frac{1}{2} & \frac{\sqrt{3}}{2} \\ \frac{-\sqrt{3}}{2} & \frac{1}{2} \end{bmatrix}$ $\quad A\begin{bmatrix} 2 \\ 1 \end{bmatrix} = \begin{bmatrix} 1 + \frac{\sqrt{3}}{2} \\ -\sqrt{3} + \frac{1}{2} \end{bmatrix}$

6. $\begin{bmatrix} 3 & 0 \\ 0 & 3 \end{bmatrix}\begin{bmatrix} x \\ y \end{bmatrix} = \begin{bmatrix} 3x \\ 3y \end{bmatrix} = \begin{bmatrix} x' \\ y' \end{bmatrix}$, so $\frac{x'^2}{9} + \frac{y'^2}{9} = 1$. Thus the images of the points on the circle $x^2 + y^2 = 1$ are the points on the circle $x^2 + y^2 = 9$.

7. $\begin{bmatrix} 0 & -1 \\ 1 & 0 \end{bmatrix}\begin{bmatrix} x \\ y \end{bmatrix} = \begin{bmatrix} -y \\ x \end{bmatrix} = \begin{bmatrix} x' \\ y' \end{bmatrix}$. $\frac{x^2}{4} + \frac{y^2}{9} = 1$, so that $\frac{y'^2}{4} + \frac{x'^2}{9} = 1$. Thus the images of the points on the ellipse $\frac{x^2}{4} + \frac{y^2}{9} = 1$ are the points on the ellipse $\frac{x^2}{9} + \frac{y^2}{4} = 1$.

8. Vertices of the image of the unit square ((1,0), (1,1), (0,1),(0,0)) are

(a) (0,1), (−1,1), (−1,0), (0,0)

(b) (2,0), (2,2), (0,2), (0,0)

(c) (3,1), (3,5), (0,4), (0,0)

(d) (4,1), (3,6), (−1,5), (0,0)

9. Vertices of the image of the unit square ((1,0), (1,1), (0,1),(0,0)) are

(a) (-2,0), (-5,4), (-3,4), (0,0)

(b) (−2,−4), (−6,−5), (−4,−1), (0,0)

(c) (0,2), (−2,2), (−2,0), (0,0)

(d) (0,−3), (3,−3), (3,0), (0,0)

10. (a) $T_2 \circ T_1(\mathbf{x}) = A_2 A_1 \mathbf{x} = \begin{bmatrix} -1 & 0 \\ 1 & 5 \end{bmatrix}\begin{bmatrix} 1 & 2 \\ 3 & 0 \end{bmatrix}\mathbf{x} = \begin{bmatrix} -1 & -2 \\ 16 & 2 \end{bmatrix}\mathbf{x}$, so

$$T_2 \circ T_1\left(\begin{bmatrix} 5 \\ 2 \end{bmatrix}\right) = \begin{bmatrix} -1 & -2 \\ 16 & 2 \end{bmatrix}\begin{bmatrix} 5 \\ 2 \end{bmatrix} = \begin{bmatrix} -9 \\ 84 \end{bmatrix}.$$

(b) $T_2 \circ T_1(\mathbf{x}) = A_2 A_1 \mathbf{x} = \begin{bmatrix} 2 & 2 \\ 1 & -1 \end{bmatrix}\begin{bmatrix} 0 & 1 & 2 \\ 3 & 4 & -1 \end{bmatrix}\mathbf{x} = \begin{bmatrix} 6 & 10 & 2 \\ -3 & -3 & 3 \end{bmatrix}\mathbf{x}$, so

$$T_2 \circ T_1\left(\begin{bmatrix} 0 \\ 1 \\ 3 \end{bmatrix}\right) = \begin{bmatrix} 6 & 10 & 2 \\ -3 & -3 & 3 \end{bmatrix}\begin{bmatrix} 0 \\ 1 \\ 3 \end{bmatrix} = \begin{bmatrix} 16 \\ 6 \end{bmatrix}.$$

(c) $T_2 \circ T_1(\mathbf{x}) = A_2 A_1 \mathbf{x} = \begin{bmatrix} 2 & 2 \\ 1 & -1 \\ 0 & 4 \end{bmatrix}\begin{bmatrix} 3 & -2 \\ 0 & 1 \end{bmatrix}\mathbf{x} = \begin{bmatrix} 6 & -2 \\ 3 & -3 \\ 0 & 4 \end{bmatrix}\mathbf{x}$, so

$$T_2 \circ T_1\left(\begin{bmatrix} -3 \\ 2 \end{bmatrix}\right) = \begin{bmatrix} 6 & -2 \\ 3 & -3 \\ 0 & 4 \end{bmatrix}\begin{bmatrix} -3 \\ 2 \end{bmatrix} = \begin{bmatrix} -22 \\ -15 \\ 8 \end{bmatrix}.$$

11. (a) $T_2 \circ T_1(\mathbf{x}) = A_2 A_1 \mathbf{x} = \begin{bmatrix} 2 & 3 \\ 0 & -4 \end{bmatrix}\begin{bmatrix} 1 & 1 \\ 1 & 1 \end{bmatrix}\mathbf{x} = \begin{bmatrix} 5 & 5 \\ -4 & -4 \end{bmatrix}\mathbf{x}$, so

$$T_2 \circ T_1\left(\begin{bmatrix} 1 \\ 3 \end{bmatrix}\right) = \begin{bmatrix} 5 & 5 \\ -4 & -4 \end{bmatrix}\begin{bmatrix} 1 \\ 3 \end{bmatrix} = \begin{bmatrix} 20 \\ -16 \end{bmatrix}.$$

(b) $T_2 \circ T_1(\mathbf{x}) = A_2 A_1 \mathbf{x} = \begin{bmatrix} 3 & 1 \\ 0 & 2 \end{bmatrix}\begin{bmatrix} 2 & 1 & 0 \\ 3 & -2 & 5 \end{bmatrix}\mathbf{x} = \begin{bmatrix} 9 & 1 & 5 \\ 6 & -4 & 10 \end{bmatrix}\mathbf{x}$, so

$$T_2 \circ T_1\left(\begin{bmatrix} 3 \\ -2 \\ 1 \end{bmatrix}\right) = \begin{bmatrix} 9 & 1 & 5 \\ 6 & -4 & 10 \end{bmatrix}\begin{bmatrix} 3 \\ -2 \\ 1 \end{bmatrix} = \begin{bmatrix} 30 \\ 36 \end{bmatrix}.$$

(c) $T_2 \circ T_1(\mathbf{x}) = A_2 A_1 \mathbf{x} = \begin{bmatrix} 3 & 0 \\ 1 & -7 \\ 2 & 5 \end{bmatrix}\begin{bmatrix} 5 & -2 \\ 3 & 6 \end{bmatrix}\mathbf{x} = \begin{bmatrix} 15 & -6 \\ -16 & -44 \\ 25 & 26 \end{bmatrix}\mathbf{x}$, so

$$T_2 \circ T_1\left(\begin{bmatrix} 4 \\ -5 \end{bmatrix}\right) = \begin{bmatrix} 15 & -6 \\ -16 & -44 \\ 25 & 26 \end{bmatrix}\begin{bmatrix} 4 \\ -5 \end{bmatrix} = \begin{bmatrix} 90 \\ 156 \\ -30 \end{bmatrix}.$$

12. (a) $\begin{bmatrix} 0.5 & 0 \\ 0 & 0.5 \end{bmatrix}\begin{bmatrix} 0 & -1 \\ 1 & 0 \end{bmatrix} = \begin{bmatrix} 0 & -0.5 \\ 0.5 & 0 \end{bmatrix}$ and $\begin{bmatrix} 0 & -0.5 \\ 0.5 & 0 \end{bmatrix}\begin{bmatrix} 2 \\ 1 \end{bmatrix} = \begin{bmatrix} -0.5 \\ 1 \end{bmatrix}$.

contraction rotation

(b) $\begin{bmatrix} 1 & 0 \\ 0 & -1 \end{bmatrix}\begin{bmatrix} 4 & 0 \\ 0 & 4 \end{bmatrix} = \begin{bmatrix} 4 & 0 \\ 0 & -4 \end{bmatrix}$ and $\begin{bmatrix} 4 & 0 \\ 0 & -4 \end{bmatrix}\begin{bmatrix} 2 \\ 1 \end{bmatrix} = \begin{bmatrix} 8 \\ -4 \end{bmatrix}$.

reflection dilation

(c) $\begin{bmatrix} -1 & 0 \\ 0 & -1 \end{bmatrix}\begin{bmatrix} 0 & 1 \\ 1 & 0 \end{bmatrix} = \begin{bmatrix} 0 & -1 \\ -1 & 0 \end{bmatrix}$ and $\begin{bmatrix} 0 & -1 \\ -1 & 0 \end{bmatrix}\begin{bmatrix} 2 \\ 1 \end{bmatrix} = \begin{bmatrix} -1 \\ -2 \end{bmatrix}$.

rotation reflection

13. $\begin{bmatrix} 2 & 0 \\ 0 & 2 \end{bmatrix}\begin{bmatrix} 0 & -1 \\ 1 & 0 \end{bmatrix} = \begin{bmatrix} 0 & -2 \\ 2 & 0 \end{bmatrix}$

dilation rotation

14. $\begin{bmatrix} r & 0 \\ 0 & r \end{bmatrix}\begin{bmatrix} \cos\theta & -\sin\theta \\ \sin\theta & \cos\theta \end{bmatrix} = \begin{bmatrix} r\cos\theta & -r\sin\theta \\ r\sin\theta & r\cos\theta \end{bmatrix}$

15. $A = \begin{bmatrix} \frac{1}{\sqrt{2}} & \frac{-1}{\sqrt{2}} \\ \frac{1}{\sqrt{2}} & \frac{1}{\sqrt{2}} \end{bmatrix}$, $A^2 = \begin{bmatrix} 0 & -1 \\ 1 & 0 \end{bmatrix}$, $A^4 = \begin{bmatrix} -1 & 0 \\ 0 & -1 \end{bmatrix}$, and $A^8 = \begin{bmatrix} 1 & 0 \\ 0 & 1 \end{bmatrix}$.

Eight successive rotations of $\frac{\pi}{4}$ gives a total rotation of 2π, taking every point back to where it started.

16. $A^2BA^2 = \begin{bmatrix} 0 & -1 \\ 1 & 0 \end{bmatrix}\begin{bmatrix} -1 & 0 \\ 0 & -1 \end{bmatrix}\begin{bmatrix} 0 & -1 \\ 1 & 0 \end{bmatrix} = \begin{bmatrix} 1 & 0 \\ 0 & 1 \end{bmatrix}$.

These three matrices A^2, B, and A^2 represent successive rotations of $\frac{\pi}{2}$, π, and $\frac{\pi}{2}$, for a total rotation of 2π, taking every point back to where it started.

17. Rotation: $T_1\left(\begin{bmatrix} x \\ y \end{bmatrix}\right) = \begin{bmatrix} \cos\theta & -\sin\theta \\ \sin\theta & \cos\theta \end{bmatrix}\begin{bmatrix} x \\ y \end{bmatrix}$, Dilation: $T_2\left(\begin{bmatrix} x \\ y \end{bmatrix}\right) = \begin{bmatrix} r & 0 \\ 0 & r \end{bmatrix}\begin{bmatrix} x \\ y \end{bmatrix}$.

Algebra - $T_2 \circ T_1 \left(\begin{bmatrix} x \\ y \end{bmatrix}\right) = \begin{bmatrix} r & 0 \\ 0 & r \end{bmatrix}\begin{bmatrix} \cos\theta & -\sin\theta \\ \sin\theta & \cos\theta \end{bmatrix}\begin{bmatrix} x \\ y \end{bmatrix} = \begin{bmatrix} r\cos\theta & -r\sin\theta \\ r\sin\theta & r\cos\theta \end{bmatrix}\begin{bmatrix} x \\ y \end{bmatrix}$

$$= \begin{bmatrix} \cos\theta & -\sin\theta \\ \sin\theta & \cos\theta \end{bmatrix}\begin{bmatrix} r & 0 \\ 0 & r \end{bmatrix}\begin{bmatrix} x \\ y \end{bmatrix} T_1 \circ T_2 \left(\begin{bmatrix} x \\ y \end{bmatrix}\right).$$

The effect of $T_2 \circ T_1$ and $T_1 \circ T_2$ are the same on all elements thus $T_2 \circ T_1 = T_1 \circ T_2$.

Geometry – $T_2 \circ T_1$: A->B->C, while $T_1 \circ T_2$: A->D->C. Route is different, but arrive at the same point. See figure below.

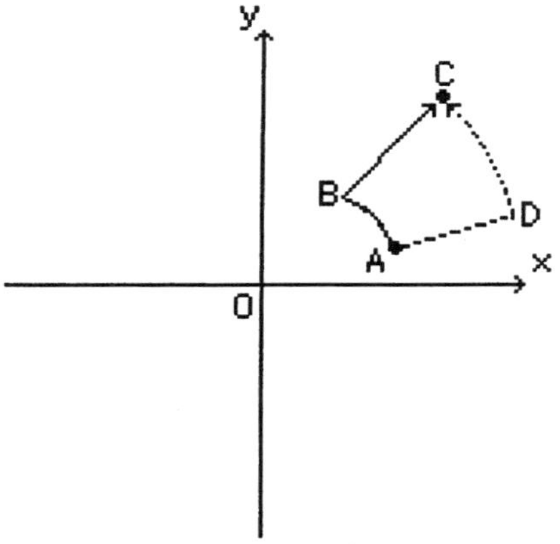

Exercise 17

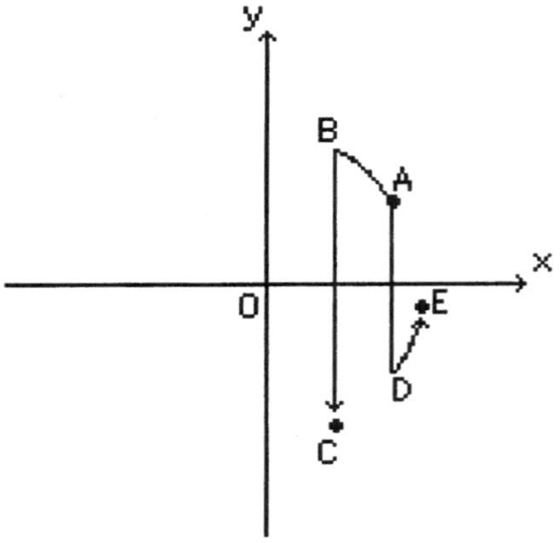

Exercise 18

18. Rotation: $T_1\left(\begin{bmatrix} x \\ y \end{bmatrix}\right) = \begin{bmatrix} \cos\theta & -\sin\theta \\ \sin\theta & \cos\theta \end{bmatrix}\begin{bmatrix} x \\ y \end{bmatrix}$, Reflection in x-axis: $T_2\left(\begin{bmatrix} x \\ y \end{bmatrix}\right) = \begin{bmatrix} 1 & 0 \\ 0 & -1 \end{bmatrix}\begin{bmatrix} x \\ y \end{bmatrix}$.

Algebra - $T_2 \circ T_1 \left(\begin{bmatrix} x \\ y \end{bmatrix}\right) = \begin{bmatrix} 1 & 0 \\ 0 & -1 \end{bmatrix}\begin{bmatrix} \cos\theta & -\sin\theta \\ \sin\theta & \cos\theta \end{bmatrix}\begin{bmatrix} x \\ y \end{bmatrix} = \begin{bmatrix} \cos\theta & -\sin\theta \\ -\sin\theta & -\cos\theta \end{bmatrix}\begin{bmatrix} x \\ y \end{bmatrix}$

while $T_1 \circ T_2 \left(\begin{bmatrix} x \\ y \end{bmatrix}\right) = \begin{bmatrix} \cos\theta & -\sin\theta \\ \sin\theta & \cos\theta \end{bmatrix}\begin{bmatrix} 1 & 0 \\ 0 & -1 \end{bmatrix}\begin{bmatrix} x \\ y \end{bmatrix} = \begin{bmatrix} \cos\theta & \sin\theta \\ \sin\theta & -\cos\theta \end{bmatrix}\begin{bmatrix} x \\ y \end{bmatrix}$

Thus $T_2 \circ T_1 \left(\begin{bmatrix} x \\ y \end{bmatrix}\right) \neq T_1 \circ T_2 \left(\begin{bmatrix} x \\ y \end{bmatrix}\right)$. The transformations are not commutative.

Geometry - $T_2 \circ T_1$: A->B->C, while $T_1 \circ T_2$: A->D->E. Route is different, but arrive at the same point. See Figure above.

19. $A^t = \begin{bmatrix} 0 & 1 \\ -1 & 0 \end{bmatrix}$. $AA^t = \begin{bmatrix} 0 & -1 \\ 1 & 0 \end{bmatrix}\begin{bmatrix} 0 & 1 \\ -1 & 0 \end{bmatrix} = I$ and $A^tA = \begin{bmatrix} 0 & 1 \\ -1 & 0 \end{bmatrix}\begin{bmatrix} 0 & -1 \\ 1 & 0 \end{bmatrix} = I$. Thus $A^{-1} = A^t$.

So A is orthogonal.

$A\mathbf{u} = \begin{bmatrix} -3 \\ 3 \end{bmatrix}$ and $A\mathbf{v} = \begin{bmatrix} -5 \\ 1 \end{bmatrix}$. $\|\mathbf{u}\|^2 = 3^2 + 3^2 = (-3)^2 + 3^2 = \|A\mathbf{u}\|^2$ and

$\|\mathbf{v}\|^2 = 1 + 5^2 = (-5)^2 + 1 = \|A\mathbf{v}\|^2$. $\mathbf{u}\cdot\mathbf{v} = 3 + 15 = 15 + 3 = A\mathbf{u}\cdot A\mathbf{v}$, so that $\dfrac{\mathbf{u}\cdot\mathbf{v}}{\|\mathbf{u}\|\,\|\mathbf{v}\|} = \dfrac{A\mathbf{u}\cdot A\mathbf{v}}{\|A\mathbf{u}\|\,\|A\mathbf{v}\|}$ and therefore the angle between the vectors **u** and **v** and the angle between the vectors A**u** and A**v** are the same.

$d\left(\begin{bmatrix}3\\3\end{bmatrix}, \begin{bmatrix}1\\5\end{bmatrix}\right) = \left\|\begin{bmatrix}2\\-2\end{bmatrix}\right\| = \left\|\begin{bmatrix}2\\2\end{bmatrix}\right\| = d\left(\begin{bmatrix}-3\\3\end{bmatrix}, \begin{bmatrix}-5\\1\end{bmatrix}\right)$.

20. Let $A = \begin{bmatrix}\cos\theta & -\sin\theta\\ \sin\theta & \cos\theta\end{bmatrix}$, the rotation matrix. Then $AA^t = \begin{bmatrix}\cos\theta & -\sin\theta\\ \sin\theta & \cos\theta\end{bmatrix}\begin{bmatrix}\cos\theta & \sin\theta\\ -\sin\theta & \cos\theta\end{bmatrix} =$

$\begin{bmatrix}\cos^2\theta+\sin^2\theta & \cos\theta\sin\theta-\sin\theta\cos\theta\\ \sin\theta\cos\theta-\cos\theta\sin\theta & \sin^2\theta+\cos^2\theta\end{bmatrix} = \begin{bmatrix}1 & 0\\ 0 & 1\end{bmatrix}$.

Similarly, $A^tA = I$. Thus $A^{-1} = A^t$. So A is orthogonal.

21. (a) It was proved in the text that $\mathbf{u}\cdot\mathbf{v} = A\mathbf{u}\cdot A\mathbf{v}$ and $\|\mathbf{u}\| = \|A\mathbf{u}\|$. Thus $\dfrac{\mathbf{u}\cdot\mathbf{v}}{\|\mathbf{u}\|\,\|\mathbf{v}\|} = \dfrac{A\mathbf{u}\cdot A\mathbf{v}}{\|A\mathbf{u}\|\,\|A\mathbf{v}\|}$, and so the angle between the vectors **u** and **v** and the angle between the vectors A**u** and A**v** are the same.

(b) $d(P,Q) = \|\mathbf{u}-\mathbf{v}\| = \|A(\mathbf{u}-\mathbf{v})\| = \|A\mathbf{u}-A\mathbf{v}\| = d(S,T)$.

22. $T(\mathbf{u}) = \mathbf{u} + \mathbf{v}$, so $\mathbf{v} = T(\mathbf{u}) - \mathbf{u} = T(1,2) - (1,2) = (2,-3) - (1,2) = (1,-5)$.

$T(3,4) = (3,4) + (1,-5) = (4,-1)$ and $T(4,6) = (4,6) + (1,-5) = (5,1)$, so image of triangle with vertices (1,2), (3,4), and (4,6) is triangle with vertices (2,−3), (4,−1), and (5,1).

23. (a) $T\left(\begin{bmatrix}x\\y\end{bmatrix}\right) = \begin{bmatrix}x\\y\end{bmatrix} + \begin{bmatrix}2\\5\end{bmatrix} = \begin{bmatrix}x+2\\y+5\end{bmatrix} = \begin{bmatrix}x'\\y'\end{bmatrix}$. $y = 3x + 1$, so $y' - 5 = 3(x' - 2) + 1$ and thus $y' = 3x'$, so the image of the line $y = 3x + 1$ is the line $y = 3x$.

(b) $T\left(\begin{bmatrix}x\\y\end{bmatrix}\right) = \begin{bmatrix}x\\y\end{bmatrix} + \begin{bmatrix}-1\\1\end{bmatrix} = \begin{bmatrix}x-1\\y+1\end{bmatrix} = \begin{bmatrix}x'\\y'\end{bmatrix}$. $y = 3x + 1$, so $y' - 1 = 3(x' + 1) + 1$ and thus $y' = 3x' + 5$, so the image of the line $y = 3x + 1$ is the line $y = 3x + 5$.

24. (a) $\begin{bmatrix}1\\0\end{bmatrix} \mapsto \begin{bmatrix}6\\4\end{bmatrix}$, $\begin{bmatrix}1\\1\end{bmatrix} \mapsto \begin{bmatrix}6\\6\end{bmatrix}$, $\begin{bmatrix}0\\1\end{bmatrix} \mapsto \begin{bmatrix}4\\6\end{bmatrix}$, and $\begin{bmatrix}0\\0\end{bmatrix} \mapsto \begin{bmatrix}4\\4\end{bmatrix}$.

$\begin{bmatrix}x\\y\end{bmatrix} \mapsto \begin{bmatrix}2x+4\\2y+4\end{bmatrix} = \begin{bmatrix}x'\\y'\end{bmatrix}$, so the image of $x^2 + y^2 = 1$ is $(x-4)^2 + (y-4)^2 = 4$.

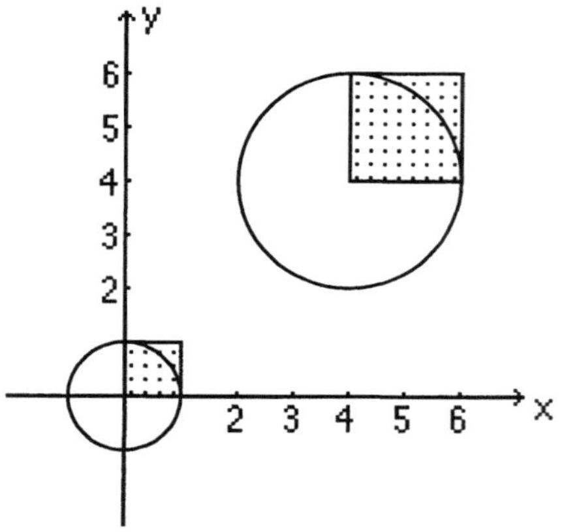

Figure for (a)

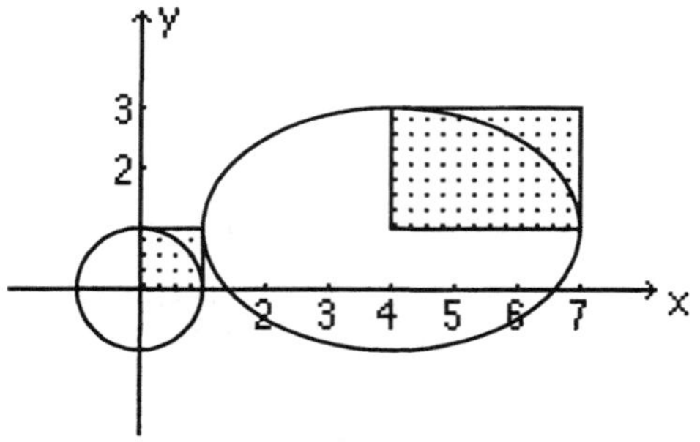

Figure for (b)

(b) $\begin{bmatrix}1\\0\end{bmatrix} \mapsto \begin{bmatrix}7\\1\end{bmatrix}$, $\begin{bmatrix}1\\1\end{bmatrix} \mapsto \begin{bmatrix}7\\3\end{bmatrix}$, $\begin{bmatrix}0\\1\end{bmatrix} \mapsto \begin{bmatrix}4\\3\end{bmatrix}$, and $\begin{bmatrix}0\\0\end{bmatrix} \mapsto \begin{bmatrix}4\\1\end{bmatrix}$.

$\begin{bmatrix}x\\y\end{bmatrix} \mapsto \begin{bmatrix}3x+4\\2y+1\end{bmatrix} = \begin{bmatrix}x'\\y'\end{bmatrix}$, so the image of $x^2 + y^2 = 1$ is $\frac{(x-4)^2}{9} + \frac{(y-1)^2}{4} = 1$.

(c) $\begin{bmatrix}1\\0\end{bmatrix} \mapsto \begin{bmatrix}\frac{1}{\sqrt{2}}+3\\ \frac{1}{\sqrt{2}}+1\end{bmatrix}$, $\begin{bmatrix}1\\1\end{bmatrix} \mapsto \begin{bmatrix}3\\ \frac{2}{\sqrt{2}}+1\end{bmatrix}$, $\begin{bmatrix}0\\1\end{bmatrix} \mapsto \begin{bmatrix}\frac{-1}{\sqrt{2}}+3\\ \frac{1}{\sqrt{2}}+1\end{bmatrix}$, and

$\begin{bmatrix}0\\0\end{bmatrix} \mapsto \begin{bmatrix}3\\1\end{bmatrix}$. $\begin{bmatrix}x\\y\end{bmatrix} \mapsto \begin{bmatrix}\frac{x}{\sqrt{2}} - \frac{y}{\sqrt{2}} + 3\\ \frac{x}{\sqrt{2}} + \frac{y}{\sqrt{2}} + 1\end{bmatrix} = \begin{bmatrix}x'\\y'\end{bmatrix}$, so the image of the circle

$x^2 + y^2 = 1$ is $\frac{(x+y-4)^2}{2} + \frac{(y-x+2)^2}{2} = 1$, which is the circle $(x-3)^2 + (y-1)^2 = 1$.

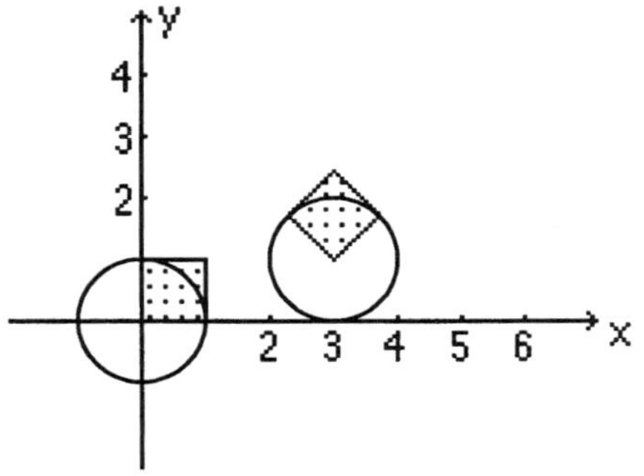

Figure for (c)

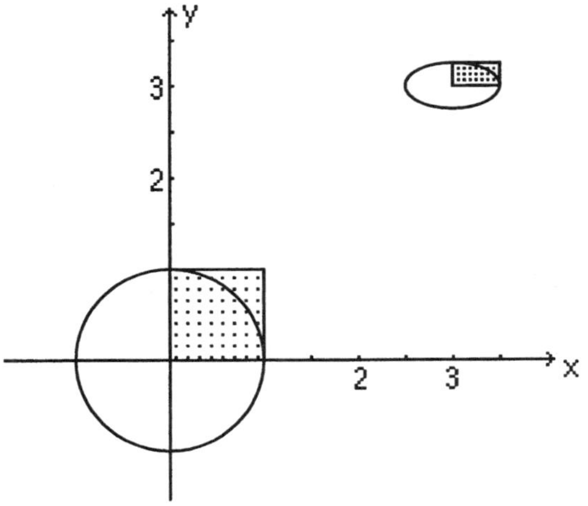

Figure for (d)

(d) $\begin{bmatrix}1\\0\end{bmatrix} \mapsto \begin{bmatrix}3.5\\3\end{bmatrix}, \begin{bmatrix}1\\1\end{bmatrix} \mapsto \begin{bmatrix}3.5\\3.25\end{bmatrix}, \begin{bmatrix}0\\1\end{bmatrix} \mapsto \begin{bmatrix}3\\3.25\end{bmatrix}$, and $\begin{bmatrix}0\\0\end{bmatrix} \mapsto \begin{bmatrix}3\\3\end{bmatrix}$.

$\begin{bmatrix}x\\y\end{bmatrix} \mapsto \begin{bmatrix}.5x+3\\.25y+3\end{bmatrix} = \begin{bmatrix}x'\\y'\end{bmatrix}$, so the image of $x^2 + y^2 = 1$ is $4(x-3)^2 + 16(y-3)^2 = 1$.

Exercise Set 2.6

1. $T((x_1, y_1)+(x_2, y_2)) = T(x_1+x_2, y_1+y_2) = (2(x_1+x_2), x_1+x_2-y_1-y_2)$
$= (2x_1+2x_2, x_1-y_1+x_2-y_2) = (2x_1, x_1-y_1) + (2x_2, x_2-y_2)$
$= T(x_1, y_1) + T(x_2, y_2)$ and $T(c(x,y)) = T(cx,cy) = (2cx, cx-cy) = c(2x, x-y) = cT(x,y)$.
Thus T is linear. $T(1,2) = (2,-1)$ and $T(-1,4) = (-2,-5)$.

2. $T((x_1, y_1)+(x_2, y_2)) = T(x_1+x_2, y_1+y_2)$
$= (3(x_1+x_2)+y_1+y_2, 2(y_1+y_2), x_1+x_2-y_1-y_2)$
$= (3x_1+3x_2+y_1+y_2, 2y_1+2y_2, x_1+x_2-y_1-y_2)$
$= (3x_1+y_1, 2y_1, x_1-y_1) + (3x_2+y_2, 2y_2, x_2-y_2) = T(x_1, y_1) + T(x_2, y_2)$,
and $T(c(x,y)) = T(cx,cy) = (3cx+cy, 2cy, cx-cy) = c(3x+y, 2y, x-y) = cT(x,y)$.
Thus T is linear. $T(1,2) = (5,4,-1)$ and $T(2,-5) = (1,-10,7)$.

3. $T((x_1, y_1, z_1)+(x_2, y_2, z_2)) = T(x_1+x_2, y_1+y_2, z_1+z_2) = (0, y_1+y_2, 0)$
$= (0, y_1, 0) + (0, y_2, 0) = T((x_1, y_1, z_1) + T(x_2, y_2, z_2)$,
and $T(c(x,y,z) = T(cx,cy,cz) = (0,cy,0) = c(0,y,0) = cT(x,y,z)$.
Thus T is linear. The image of (x,y,z) under T is the projection of (x,y,z) on the y axis.

4. (a) $T(c(x,y,z)) = T(cx,cy,cz) = (3cx,(cy)^2) = (3cx,c^2y^2)=c(3x,cy^2) \neq c(3x,y^2)=cT(x,y,z)$.

   Thus scalar multiplication not preserved. T is not linear.

   (b) $T(c(x,y,z)) = T(cx,cy,cz) = (cx+2,4cy) \neq c(x+2,4y) = cT(x,y,z)$, so T is not linear.

5. $T(c(x,y)) = T(cx,cy) = cx + a \neq c(x+a) = cT(x,y)$, so T is not linear.

6. $T((x_1,y_1,z_1)+(x_2,y_2,z_2)) = T(x_1+x_2,y_1+y_2,z_1+z_2) = (2(x_1+x_2),y_1+y_2)$
   $= (2x_1+2x_2,y_1+y_2) = (2x_1,y_1) = (2x_2,y_2) = T((x_1,y_1,z_1)+T(x_2,y_2,z_2)$,
   and $T(c(x,y,z) = T(cx,cy,cz) = (2cx,cy) = c(2x,y) = cT(x,y,z)$. Thus T is linear.

7. $T((x_1,y_1)+(x_2,y_2)) = T(x_1+x_2,y_1+y_2) = x_1+x_2-y_1-y_2 = x_1-y_1+x_2-y_2$
   $= T(x_1,y_1) + T(x_2,y_2)$
   and $T(c(x,y)) = T(cx,cy) = cx-cy = c(x-y) = cT(x,y)$. Thus T is linear.

8. (a) $T((x_1,y_1)+(x_2,y_2)) = T(x_1+x_2,y_1+y_2) = (x_1+x_2,y_1+y_2,0)$
   $= (x_1,y_1,0) + (x_2,y_2,0) = T(x_1,y_1) + T(x_2,y_2)$
   and $T(c(x,y)) = T(cx,cy) = (cx,cy,0) = c(x,y,0) = cT(x,y)$. Thus T is linear.

   (b) $T((x_1,y_1)+(x_2,y_2)) = T(x_1+x_2,y_1+y_2) = (x_1+x_2,y_1+y_2,1)$
   $\neq (x_1,y_1,1) + (x_2,y_2,1) = T(x_1,y_1) + T(x_2,y_2)$. Thus T is not linear.

9. $T(x_1+x_2) = (x_1+x_2, 2(x_1+x_2), 3(x_1+x_2)) = (x_1+x_2, 2x_1+2x_2, 3x_1+3x_2)$
   $= (x_1,2x_1,3x_1) + (x_2,2x_2,3x_2) = T(x_1)+T(x_2)$,
   and $T(cx) = (cx,2cx,3cx) = c(x,2x,3x) = cT(x)$. Thus T is linear.

10. If $c \neq 0$ or 1, $T(c(x,y)) = T(cx,cy) = ((cx)^2,cy) = (c^2x^2,cy) = c(cx^2,y)$
    $\neq c(x^2,y) = cT(x,y)$, so T is not linear.

11. $T((x_1, y_1, z_1)+(x_2, y_2, z_2)) = T(x_1+x_2, y_1+y_2, z_1+z_2)$
$= (x_1+x_2+2(y_1+y_2), x_1+x_2+y_1+y_2+z_1+z_2, 3(z_1+z_2))$
$= (x_1+x_2+2y_1+2y_2, x_1+x_2+y_1+y_2+z_1+z_2, 3z_1+3z_2)$
$= (x_1+2y_1, x_1+y_1+z_1, 3z_1) + (x_2+2y_2, x_2+y_2+z_2, 3z_2)$
$= T((x_1, y_1, z_1)+T(x_2, y_2, z_2),$
and $T(c(x,y,z) = T(cx,cy,cz) = (cx+2cy, cx+cy+cz, 3cz) = c(x+2y, x+y+z, 3z) = cT(x,y,z)$.
Thus T is linear.

12. (a) $\begin{bmatrix} 1 \\ 0 \end{bmatrix} \mapsto \begin{bmatrix} 2 \\ 1 \end{bmatrix}$ and $\begin{bmatrix} 0 \\ 1 \end{bmatrix} \mapsto \begin{bmatrix} 0 \\ -1 \end{bmatrix}$, so $A = \begin{bmatrix} 2 & 0 \\ 1 & -1 \end{bmatrix}$.

(b) $\begin{bmatrix} 1 \\ 0 \end{bmatrix} \mapsto \begin{bmatrix} 1 \\ 1 \end{bmatrix}$ and $\begin{bmatrix} 0 \\ 1 \end{bmatrix} \mapsto \begin{bmatrix} -1 \\ 1 \end{bmatrix}$, so $A = \begin{bmatrix} 1 & -1 \\ 1 & 1 \end{bmatrix}$.

(c) $\begin{bmatrix} 1 \\ 0 \end{bmatrix} \mapsto \begin{bmatrix} 2 \\ 0 \end{bmatrix}$ and $\begin{bmatrix} 0 \\ 1 \end{bmatrix} \mapsto \begin{bmatrix} -5 \\ 3 \end{bmatrix}$, so $A = \begin{bmatrix} 2 & -5 \\ 0 & 3 \end{bmatrix}$.

(d) $\begin{bmatrix} 1 \\ 0 \end{bmatrix} \mapsto \begin{bmatrix} 0 \\ -3 \end{bmatrix}$ and $\begin{bmatrix} 0 \\ 1 \end{bmatrix} \mapsto \begin{bmatrix} 2 \\ 0 \end{bmatrix}$, so $A = \begin{bmatrix} 0 & 2 \\ -3 & 0 \end{bmatrix}$.

13. (a) $\begin{bmatrix} 1 \\ 0 \end{bmatrix} \mapsto \begin{bmatrix} 1 \\ 0 \\ 1 \end{bmatrix}$ and $\begin{bmatrix} 0 \\ 1 \end{bmatrix} \mapsto \begin{bmatrix} 0 \\ 1 \\ 1 \end{bmatrix}$, so $A = \begin{bmatrix} 1 & 0 \\ 0 & 1 \\ 1 & 1 \end{bmatrix}$. $A\begin{bmatrix} x \\ y \end{bmatrix} = \begin{bmatrix} x \\ y \\ x+y \end{bmatrix}$.

(b) $\begin{bmatrix} 1 \\ 0 \end{bmatrix} \mapsto \begin{bmatrix} 1 \\ 2 \\ 3 \end{bmatrix}$ and $\begin{bmatrix} 0 \\ 1 \end{bmatrix} \mapsto \begin{bmatrix} 0 \\ 0 \\ 0 \end{bmatrix}$, so $A = \begin{bmatrix} 1 & 0 \\ 2 & 0 \\ 3 & 0 \end{bmatrix}$. $A\begin{bmatrix} x \\ y \end{bmatrix} = \begin{bmatrix} x \\ 2x \\ 3x \end{bmatrix}$.

(c) $\begin{bmatrix} 1 \\ 0 \\ 0 \end{bmatrix} \mapsto \begin{bmatrix} 2 \\ 1 \end{bmatrix}, \begin{bmatrix} 0 \\ 1 \\ 0 \end{bmatrix} \mapsto \begin{bmatrix} 0 \\ 1 \end{bmatrix}$, and $\begin{bmatrix} 0 \\ 0 \\ 1 \end{bmatrix} \mapsto \begin{bmatrix} 0 \\ 0 \end{bmatrix}$, so $A = \begin{bmatrix} 2 & 0 & 0 \\ 1 & 1 & 0 \end{bmatrix}$. $A\begin{bmatrix} x \\ y \\ z \end{bmatrix} = \begin{bmatrix} 2x \\ x+y \end{bmatrix}$.

(d) $\begin{bmatrix} 1 \\ 0 \\ 0 \end{bmatrix} \mapsto \begin{bmatrix} 1 \\ 0 \end{bmatrix}, \begin{bmatrix} 0 \\ 1 \\ 0 \end{bmatrix} \mapsto \begin{bmatrix} 0 \\ 1 \end{bmatrix}$, and $\begin{bmatrix} 0 \\ 0 \\ 1 \end{bmatrix} \mapsto \begin{bmatrix} 2 \\ 3 \end{bmatrix}$, so $A = \begin{bmatrix} 1 & 0 & 2 \\ 0 & 1 & 3 \end{bmatrix}$. $A\begin{bmatrix} x \\ y \\ z \end{bmatrix} = \begin{bmatrix} x+2z \\ y+3z \end{bmatrix}$.

14. We see that (1,0)->($\cos\theta$, $\sin\theta$), and (0,1) ->($-\sin\theta$, $\cos\theta$), below. Write these elements in

column form. $T\left(\begin{bmatrix}1\\0\end{bmatrix}\right) = \begin{bmatrix}\cos\theta\\ \sin\theta\end{bmatrix}$ and $T\left(\begin{bmatrix}0\\1\end{bmatrix}\right) = \begin{bmatrix}-\sin\theta\\ \cos\theta\end{bmatrix}$, so $A = \begin{bmatrix}\cos\theta & -\sin\theta\\ \sin\theta & \cos\theta\end{bmatrix}$.

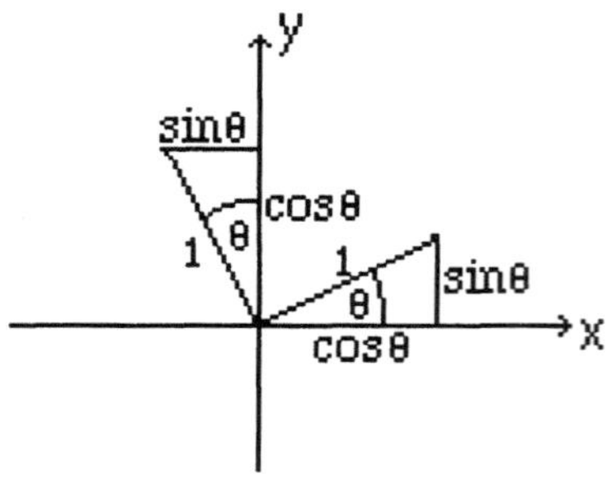

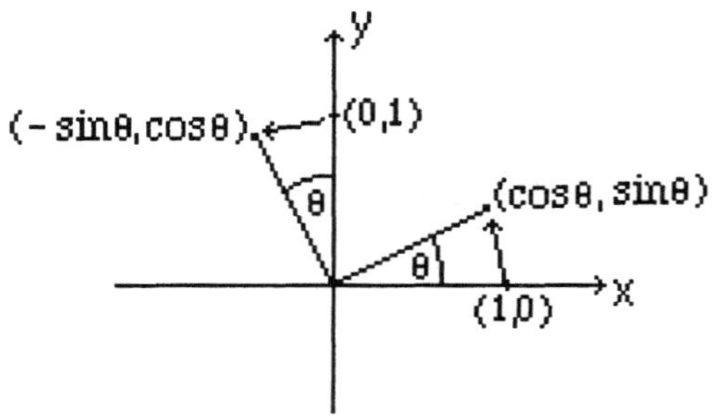

15. Plot a few points to see that T defines a reflection in the line y=x.

$\begin{bmatrix}1\\0\end{bmatrix} \mapsto \begin{bmatrix}0\\1\end{bmatrix}$ and $\begin{bmatrix}0\\1\end{bmatrix} \mapsto \begin{bmatrix}1\\0\end{bmatrix}$, so $A = \begin{bmatrix}0 & 1\\ 1 & 0\end{bmatrix}$. $A\begin{bmatrix}-3\\5\end{bmatrix} = \begin{bmatrix}5\\-3\end{bmatrix}$.

16. $\begin{bmatrix}1\\0\end{bmatrix} \mapsto \begin{bmatrix}1\\0\end{bmatrix}$ and $\begin{bmatrix}0\\1\end{bmatrix} \mapsto \begin{bmatrix}0\\0\end{bmatrix}$, so $A = \begin{bmatrix}1 & 0\\ 0 & 0\end{bmatrix}$.

17. $T\left(\begin{bmatrix}x\\y\end{bmatrix}\right) = \begin{bmatrix}0\\y\end{bmatrix}$, so $\begin{bmatrix}1\\0\end{bmatrix} \mapsto \begin{bmatrix}0\\0\end{bmatrix}$ and $\begin{bmatrix}0\\1\end{bmatrix} \mapsto \begin{bmatrix}0\\1\end{bmatrix}$, and $A = \begin{bmatrix}0 & 0\\ 0 & 1\end{bmatrix}$.

18. $T\left(\begin{bmatrix}x\\y\end{bmatrix}\right) = \begin{bmatrix}\frac{x+y}{2}\\ \frac{x+y}{2}\end{bmatrix}$, so $\begin{bmatrix}1\\0\end{bmatrix} \mapsto \begin{bmatrix}\frac{1}{2}\\ \frac{1}{2}\end{bmatrix}$ and $\begin{bmatrix}0\\1\end{bmatrix} \mapsto \begin{bmatrix}\frac{1}{2}\\ \frac{1}{2}\end{bmatrix}$, and $A = \begin{bmatrix}\frac{1}{2} & \frac{1}{2}\\ \frac{1}{2} & \frac{1}{2}\end{bmatrix}$.

$\begin{bmatrix}4\\2\end{bmatrix} \mapsto \begin{bmatrix}3\\3\end{bmatrix}$.

19. $\begin{bmatrix}1\\0\end{bmatrix} \mapsto \begin{bmatrix}a\\0\end{bmatrix}$ and $\begin{bmatrix}0\\1\end{bmatrix} \mapsto \begin{bmatrix}0\\b\end{bmatrix}$, so $A = \begin{bmatrix}a & 0\\ 0 & b\end{bmatrix}$.

If a = 3 and b = 2, the unit square becomes a 3x2 rectangle.

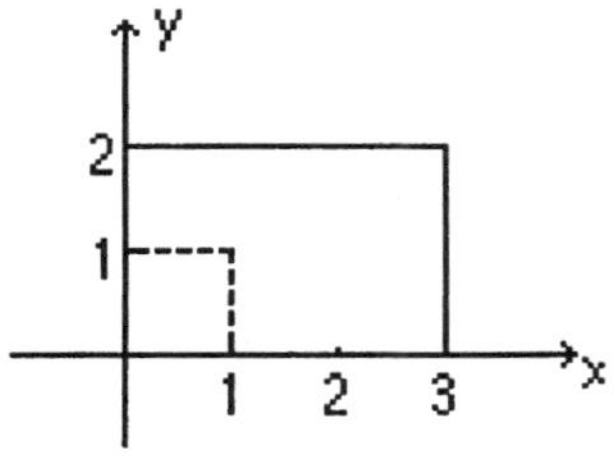

20. $\begin{bmatrix} 2 & 0 \\ 0 & 3 \end{bmatrix}\begin{bmatrix} x \\ y \end{bmatrix} = \begin{bmatrix} 2x \\ 3y \end{bmatrix} = \begin{bmatrix} x' \\ y' \end{bmatrix}$. $y = 2x$, so $\frac{y'}{3} = 2\frac{x'}{2}$, and $y' = 3x'$. Thus the images of

the points on the line $y = 2x$ are the points on the line $y = 3x$.

21. $\begin{bmatrix} 4 & 0 \\ 0 & 3 \end{bmatrix}\begin{bmatrix} x \\ y \end{bmatrix} = \begin{bmatrix} 4x \\ 3y \end{bmatrix} = \begin{bmatrix} x' \\ y' \end{bmatrix}$. $x^2 + y^2 = 1$, so $\left(\frac{x'}{4}\right)^2 + \left(\frac{y'}{3}\right)^2 = 1$. Thus the images of the points on the circle $x^2 + y^2 = 1$ are the points on the ellipse $\frac{x^2}{16} + \frac{y^2}{9} = 1$.

22. $\begin{bmatrix} 1 \\ 0 \end{bmatrix} \mapsto \begin{bmatrix} 1 \\ 0 \end{bmatrix}$ and $\begin{bmatrix} 0 \\ 1 \end{bmatrix} \mapsto \begin{bmatrix} c \\ 1 \end{bmatrix}$, so $A = \begin{bmatrix} 1 & c \\ 0 & 1 \end{bmatrix}$. If $c = 2$, then

$\begin{bmatrix} 1 \\ 0 \end{bmatrix} \mapsto \begin{bmatrix} 1 \\ 0 \end{bmatrix}$, $\begin{bmatrix} 1 \\ 1 \end{bmatrix} \mapsto \begin{bmatrix} 3 \\ 1 \end{bmatrix}$, $\begin{bmatrix} 0 \\ 1 \end{bmatrix} \mapsto \begin{bmatrix} 2 \\ 1 \end{bmatrix}$ and $\begin{bmatrix} 0 \\ 0 \end{bmatrix} \mapsto \begin{bmatrix} 0 \\ 0 \end{bmatrix}$.

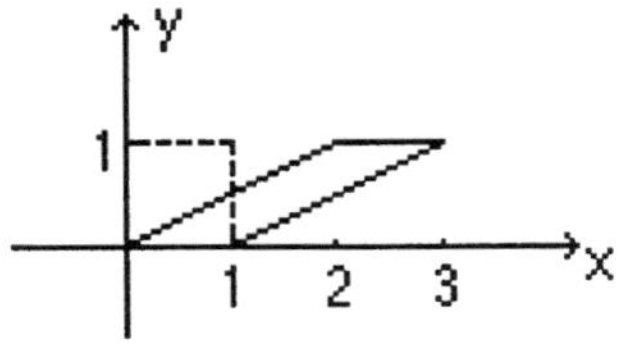

sketch for Exercise 22

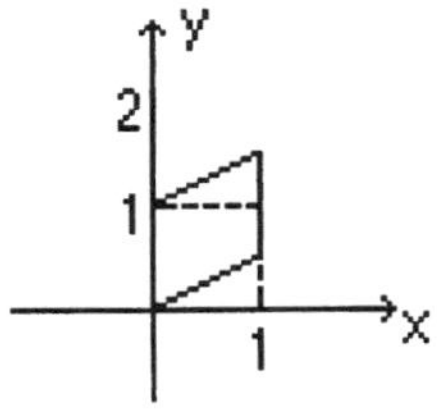

sketch for Exercise 23

23. $T\left(\begin{bmatrix} x \\ y \end{bmatrix}\right) = \begin{bmatrix} x \\ y+cx \end{bmatrix}$, so $\begin{bmatrix} 1 \\ 0 \end{bmatrix} \mapsto \begin{bmatrix} 1 \\ c \end{bmatrix}$ and $\begin{bmatrix} 0 \\ 1 \end{bmatrix} \mapsto \begin{bmatrix} 0 \\ 1 \end{bmatrix}$, and $A = \begin{bmatrix} 1 & 0 \\ c & 1 \end{bmatrix}$.

If $c = .5$, then $\begin{bmatrix} 1 \\ 0 \end{bmatrix} \mapsto \begin{bmatrix} 1 \\ .5 \end{bmatrix}$, $\begin{bmatrix} 1 \\ 1 \end{bmatrix} \mapsto \begin{bmatrix} 1 \\ 1.5 \end{bmatrix}$, $\begin{bmatrix} 0 \\ 1 \end{bmatrix} \mapsto \begin{bmatrix} 0 \\ 1 \end{bmatrix}$ and $\begin{bmatrix} 0 \\ 0 \end{bmatrix} \mapsto \begin{bmatrix} 0 \\ 0 \end{bmatrix}$.

24. $\begin{bmatrix} 1 & 5 \\ 0 & 1 \end{bmatrix}\begin{bmatrix} x \\ y \end{bmatrix} = \begin{bmatrix} x+5y \\ y \end{bmatrix} = \begin{bmatrix} x' \\ y' \end{bmatrix}$. $y = 3x$, so $y' = 3(x' - 5y')$ and $16y' = 3x'$. Thus the

images of the points on the line y = 3x are the points on the line 16y = 3x.

25. (a) $\begin{bmatrix} 1 & 2 \\ 0 & 1 \end{bmatrix}\begin{bmatrix} 3 & 0 \\ 0 & 3 \end{bmatrix} = \begin{bmatrix} 3 & 6 \\ 0 & 3 \end{bmatrix}$ and $\begin{bmatrix} 3 \\ 2 \end{bmatrix} \mapsto \begin{bmatrix} 21 \\ 6 \end{bmatrix}$.

shear dilation

(b) $\begin{bmatrix} 0 & 1 \\ 1 & 0 \end{bmatrix}\begin{bmatrix} 3 & 0 \\ 0 & 2 \end{bmatrix} = \begin{bmatrix} 0 & 2 \\ 3 & 0 \end{bmatrix}$ and $\begin{bmatrix} 3 \\ 2 \end{bmatrix} \mapsto \begin{bmatrix} 4 \\ 9 \end{bmatrix}$.

reflection scaling

(c) $\begin{bmatrix} 0 & -1 \\ 1 & 0 \end{bmatrix}\begin{bmatrix} 1 & 3 \\ 0 & 1 \end{bmatrix}\begin{bmatrix} 2 & 0 \\ 0 & 2 \end{bmatrix} = \begin{bmatrix} 0 & -2 \\ 2 & 6 \end{bmatrix}$ and $\begin{bmatrix} 3 \\ 2 \end{bmatrix} \mapsto \begin{bmatrix} -4 \\ 18 \end{bmatrix}$.

rotation shear dilation

26. $\begin{bmatrix} a & b \\ c & d \end{bmatrix}\begin{bmatrix} 1 \\ 2 \end{bmatrix} = \begin{bmatrix} 7 \\ 3 \end{bmatrix}$, so a + 2b = 7 and c + 2d = 3, and $\begin{bmatrix} a & b \\ c & d \end{bmatrix}\begin{bmatrix} -1 \\ 1 \end{bmatrix} = \begin{bmatrix} 2 \\ 3 \end{bmatrix}$, so

-a +b = 2 and -c +d = 3. Thus a = 1, b = 3, c = -1, and d = 2, and the matrix is

$\begin{bmatrix} a & b \\ c & d \end{bmatrix} = \begin{bmatrix} 1 & 3 \\ -1 & 2 \end{bmatrix}$.

27. The pairs that commute are D and R, D and F, D and S, and D and H.

28. If $\begin{bmatrix} x \\ y \\ z \end{bmatrix} \mapsto \begin{bmatrix} -y \\ x \\ z \end{bmatrix}$ then $\begin{bmatrix} 1 \\ 0 \\ 0 \end{bmatrix} \mapsto \begin{bmatrix} 0 \\ 1 \\ 0 \end{bmatrix}$, $\begin{bmatrix} 0 \\ 1 \\ 0 \end{bmatrix} \mapsto \begin{bmatrix} -1 \\ 0 \\ 0 \end{bmatrix}$, and $\begin{bmatrix} 0 \\ 0 \\ 1 \end{bmatrix} \mapsto \begin{bmatrix} 0 \\ 0 \\ 1 \end{bmatrix}$, so that

$A = \begin{bmatrix} 0 & -1 & 0 \\ 1 & 0 & 0 \\ 0 & 0 & 1 \end{bmatrix}$. If $\begin{bmatrix} x \\ y \\ z \end{bmatrix} \mapsto \begin{bmatrix} y \\ -x \\ z \end{bmatrix}$ then $A = \begin{bmatrix} 0 & 1 & 0 \\ -1 & 0 & 0 \\ 0 & 0 & 1 \end{bmatrix}$.

29. $\begin{bmatrix} 3 & 0 & 0 \\ 0 & 3 & 0 \\ 0 & 0 & 3 \end{bmatrix}$

30. Let $\theta = \frac{\pi}{2}$, h = 5, and k = 1 in the matrix $\begin{bmatrix} \cos\theta & -\sin\theta & -h\cos\theta+k\sin\theta+h \\ \sin\theta & \cos\theta & -h\sin\theta-k\cos\theta+k \\ 0 & 0 & 1 \end{bmatrix}$.

The resulting matrix is $\begin{bmatrix} 0 & -1 & 6 \\ 1 & 0 & -4 \\ 0 & 0 & 1 \end{bmatrix}$, and $\begin{bmatrix} 1 \\ 0 \\ 1 \end{bmatrix} \mapsto \begin{bmatrix} 6 \\ -3 \\ 1 \end{bmatrix}$, $\begin{bmatrix} 1 \\ 1 \\ 1 \end{bmatrix} \mapsto \begin{bmatrix} 5 \\ -3 \\ 1 \end{bmatrix}$,

$\begin{bmatrix} 0 \\ 1 \\ 1 \end{bmatrix} \mapsto \begin{bmatrix} 5 \\ -4 \\ 1 \end{bmatrix}$, and $\begin{bmatrix} 0 \\ 0 \\ 1 \end{bmatrix} \mapsto \begin{bmatrix} 6 \\ -4 \\ 1 \end{bmatrix}$, so that the image of the unit square is the square

with vertices (6,−3), (5,−3), (5,−4), and (6,−4).

31. $T^{-1} = \begin{bmatrix} 1 & 0 & -h \\ 0 & 1 & -k \\ 0 & 0 & 1 \end{bmatrix}$.

32. $S^{-1} = \begin{bmatrix} 1/c & 0 & 0 \\ 0 & 1/d & 0 \\ 0 & 0 & 1 \end{bmatrix}$.

33. $SRT = \begin{bmatrix} 3 & 0 & 0 \\ 0 & 5 & 0 \\ 0 & 0 & 1 \end{bmatrix}\begin{bmatrix} 0 & 1 & 0 \\ -1 & 0 & 0 \\ 0 & 0 & 1 \end{bmatrix}\begin{bmatrix} 1 & 0 & 4 \\ 0 & 1 & -3 \\ 0 & 0 & 1 \end{bmatrix} = \begin{bmatrix} 0 & 3 & -9 \\ -5 & 0 & -20 \\ 0 & 0 & 1 \end{bmatrix}$.

$\begin{bmatrix} 1 \\ 6 \\ 1 \end{bmatrix} \mapsto \begin{bmatrix} 9 \\ -25 \\ 1 \end{bmatrix}$, $\begin{bmatrix} 3 \\ 0 \\ 1 \end{bmatrix} \mapsto \begin{bmatrix} -9 \\ -35 \\ 1 \end{bmatrix}$, and $\begin{bmatrix} 4 \\ 6 \\ 1 \end{bmatrix} \mapsto \begin{bmatrix} 9 \\ -40 \\ 1 \end{bmatrix}$, so that the image of the

given triangle is the triangle with vertices (9,−25), (−9,−35), and (9,−40).

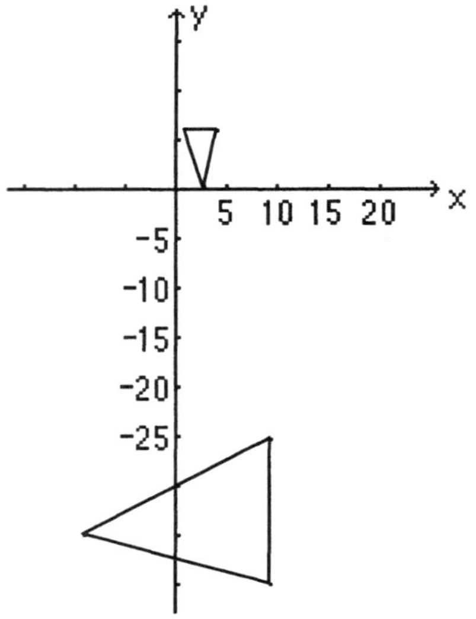

34. (a) For any scalar c and linear transformation T, $T(c\mathbf{v}) = cT(\mathbf{v})$, so with $c = -1$,

$$T(-\mathbf{v}) = T((-1)\mathbf{v}) = (-1)T(\mathbf{v}) = -T(\mathbf{v}).$$

(b) $T(\mathbf{v}-\mathbf{w}) = T(\mathbf{v}+(-1)\mathbf{w}) = T(\mathbf{v})+T((-1)\mathbf{w}) = T(\mathbf{v}) +(-1)T(\mathbf{w}) = T(\mathbf{v}) -T(\mathbf{w})$.

35. If T is linear, then $T(a\mathbf{u} + b\mathbf{v}) = T(a\mathbf{u}) + T(b\mathbf{v}) = aT(\mathbf{u}) + bT(\mathbf{v})$.

If $T(a\mathbf{u} + b\mathbf{v}) = aT(\mathbf{u}) + bT(\mathbf{v})$, then on letting $a = b = 1$, $T(\mathbf{u} + \mathbf{v}) = T(\mathbf{u}) + T(\mathbf{v})$. Further, on letting $c=a$ and $b=0$, $T(c\mathbf{u}) = cT(\mathbf{u})$, so T is linear.

36. $T(c_1\mathbf{v}_1) = c_1T(\mathbf{v}_1)$ and $T(c_1\mathbf{v}_1 + c_2\mathbf{v}_2) = T(c_1\mathbf{v}_1)+T(c_2\mathbf{v}_2) = c_1T(\mathbf{v}_1)+ c_2T(\mathbf{v}_2)$. Thus

$$T(c_1\mathbf{v}_1 + c_2\mathbf{v}_2 + c_3\mathbf{v}_3) = T(c_1\mathbf{v}_1 + c_2\mathbf{v}_2)+T(c_3\mathbf{v}_3) = c_1T(\mathbf{v}_1)+ c_2T(\mathbf{v}_2)+ c_3T(\mathbf{v}_3).$$

Continuing thus, $T(c_1\mathbf{v}_1 +\ldots+ c_{m-1}\mathbf{v}_{m-1} + c_m\mathbf{v}_m) = T(c_1\mathbf{v}_1 +\ldots+ c_{m-1}\mathbf{v}_{m-1})+ T(c_m\mathbf{v}_m) = c_1T(\mathbf{v}_1) +\ldots+ c_{m-1}T(\mathbf{v}_{m-1})+ c_mT(\mathbf{v}_m)$.

Proof by induction: $T(c_1\mathbf{v}_1) = c_1T(\mathbf{v}_1)$ from the definition of linear transformation.

If $T(c_1\mathbf{v}_1 +\ldots+ c_{m-1}\mathbf{v}_{m-1}) = c_1T(\mathbf{v}_1) +\ldots+ c_{m-1}T(\mathbf{v}_{m-1})$, then $T(c_1\mathbf{v}_1 +\ldots+ c_{m-1}\mathbf{v}_{m-1} + c_m\mathbf{v}_m)$

$$= T(c_1\mathbf{v}_1 +\ldots+ c_{m-1}\mathbf{v}_{m-1})+T(c_m\mathbf{v}_m) = c_1T(\mathbf{v}_1) +\ldots+ c_{m-1}T(\mathbf{v}_{m-1})+ c_mT(\mathbf{v}_m).$$

37. Let U, V, and W be vector spaces and $T_1:U \rightarrow V$, $T_2:V \rightarrow W$ be linear transformations. Let **u** and **v** be vectors in U and c be a scalar. We use the linearity of T_1 and T_2 to get

$$T_2 \circ T_1(\mathbf{u}+\mathbf{v}) = T_2(T_1(\mathbf{u}+\mathbf{v})) = T_2(T_1(\mathbf{u}) + T_1(\mathbf{v}))$$
$$= T_2(T_1(\mathbf{u})) + T_2(T_1(\mathbf{v})) = T_2 \circ T_1(\mathbf{u}) + T_2 \circ T_1(\mathbf{v})$$

$T_2 \circ T_1(c\mathbf{u}) = T_2(T_1(c\mathbf{u})) = T_2(cT_1(\mathbf{u})) = cT_2(T_1(\mathbf{u})) = cT_2 \circ T_1(\mathbf{u})$. Thus $T_2 \circ T_1$ is linear.

38. (a) True: $T((x_1, y_1)+(x_2, y_2)) = T(x_1 + x_2, y_1 + y_2) = (x_1 + x_2, y_1 + y_2)$
$= (x_1, y_1)+(x_2, y_2)$.
$T(c(x_1, y_1)) = T(cx_1, cy_1) = (cx_1, cy_1) = c(x_1, y_1)$.
Thus T is linear. T is called the identity transformation.
If vectors are written in column matrix form it is defined by the Identity matrix.

(b) False: e.g., $T(x,y)=x+y$ is linear mapping of $\mathbf{R}^2$ to $\mathbf{R}$. $T(1,2) = T(2,1) = 3$.

(c) False: Any 2x3 matrix defines a linear transformation from $\mathbf{R}^2$ to $\mathbf{R}^3$.

39. Let $T(\mathbf{u}) = \mathbf{u} + \mathbf{w}$, $\mathbf{w} \neq 0$. Then $T(\mathbf{u} + \mathbf{v}) = \mathbf{u} + \mathbf{v} + \mathbf{w}$ and
$T(\mathbf{u})+T(\mathbf{v}) = \mathbf{u} + \mathbf{w} + \mathbf{v} + \mathbf{w} = \mathbf{u} + \mathbf{v} + 2\mathbf{w}$. Thus $T(\mathbf{u} + \mathbf{v}) \neq T(\mathbf{u})+T(\mathbf{v})$.
Let $T(\mathbf{u}) = A\mathbf{u} + \mathbf{w}$, $\mathbf{w} \neq 0$. Then $T(\mathbf{u} + \mathbf{v}) = A(\mathbf{u} + \mathbf{v}) + \mathbf{w}$.
But $T(\mathbf{u}) + T(\mathbf{v}) = A\mathbf{u} + \mathbf{w} + A\mathbf{v} + \mathbf{w} = A(\mathbf{u} + \mathbf{v}) + 2\mathbf{w}$. Thus $T(\mathbf{u} + \mathbf{v}) \neq T(\mathbf{u}) + T(\mathbf{v})$.

Exercise Set 2.7

(Students computing matrix inverses by hand may want to convert the decimal entries of the matrices to fractions, and then do the computation. We give results in fraction form for the convenience of these students. Students using calculators or computers should have no difficulty checking these answers.)

1. (a) $a_{32} = .25$ (b) $a_{21} = .40$ (c) electrical industry ($a_{43} = .30$)

(d) steel industry ($a_{32} = .25$) (e) steel industry ($a_{21} = .40$)

2. $(I - A)^{-1} = \begin{bmatrix} 15/8 & 5/4 \\ 5/6 & 5/3 \end{bmatrix}$. We do all the multiplications at once.

$\begin{bmatrix} 15/8 & 5/4 \\ 5/6 & 5/3 \end{bmatrix}\begin{bmatrix} 24 & 8 & 0 \\ 12 & 6 & 12 \end{bmatrix} = \begin{bmatrix} 60 & 45/2 & 15 \\ 40 & 50/3 & 20 \end{bmatrix}$, so the values of X, in turn, are

$\begin{bmatrix} 60 \\ 40 \end{bmatrix}$, $\begin{bmatrix} 45/2 \\ 50/3 \end{bmatrix}$, and $\begin{bmatrix} 15 \\ 20 \end{bmatrix}$.

3. $(I - A)^{-1} = \begin{bmatrix} 4/3 & 2/3 \\ 1/2 & 3/2 \end{bmatrix}$.

$\begin{bmatrix} 4/3 & 2/3 \\ 1/2 & 3/2 \end{bmatrix}\begin{bmatrix} 6 & 18 & 24 \\ 12 & 6 & 12 \end{bmatrix} = \begin{bmatrix} 16 & 28 & 40 \\ 21 & 18 & 30 \end{bmatrix}$, so the values of X, in turn, are

$\begin{bmatrix} 16 \\ 21 \end{bmatrix}$, $\begin{bmatrix} 28 \\ 18 \end{bmatrix}$, and $\begin{bmatrix} 40 \\ 30 \end{bmatrix}$.

4. $(I - A)^{-1} = \begin{bmatrix} 15/7 & 10/7 \\ 5/6 & 5/3 \end{bmatrix}$.

$\begin{bmatrix} 15/7 & 10/7 \\ 5/6 & 5/3 \end{bmatrix}\begin{bmatrix} 42 & 0 & 14 & 42 \\ 84 & 10 & 7 & 42 \end{bmatrix} = \begin{bmatrix} 210 & 100/7 & 40 & 150 \\ 175 & 50/3 & 70/3 & 105 \end{bmatrix}$, so the values of X,

in turn, are $\begin{bmatrix} 210 \\ 175 \end{bmatrix}$, $\begin{bmatrix} 100/7 \\ 50/3 \end{bmatrix}$, $\begin{bmatrix} 40 \\ 70/3 \end{bmatrix}$, and $\begin{bmatrix} 150 \\ 105 \end{bmatrix}$.

5. $(I - A)^{-1} = \begin{bmatrix} 5/4 & 5/8 & 5/8 \\ 0 & 2 & 1 \\ 0 & 1 & 3 \end{bmatrix}$.

$\begin{bmatrix} 5/4 & 5/8 & 5/8 \\ 0 & 2 & 1 \\ 0 & 1 & 3 \end{bmatrix}\begin{bmatrix} 4 & 0 & 8 \\ 8 & 8 & 24 \\ 8 & 16 & 8 \end{bmatrix} = \begin{bmatrix} 15 & 15 & 30 \\ 24 & 32 & 56 \\ 32 & 56 & 48 \end{bmatrix}$, so the values of X, in turn, are

$\begin{bmatrix} 15 \\ 24 \\ 32 \end{bmatrix}$, $\begin{bmatrix} 15 \\ 32 \\ 56 \end{bmatrix}$, and $\begin{bmatrix} 30 \\ 56 \\ 48 \end{bmatrix}$.

6. $(I - A)^{-1} = \begin{bmatrix} 25/12 & 5/6 & 5/6 \\ 10/3 & 10/3 & 10/3 \\ 35/18 & 10/9 & 25/9 \end{bmatrix}$.

$\begin{bmatrix} 25/12 & 5/6 & 5/6 \\ 10/3 & 10/3 & 10/3 \\ 35/18 & 10/9 & 25/9 \end{bmatrix}\begin{bmatrix} 36 & 36 & 36 & 0 \\ 72 & 0 & 0 & 18 \\ 36 & 18 & 0 & 18 \end{bmatrix} = \begin{bmatrix} 165 & 90 & 75 & 30 \\ 480 & 180 & 120 & 120 \\ 250 & 120 & 70 & 70 \end{bmatrix}$, so the values of X, in turn, are $\begin{bmatrix} 165 \\ 480 \\ 250 \end{bmatrix}$, $\begin{bmatrix} 90 \\ 180 \\ 120 \end{bmatrix}$, $\begin{bmatrix} 75 \\ 120 \\ 70 \end{bmatrix}$, and $\begin{bmatrix} 30 \\ 120 \\ 70 \end{bmatrix}$.

7. $(I - A)X = \begin{bmatrix} .8 & -.4 \\ -.5 & .9 \end{bmatrix}\begin{bmatrix} 8 \\ 10 \end{bmatrix} = \begin{bmatrix} 2.4 \\ 5 \end{bmatrix} = D.$

8. $(I - A)X = \begin{bmatrix} .9 & -.2 & -.3 \\ 0 & .9 & -.4 \\ -.5 & -.4 & .8 \end{bmatrix}\begin{bmatrix} 10 \\ 10 \\ 20 \end{bmatrix} = \begin{bmatrix} 1 \\ 1 \\ 7 \end{bmatrix} = D.$

9. $(I - A)X = \begin{bmatrix} .9 & -.1 & -.2 \\ -.2 & .9 & -.3 \\ -.4 & -.3 & .85 \end{bmatrix}\begin{bmatrix} 6 \\ 4 \\ 5 \end{bmatrix} = \begin{bmatrix} 4 \\ .9 \\ .65 \end{bmatrix} = D.$

10. We assume here that units are monetary values. a_{ij} is the amount of commodity i required to produce one unit of commodity j. Thus no a_{ij} can be negative. If the amount of commodity i required for one unit of commodity j were greater than 1, the value going into commodity j would be greater than the final value, an economically infeasible situation.

11. The sum of the elements of column j of the input-output matrix is the sum of the amounts of all the commodities required to produce one unit of commodity j. This sum will certainly be greater than or equal to zero since all elements are, and in an economically feasible situation, the sum of all that goes into producing one unit of commodity j should be worth less than one unit of commodity j.

12. (a) $A = \begin{bmatrix} a & 1-b \\ 1-a & b \end{bmatrix}$. $I - A = \begin{bmatrix} 1 & 0 \\ 0 & 1 \end{bmatrix} - \begin{bmatrix} a & 1-b \\ 1-a & b \end{bmatrix} = \begin{bmatrix} 1-a & 1-b \\ 1-a & 1-b \end{bmatrix}$.

$$[I-A : I] = \begin{bmatrix} 1-a & 1-b & 1 & 0 \\ 1-a & 1-b & 0 & 1 \end{bmatrix} \approx \begin{bmatrix} 1 & (1-b)/(1-a) & 1/(1-a) & 0 \\ 1-a & 1-b & 0 & 1 \end{bmatrix}$$

$\approx \begin{bmatrix} 1 & (1-b)/(1-a) & 1/(1-a) & 0 \\ 0 & 0 & -1 & 1 \end{bmatrix}$. This matrix cannot be transformed into the form [I : B]

(b) $A = \begin{bmatrix} .2 & .7 \\ .8 & .3 \end{bmatrix}$. $I - A = \begin{bmatrix} 1 & 0 \\ 0 & 1 \end{bmatrix} - \begin{bmatrix} .2 & .7 \\ .8 & .3 \end{bmatrix} = \begin{bmatrix} .8 & .7 \\ .8 & .7 \end{bmatrix}$.

$$[I-A : I] = \begin{bmatrix} .8 & .7 & 1 & 0 \\ .8 & .7 & 0 & 1 \end{bmatrix} \approx \begin{bmatrix} 1 & .875 & .125 & 0 \\ .8 & .7 & 0 & 1 \end{bmatrix} \approx \begin{bmatrix} 1 & .875 & .125 & 0 \\ 0 & 0 & -1 & 1 \end{bmatrix}.$$

Matrix A has no inverse.

(c) The implication is that the equation $X = (I - A)^{-1}D$ has no solution for any value of D where D is non-zero. There are no industrial outputs that can meet the demands of the open sector. All the outputs must go into the industries themselves.

Exercise Set 2.8

1. (a) stochastic

(b) The elements in column 2 are not numbers between zero and 1, so the matrix is not stochastic.

(c) The elements in column 2 do not add up to 1, so the matrix is not stochastic.

(d) stochastic (e) stochastic

(f) There is a negative number in column 3, so the matrix is not stochastic.

2. $$\begin{bmatrix} x & y \\ 1-x & 1-y \end{bmatrix}\begin{bmatrix} u & w \\ 1-u & 1-w \end{bmatrix} = \begin{bmatrix} xu+y(1-u) & xw+y(1-w) \\ (1-x)u+(1-y)(1-u) & (1-x)w+(1-y)(1-w) \end{bmatrix}.$$

The elements of the product are all greater than or equal to zero and the sum of the terms in each column is 1, so the matrix is stochastic.

3. $\begin{bmatrix} 1 & 0 \\ 0 & 1 \end{bmatrix}$ and $\begin{bmatrix} 1/2 & 1/4 & 1/4 \\ 1/4 & 1/2 & 1/4 \\ 1/4 & 1/4 & 1/2 \end{bmatrix}$ are doubly stochastic matrices.

If A and B are doubly stochastic matrices, then AB is stochastic, and A^t and B^t are stochastic, so $B^t A^t = (AB)^t$ is stochastic. Thus AB is doubly stochastic.

4. (a) $p_{21} = 0.1$. (b) $p_{14} = 0.05$.

(c) Largest element in column 5 of P, $p_{25} = 0.35$. Vacant land in 2000 has highest probability (0.35) of becoming office space in 2005.

(d) Largest diagonal element of P is $p_{33}=0.5$. Commercial is the most stable use of land.

5. (a) $P^2 = \begin{bmatrix} .96 & .01 \\ .04 & .99 \end{bmatrix}^2 = \begin{bmatrix} .922 & .0195 \\ .0780 & .9805 \end{bmatrix}$, so the probability of moving from city to suburb in two years is .0780.

(b) $P^3 = \begin{bmatrix} .96 & .01 \\ .04 & .99 \end{bmatrix}^3 = \begin{bmatrix} .8859 & .028525 \\ .1141 & .971475 \end{bmatrix}$, so the probability of moving from suburb to city in three years is .028525.

6. 2008: $\begin{bmatrix} .99 & .02 \\ .01 & .98 \end{bmatrix}\begin{bmatrix} 245 \\ 52 \end{bmatrix} = \begin{bmatrix} 243.59 \\ 53.41 \end{bmatrix}$. $\begin{matrix} Metro: 243.59 million \\ Nonmetro: 53.41 million \end{matrix}$

2009: $\begin{bmatrix} .99 & .02 \\ .01 & .98 \end{bmatrix}\begin{bmatrix} 248.54 \\ 53.46 \end{bmatrix} = \begin{bmatrix} 242.2223 \\ 54.7777 \end{bmatrix}$. $\begin{matrix} Metro: 242.2223 million \\ Nonmetro: 54.7777 million \end{matrix}$

2010: $\begin{bmatrix} .99 & .02 \\ .01 & .98 \end{bmatrix}\begin{bmatrix} 242.2223 \\ 54..7777 \end{bmatrix} = \begin{bmatrix} 240.8956 \\ 56.1044 \end{bmatrix}$. $\begin{matrix} Metro: 240.8956 million \\ Nonmetro: 56.1044 million \end{matrix}$

$\begin{bmatrix} .99 & .02 \\ .01 & .98 \end{bmatrix}^3 = \begin{bmatrix} .9709 & .0582 \\ .0291 & .9418 \end{bmatrix}$; the probability that a person living in a metropolitan area in 2007 will still be living in a metropolitan area in 2010 is .9709.

7. 2008: $\begin{bmatrix} .96 & .01 & .015 \\ .03 & .98 & .005 \\ .01 & .01 & .98 \end{bmatrix}\begin{bmatrix} 82 \\ 163 \\ 52 \end{bmatrix} = \begin{bmatrix} 81.13 \\ 162.46 \\ 53.41 \end{bmatrix}$.

City: 81.13 million, Suburb: 162.46 million, Nonmetro: 53.41 million

2009: $\begin{bmatrix} .96 & .01 & .015 \\ .03 & .98 & .005 \\ .01 & .01 & .98 \end{bmatrix}\begin{bmatrix} 81.13 \\ 162.46 \\ 53.41 \end{bmatrix} = \begin{bmatrix} 80.3106 \\ 161.9118 \\ 54.7777 \end{bmatrix}$.

City: 80.3106 million, Suburb: 161.9118 million, Nonmetro: 54.7777 million

2010: $\begin{bmatrix} .96 & .01 & .015 \\ .03 & .98 & .005 \\ .01 & .01 & .98 \end{bmatrix}\begin{bmatrix} 80.3106 \\ 161.9118 \\ 54.7777 \end{bmatrix} = \begin{bmatrix} 79.5389 \\ 161.3567 \\ 56.1044 \end{bmatrix}$.

City: 79.5389 million, Suburb: 161.3567 million, Nonmetro: 56.1044 million

$\begin{bmatrix} .96 & .01 & .015 \\ .03 & .98 & .005 \\ .01 & .01 & .98 \end{bmatrix}^2 = \begin{bmatrix} .9221 & .0196 & .0292 \\ .0583 & .9608 & .0103 \\ .0197 & .0197 & .9606 \end{bmatrix}$, so the probability that a person living in the city in 2007 will be living in a nonmetropolitan area in 2009 is .0197.

8. $X_1 = 1.01\begin{bmatrix} .96 & .01 \\ .04 & .99 \end{bmatrix}\begin{bmatrix} 82 \\ 163 \end{bmatrix} = \begin{bmatrix} .9696 & .0101 \\ .0404 & .9999 \end{bmatrix}\begin{bmatrix} 82 \\ 163 \end{bmatrix}$. Likewise

$X_2 = \begin{bmatrix} .9696 & .0101 \\ .0404 & .9999 \end{bmatrix}X_1 = \begin{bmatrix} .9696 & .0101 \\ .0404 & .9999 \end{bmatrix}^2\begin{bmatrix} 82 \\ 163 \end{bmatrix}$

and in general $X_n = \begin{bmatrix} .9696 & .0101 \\ .0404 & .9999 \end{bmatrix}^n\begin{bmatrix} 73 \\ 157 \end{bmatrix}$.

$X_3 = \begin{bmatrix} .9696 & .0101 \\ .0404 & .9999 \end{bmatrix}^3\begin{bmatrix} 82 \\ 163 \end{bmatrix} = \begin{bmatrix} 79.6354 \\ 172.7883 \end{bmatrix}$, so the predicted 2010 city population is 79.6354 million and suburban population is 172.7883 million.

9. $X_1 = \begin{bmatrix} 1.012 & 0 \\ 0 & 1.008 \end{bmatrix}\begin{bmatrix} .96 & .01 \\ .04 & .99 \end{bmatrix}\begin{bmatrix} 82 \\ 163 \end{bmatrix} = \begin{bmatrix} .9715 & .0101 \\ .0403 & .9980 \end{bmatrix}\begin{bmatrix} 82 \\ 163 \end{bmatrix}$.

Likewise $X_2 = \begin{bmatrix} .9715 & .0101 \\ .0403 & .9980 \end{bmatrix}X_1 = \begin{bmatrix} .9715 & .0101 \\ .0403 & .9980 \end{bmatrix}^2\begin{bmatrix} 82 \\ 163 \end{bmatrix}$

and in general $X_n = \begin{bmatrix} .9715 & .0101 \\ .0403 & .9980 \end{bmatrix}^n\begin{bmatrix} 82 \\ 163 \end{bmatrix}$.

$X_3=\begin{bmatrix} .9715 & .0101 \\ .0403 & .9980 \end{bmatrix}^3 \begin{bmatrix} 82 \\ 163 \end{bmatrix} = \begin{bmatrix} 80.0754 \\ 171.8365 \end{bmatrix}$, so the predicted 2010 city

population is 80.0754 million and the suburban population is 171.8365 million.

10. $X_{n-1}=A^{-1}X_n$. $A^{-1}=\begin{bmatrix} 1.0421 & -0.0105 \\ -0.0421 & 1.0105 \end{bmatrix}$.

$X_{2006}=\begin{bmatrix} 83.7407 \\ 161.2593 \end{bmatrix}$, $X_{2005}=\begin{bmatrix} 85.5730 \\ 159.4270 \end{bmatrix}$, $X_{2004}=\begin{bmatrix} 87.5016 \\ 157.4984 \end{bmatrix}$ *city* *suburb*

Not stochastic, A^{-1} has negative numbers, but sum of a column is 1.

11. (a) .2 (b) $\begin{bmatrix} 1 & .2 \\ 0 & .8 \end{bmatrix} \begin{bmatrix} 10000 \\ 20000 \end{bmatrix} = \begin{bmatrix} 14000 \\ 16000 \end{bmatrix}$ white collar, manual

12. (a) .6 (b) next generation: $\begin{bmatrix} 1 & .4 \\ 0 & .6 \end{bmatrix} \begin{bmatrix} 10000 \\ 1000 \end{bmatrix} = \begin{bmatrix} 10400 \\ 600 \end{bmatrix}$ nonfarming, farming

$\begin{bmatrix} 1 & .4 \\ 0 & .6 \end{bmatrix}^4 \begin{bmatrix} 10000 \\ 1000 \end{bmatrix} = \begin{bmatrix} 10870.4 \\ 129.6 \end{bmatrix}$, so the distribution after four generations will

be 10870 nonfarming and 130 farming.

(c) $\begin{bmatrix} 1 & .4 \\ 0 & .6 \end{bmatrix}^2 = \begin{bmatrix} 1 & .64 \\ 0 & .36 \end{bmatrix}$, so the probability the grandson will be a farmer is .36.

13. $\begin{bmatrix} .8 & .4 \\ .2 & .6 \end{bmatrix}^4 \begin{bmatrix} 40000 \\ 50000 \end{bmatrix} = \begin{bmatrix} 59488 \\ 30512 \end{bmatrix}$ small, large

14. $\begin{bmatrix} .8 & .2 & .6 \\ .15 & .7 & .3 \\ .05 & .1 & .1 \end{bmatrix} \begin{bmatrix} 2.5 \\ 1.5 \\ .25 \end{bmatrix} = \begin{bmatrix} 2.45 \\ 1.5 \\ .3 \end{bmatrix}$ Democrats, Republicans, Independents

15. AA Aa aa after one generation

$$\begin{bmatrix} 1/2 & 1/4 & 0 \\ 1/2 & 1/2 & 1/2 \\ 0 & 1/4 & 1/2 \end{bmatrix} \begin{matrix} AA \\ Aa \\ aa \end{matrix} \qquad \begin{bmatrix} 1/2 & 1/4 & 0 \\ 1/2 & 1/2 & 1/2 \\ 0 & 1/4 & 1/2 \end{bmatrix} \begin{bmatrix} 1/3 \\ 1/3 \\ 1/3 \end{bmatrix} = \begin{bmatrix} 1/4 \\ 1/2 \\ 1/4 \end{bmatrix} \begin{matrix} AA \\ Aa \\ aa \end{matrix}$$

after two generations

$$\begin{bmatrix} 1/2 & 1/4 & 0 \\ 1/2 & 1/2 & 1/2 \\ 0 & 1/4 & 1/2 \end{bmatrix} \begin{bmatrix} 1/4 \\ 1/2 \\ 1/4 \end{bmatrix} = \begin{bmatrix} 1/4 \\ 1/2 \\ 1/4 \end{bmatrix} \begin{matrix} AA \\ Aa \\ aa \end{matrix}$$

after three generations

$$\begin{bmatrix} 1/2 & 1/4 & 0 \\ 1/2 & 1/2 & 1/2 \\ 0 & 1/4 & 1/2 \end{bmatrix} \begin{bmatrix} 1/4 \\ 1/2 \\ 1/4 \end{bmatrix} = \begin{bmatrix} 1/4 \\ 1/2 \\ 1/4 \end{bmatrix} \begin{matrix} AA \\ Aa \\ aa \end{matrix}$$

16. P is stochastic. Let $P = \begin{bmatrix} x & y \\ 1-x & 1-y \end{bmatrix}$, where $0 \le x \le 1$, $0 \le y \le 1$. Since P is also symmetric, y=1-x. Thus 1-y=x. P is of the form $\begin{bmatrix} x & 1-x \\ 1-x & x \end{bmatrix}$. Probability of staying in state 1 is same as probability of staying in state 2. Also, naturally, probability of moving from state 1 to state 2 is same as that of moving from state 2 to state 1.

Exercise Set 2.9

1. adjacency matrix / distance matrix

(a) $\begin{bmatrix} 0 & 1 & 0 \\ 1 & 0 & 1 \\ 1 & 0 & 0 \end{bmatrix}$ $\qquad$ $\begin{bmatrix} 0 & 1 & 2 \\ 1 & 0 & 1 \\ 1 & 2 & 0 \end{bmatrix}$

adjacency matrix / distance matrix

(b) $\begin{bmatrix} 0 & 1 & 0 & 0 & 0 \\ 1 & 0 & 1 & 0 & 0 \\ 0 & 0 & 0 & 1 & 1 \\ 0 & 1 & 0 & 0 & 0 \\ 0 & 0 & 0 & 0 & 0 \end{bmatrix}$ $\qquad$ $\begin{bmatrix} 0 & 1 & 2 & 3 & 3 \\ 1 & 0 & 1 & 2 & 2 \\ 3 & 2 & 0 & 1 & 1 \\ 2 & 1 & 2 & 0 & 3 \\ x & x & x & x & 0 \end{bmatrix}$

(c) $\begin{bmatrix} 0 & 1 & 1 & 1 & 1 \\ 0 & 0 & 0 & 0 & 0 \\ 0 & 0 & 0 & 0 & 0 \\ 0 & 0 & 0 & 0 & 0 \\ 0 & 0 & 0 & 0 & 0 \end{bmatrix}$ $\qquad$ $\begin{bmatrix} 0 & 1 & 1 & 1 & 1 \\ x & 0 & x & x & x \\ x & x & 0 & x & x \\ x & x & x & 0 & x \\ x & x & x & x & 0 \end{bmatrix}$

(d) $\begin{bmatrix} 0 & 0 & 1 & 1 & 0 \\ 1 & 0 & 1 & 0 & 1 \\ 0 & 1 & 0 & 1 & 0 \\ 0 & 0 & 0 & 0 & 1 \\ 1 & 0 & 0 & 0 & 0 \end{bmatrix}$ $\begin{bmatrix} 0 & 2 & 1 & 1 & 2 \\ 1 & 0 & 1 & 2 & 1 \\ 2 & 1 & 0 & 1 & 2 \\ 2 & 4 & 3 & 0 & 1 \\ 1 & 3 & 2 & 2 & 0 \end{bmatrix}$

2. (a) 2 (b) undefined (c) undefined (d) 4

3. (a) (b) (c) (d) (e)

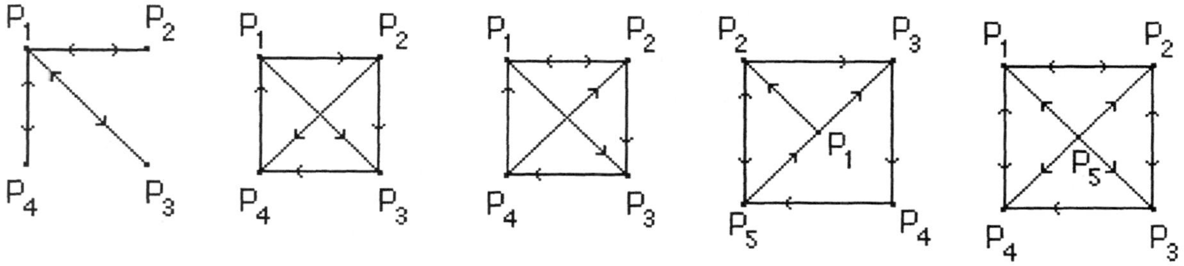

4. adjacency matrix distance matrix

$\begin{bmatrix} 0 & 1 & 0 & 0 & 0 & 0 \\ 0 & 0 & 1 & 0 & 1 & 0 \\ 0 & 0 & 0 & 1 & 0 & 0 \\ 0 & 0 & 0 & 0 & 1 & 0 \\ 0 & 1 & 0 & 0 & 0 & 1 \\ 1 & 0 & 0 & 0 & 0 & 0 \end{bmatrix}$ $\begin{bmatrix} 0 & 1 & 2 & 3 & 2 & 3 \\ 3 & 0 & 1 & 2 & 1 & 2 \\ 4 & 3 & 0 & 1 & 2 & 3 \\ 3 & 2 & 3 & 0 & 1 & 2 \\ 2 & 1 & 2 & 3 & 0 & 1 \\ 1 & 2 & 3 & 4 & 3 & 0 \end{bmatrix}$

5.

	R	B	D	G	I
Raccoon	0	0	0	0	1
Bird	0	0	0	1	1
Deer	0	0	0	1	0
Grass	0	0	0	0	0
Insect	0	0	0	1	0

6.

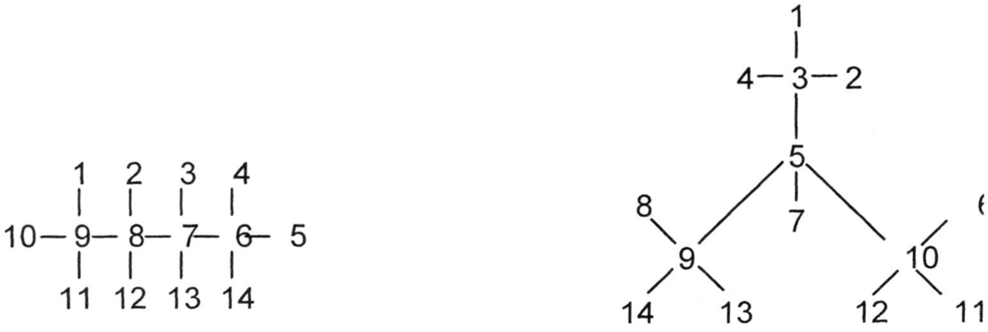

Butane

$$\begin{bmatrix}
0&0&0&0&0&0&0&0&1&0&0&0&0&0\\
0&0&0&0&0&0&0&1&0&0&0&0&0&0\\
0&0&0&0&0&0&1&0&0&0&0&0&0&0\\
0&0&0&0&0&1&0&0&0&0&0&0&0&0\\
0&0&0&0&0&1&0&0&0&0&0&0&0&0\\
0&0&0&1&1&0&1&0&0&0&0&0&0&1\\
0&0&1&0&0&1&0&1&0&0&0&0&1&0\\
0&1&0&0&0&0&1&0&1&0&0&1&0&0\\
1&0&0&0&0&0&0&1&0&1&1&0&0&0\\
0&0&0&0&0&0&0&0&1&0&0&0&0&0\\
0&0&0&0&0&0&0&0&1&0&0&0&0&0\\
0&0&0&0&0&0&0&1&0&0&0&0&0&0\\
0&0&0&0&0&0&1&0&0&0&0&0&0&0\\
0&0&0&0&0&1&0&0&0&0&0&0&0&0
\end{bmatrix}$$

Isobutane

$$\begin{bmatrix}
0&0&1&0&0&0&0&0&0&0&0&0&0&0\\
0&0&1&0&0&0&0&0&0&0&0&0&0&0\\
1&1&0&1&1&0&0&0&0&0&0&0&0&0\\
0&0&1&0&0&0&0&0&0&0&0&0&0&0\\
0&0&1&0&0&0&1&0&1&1&0&0&0&0\\
0&0&0&0&0&0&0&0&0&1&0&0&0&0\\
0&0&0&0&1&0&0&0&0&0&0&0&0&0\\
0&0&0&0&0&0&0&0&1&0&0&0&0&0\\
0&0&0&0&1&0&0&1&0&0&0&0&1&1\\
0&0&0&0&1&1&0&0&0&0&1&1&0&0\\
0&0&0&0&0&0&0&0&0&1&0&0&0&0\\
0&0&0&0&0&0&0&0&0&1&0&0&0&0\\
0&0&0&0&0&0&0&0&1&0&0&0&0&0\\
0&0&0&0&0&0&0&0&1&0&0&0&0&0
\end{bmatrix}$$

7.

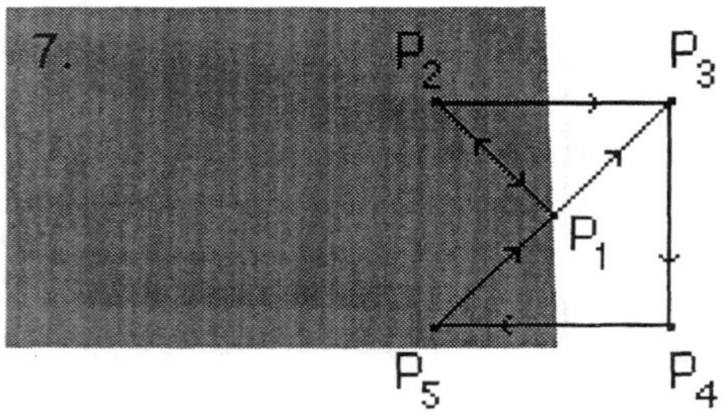

(a) $P_2 \rightarrow P_3 \rightarrow P_4 \rightarrow P_5$

length = 3

(b) $P_3 \rightarrow P_4 \rightarrow P_5 \rightarrow P_1 \rightarrow P_2$

length = 4

distance matrix

$$\begin{bmatrix}
0&1&1&2&3\\
1&0&1&2&3\\
3&4&0&1&2\\
2&3&3&0&1\\
1&2&2&3&0
\end{bmatrix}$$

8. (a) $(a_{22})^2 = 1$. There is one 2-path from P_2 to P_2.

$(a_{24})^2 = 0$. There is no 2-path from P_2 to P_4.

$(a_{31})^2 = 1$. There is one 2-path from P_3 to P_1.

$(a_{42})^2 = 1$. There is one 2-path from P_4 to P_2.

$(a_{12})^3 = 1$. There is one 3-path from P_1 to P_2.

$(a_{24})^3 = 0$. There is no 3-path from P_2 to P_4.

$(a_{32})^3 = 1$. There is one 3-path from P_3 to P_2.

$(a_{41})^3 = 1$. There is one 3-path from P_4 to P_1.

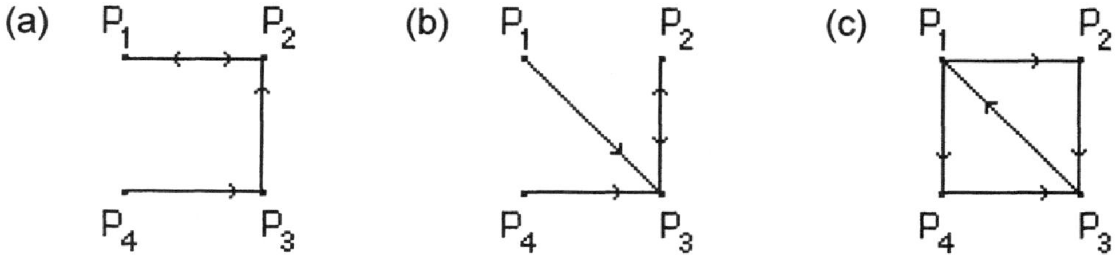

(b) $(a_{21})^2 = 0$. There is no 2-path from P_2 to P_1.

$(a_{33})^2 = 1$. There is one 2-path from P_3 to P_3.

$(a_{42})^2 = 1$. There is one 2-path from P_4 to P_2.

$(a_{13})^3 = 1$. There is one 3-path from P_1 to P_3.

$(a_{32})^3 = 1$. There is one 3-path from P_3 to P_2.

$(a_{34})^3 = 0$. There is no 3-path from P_3 to P_4.

(c) $(a_{13})^2 = 2$. There are two 2-paths from P_1 to P_3.

$(a_{21})^2 = 1$. There is one 2-path from P_2 to P_1.

$(a_{34})^2 = 1$. There is one 2-path from P_3 to P_4.

$(a_{11})^3 = 2$. There are two 3-paths from P_1 to P_1.

$(a_{24})^3 = 1$. There is one 3-path from P_2 to P_4.

$(a_{33})^3 = 2$. There are two 3-paths from P_3 to P_3.

$(a_{42})^3 = 1$. There is one 3-path from P_4 to P_2.

(d) $(a_{14})^2 = 2$. There are two 2-paths from P_1 to P_4.

$(a_{23})^2 = 1$. There is one 2-path from P_2 to P_3.

$(a_{53})^2 = 1$. There is one 2-path from P_5 to P_3.

$(a_{13})^3 = 2$. There are two 3-paths from P_1 to P_3.

$(a_{31})^3 = 0$. There is no 3-path from P_3 to P_1.

(d)

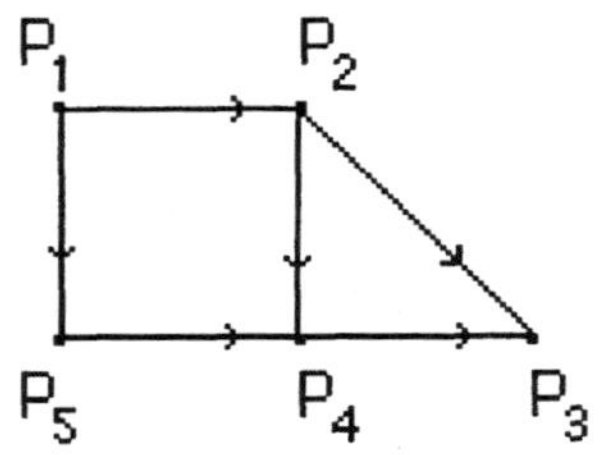

(e)

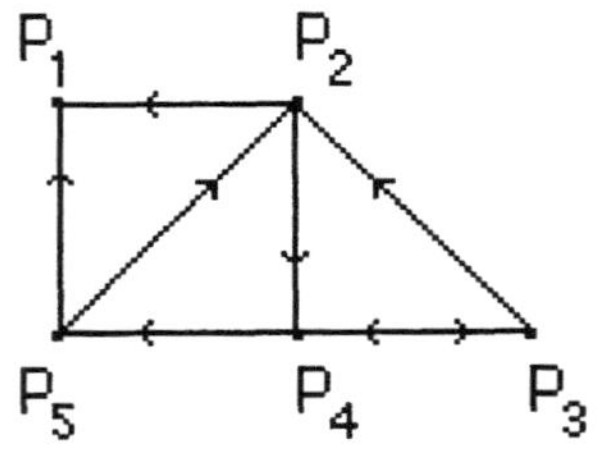

(e) $(a_{25})^2 = 1$. There is one 2-path from P_2 to P_5.
$(a_{31})^2 = 1$. There is one 2-path from P_3 to P_1.
$(a_{42})^2 = 2$. There are two 2-paths from P_4 to P_2.
$(a_{51})^2 = 1$. There is one 2-path from P_5 to P_1.
$(a_{53})^2 = 0$. There is no 2-path from P_5 to P_3.
$(a_{54})^2 = 1$. There is one 2-path from P_5 to P_4.
$(a_{21})^3 = 1$. There is one 3-path from P_2 to P_1.
$(a_{22})^3 = 2$. There are two 3-paths from P_2 to P_2.
$(a_{32})^3 = 2$. There are two 3-paths from P_3 to P_2.
$(a_{34})^3 = 1$. There is one 3-path from P_3 to P_4.
$(a_{35})^3 = 1$. There is one 3-path from P_3 to P_5.
$(a_{41})^3 = 2$. There are two 3-paths from P_4 to P_1.
$(a_{53})^3 = 1$. There is one 3-path from P_5 to P_3.

9. (a) There are no arcs from P_3 to any other vertex.
(b) No arcs lead to P_4.
(c) There are three arcs from P_5.
(d) Two arcs lead to P_2.
(e) There are no 3-paths from P_2.
(f) No 4-paths lead to P_3.

10. (a) There are no arcs from P_2 to any other vertex.
(b) No arcs lead to P_3.

(c) There is an arc from P_4 to every other vertex.
(d) There is an arc from every other vertex to P_5.
(e) There are five arcs from P_3.
(f) Four arcs lead to P_2.
(g) There are seven arcs in the digraph.
(h) There are three 4-paths from P_2.
(i) Four 5-paths lead to P_3.
(j) There are no 2-paths from P_4.
(k) No 4-paths lead to P_3.

11. From A^2 it is known that there is one 2-path from P_1 to P_2, one 2-path from P_2 to P_3, one 2-path from P_3 to P_4, and one 2-path from P_4 to P_2. The possible 2-paths are

$P_1 \to P_3 \to P_2$ or $P_1 \to P_4 \to P_2$

$P_2 \to P_1 \to P_3$ or $P_2 \to P_4 \to P_3$

$P_3 \to P_1 \to P_4$ or $P_3 \to P_2 \to P_4$

$P_4 \to P_1 \to P_2$ or $P_4 \to P_3 \to P_2$

The only combination of these that does not produce additional 2-paths is $P_1 \to P_3 \to P_2$, $P_2 \to P_4 \to P_3$, $P_3 \to P_2 \to P_4$, $P_4 \to P_3 \to P_2$.

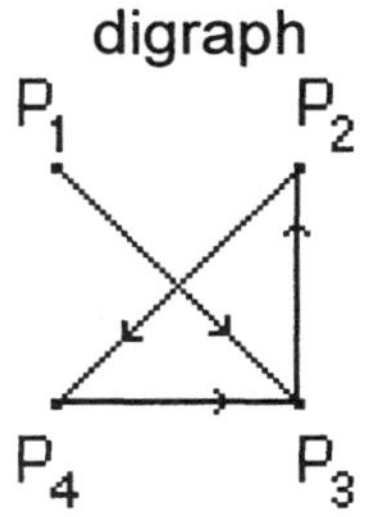

$$A^3 = \begin{bmatrix} 0 & 0 & 0 & 1 \\ 0 & 1 & 0 & 0 \\ 0 & 0 & 1 & 0 \\ 0 & 0 & 0 & 1 \end{bmatrix}$$

12. There is a 2-path from P_1 to P_3 and a 2-path from P_1 to P_4. The possible 2-paths are

$P_1 \to P_2 \to P_3$ or $P_1 \to P_4 \to P_3$

$P_1 \to P_2 \to P_4$ or $P_1 \to P_3 \to P_4$

The only combination of these that does not produce additional 2-paths is $P_1 \to P_2 \to P_3$ and $P_1 \to P_2 \to P_4$. The dashed lines in the digraph indicate arcs that do not cause additional 2-paths and therefore may be in the digraph.

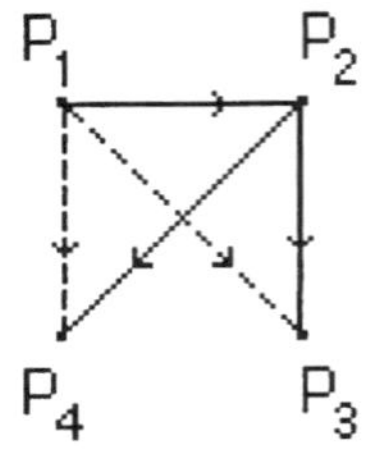

13\. (a)

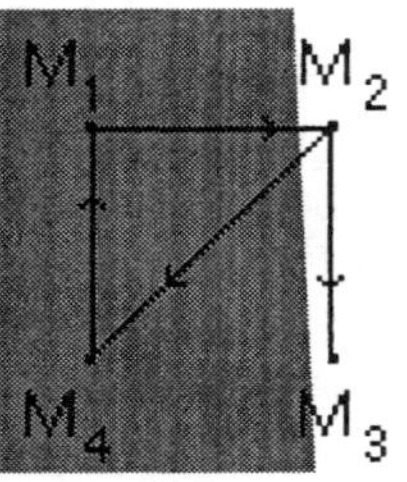

$$D = \begin{bmatrix} 0 & 1 & 2 & 2 \\ 2 & 0 & 1 & 1 \\ x & x & 0 & x \\ 1 & 2 & 3 & 0 \end{bmatrix} \begin{matrix} 5 \\ 4 \\ 3x \\ 6 \end{matrix}$$

most to least influential:
M_2, M_1, M_4, M_3

(b)

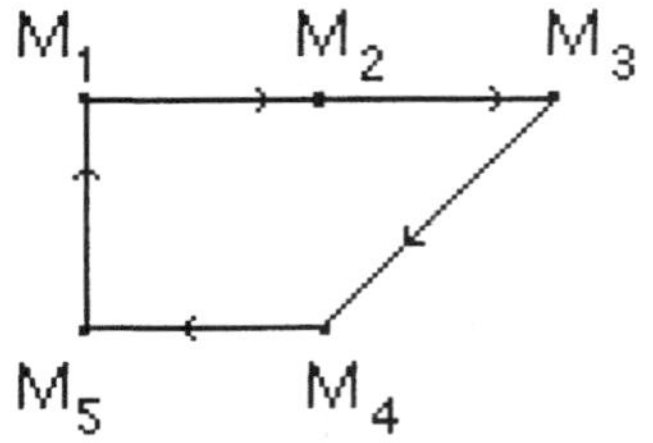

$$D = \begin{bmatrix} 0 & 1 & 2 & 3 & 4 \\ 4 & 0 & 1 & 2 & 3 \\ 3 & 4 & 0 & 1 & 2 \\ 2 & 3 & 4 & 0 & 1 \\ 1 & 2 & 3 & 4 & 0 \end{bmatrix}$$

All are equally influential.

(c)

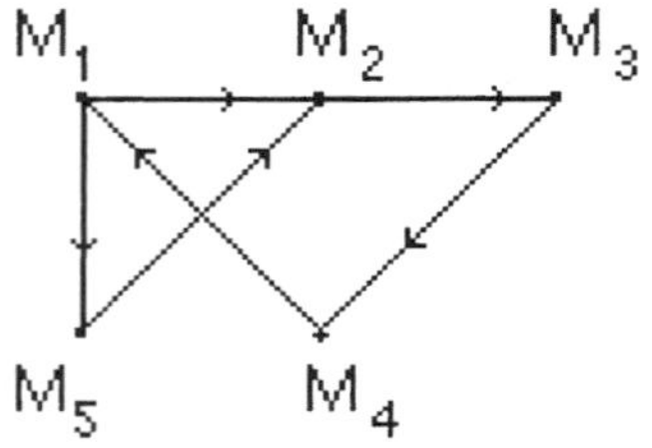

(d)

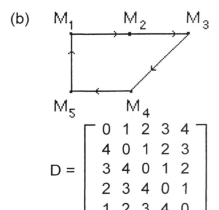

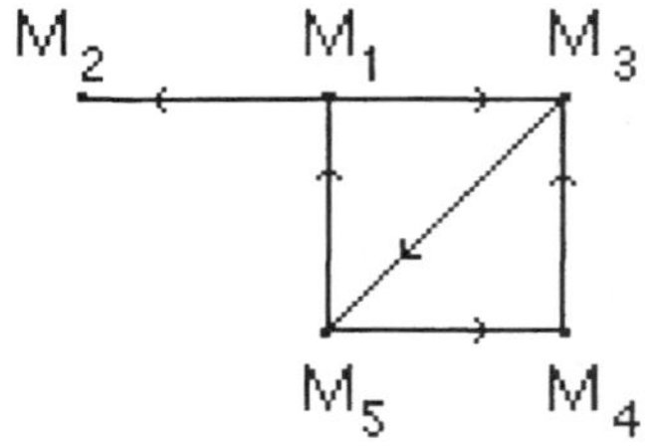

(continued on next page)

(c) $$D = \begin{bmatrix} 0 & 1 & 2 & 3 & 1 \\ 3 & 0 & 1 & 2 & 4 \\ 2 & 3 & 0 & 1 & 3 \\ 1 & 2 & 3 & 0 & 2 \\ 4 & 1 & 2 & 3 & 0 \end{bmatrix} \begin{matrix} 7 \\ 10 \\ 9 \\ 8 \\ 10 \end{matrix}$$

most to least influential:

(d) $$D = \begin{bmatrix} 0 & 1 & 1 & 3 & 2 \\ x & 0 & x & x & x \\ 2 & 3 & 0 & 2 & 1 \\ 3 & 4 & 1 & 0 & 2 \\ 1 & 2 & 2 & 1 & 0 \end{bmatrix} \begin{matrix} 7 \\ 4x \\ 8 \\ 9 \\ 6 \end{matrix}$$

most to least influential:

M_1 , M_4 , M_3 , with M_2 and M_5 equally uninfluential $\qquad$ M_5 , M_1 , M_3 , M_4 , M_2

14. (a)

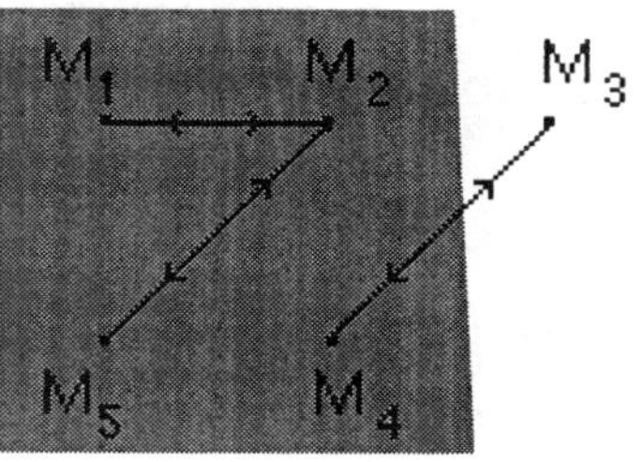

(b)

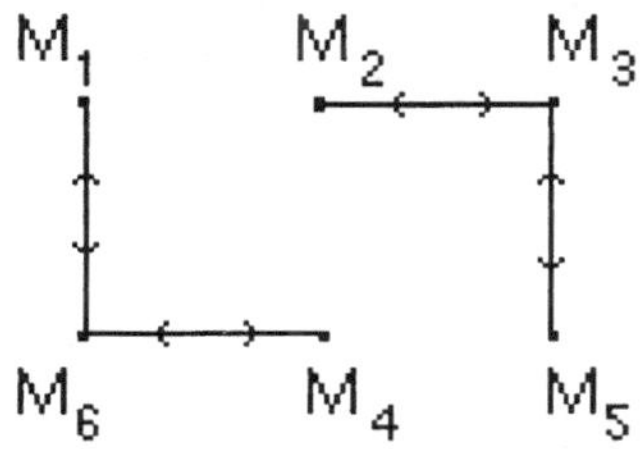

A symmetric matrix implies that all direct friendships are mutual.

15.

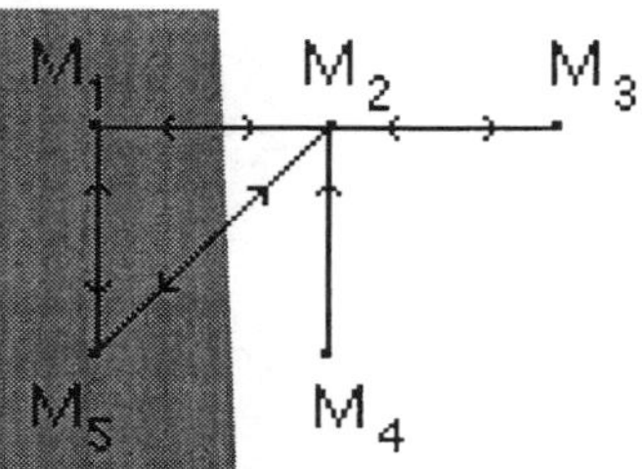

In this digraph there is one clique:

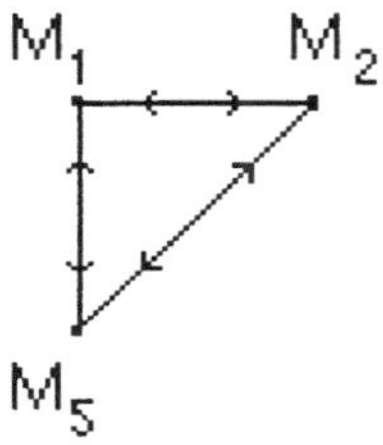

16. The number of rows and the number of columns in the adjacency matrix are both equal to the number of vertices in the digraph and are therefore the same.

17. If the digraph of A contains an arc from P_i to P_j , then the digraph of A^t contains an arc from P_j to P_i . If the digraph of A does not contain an arc from P_i to P_j , then the digraph of A^t does not contain an arc from P_j to P_i .

18. There is an arc from P_i to P_j if and only if there is an arc from P_j to P_i .

19. If a path from P_i to P_j contains P_k twice,

$$P_i \to \ldots \to P_k \to \ldots \to P_k \to \ldots \to P_j ,$$

then there is a shorter path from P_i to P_j obtained by omitting the path from P_k to P_k .

20. n - 1

21. $c_{ij} = a_{i1}a_{j1} + a_{i2}a_{j2} + \ldots + a_{in}a_{jn}$. $a_{ik}a_{jk} = 1$ if station k can receive messages directly from both station i and station j and $a_{ik}a_{jk} = 0$ otherwise. Thus c_{ij} is the number of stations that can receive messages directly from both station i and station j.

22. (a) $\begin{bmatrix} 1 & 1 & 0 & 0 \\ 1 & 1 & 0 & 0 \\ 1 & 1 & 1 & 0 \\ 1 & 1 & 1 & 1 \end{bmatrix}$ (b) $\begin{bmatrix} 1 & 1 & 1 & 0 \\ 0 & 1 & 1 & 0 \\ 0 & 1 & 1 & 0 \\ 0 & 1 & 1 & 1 \end{bmatrix}$ (c) $\begin{bmatrix} 1 & 1 & 1 & 1 \\ 1 & 1 & 1 & 1 \\ 1 & 1 & 1 & 1 \\ 1 & 1 & 1 & 1 \end{bmatrix}$

(d) $\begin{bmatrix} 1 & 1 & 1 & 1 & 1 \\ 0 & 1 & 1 & 1 & 0 \\ 0 & 0 & 1 & 0 & 0 \\ 0 & 0 & 1 & 1 & 0 \\ 0 & 0 & 1 & 1 & 1 \end{bmatrix}$ (e) $\begin{bmatrix} 1 & 0 & 0 & 0 & 0 \\ 1 & 1 & 1 & 1 & 1 \\ 1 & 1 & 1 & 1 & 1 \\ 1 & 1 & 1 & 1 & 1 \\ 1 & 1 & 1 & 1 & 1 \end{bmatrix}$

23. For each vertex, the adjacency matrix gives the vertices reachable by arcs or 1-paths, its square gives all vertices reachable by 2-paths, etc. Since there are n vertices, any vertex can be reached by a path of length at most n - 1 or else it cannot be reached at all. Thus all the information needed will be contained in the first n - 1 powers of the adjacency matrix.

24. (a) Yes. If P_j is reachable from P_i, then P_i is reachable from P_j by reversing the path.

(b) No. The digraph that has adjacency matrix $\begin{bmatrix} 0 & 1 & 0 \\ 0 & 0 & 1 \\ 1 & 1 & 0 \end{bmatrix}$ has reachability matrix $\begin{bmatrix} 1 & 1 & 1 \\ 1 & 1 & 1 \\ 1 & 1 & 1 \end{bmatrix}$.

25. r(i) is the number of stations that can receive a message from station i. c(j) is the number of stations that can send a message to station j.

26. $(r_{ij})^2 = r_{i1}r_{1j} + r_{i2}r_{2j} + \ldots + r_{in}r_{nj}$ is the number of vertices that can be reached from vertex i and that can reach vertex j; that is, the number of vertices that could be on paths from vertex i to vertex j.

27. The (i,j)th element of R^t is 1 if vertex i can be reached from vertex j or if i = j. It is zero if vertex i cannot be reached from vertex j.

28. (a) and (b) No. R has a 1 in every diagonal position and A has a zero in every diagonal position, and every 1 in A causes at least one 1 in R, so there cannot be as many ones in A as in R.

(c) Yes. If $A = \begin{bmatrix} 0 & 1 & 0 \\ 0 & 0 & 1 \\ 0 & 0 & 0 \end{bmatrix}$, then $R = \begin{bmatrix} 1 & 1 & 1 \\ 0 & 1 & 1 \\ 0 & 0 & 1 \end{bmatrix}$. In fact R will always contain more 1s than A.

Chapter 2 Review Exercises

1. (a) $2AB = \begin{bmatrix} 28 & 0 \\ 100 & -6 \end{bmatrix}$. (b) AB + C does not exist.

(c) $BA + AB = \begin{bmatrix} 28 & 0 \\ 69 & -6 \end{bmatrix}$. (d) $AD - 3D = \begin{bmatrix} -6 \\ 26 \end{bmatrix}$.

(e) $AC + BC = \begin{bmatrix} 54 & -9 & 27 \\ 46 & -6 & 14 \end{bmatrix}$. (f) 2DA + B does not exist.

2. (a) 2x2 (b) 2x3 (c) 2x2 (d) Does not exist. (e) 3x2

(f) Does not exist. (g) 2x2

3. (a) $d_{12} = 2(1x2 + (-3)x0) - 3(-4) = 16$.

(b) $d_{23} = 2(0x(-3) + 4x(-1)) - 3x0 = -8$.

4. (a) $AB_1 = \begin{bmatrix} 3 & 1 \\ 7 & 2 \end{bmatrix}\begin{bmatrix} 1 \\ 2 \end{bmatrix} = \begin{bmatrix} 5 \\ 11 \end{bmatrix}$, $AB_2 = \begin{bmatrix} 3 & 1 \\ 7 & 2 \end{bmatrix}\begin{bmatrix} 3 \\ 0 \end{bmatrix} = \begin{bmatrix} 9 \\ 21 \end{bmatrix}$,

$AB_3 = \begin{bmatrix} 3 & 1 \\ 7 & 2 \end{bmatrix}\begin{bmatrix} 6 \\ -1 \end{bmatrix} = \begin{bmatrix} 17 \\ 40 \end{bmatrix}$. $AC = \begin{bmatrix} 5 & 9 & 17 \\ 11 & 21 & 40 \end{bmatrix}$.

(b) $PQ = 2\begin{bmatrix}1\\5\end{bmatrix} - 3\begin{bmatrix}2\\-1\end{bmatrix} + 5\begin{bmatrix}3\\4\end{bmatrix}$

(c) $B = \left[\begin{array}{c|c|c} 2 & 4 & 2 \\ -1 & 3 & 7 \\ \hline 0 & 1 & -2 \end{array}\right], \left[\begin{array}{c|cc} 2 & 4 & 2 \\ -1 & 3 & 7 \\ \hline 0 & 1 & -2 \end{array}\right], \left[\begin{array}{cc|c} 2 & 4 & 2 \\ -1 & 3 & 7 \\ \hline 0 & 1 & -2 \end{array}\right]$, or $\left[\begin{array}{ccc} 2 & 4 & 2 \\ -1 & 3 & 7 \\ \hline 0 & 1 & -2 \end{array}\right]$

5. (a) $(A^t)^2 = (A^2)^t = \begin{bmatrix}9 & 5\\0 & 4\end{bmatrix}^t = \begin{bmatrix}9 & 0\\5 & 4\end{bmatrix}$.

(b) $A^t - B^2 = \begin{bmatrix}3 & 0\\1 & 2\end{bmatrix} - \begin{bmatrix}7 & -1\\-3 & 4\end{bmatrix} = \begin{bmatrix}-4 & 1\\4 & -2\end{bmatrix}$.

(c) $AB^3 - 2C^2 = \begin{bmatrix}3 & 1\\0 & 2\end{bmatrix}\begin{bmatrix}-17 & 6\\18 & 1\end{bmatrix} - 2\begin{bmatrix}3 & 2\\6 & 7\end{bmatrix} = \begin{bmatrix}-39 & 15\\24 & -12\end{bmatrix}$.

(d) $A^2 - 3A + 4I_2 = \begin{bmatrix}9 & 5\\0 & 4\end{bmatrix} - \begin{bmatrix}9 & 3\\0 & 6\end{bmatrix} + \begin{bmatrix}4 & 0\\0 & 4\end{bmatrix} = \begin{bmatrix}4 & 2\\0 & 2\end{bmatrix}$.

6. e.g., $X_3 = \begin{bmatrix}5\\3\\-1\\2\end{bmatrix} + \begin{bmatrix}3\\-1\\1\\2\end{bmatrix} = \begin{bmatrix}8\\2\\0\\4\end{bmatrix}$. $X_4 = 2\begin{bmatrix}5\\3\\-1\\2\end{bmatrix} = \begin{bmatrix}10\\6\\-2\\4\end{bmatrix}$. $X_5 = 3\begin{bmatrix}3\\-1\\1\\2\end{bmatrix} = \begin{bmatrix}9\\-3\\3\\6\end{bmatrix}$. $X_6 = X_4 + X_5 = \begin{bmatrix}10\\6\\-2\\4\end{bmatrix} + \begin{bmatrix}9\\-3\\3\\6\end{bmatrix} = \begin{bmatrix}19\\3\\1\\10\end{bmatrix}$.

The linear combinations $a\begin{bmatrix}5\\3\\-1\\2\end{bmatrix} + b\begin{bmatrix}3\\-1\\1\\2\end{bmatrix}$ are solutions for all values of a and b.

Want x=1,y=9 in solution. Thus 5a+3b=1. 3a-b=9. Gives a=2, b=-3.

The solution is $2\begin{bmatrix}5\\3\\-1\\2\end{bmatrix} - 3\begin{bmatrix}3\\-1\\1\\2\end{bmatrix} = \begin{bmatrix}1\\9\\-5\\-2\end{bmatrix}$.

7. $\begin{bmatrix} 1 & 1 & 2 & 2 & 0 \\ 1 & 2 & 6 & 1 & 0 \\ 3 & 2 & 2 & 7 & 0 \end{bmatrix} \underset{\substack{R2-R1 \\ R3-3R1}}{\sim} \begin{bmatrix} 1 & 1 & 2 & 2 & 0 \\ 0 & 1 & 4 & -1 & 0 \\ 0 & -1 & -4 & 1 & 0 \end{bmatrix} \underset{\substack{R1-R2 \\ R3+R2}}{\sim} \begin{bmatrix} 1 & 0 & -2 & 3 & 0 \\ 0 & 1 & 4 & -1 & 0 \\ 0 & 0 & 0 & 0 & 0 \end{bmatrix}$

Thus $x_1 - 2x_3 + 3x_4 = 0$ and $x_2 + 4x_3 - x_4 = 0$.
General solution is $x_1 = 2r - 3s$, $x_2 = -4r + s$, $x_3 = r$, $x_4 = s$.
Write as (2r - 3s, -4r + s, r, s), or r(2, -4, 1, 0) + s(-3, 1, 0, 1).
Subspace has basis {(2, -4, 1, 0), (-3, 1, 0, 1)}. Dimension is 2.

8. (a) Corresponding system of homogeneous linear equations:
$$\begin{aligned} x_1 + 2x_2 - 8x_3 &= 0 \\ x_2 - 3x_3 &= 0 \\ x_1 + x_2 - 5x_3 &= 0 \end{aligned}$$
General solution of nonhomogeneous system is (2r+4, 3r+2, r) = (2r, 3r, r) + (4, 2, 0).
Thus general solution of homogeneous system is (2r, 3r, r). (Check that it satisfies equs.)

(b) Homog system: Write general solution as r(2, 3, 1). Basis {(2, 3, 1)}.
Subspace of solutions is a line through the origin defined by the vector (2, 3, 1).

(c) Nonhomog system: (2r+4, 3r+2, r) = r(2, 3, 1) + (4, 2, 0). Solution is the line defined by the vector (2, 3, 1) slid in a manner defined by the vector (4, 2, 0). i.e. it is the line through the point (4, 2, 0) parallel to the line defined by the vector (2, 3, 1).

9. (a) $\begin{bmatrix} 1 & 4 & 1 & 0 \\ 2 & -1 & 0 & 1 \end{bmatrix} \underset{R2+(-2)R1}{\approx} \begin{bmatrix} 1 & 4 & 1 & 0 \\ 0 & -9 & -2 & 1 \end{bmatrix} \underset{(-1/9)R2}{\approx} \begin{bmatrix} 1 & 4 & 1 & 0 \\ 0 & 1 & 2/9 & -1/9 \end{bmatrix}$

$\underset{R1+(-4)R2}{\approx} \begin{bmatrix} 1 & 0 & 1/9 & 4/9 \\ 0 & 1 & 2/9 & -1/9 \end{bmatrix}$, so the inverse is $\begin{bmatrix} 1/9 & 4/9 \\ 2/9 & -1/9 \end{bmatrix}$.

(b) $\begin{bmatrix} 0 & 3 & 3 & 1 & 0 & 0 \\ 1 & 2 & 3 & 0 & 1 & 0 \\ 1 & 4 & 6 & 0 & 0 & 1 \end{bmatrix} \underset{R1 \Leftrightarrow R2}{\approx} \begin{bmatrix} 1 & 2 & 3 & 0 & 1 & 0 \\ 0 & 3 & 3 & 1 & 0 & 0 \\ 1 & 4 & 6 & 0 & 0 & 1 \end{bmatrix}$

$\underset{R3+(-1)R1}{\approx} \begin{bmatrix} 1 & 2 & 3 & 0 & 1 & 1 \\ 0 & 3 & 3 & 1 & 0 & 0 \\ 0 & 2 & 3 & 0 & -1 & 1 \end{bmatrix} \underset{(1/3)R2}{\approx} \begin{bmatrix} 1 & 2 & 3 & 0 & 1 & 0 \\ 0 & 1 & 1 & 1/3 & 0 & 0 \\ 0 & 2 & 3 & 0 & -1 & 1 \end{bmatrix}$

$$\approx_{\substack{R1+(-2)R2\\R3+(-2)R2}} \begin{bmatrix} 1 & 0 & 1 & -2/3 & 1 & 0 \\ 0 & 1 & 1 & 1/3 & 0 & 0 \\ 0 & 0 & 1 & -2/3 & -1 & 1 \end{bmatrix} \approx_{\substack{R1+(-1)R3\\R2+(-1)R3}} \begin{bmatrix} 1 & 0 & 0 & 0 & 2 & -1 \\ 0 & 1 & 0 & 1 & 1 & -1 \\ 0 & 0 & 1 & -2/3 & -1 & 1 \end{bmatrix},$$

so the inverse is $\begin{bmatrix} 0 & 2 & -1 \\ 1 & 1 & -1 \\ -2/3 & -1 & 1 \end{bmatrix}$.

(c) $$\begin{bmatrix} 1 & 2 & 3 & 1 & 0 & 0 \\ 2 & 5 & 3 & 0 & 1 & 0 \\ 1 & 0 & 8 & 0 & 0 & 1 \end{bmatrix} \approx_{\substack{R2+(-2)R1\\R3+(-1)R1}} \begin{bmatrix} 1 & 2 & 3 & 1 & 0 & 0 \\ 0 & 1 & -3 & -2 & 1 & 0 \\ 0 & -2 & 5 & -1 & 0 & 1 \end{bmatrix}$$

$$\approx_{\substack{R1+(-2)R2\\R3+(2)R2}} \begin{bmatrix} 1 & 0 & 9 & 5 & -2 & 0 \\ 0 & 1 & -3 & -2 & 1 & 0 \\ 0 & 0 & -1 & -5 & 2 & 1 \end{bmatrix} \approx_{(-1)R3} \begin{bmatrix} 1 & 0 & 9 & 5 & -2 & 0 \\ 0 & 1 & -3 & -2 & 1 & 0 \\ 0 & 0 & 1 & 5 & -2 & -1 \end{bmatrix}$$

$$\approx_{\substack{R1+(-9)R3\\R2+(3)R3}} \begin{bmatrix} 1 & 0 & 0 & -40 & 16 & 9 \\ 0 & 1 & 0 & 13 & -5 & -3 \\ 0 & 0 & 1 & 5 & -2 & -1 \end{bmatrix},$$ so the inverse is $\begin{bmatrix} -40 & 16 & 9 \\ 13 & -5 & -3 \\ 5 & -2 & -1 \end{bmatrix}$.

10. The inverse of the coefficient matrix is $\begin{bmatrix} 14 & -8 & -1 \\ -17 & 10 & 1 \\ -19 & 11 & 1 \end{bmatrix}$.

$$\begin{bmatrix} x_1 \\ x_2 \\ x_3 \end{bmatrix} = \begin{bmatrix} 14 & -8 & -1 \\ -17 & 10 & 1 \\ -19 & 11 & 1 \end{bmatrix} \begin{bmatrix} 1 \\ 5 \\ 7 \end{bmatrix} = \begin{bmatrix} -33 \\ 40 \\ 43 \end{bmatrix}.$$

11. $A^{-1} = \frac{1}{3}\begin{bmatrix} 5 & -6 \\ -2 & 3 \end{bmatrix}$, $A = 3\begin{bmatrix} 5 & -6 \\ -2 & 3 \end{bmatrix}^{-1} = 3\begin{bmatrix} 1 & 2 \\ 2/3 & 5/3 \end{bmatrix} = \begin{bmatrix} 3 & 6 \\ 2 & 5 \end{bmatrix}$.

12. The (i,j)th element of A(BC) is $A_i(BC_j)$, where A_i is the ith row of A and C_j is the jth column of C. Likewise, the (i,j)th element of (AB)C is $(A_i B)C_j$. We show that these elements are the same.

$$A_i(BC_j) = [\,a_{i1} \quad a_{i2} \quad \ldots\ a_{in}]\begin{bmatrix} b_{11}c_{1j} + b_{12}c_{2j} + \ldots + b_{1r}c_{rj} \\ b_{21}c_{1j} + b_{22}c_{2j} + \ldots + b_{2r}c_{rj} \\ \vdots \\ b_{n1}c_{1j} + b_{n2}c_{2j} + \ldots + b_{nr}c_{rj} \end{bmatrix}$$

$= a_{i1}(b_{11}c_{ij} + b_{12}c_{2j} + \ldots + b_{1r}c_{rj}) + a_{i2}(b_{21}c_{1j} + b_{22}c_{2j} + \ldots + b_{2r}c_{rj}) + \ldots$
$+ a_{in}(b_{n1}c_{1j} + b_{n2}c_{2j} + \ldots + b_{nr}c_{rj})$

$= (a_{i1}b_{11} + a_{i2}b_{21} + \ldots a_{in}b_{n1})c_{1j} + (a_{i1}b_{12} + a_{i2}b_{22} + a_{in}b_{n2})c_{2j} + \ldots$
$+ (a_{i1}b_{1r} + a_{i2}b_{2r} + \ldots + a_{in}b_{nr})c_{rj}$

$= (A_i\, B)C_j$.

13. (a) R1<->R3: $\begin{bmatrix} 0 & 0 & 1 \\ 0 & 1 & 0 \\ 1 & 0 & 0 \end{bmatrix}$. -4R2+R1: $\begin{bmatrix} 1 & -4 & 0 \\ 0 & 1 & 0 \\ 0 & 0 & 1 \end{bmatrix}$.

(b) Add 2 times row1 to row 3. Interchange rows 2 and 3.

14. $(cA)^n = (cA)(cA)\ldots(cA) = c^n A^n$.

15. The ith diagonal element of AA^t is

$$a_{i1}\, a_{i1} + a_{i2}\, a_{i2} + \ldots + a_{in}a_{in} = (a_{i1})^2 + (a_{i2})^2 + \ldots + (a_{in})^2 .$$

Since each term is a square, the sum can equal zero only if each term is zero, i.e., if each $a_{ij} = 0$. Thus, if $AA^t = O$ then $A = O$.

16. If A is a symmetric matrix then $A = A^t$, so $AA^t = A^2 = A^t A$ and A is normal.

17. If $A = A^2$ then $A^t = (A^2)^t = (A^t)^2$, so A^t is idempotent.

18. From Exercise 7 in Section 2.3, $(A^t)^n = (A^n)^t$. If $n < p$, $A^n \neq O$, so $(A^n)^t \neq O$, so $(A^t)^n \neq O$. $A^p = O$ so $(A^p)^t = (A^t)^p = O$. Thus A^t is nilpotent with degree of nilpotency= p.

19. $A = A^t$, so $A^{-1} = (A^t)^{-1} = (A^{-1})^t$, so A^{-1} is symmetric.

20. If row i of A is all zeros, then row i of AB is all zeros for any matrix B. The ith diagonal term of I_n is 1, so there is no matrix B for which $AB = I_n$. Likewise, if column j of A is all zeros, then column j of BA is all zeros for any B, so there is no matrix B with $BA = I_n$.

21. $\begin{bmatrix} 1 & 2 & -3 \\ 0 & 4 & 1 \end{bmatrix}\begin{bmatrix} 3 \\ 0 \\ 1 \end{bmatrix} = \begin{bmatrix} 0 \\ 1 \end{bmatrix}$. $\begin{bmatrix} 1 & 2 & -3 \\ 0 & 4 & 1 \end{bmatrix}\begin{bmatrix} -1 \\ 4 \\ 2 \end{bmatrix} = \begin{bmatrix} 1 \\ 18 \end{bmatrix}$.

22. (a) $T((x_1, y_1)+(x_2, y_2)) = T(x_1+x_2, y_1+y_2) = (2(x_1+x_2), y_1+y_2, y_1+y_2-x_1-x_2)$

$= (2x_1+2x_2, y_1+y_2, y_1-x_1+y_2-x_2) = (2x_1, y_1, y_1-x_1) + (2x_2, y_2, y_2-x_2)$
$= T(x_1, y_1) + T(x_2, y_2)$

and $T(c(x,y)) = T(cx,cy) = (2cx,cy,cy-cx) = c(2x,y,y-x) = cT(x,y)$, so T is linear.

(b) $T(c(x,y)) = T(cx,cy) = (cx+cy, 2cy+3)$. $cT(x,y) = c(x+y, 2y+3) = (cx+cy, 2cy+3c)$

$\neq T(c(x,y))$, so T is not linear.

23. $\begin{bmatrix} a & b \\ c & d \end{bmatrix}\begin{bmatrix} 1 \\ 2 \end{bmatrix} = \begin{bmatrix} 5 \\ 1 \end{bmatrix}$, so $a + 2b = 5$ and $c + 2d = 1$, and $\begin{bmatrix} a & b \\ c & d \end{bmatrix}\begin{bmatrix} 3 \\ -2 \end{bmatrix} = \begin{bmatrix} -1 \\ 11 \end{bmatrix}$, so

$3a - 2b = -1$ and $3c - 2d = 11$. Thus $a = 1$, $b = 2$, $c = 3$, and $d = -1$, and the matrix is

$\begin{bmatrix} a & b \\ c & d \end{bmatrix} = \begin{bmatrix} 1 & 2 \\ 3 & -1 \end{bmatrix}$.

24. $$\begin{bmatrix} 3 & 0 \\ 0 & 3 \end{bmatrix}\begin{bmatrix} \frac{\sqrt{3}}{2} & \frac{-1}{2} \\ \frac{1}{2} & \frac{\sqrt{3}}{2} \end{bmatrix} = \begin{bmatrix} \frac{3\sqrt{3}}{2} & \frac{-3}{2} \\ \frac{3}{2} & \frac{3\sqrt{3}}{2} \end{bmatrix}.$$

25. $\begin{bmatrix}1\\0\end{bmatrix} \mapsto \begin{bmatrix}0\\-1\end{bmatrix}$ and $\begin{bmatrix}0\\1\end{bmatrix} \mapsto \begin{bmatrix}-1\\0\end{bmatrix}$, so $A = \begin{bmatrix}0 & -1\\-1 & 0\end{bmatrix}$.

26. $\begin{bmatrix}1\\0\end{bmatrix} \mapsto \begin{bmatrix}\frac{1}{2}\\ \frac{-1}{2}\end{bmatrix}$ and $\begin{bmatrix}0\\1\end{bmatrix} \mapsto \begin{bmatrix}\frac{-1}{2}\\ \frac{1}{2}\end{bmatrix}$, so $A = \begin{bmatrix}\frac{1}{2} & \frac{-1}{2}\\ \frac{-1}{2} & \frac{1}{2}\end{bmatrix}$.

27. $\begin{bmatrix}5 & 0\\0 & 2\end{bmatrix}\begin{bmatrix}x\\y\end{bmatrix} = \begin{bmatrix}5x\\2y\end{bmatrix} = \begin{bmatrix}x'\\y'\end{bmatrix}$. $y = -5x + 1$, so $\frac{y'}{2} = -5\frac{x'}{5} + 1$. Thus the images of the points on the line $y = -5x + 1$ are the points on the line $\frac{y}{2} = -x + 1$.

28. $\begin{bmatrix}1 & 0\\3 & 1\end{bmatrix}\begin{bmatrix}x\\y\end{bmatrix} = \begin{bmatrix}x\\3x+y\end{bmatrix} = \begin{bmatrix}x'\\y'\end{bmatrix}$. $y = 2x + 3$, so $y' - 3x' = 2x' + 3$ and $y' = 5x' + 3$. Thus the images of the points on the line $y = 2x + 3$ are the points on the line $y = 5x + 3$.

29. $\begin{bmatrix}-1 & 0\\0 & 1\end{bmatrix}\begin{bmatrix}2 & 0\\0 & 1\end{bmatrix}\begin{bmatrix}1 & 0\\3 & 1\end{bmatrix} = \begin{bmatrix}-2 & 0\\3 & 1\end{bmatrix}$.

30. $A + B = \begin{bmatrix}5+i & 5-5i\\6+10i & -3+i\end{bmatrix}$. $AB = \begin{bmatrix}35+24i & -3+6i\\7+6i & 12-6i\end{bmatrix}$.

$\overline{A} = \begin{bmatrix}2 & 4+3i\\4-3i & -1\end{bmatrix}$, so $A^* = \begin{bmatrix}2 & 4-3i\\4+3i & -1\end{bmatrix} = A$; i.e., A is hermitian.

31. If A is a real symmetric matrix, then $\overline{A} = A = A^t$ so $A^* = A^t = A$. Thus A is hermitian.

32. $G = AA^t = \begin{bmatrix}1 & 1 & 0 & 0 & 0\\1 & 2 & 0 & 0 & 1\\0 & 0 & 2 & 1 & 1\\0 & 0 & 1 & 1 & 0\\0 & 1 & 1 & 0 & 2\end{bmatrix}$ $P = A^tA = \begin{bmatrix}2 & 1 & 0 & 0\\1 & 2 & 1 & 0\\0 & 1 & 2 & 1\\0 & 0 & 1 & 2\end{bmatrix}$

$g_{12} = 1, g_{13} = 0, g_{14} = 0, g_{15} = 0,$
$g_{23} = 0, g_{24} = 0, g_{25} = 1,$
$g_{34} = 1, g_{35} = 1, g_{45} = 0,$
so $1 \to 2 \to 5 \to 3 \to 4$
or $4 \to 3 \to 5 \to 2 \to 1$.

$p_{12} = 1, p_{13} = 0, p_{14} = 0,$
$p_{23} = 1, p_{24} = 0, p_{34} = 1,$
so $1 \to 2 \to 3 \to 4$ or $4 \to 3 \to 2 \to 1$.

33. $P^2 = \begin{bmatrix} .9 & .25 \\ .1 & .75 \end{bmatrix}\begin{bmatrix} .9 & .25 \\ .1 & .75 \end{bmatrix} = \begin{bmatrix} .835 & .4125 \\ .165 & .5875 \end{bmatrix}$.

$\begin{bmatrix} .835 & .4125 \\ .165 & .5875 \end{bmatrix}\begin{bmatrix} 300000 \\ 750000 \end{bmatrix} = \begin{bmatrix} 559875 \\ 490125 \end{bmatrix}$, so in two generations the distribution will be 559,875 college-educated and 490,125 noncollege-educated.

The probability is .4125 that a couple with no college education will have at least one grandchild with a college education.

34. (a) No arcs lead to vertex 4.
(b) There are two arcs from vertex 3.
(c) There are four 3-paths from vertex 2.
(d) No 2-paths lead to vertex 3.
(e) There are two 3-paths from vertex 4 to vertex 4.
(f) There are three pairs of vertices joined by 4-paths.

Chapter 3

Exercise Set 3.1

1. (a) $\begin{vmatrix} 2 & 1 \\ 3 & 5 \end{vmatrix} = (2x5) - (1x3) = 7.$ (b) $\begin{vmatrix} 3 & -2 \\ 1 & 2 \end{vmatrix} = (3x2) - (-2x1) = 8.$

   (c) $\begin{vmatrix} 4 & 1 \\ -2 & 3 \end{vmatrix} = (4x3) - (1x-2) = 14.$ (d) $\begin{vmatrix} 5 & -2 \\ -3 & -4 \end{vmatrix} = (5x-4) - (-2x-3) = -26.$

2. (a) $\begin{vmatrix} 1 & -5 \\ 0 & 3 \end{vmatrix} = (1x3) - (-5x0) = 3.$ (b) $\begin{vmatrix} 1 & 2 \\ 3 & 4 \end{vmatrix} = (1x4) - (2x3) = -2.$

   (c) $\begin{vmatrix} -3 & 1 \\ 2 & -5 \end{vmatrix} = (-3x-5) - (1x2) = 13.$ (d) $\begin{vmatrix} 3 & -2 \\ -1 & 0 \end{vmatrix} = (3x0) - (-2x-1) = -2.$

3. (a) $M_{11} = \begin{vmatrix} 0 & 6 \\ 1 & -4 \end{vmatrix} = (0x-4) - (6x1) = -6.$ $C_{11} = (-1)^{1+1} M_{11} = (-1)^2(-6) = -6.$

   (b) $M_{21} = \begin{vmatrix} 2 & -3 \\ 1 & -4 \end{vmatrix} = (2x-4) - (-3x1) = -5.$ $C_{21} = (-1)^{2+1} M_{21} = (-1)^3(-5) = 5.$

   (c) $M_{23} = \begin{vmatrix} 1 & 2 \\ 7 & 1 \end{vmatrix} = (1x1) - (2x7) = -13.$ $C_{23} = (-1)^{2+3} M_{23} = (-1)^5(-13) = 13.$

   (d) $M_{33} = \begin{vmatrix} 1 & 2 \\ 5 & 0 \end{vmatrix} = (1x0) - (2x5) = -10.$ $C_{33} = (-1)^{3+3} M_{33} = (-1)^6(-10) = -10.$

4. (a) $M_{13} = \begin{vmatrix} -2 & 3 \\ 0 & -6 \end{vmatrix} = (-2x-6) - (3x0) = 12.$ $C_{13} = (-1)^{1+3} M_{13} = (-1)^4(12) = 12.$

   (b) $M_{22} = \begin{vmatrix} 5 & 1 \\ 0 & 2 \end{vmatrix} = (5x2) - (1x0) = 10.$ $C_{22} = (-1)^{2+2} M_{22} = (-1)^4(10) = 10.$

   (c) $M_{31} = \begin{vmatrix} 0 & 1 \\ 3 & 7 \end{vmatrix} = (0x7) - (1x3) = -3.$ $C_{31} = (-1)^{3+1} M_{31} = (-1)^4(-3) = -3.$

(d) $M_{33} = \begin{vmatrix} 5 & 0 \\ -2 & 3 \end{vmatrix} = (5x3) - (0x-2) = 15.$ $\quad C_{33} = (-1)^{3+3} M_{33} = (-1)^6(15) = 15.$

5. (a) $M_{12} = \begin{vmatrix} 8 & 2 & 1 \\ 4 & -5 & 0 \\ 1 & 8 & 2 \end{vmatrix}$

$= (8x-5x2) + (2x0x1) + (1x4x8) - (1x-5x1) - (8x0x8) - (2x4x2)$

$= -80 + 0 + 32 - (-5) - 0 - 16 = -59.$

$C_{12} = (-1)^{1+2} M_{12} = (-1)^3(-59) = 59.$

(b) $M_{24} = \begin{vmatrix} 2 & 0 & 1 \\ 4 & -3 & -5 \\ 1 & 4 & 8 \end{vmatrix}$

$= (2x-3x8) + (0x-5x1) + (1x4x4) - (1x-3x1) - (2x-5x4) - (0x4x8)$

$= -48 + 0 + 16 - (-3) - (-40) - 0 = 11.$

$C_{24} = (-1)^{2+4} M_{24} = (-1)^6(11) = 11.$

(c) $M_{33} = \begin{vmatrix} 2 & 0 & -5 \\ 8 & -1 & 1 \\ 1 & 4 & 2 \end{vmatrix}$

$= (2x-1x2) + (0x1x1) + (-5x8x4) - (-5x-1x1) - (2x1x4) - (0x8x2)$

$= -4 + 0 + (-160) - 5 - 8 - 0 = -177.$

$C_{33} = (-1)^{3+3} M_{33} = (-1)^6(-177) = -177.$

(d) $M_{43} = \begin{vmatrix} 2 & 0 & -5 \\ 8 & -1 & 1 \\ 4 & -3 & 0 \end{vmatrix}$

$= (2x-1x0) + (0x1x4) + (-5x8x-3) - (-5x-1x4) - (2x1x-3) - (0x8x0)$

$= 0 + 0 + 120 - 20 - (-6) - 0 = 106.$

$C_{43} = (-1)^{4+3} M_{43} = (-1)^7(106) = -106.$

6\. (a) $\begin{vmatrix} 1 & 2 & 4 \\ 4 & -1 & 5 \\ -2 & 2 & 1 \end{vmatrix} = 1\begin{vmatrix} -1 & 5 \\ 2 & 1 \end{vmatrix} - 2\begin{vmatrix} 4 & 5 \\ -2 & 1 \end{vmatrix} + 4\begin{vmatrix} 4 & -1 \\ -2 & 2 \end{vmatrix}$

= [(−1x1) − (5x2)] − 2[(4x1) − (5x−2)] + 4[(4x2) − (−1x−2)]

= −11 − 2(14) + 4(6) = −15.

"diagonals" method:

(1x−1x1) + (2x5x−2) + (4x4x2) − (4x−1x−2) − (1x5x2) − (2x4x1)

= −1 + (−20) + 32 − 8 − 10 − 8 = −15.

(b) $\begin{vmatrix} 3 & 1 & 4 \\ -7 & -2 & 1 \\ 9 & 1 & -1 \end{vmatrix} = 3\begin{vmatrix} -2 & 1 \\ 1 & -1 \end{vmatrix} - \begin{vmatrix} -7 & 1 \\ 9 & -1 \end{vmatrix} + 4\begin{vmatrix} -7 & -2 \\ 9 & 1 \end{vmatrix}$

= 3[(−2x−1) − (1x1)] − [(−7x−1) − (1x9)] + 4[(−7x1) − (−2x9)]

= 3 − (−2) + 4(11) = 49.

"diagonals" method:

(3x−2x−1) + (1x1x9) + (4x−7x1) − (4x−2x9) − (3x1x1) − (1x−7x−1)

= 6 + 9 + (−28) − (−72) − 3 − 7 = 49.

(c) $\begin{vmatrix} 4 & 1 & -2 \\ 5 & 3 & -1 \\ 2 & 4 & 1 \end{vmatrix} = 4\begin{vmatrix} 3 & -1 \\ 4 & 1 \end{vmatrix} - \begin{vmatrix} 5 & -1 \\ 2 & 1 \end{vmatrix} + (-2)\begin{vmatrix} 5 & 3 \\ 2 & 4 \end{vmatrix}$

= 4[(3x1) − (−1x4)] − [(5x1) − (−1x2)] + (−2)[(5x4) − (3x2)]
= 4(7)− 7 + (−2)(14) = −7.

"diagonals" method:
(4x3x1) + (1x−1x2) + (−2x5x4) − (−2x3x2) − (4x−1x4) − (1x5x1)

= 12 + (−2) + (−40) − (−12) − (−16) − 5 = −7.

7\. (a) $\begin{vmatrix} 2 & 0 & 7 \\ 8 & -1 & -2 \\ 5 & 6 & 1 \end{vmatrix} = 2\begin{vmatrix} -1 & -2 \\ 6 & 1 \end{vmatrix} - 0\begin{vmatrix} 8 & -2 \\ 5 & 1 \end{vmatrix} + 7\begin{vmatrix} 8 & -1 \\ 5 & 6 \end{vmatrix}$

$= 2[(-1\text{x}1) - (-2\text{x}6)] - 0 + 7[(8\text{x}6) - (-1\text{x}5)] = 2(11) - 0 + 7(53) = 393.$

"diagonals" method:

$(2\text{x}-1\text{x}1) + (0\text{x}-2\text{x}5) + (7\text{x}8\text{x}6) - (7\text{x}-1\text{x}5) - (2\text{x}-2\text{x}6) - (0\text{x}8\text{x}1)$

$= -2 + 0 + (336) - (-35) - (-24) - 0 = 393.$

(b) $\begin{vmatrix} 2 & -1 & 3 \\ 4 & 0 & 2 \\ 1 & 1 & 1 \end{vmatrix} = 2\begin{vmatrix} 0 & 2 \\ 1 & 1 \end{vmatrix} - (-1)\begin{vmatrix} 4 & 2 \\ 1 & 1 \end{vmatrix} + 3\begin{vmatrix} 4 & 0 \\ 1 & 1 \end{vmatrix}$

$= 2[(0\text{x}1) - (2\text{x}1)] - (-1)[(4\text{x}1) - (2\text{x}1)] + 3[(4\text{x}1) - (0\text{x}1)]$

$= 2(-2) - (-1)(2) + 3(4) = 10.$

"diagonals" method:

$(2\text{x}0\text{x}1) + (-1\text{x}2\text{x}1) + (3\text{x}4\text{x}1) - (3\text{x}0\text{x}1) - (2\text{x}2\text{x}1) - (-1\text{x}4\text{x}1)$

$= 0 + (-2) + 12 - 0 - (4) - (-4) = 10.$

(c) $\begin{vmatrix} 0 & 0 & 5 \\ 1 & 1 & 1 \\ 2 & 2 & 2 \end{vmatrix} = 0\begin{vmatrix} 1 & 1 \\ 2 & 2 \end{vmatrix} - 0\begin{vmatrix} 1 & 1 \\ 2 & 2 \end{vmatrix} + 5\begin{vmatrix} 1 & 1 \\ 2 & 2 \end{vmatrix}$

$= 0 - 0 + 5[(1\text{x}2) - (1\text{x}2)] = 0$

"diagonals" method:

$(0\text{x}1\text{x}2) + (0\text{x}1\text{x}2) + (5\text{x}1\text{x}2) - (5\text{x}1\text{x}2) - (0\text{x}1\text{x}2) - (0\text{x}1\text{x}2) = 0.$

8. (a) $\begin{vmatrix} 0 & 3 & 2 \\ 1 & 5 & 7 \\ -2 & -6 & -1 \end{vmatrix}$

using row 2:

$= -\begin{vmatrix} 3 & 2 \\ -6 & -1 \end{vmatrix} + 5\begin{vmatrix} 0 & 2 \\ -2 & -1 \end{vmatrix} - 7\begin{vmatrix} 0 & 3 \\ -2 & -6 \end{vmatrix} = -9 + 5\text{x}4 - 7\text{x}6 = -31.$

using column 1:

$= 0\begin{vmatrix} 5 & 7 \\ -6 & -1 \end{vmatrix} - \begin{vmatrix} 3 & 2 \\ -6 & -1 \end{vmatrix} + (-2)\begin{vmatrix} 3 & 2 \\ 5 & 7 \end{vmatrix} = 0 - 9 + (-2)(11) = -31.$

(b) $\begin{vmatrix} 4 & 2 & 1 \\ -6 & 3 & -2 \\ 7 & 1 & -1 \end{vmatrix}$

using row 3:

$= 7\begin{vmatrix} 2 & 1 \\ 3 & -2 \end{vmatrix} - \begin{vmatrix} 4 & 1 \\ -6 & -2 \end{vmatrix} + (-1)\begin{vmatrix} 4 & 2 \\ -6 & 3 \end{vmatrix} = -49 - (-2) + (-24) = -71.$

using column 2:

$= -2\begin{vmatrix} -6 & -2 \\ 7 & -1 \end{vmatrix} + 3\begin{vmatrix} 4 & 1 \\ 7 & -1 \end{vmatrix} - \begin{vmatrix} 4 & 1 \\ -6 & -2 \end{vmatrix} = -40 + (-33) - (-2) = -71.$

(c) $\begin{vmatrix} 5 & -1 & 2 \\ 3 & 0 & 6 \\ -4 & 3 & 1 \end{vmatrix}$

using row 1:

$= 5\begin{vmatrix} 0 & 6 \\ 3 & 1 \end{vmatrix} - (-1)\begin{vmatrix} 3 & 6 \\ -4 & 1 \end{vmatrix} + 2\begin{vmatrix} 3 & 0 \\ -4 & 3 \end{vmatrix} = -90 - (-27) + 18 = -45.$

using row 3:

$= (-4)\begin{vmatrix} -1 & 2 \\ 0 & 6 \end{vmatrix} - 3\begin{vmatrix} 5 & 2 \\ 3 & 6 \end{vmatrix} + \begin{vmatrix} 5 & -1 \\ 3 & 0 \end{vmatrix} = 24 - (72) + (3) = -45.$

(d) $\begin{vmatrix} 6 & 3 & 0 \\ -2 & -1 & 5 \\ 4 & 6 & -2 \end{vmatrix}$

using column 2:

$= -3\begin{vmatrix} -2 & 5 \\ 4 & -2 \end{vmatrix} + (-1)\begin{vmatrix} 6 & 0 \\ 4 & -2 \end{vmatrix} - 6\begin{vmatrix} 6 & 0 \\ -2 & 5 \end{vmatrix} = 48 + 12 - 180 = -120.$

using column 3:

$= 0\begin{vmatrix} -2 & -1 \\ 4 & 6 \end{vmatrix} - 5\begin{vmatrix} 6 & 3 \\ 4 & 6 \end{vmatrix} + (-2)\begin{vmatrix} 6 & 3 \\ -2 & -1 \end{vmatrix} = 0 - 120 + 0 = -120.$

9. (a) $\begin{vmatrix} 1 & 3 & -1 \\ 2 & 0 & 5 \\ 1 & 4 & 3 \end{vmatrix}$

using row 2:

$= -2\begin{vmatrix} 3 & -1 \\ 4 & 3 \end{vmatrix} + 0\begin{vmatrix} 1 & -1 \\ 1 & 3 \end{vmatrix} - 5\begin{vmatrix} 1 & 3 \\ 1 & 4 \end{vmatrix} = -26 + 0 - 5 = -31.$

using column 1:

$$= \begin{vmatrix} 0 & 5 \\ 4 & 3 \end{vmatrix} - 2\begin{vmatrix} 3 & -1 \\ 4 & 3 \end{vmatrix} + \begin{vmatrix} 3 & -1 \\ 0 & 5 \end{vmatrix} = -20 - 26 + 15 = -31.$$

(b) $\begin{vmatrix} -2 & -1 & 1 \\ 9 & 3 & 2 \\ 4 & 0 & 0 \end{vmatrix}$

using row 1:

$$= (-2)\begin{vmatrix} 3 & 2 \\ 0 & 0 \end{vmatrix} - (-1)\begin{vmatrix} 9 & 2 \\ 4 & 0 \end{vmatrix} + \begin{vmatrix} 9 & 3 \\ 4 & 0 \end{vmatrix} = 0 - 8 + (-12) = -20.$$

using column 3:

$$= \begin{vmatrix} 9 & 3 \\ 4 & 0 \end{vmatrix} - 2\begin{vmatrix} -2 & -1 \\ 4 & 0 \end{vmatrix} + 0\begin{vmatrix} -2 & -1 \\ 9 & 3 \end{vmatrix} = -12 - 8 + 0 = -20.$$

(c) $\begin{vmatrix} 1 & 0 & 2 \\ 3 & -2 & 1 \\ 4 & 0 & 2 \end{vmatrix}$

using column 1:

$$= \begin{vmatrix} -2 & 1 \\ 0 & 2 \end{vmatrix} - 3\begin{vmatrix} 0 & 2 \\ 0 & 2 \end{vmatrix} + 4\begin{vmatrix} 0 & 2 \\ -2 & 1 \end{vmatrix} = -4 - 0 + 16 = 12.$$

using column 2:

$$= -0\begin{vmatrix} 3 & 1 \\ 4 & 2 \end{vmatrix} + (-2)\begin{vmatrix} 1 & 2 \\ 4 & 2 \end{vmatrix} - 0\begin{vmatrix} 1 & 2 \\ 3 & 1 \end{vmatrix} = 0 + 12 - 0 = 12.$$

(d) $\begin{vmatrix} 1 & 2 & 3 \\ -1 & -4 & 0 \\ 0 & 0 & 4 \end{vmatrix}$

using row 3:

$$= 0\begin{vmatrix} 2 & 3 \\ -4 & 0 \end{vmatrix} - 0\begin{vmatrix} 1 & 3 \\ -1 & 0 \end{vmatrix} + 4\begin{vmatrix} 1 & 2 \\ -1 & -4 \end{vmatrix} = 0 - 0 + (-8) = -8.$$

using column 1:

$$= \begin{vmatrix} -4 & 0 \\ 0 & 4 \end{vmatrix} - (-1)\begin{vmatrix} 2 & 3 \\ 0 & 4 \end{vmatrix} + 0\begin{vmatrix} 2 & 3 \\ -4 & 0 \end{vmatrix} = -16 - (-8) + 0 = -8.$$

10. (a) Using column 3, $\begin{vmatrix} 1 & -2 & 3 \\ 1 & 4 & 0 \\ 2 & -1 & 0 \end{vmatrix} = 3\begin{vmatrix} 1 & 4 \\ 2 & -1 \end{vmatrix} = -27.$

(b) Using row 2, $\begin{vmatrix} 3 & -1 & 2 \\ 0 & 4 & 0 \\ -5 & 1 & 9 \end{vmatrix} = 4\begin{vmatrix} 3 & 2 \\ -5 & 9 \end{vmatrix} = 148.$

(c) Using row 3, $\begin{vmatrix} 9 & 2 & 1 \\ -3 & 2 & 6 \\ 0 & 0 & -3 \end{vmatrix} = (-3)\begin{vmatrix} 9 & 2 \\ -3 & 2 \end{vmatrix} = -72.$

11. (a) Using column 4, $\begin{vmatrix} 1 & -2 & 3 & 0 \\ 4 & 0 & 5 & 0 \\ 7 & -3 & 8 & 4 \\ -3 & 0 & 4 & 0 \end{vmatrix} = -4\begin{vmatrix} 1 & -2 & 3 \\ 4 & 0 & 5 \\ -3 & 0 & 4 \end{vmatrix}$

(using column 2 of the 3x3 matrix) $= -4(-(-2))\begin{vmatrix} 4 & 5 \\ -3 & 4 \end{vmatrix} = -8\text{x}31 = -248.$

(b) Using row 3, $\begin{vmatrix} 1 & 4 & 5 & 9 \\ 2 & 3 & -7 & 1 \\ 0 & 0 & 0 & -3 \\ 0 & 1 & 0 & 8 \end{vmatrix} = -(-3)\begin{vmatrix} 1 & 4 & 5 \\ 2 & 3 & -7 \\ 0 & 1 & 0 \end{vmatrix}$

(using row 3 of the 3x3 matrix) $= -(-3)(-1)\begin{vmatrix} 1 & 5 \\ 2 & -7 \end{vmatrix} = -3\text{x}(-17) = 51.$

(c) Using column 2, $\begin{vmatrix} 9 & 3 & 7 & -8 \\ 1 & 0 & 4 & 2 \\ 1 & 0 & 0 & -1 \\ -2 & 0 & -1 & 3 \end{vmatrix} = -3\begin{vmatrix} 1 & 4 & 2 \\ 1 & 0 & -1 \\ -2 & -1 & 3 \end{vmatrix}$

(using row 2 of the 3x3 matrix) $= -3\left(-\begin{vmatrix} 4 & 2 \\ -1 & 3 \end{vmatrix} - (-1)\begin{vmatrix} 1 & 4 \\ -2 & -1 \end{vmatrix}\right)$

$= -3(-14 - (-7)) = 21.$

12. $\begin{vmatrix} x+1 & x \\ 3 & x-2 \end{vmatrix} = (x+1)(x-2) - 3x = x^2 - x - 2 - 3x = 3$, so $x^2 - 4x - 5 = 0$.

$(x-5)(x+1) = 0$, so there are two solutions, $x = 5$ and $x = -1$.

13. $\begin{vmatrix} 2x & -3 \\ x-1 & x+2 \end{vmatrix} = (2x)(x+2) - (-3)(x-1) = 2x^2 + 4x - (-3x) - 3 = 1$, so $2x^2 + 7x - 4 = 0$.

$(2x-1)(x+4) = 0$, so there are two solutions, $x = 1/2$ and $x = -4$.

14. $\begin{vmatrix} x-1 & -2 \\ x-2 & x-1 \end{vmatrix} = (x-1)(x-1) - (-2)(x-2) = x^2 - 2x + 1 - (-2x) - 4 = 0$,

so $x^2 - 3 = 0$, and there are two solutions, $\sqrt{3}$ and $-\sqrt{3}$.

15. $\begin{vmatrix} x & 0 & 2 \\ 2x & x-1 & 4 \\ -x & x-1 & x+1 \end{vmatrix} = x(x-1)(x+1) + 0 + 2(2x)(x-1) - 2(x-1)(-x) - x(4)(x-1) - 0$

$= x^3 - x + 4x^2 - 4x + 2x^2 - 2x - 4x^2 + 4x = x^3 + 2x^2 - 3x = x(x+3)(x-1) = 0$, so there are three solutions, $x = 0$, $x = -3$, and $x = 1$.

16. The cofactor expansion of each determinant using the third column gives $-3\begin{vmatrix} 4 & -1 \\ 2 & 1 \end{vmatrix}$.

17. The cofactor expansion using the third row is independent of the numbers a and b.

18. (a) $4213 \to 4123 \to 1423 \to 1243 \to 1234$; even

(b) $3142 \to 1342 \to 1324 \to 1234$; odd

(c) $3214 \to 3124 \to 1324 \to 1234$; odd

(d) $2413 \to 2143 \to 1243 \to 1234$; odd

(e) $4321 \to 4312 \to 3412 \to 3142 \to 1342 \to 1324 \to 1234$; even

19. (a) $35241 \to 32541 \to 32451 \to 32415 \to 32145 \to 31245 \to 13245 \to 12345$; odd

(b) $43152 \to 41352 \to 14352 \to 13452 \to 13425 \to 13245 \to 12345$; even

(c) $54312 \to 45312 \to 43512 \to 43152 \to 43125 \to 34125 \to 31425 \to 31245$
$\to 13245 \to 12345$; odd

(d) 25143 $\rightarrow$ 21543 $\rightarrow$ 12543 $\rightarrow$ 12453 $\rightarrow$ 12435 $\rightarrow$ 12345; odd

(e) 32514 $\rightarrow$ 23514 $\rightarrow$ 23154 $\rightarrow$ 21354 $\rightarrow$ 12354 $\rightarrow$ 12345; odd

20.

1234	even	2134 odd
1243	odd	2143 even
1324	odd	2314 even
1342	even	2341 odd
1423	even	2413 odd
1432	odd	2431 even
3124	even	4123 odd
3142	odd	4132 even
3214	odd	4213 even
3241	even	4231 odd
3412	even	4312 odd
3421	odd	4321 even

21. Cofactor expansion gives $|A| = a_{11}C_{11} + a_{12}C_{12} + a_{13}C_{13}$
$= a_{11}(a_{22}a_{33} - a_{23}a_{32}) - a_{12}(a_{21}a_{33} - a_{23}a_{31}) + a_{13}(a_{21}a_{32} - a_{22}a_{31})$
$= a_{11}a_{22}a_{33} - a_{11}a_{23}a_{32} - a_{12}a_{21}a_{33} + a_{12}a_{23}a_{31} + a_{13}a_{21}a_{32} - a_{13}a_{22}a_{31}$.

The permutations of 1,2,3 are 123(even), 132(odd), 213(odd), 231(even), 312(even), and 321(odd), so the second definition gives $|A| = a_{11}a_{22}a_{33} - a_{11}a_{23}a_{32} - a_{12}a_{21}a_{33} + a_{12}a_{23}a_{31} + a_{13}a_{21}a_{32} - a_{13}a_{22}a_{31}$. The two expressions are the same.

22. (a) (1)(-1)(1)-(1)(5(2)-(2)(4)(1)+(2)(5)(-2)+(4)(4)(2)-(4)(-1)(-2) = -15.

(b) (3)(-2)(-1)-(3)(1)(1)-(1)(-7)(-1)+(1)(1)(9)+(4)(-7)(1)-(4)(-2)(9) = 49.

(c) (4)(3)(1)-(4)(-1)(4)-(1)(5)(1)+(1)(-1)(2)+(-2)(5)(4)-(-2)(3)(2) = -7.

<u>Exercise Set 3.2</u>

1. (a) $\begin{vmatrix} 1 & 2 & 3 \\ 2 & 4 & 1 \\ 1 & 1 & 1 \end{vmatrix} \underset{C2+(-2)C1}{\approx} \begin{vmatrix} 1 & 0 & 3 \\ 2 & 0 & 1 \\ 1 & -1 & 1 \end{vmatrix} = -(-1)\begin{vmatrix} 1 & 3 \\ 2 & 1 \end{vmatrix} = -5.$

(b) $\begin{vmatrix} 0 & 1 & 5 \\ 1 & 1 & 6 \\ 2 & 2 & 7 \end{vmatrix} \underset{C2+(-1)C1}{=} \begin{vmatrix} 0 & 1 & 5 \\ 1 & 0 & 6 \\ 2 & 0 & 7 \end{vmatrix} = -\begin{vmatrix} 1 & 6 \\ 2 & 7 \end{vmatrix} = 5.$

(c) $\begin{vmatrix} 2 & 1 & -1 \\ 3 & -1 & 1 \\ 1 & 4 & -4 \end{vmatrix} \underset{C2+C3}{=} \begin{vmatrix} 2 & 0 & -1 \\ 3 & 0 & 1 \\ 1 & 0 & -4 \end{vmatrix} = 0.$

(d) $\begin{vmatrix} 3 & -1 & 0 \\ 4 & 2 & 1 \\ 1 & 1 & 2 \end{vmatrix} \underset{R3+(-2)R2}{=} \begin{vmatrix} 3 & -1 & 0 \\ 4 & 2 & 1 \\ -7 & -3 & 0 \end{vmatrix} = -\begin{vmatrix} 3 & -1 \\ -7 & -3 \end{vmatrix} = 16.$

2. (a) $\begin{vmatrix} 2 & -1 & 2 \\ 1 & 2 & -4 \\ 3 & 1 & 2 \end{vmatrix} \underset{C3+(2)C2}{=} \begin{vmatrix} 2 & -1 & 0 \\ 1 & 2 & 0 \\ 3 & 1 & 4 \end{vmatrix} = 4\begin{vmatrix} 2 & -1 \\ 1 & 2 \end{vmatrix} = 20.$

(b) $\begin{vmatrix} 5 & 1 & 3 \\ 1 & 2 & 4 \\ -1 & 1 & -4 \end{vmatrix} \underset{R2+R3}{=} \begin{vmatrix} 5 & 1 & 3 \\ 0 & 3 & 0 \\ -1 & 1 & -4 \end{vmatrix} = 3\begin{vmatrix} 5 & 3 \\ -1 & -4 \end{vmatrix} = -51.$

(c) $\begin{vmatrix} 1 & -2 & 3 \\ -3 & 6 & -9 \\ 4 & 5 & 7 \end{vmatrix} \underset{R2+(3)R1}{=} \begin{vmatrix} 1 & -2 & 3 \\ 0 & 0 & 0 \\ 4 & 5 & 7 \end{vmatrix} = 0.$

(d) $\begin{vmatrix} -1 & 3 & 2 \\ 2 & 5 & -4 \\ 4 & 1 & -8 \end{vmatrix} \underset{C3+(2)C1}{=} \begin{vmatrix} -1 & 3 & 0 \\ 2 & 5 & 0 \\ 4 & 1 & 0 \end{vmatrix} = 0.$

3. (a) The given matrix can be obtained from A by multiplying the third row by 2, so its determinant is 2 |A| = −4.

(b) The given matrix can be obtained from A by interchanging rows 1 and 2, so its determinant is −|A| = 2.

(c) The given matrix can be obtained from A by adding twice row 1 to row 2, so its determinant is |A| = −2.

4. (a) The given matrix can be obtained from A by interchanging columns 2 and 3, so its determinant is $-|A| = -5$.

   (b) The given matrix can be obtained from A by adding -2 times column 3 to column 2, so its determinant is $|A| = 5$.

   (c) The given matrix is the transpose of A, and $|A^t| = |A| = 5$.

5. The second answer is correct. 2R1 - R3 does not preserve the value of the determinant. It multiplies the determinant by 2 before subtracting R3.

6. (a) Row 3 is all zeros.
   (b) Columns 2 and 3 are equal.
   (c) Row 3 is -3 times row 1.
   (d) Row 3 is all zeros.

7. (a) Column 3 is all zeros.
   (b) Row 3 is all zeros.
   (c) Row 3 is 3 times row 1.
   (d) Column 3 is $-3/2$ times column 1.

8. (a) $|2A| = (2)^2 |A| = 12$.
   (b) $|3A^t| = (3)^2 |A^t| = 9|A| = 27$.
   (c) $|A^2| = |A||A| = 9$.
   (d) $|(A^t)^2| = |A^t||A^t| = |A||A| = 9$.
   (e) $|(A^2)^t| = |A^2| = 9$.
   (f) $|4A^{-1}| = (4)^2 |A^{-1}| = 16/|A| = 16/3$.

9. (a) $|AB| = |A||B| = -6$.
   (b) $|AA^t| = |A||A^t| = |A||A| = 9$.
   (c) $|A^t B| = |A^t||B| = |A||B| = -6$.
   (d) $|3A^2 B| = (3)^3 |A||A||B| = 486$.
   (e) $|2AB^{-1}| = (2)^3 |A||B^{-1}| = 8|A|/|B| = -12$. (f) $|(A^2 B^{-1})^t| = |A^2 B^{-1}| = |A||A|/|B| = 9/2$.

10. (a) The given matrix can be obtained from A by interchanging rows 1 and 2 and then interchanging rows 2 and 3. Thus its determinant is $(-1)(-1)|A| = 3$.

    (b) The given matrix can be obtained from A by interchanging rows 1 and 2 and then taking the transpose. Thus its determinant is $(-1)|A| = -3$.

(c) The given matrix can be obtained from A by interchanging rows 1 and 2 and then interchanging columns 2 and 3 in the resulting matrix. Thus the determinant of the given matrix is (−1)(−1)|A| = 3.

(d) The given matrix can be obtained from A by interchanging rows 1 and 2 of A and multiplying row 3 by 2. Thus the determinant is (−1)(2)|A| = −6.

11. $$\begin{vmatrix} a+b & c+d & e+f \\ p & q & r \\ u & v & w \end{vmatrix} = (a+b)\begin{vmatrix} q & r \\ v & w \end{vmatrix} - (c+d)\begin{vmatrix} p & r \\ u & w \end{vmatrix} + (e+f)\begin{vmatrix} p & q \\ u & v \end{vmatrix}$$

$$= a\begin{vmatrix} q & r \\ v & w \end{vmatrix} - c\begin{vmatrix} p & r \\ u & w \end{vmatrix} + e\begin{vmatrix} p & q \\ u & v \end{vmatrix} + b\begin{vmatrix} q & r \\ v & w \end{vmatrix} - d\begin{vmatrix} p & r \\ u & w \end{vmatrix} + f\begin{vmatrix} p & q \\ u & v \end{vmatrix}$$

$$= \begin{vmatrix} a & c & e \\ p & q & r \\ u & v & w \end{vmatrix} + \begin{vmatrix} b & d & f \\ p & q & r \\ u & v & w \end{vmatrix}.$$

12. (a) 3x−1x4 = −12 (b) 2x3x5x−2 = −60 (c) 3x−2x5x1 = −30

13. (a) $$\begin{vmatrix} 1 & 0 & -1 \\ 2 & 1 & 2 \\ -1 & 1 & 1 \end{vmatrix} \underset{\substack{R2+(-2)R1 \\ R3+R1}}{=} \begin{vmatrix} 1 & 0 & -1 \\ 0 & 1 & 4 \\ 0 & 1 & 0 \end{vmatrix} \underset{R3+(-1)R2}{=} \begin{vmatrix} 1 & 0 & -1 \\ 0 & 1 & 4 \\ 0 & 0 & -4 \end{vmatrix} = -4.$$

(b) $$\begin{vmatrix} 1 & -2 & 3 \\ -1 & 2 & 1 \\ 2 & 1 & 3 \end{vmatrix} \underset{\substack{R2+R1 \\ R3+(-2)R1}}{=} \begin{vmatrix} 1 & -2 & 3 \\ 0 & 0 & 4 \\ 0 & 5 & -3 \end{vmatrix} \underset{R2\leftrightarrow R3}{=} -\begin{vmatrix} 1 & -2 & 3 \\ 0 & 5 & -3 \\ 0 & 0 & 4 \end{vmatrix} = -20.$$

(c) $$\begin{vmatrix} 2 & 3 & 8 \\ -2 & -3 & 4 \\ 4 & 6 & -2 \end{vmatrix} \underset{\substack{R2+R1 \\ R3+(-2)R1}}{=} \begin{vmatrix} 2 & 3 & 8 \\ 0 & 0 & 12 \\ 0 & 0 & -18 \end{vmatrix} = 0.$$

(d) $$\begin{vmatrix} 2 & -1 & 4 \\ -2 & 1 & -3 \\ 0 & 3 & -2 \end{vmatrix} \underset{R2+R1}{=} \begin{vmatrix} 2 & -1 & 4 \\ 0 & 0 & 1 \\ 0 & 3 & -2 \end{vmatrix} \underset{R2\leftrightarrow R3}{=} -\begin{vmatrix} 2 & -1 & 4 \\ 0 & 3 & -2 \\ 0 & 0 & 1 \end{vmatrix} = -6.$$

14. (a) $$\begin{vmatrix} 2 & 1 & 3 & 1 \\ -2 & 3 & -1 & 2 \\ 2 & 1 & 2 & 3 \\ -4 & -2 & 0 & -1 \end{vmatrix} \underset{\substack{R2+R1 \\ R3+(-1)R1 \\ R4+(2)R1}}{=} \begin{vmatrix} 2 & 1 & 3 & 1 \\ 0 & 4 & 2 & 3 \\ 0 & 0 & -1 & 2 \\ 0 & 0 & 6 & 1 \end{vmatrix} \underset{R4+(6)R3}{=} \begin{vmatrix} 2 & 1 & 3 & 1 \\ 0 & 4 & 2 & 3 \\ 0 & 0 & -1 & 2 \\ 0 & 0 & 0 & 13 \end{vmatrix} = -104.$$

(b) $$\begin{vmatrix} 1 & 2 & -1 & 0 \\ 1 & 4 & 2 & 1 \\ -1 & 2 & 6 & 6 \\ 2 & 2 & -4 & -2 \end{vmatrix} \underset{\substack{R2+(-1)R1 \\ R3+R1 \\ R4+(-2)R1}}{=} \begin{vmatrix} 1 & 2 & -1 & 0 \\ 0 & 2 & 3 & 1 \\ 0 & 4 & 5 & 6 \\ 0 & -2 & -2 & -2 \end{vmatrix} \underset{\substack{R3+(-2)R2 \\ R4+R2}}{=} \begin{vmatrix} 1 & 2 & -1 & 0 \\ 0 & 2 & 3 & 1 \\ 0 & 0 & -1 & 4 \\ 0 & 0 & 1 & -1 \end{vmatrix}$$

$$\underset{R4+R3}{=} \begin{vmatrix} 1 & 2 & -1 & 0 \\ 0 & 2 & 3 & 1 \\ 0 & 0 & -1 & 4 \\ 0 & 0 & 0 & 3 \end{vmatrix} = -6.$$

(c) $$\begin{vmatrix} 1 & -1 & 0 & 2 \\ -1 & 1 & 0 & 0 \\ 2 & -2 & 0 & 1 \\ 3 & 1 & 5 & -1 \end{vmatrix} \underset{\substack{R2+R1 \\ R3+(-2)R1 \\ R4+(-3)R1}}{=} \begin{vmatrix} 1 & -1 & 0 & 2 \\ 0 & 0 & 0 & 2 \\ 0 & 0 & 0 & -3 \\ 0 & 4 & 5 & -7 \end{vmatrix} \underset{R2\leftrightarrow R4}{=} -\begin{vmatrix} 1 & -1 & 0 & 2 \\ 0 & 4 & 5 & -7 \\ 0 & 0 & 0 & -3 \\ 0 & 0 & 0 & 2 \end{vmatrix} = 0.$$

15. Expand by row (or column) 1 at each stage.

$$\begin{vmatrix} a_{11} & 0 & 0 & \dots & 0 \\ 0 & a_{22} & 0 & \dots & 0 \\ 0 & 0 & a_{33} & \dots & 0 \\ \vdots & \vdots & \vdots & & \vdots \\ 0 & 0 & 0 & \dots & a_{nn} \end{vmatrix} = a_{11} \begin{vmatrix} a_{22} & 0 & \dots & 0 \\ 0 & a_{33} & \dots & 0 \\ \vdots & \vdots & & \vdots \\ 0 & 0 & \dots & a_{nn} \end{vmatrix} = a_{11} a_{22} \begin{vmatrix} a_{33} & \dots & 0 \\ \vdots & & \vdots \\ 0 & \dots & a_{nn} \end{vmatrix}$$

$$= \dots = a_{11} a_{22} a_{33} \dots a_{nn}$$

16. Suppose row i of B is row i of A plus k times row j of A, where k is any real number. The cofactors of the ith rows of A and B are the same since only the elements in the

ith rows are different.

$|B| = b_{i1}C_{i1} + b_{i2}C_{i2} + \ldots + b_{in}C_{in} = (a_{i1} + ka_{j1})C_{i1} + (a_{i2} + ka_{j2})C_{i2} + \ldots + (a_{in} + ka_{jn})C_{in}$

$= a_{i1}C_{i1} + a_{i2}C_{i2} + \ldots + a_{in}C_{in} + k(a_{j1}C_{i1} + a_{j2}C_{i2} + \ldots + a_{jn}C_{in}) = |A| + k|D|$,
where D is the matrix obtained from A by replacing the ith row of A by the jth row of A. Thus the ith and jth rows of D are equal, so $|D| = 0$ and $|B| = |A|$.
The proof for columns is similar.

17. Suppose row j is k times row i. Let B be the matrix obtained from A by interchanging rows i and j. Call the cofactors of the ith row of A $C_{i1}, C_{i2}, \ldots, C_{in}$ and the cofactors of the ith row of B $D_{i1}, D_{i2}, \ldots, D_{in}$. Each cofactor of A is k times the corresponding cofactor of B, because row j in A is k times row j in B.

$|A| = a_{i1}C_{i1} + a_{i2}C_{i2} + \ldots + a_{in}C_{in} = a_{i1}kD_{i1} + a_{i2}kD_{i2} + \ldots + a_{in}kD_{in}$
and $|B| = a_{j1}D_{i1} + a_{j2}D_{i2} + \ldots + a_{jn}D_{in} = ka_{i1}D_{i1} + ka_{i2}D_{i2} + \ldots + ka_{in}D_{in}$,
so $|A| = |B|$.

But $|B| = -|A|$ because B was obtained from A by interchanging two rows, so $|A| = -|A|$, and therefore $|A| = 0$.

The proof for proportional columns is similar.

18. Suppose the sum of the elements in each column is zero.

$$|A| = \begin{vmatrix} a_{11} & \cdots & a_{1n} \\ a_{21} & \cdots & a_{2n} \\ a_{31} & \cdots & a_{3n} \\ \vdots & & \vdots \\ a_{n1} & \cdots & a_{nn} \end{vmatrix} \underset{R1+R2}{=} \begin{vmatrix} a_{11}+a_{21} & \cdots & a_{1n}+a_{2n} \\ a_{21} & \cdots & a_{2n} \\ a_{31} & \cdots & a_{3n} \\ \vdots & & \vdots \\ a_{n1} & \cdots & a_{nn} \end{vmatrix} \underset{R1+R3}{=} \begin{vmatrix} a_{11}+a_{21}+a_{31} & \cdots & a_{1n}+a_{2n}+a_{3n} \\ a_{21} & \cdots & a_{2n} \\ a_{31} & \cdots & a_{3n} \\ \vdots & & \vdots \\ a_{n1} & \cdots & a_{nn} \end{vmatrix}$$

$$\underset{R1+R4}{=} \cdots \underset{R1+Rn}{=} \begin{vmatrix} a_{11}+a_{21}+a_{31}+\ldots+a_{n1} & \cdots & a_{1n}+a_{2n}+a_{3n}+\ldots+a_{nn} \\ a_{21} & \cdots & a_{2n} \\ a_{31} & \cdots & a_{3n} \\ \vdots & & \vdots \\ a_{n1} & \cdots & a_{nn} \end{vmatrix} = 0,$$

because the first row is all zeros.

19. $|A^n| = |A||A|...|A| = |A|^n$. If $A^n = O$, then $|A^n| = 0$, so $|A|^n = 0$, but that means $|A| = 0$.

20. $|AB| = |A||B| = |B||A| = |BA|$.

21. $|ABC| = |AB||C| = |A||B||C|$.

22. Let $A = \begin{bmatrix} 2 & 1 \\ 0 & 0 \end{bmatrix}$ and let $B = \begin{bmatrix} 0 & 0 \\ 4 & -1 \end{bmatrix}$. Then $A + B = \begin{bmatrix} 2 & 1 \\ 4 & -1 \end{bmatrix}$.

$|A| = |B| = 0$ and $|A+B| = -6$, so $|A+B| \neq |A| + |B|$ for this example.

23. If B is obtained from A using an elementary row operation, then $|B| = |A|$ or $|B| = -|A|$ or $|B| = c|A|$. Since $c \neq 0$ by definition, $|B| \neq 0$ if and only if $|A| \neq 0$.

24. (a) Let E be the reduced echelon form of a 2x2 matrix A and suppose E has no zero rows. Since E is a reduced echelon matrix, the first position in row 2 that can be nonzero is the (2,2) position, and it must be 1 in order for E to have no zero rows. This means the (1,2) position is zero and the (1,1) position is 1. So $E = I_2$.

(b) Let E be the reduced echelon form of a 3x3 matrix A and suppose E has no zero rows. Since E is a reduced echelon matrix, the first position in row 3 that can be nonzero is the (3,3) position, and it must be 1 in order for E to have no zero rows. This means the (1,3) and (2,3) positions are zero and the (2,2) position is 1, since it is now both the first and last position in row 2 that can be nonzero. Therefore the (1,2) position is zero and the (1,1) position is 1. So $E = I_3$.

(c) Let E be the reduced echelon form of an nxn matrix A and suppose E has no zero rows. Since E is a reduced echelon matrix, the first position in row n that can be nonzero is the (n,n) position, and it must be 1 in order for E to have no zero rows. This means the (1,n), (2,n),. . .,(n−1,n) positions are zero and the (n−1,n−1) position is 1, since it is now both the first and last position in row n−1 that can be nonzero. Therefore the (1,n−1), (2,n−1), . . . , (n−2,n−1) positions are zero and the (n−2,n−2) position is 1, ... Continuing in this way, we see that all diagonal elements of E must be 1 to avoid having a zero row, and therefore all other elements of E are zero. Thus $E = I_n$.

(d) E is obtained from A by performing a sequence of elementary row operations.

$|A| = (-1)^k c|E|$, where k is the number of row interchanges required and c is 1 divided by the product of all constants multiplied by rows. Thus if $|A| \neq 0$ then $|E| \neq 0$, but if $|E| \neq 0$, then E has no zero rows, so from part (c), $E = I_n$. On the other hand, if $E = I_n$, then $|E| \neq 0$, so $|A| \neq 0$.

25. Let $A=\begin{bmatrix} 1 & 0 \\ 0 & 1 \end{bmatrix}$ and $B=\begin{bmatrix} -1 & 0 \\ 0 & -1 \end{bmatrix}$. Then |A|=1, |B|=1. A and B are nonsingular. $A+B=\begin{bmatrix} 0 & 0 \\ 0 & 0 \end{bmatrix}$.

A+B is singular. S is not closed under addition. Further, $0A=\begin{bmatrix} 0 & 0 \\ 0 & 0 \end{bmatrix}$. 0A is singular. S is not closed under scalar multiplication.

Exercise Set 3.3

1. (a) The determinant is 5. The matrix is invertible.

   (b) The determinant is zero. The matrix is singular. The inverse does not exist.

   (c) The determinant is zero. The matrix is singular. The inverse does not exist.

   (d) The determinant is 1. The matrix is invertible.

2. (a) The determinant is −6. The matrix is invertible.

   (b) The determinant is zero. The matrix is singular. The inverse does not exist.

   (c) The determinant is 7. The matrix is invertible.

   (d) The determinant is zero. The matrix is singular. The inverse does not exist.

3. (a) The determinant is 18. The matrix is invertible.

   (b) The determinant is zero. The matrix is singular. The inverse does not exist.

   (c) The determinant is −105. The matrix is invertible.

   (d) The determinant is zero. The matrix is singular. The inverse does not exist.

4. (a) The determinant is zero. The matrix is singular. The inverse does not exist.

   (b) The determinant is 42. The matrix is invertible.

   (c) The determinant is −27. The matrix is invertible.

   (d) The determinant is zero. The matrix is singular. The inverse does not exist.

5. (a) The determinant is −10.

$$\begin{bmatrix} 1 & 4 \\ 3 & 2 \end{bmatrix}^{-1} = \frac{-1}{10}\begin{bmatrix} 2 & -4 \\ -3 & 1 \end{bmatrix}.$$

   (b) The determinant is 1.

$$\begin{bmatrix} -2 & -1 \\ 7 & 3 \end{bmatrix}^{-1} = \begin{bmatrix} 3 & 1 \\ -7 & -2 \end{bmatrix}.$$

   (c) The determinant is zero. The inverse does not exist.

   (d) The determinant is 2.

$$\begin{bmatrix} 2 & 1 \\ 4 & 3 \end{bmatrix}^{-1} = \frac{1}{2}\begin{bmatrix} 3 & -1 \\ -4 & 2 \end{bmatrix}.$$

6. (a) The determinant is −3.

$$\begin{bmatrix} 1 & 2 & 3 \\ 0 & 1 & 2 \\ 4 & 5 & 3 \end{bmatrix}^{-1} = \frac{-1}{3}\begin{bmatrix} \begin{vmatrix} 1 & 2 \\ 5 & 3 \end{vmatrix} & -\begin{vmatrix} 2 & 3 \\ 5 & 3 \end{vmatrix} & \begin{vmatrix} 2 & 3 \\ 1 & 2 \end{vmatrix} \\ -\begin{vmatrix} 0 & 2 \\ 4 & 3 \end{vmatrix} & \begin{vmatrix} 1 & 3 \\ 4 & 3 \end{vmatrix} & -\begin{vmatrix} 1 & 3 \\ 0 & 2 \end{vmatrix} \\ \begin{vmatrix} 0 & 1 \\ 4 & 5 \end{vmatrix} & -\begin{vmatrix} 1 & 2 \\ 4 & 5 \end{vmatrix} & \begin{vmatrix} 1 & 2 \\ 0 & 1 \end{vmatrix} \end{bmatrix} = \frac{-1}{3}\begin{bmatrix} -7 & 9 & 1 \\ 8 & -9 & -2 \\ -4 & 3 & 1 \end{bmatrix}.$$

   (b) The determinant is −3.

$$\begin{bmatrix} 0 & 3 & 3 \\ 1 & 2 & 3 \\ 1 & 4 & 6 \end{bmatrix}^{-1} = \frac{-1}{3}\begin{bmatrix} \begin{vmatrix} 2 & 3 \\ 4 & 6 \end{vmatrix} & -\begin{vmatrix} 3 & 3 \\ 4 & 6 \end{vmatrix} & \begin{vmatrix} 3 & 3 \\ 2 & 3 \end{vmatrix} \\ -\begin{vmatrix} 1 & 3 \\ 1 & 6 \end{vmatrix} & \begin{vmatrix} 0 & 3 \\ 1 & 6 \end{vmatrix} & -\begin{vmatrix} 0 & 3 \\ 1 & 3 \end{vmatrix} \\ \begin{vmatrix} 1 & 2 \\ 1 & 4 \end{vmatrix} & -\begin{vmatrix} 0 & 3 \\ 1 & 4 \end{vmatrix} & \begin{vmatrix} 0 & 3 \\ 1 & 2 \end{vmatrix} \end{bmatrix} = \frac{-1}{3}\begin{bmatrix} 0 & -6 & 3 \\ -3 & -3 & 3 \\ 2 & 3 & -3 \end{bmatrix}.$$

(c) The determinant is −4.

$$\begin{bmatrix} 1 & 2 & -1 \\ 2 & 4 & -3 \\ 1 & -2 & 0 \end{bmatrix}^{-1} = \frac{-1}{4}\begin{bmatrix} -6 & 2 & -2 \\ -3 & 1 & 1 \\ -8 & 4 & 0 \end{bmatrix}.$$

(d) The determinant is zero. The inverse does not exist.

7. (a) The determinant is −1.

$$\begin{bmatrix} 5 & 2 & 4 \\ 2 & 1 & 2 \\ 4 & 2 & 3 \end{bmatrix}^{-1} = -\begin{bmatrix} \begin{vmatrix} 1 & 2 \\ 2 & 3 \end{vmatrix} & -\begin{vmatrix} 2 & 4 \\ 2 & 3 \end{vmatrix} & \begin{vmatrix} 2 & 4 \\ 1 & 2 \end{vmatrix} \\ -\begin{vmatrix} 2 & 2 \\ 4 & 3 \end{vmatrix} & \begin{vmatrix} 5 & 4 \\ 4 & 3 \end{vmatrix} & -\begin{vmatrix} 5 & 4 \\ 2 & 2 \end{vmatrix} \\ \begin{vmatrix} 2 & 1 \\ 4 & 2 \end{vmatrix} & -\begin{vmatrix} 5 & 2 \\ 4 & 2 \end{vmatrix} & \begin{vmatrix} 5 & 2 \\ 2 & 1 \end{vmatrix} \end{bmatrix} = -\begin{bmatrix} -1 & 2 & 0 \\ 2 & -1 & -2 \\ 0 & -2 & 1 \end{bmatrix}.$$

(b) The determinant is 2.

$$\begin{bmatrix} -3 & -2 & -5 \\ 3 & 4 & 3 \\ 1 & 1 & 1 \end{bmatrix}^{-1} = \frac{1}{2}\begin{bmatrix} 1 & -3 & 14 \\ 0 & 2 & -6 \\ -1 & 1 & -6 \end{bmatrix}.$$

(c) The determinant is zero. The inverse does not exist.

(d) The determinant is −3.

$$\begin{bmatrix} 2 & 2 & 1 \\ 4 & -1 & 4 \\ 7 & 4 & 5 \end{bmatrix}^{-1} = \frac{-1}{3}\begin{bmatrix} -21 & -6 & 9 \\ 8 & 3 & -4 \\ 23 & 6 & -10 \end{bmatrix}.$$

8. (a) $x_1 = \dfrac{\begin{vmatrix} 8 & 2 \\ 19 & 5 \end{vmatrix}}{\begin{vmatrix} 1 & 2 \\ 2 & 5 \end{vmatrix}} = \dfrac{2}{1} = 2.$ $\quad x_2 = \dfrac{\begin{vmatrix} 1 & 8 \\ 2 & 19 \end{vmatrix}}{\begin{vmatrix} 1 & 2 \\ 2 & 5 \end{vmatrix}} = \dfrac{3}{1} = 3.$

(b) $x_1 = \frac{\begin{vmatrix} 3 & 2 \\ -1 & 1 \end{vmatrix}}{\begin{vmatrix} 1 & 2 \\ 3 & 1 \end{vmatrix}} = \frac{5}{-5} = -1.$ $x_2 = \frac{\begin{vmatrix} 1 & 3 \\ 3 & -1 \end{vmatrix}}{\begin{vmatrix} 1 & 2 \\ 3 & 1 \end{vmatrix}} = \frac{-10}{-5} = 2.$

(c) $x_1 = \frac{\begin{vmatrix} 11 & 3 \\ -1 & 1 \end{vmatrix}}{\begin{vmatrix} 1 & 3 \\ -2 & 1 \end{vmatrix}} = \frac{14}{7} = 2.$ $x_2 = \frac{\begin{vmatrix} 1 & 11 \\ -2 & -1 \end{vmatrix}}{\begin{vmatrix} 1 & 3 \\ -2 & 1 \end{vmatrix}} = \frac{21}{7} = 3.$

9. (a) $x_1 = \frac{\begin{vmatrix} -1 & 1 \\ 3 & 1 \end{vmatrix}}{\begin{vmatrix} 3 & 1 \\ 1 & 1 \end{vmatrix}} = \frac{-4}{2} = -2.$ $x_2 = \frac{\begin{vmatrix} 3 & -1 \\ 1 & 3 \end{vmatrix}}{\begin{vmatrix} 3 & 1 \\ 1 & 1 \end{vmatrix}} = \frac{10}{2} = 5.$

(b) $x_1 = \frac{\begin{vmatrix} 11 & 2 \\ 14 & 3 \end{vmatrix}}{\begin{vmatrix} 3 & 2 \\ 2 & 3 \end{vmatrix}} = \frac{5}{5} = 1.$ $x_2 = \frac{\begin{vmatrix} 3 & 11 \\ 2 & 14 \end{vmatrix}}{\begin{vmatrix} 3 & 2 \\ 2 & 3 \end{vmatrix}} = \frac{20}{5} = 4.$

(c) $x_1 = \frac{\begin{vmatrix} -1 & 1 \\ 10 & 2 \end{vmatrix}}{\begin{vmatrix} 2 & 1 \\ -2 & 2 \end{vmatrix}} = \frac{-12}{6} = -2.$ $x_2 = \frac{\begin{vmatrix} 2 & -1 \\ -2 & 10 \end{vmatrix}}{\begin{vmatrix} 2 & 1 \\ -2 & 2 \end{vmatrix}} = \frac{18}{6} = 3.$

10. (a) $|A| = \begin{vmatrix} 1 & 3 & 4 \\ 2 & 6 & 9 \\ 3 & 1 & -2 \end{vmatrix} = 8, |A_1| = \begin{vmatrix} 3 & 3 & 4 \\ 5 & 6 & 9 \\ 7 & 1 & -2 \end{vmatrix} = 8, |A_2| = \begin{vmatrix} 1 & 3 & 4 \\ 2 & 5 & 9 \\ 3 & 7 & -2 \end{vmatrix} = 16,$

$|A_3| = \begin{vmatrix} 1 & 3 & 3 \\ 2 & 6 & 5 \\ 3 & 1 & 7 \end{vmatrix} = -8$, so $x_1 = \frac{8}{8} = 1, x_2 = \frac{16}{8} = 2, x_3 = \frac{-8}{8} = -1.$

(b) $|A| = \begin{vmatrix} 1 & 2 & 1 \\ 1 & 3 & -1 \\ 1 & 4 & -1 \end{vmatrix} = 2$, $|A_1| = \begin{vmatrix} 9 & 2 & 1 \\ 4 & 3 & -1 \\ 7 & 4 & -1 \end{vmatrix} = -2$, $|A_2| = \begin{vmatrix} 1 & 9 & 1 \\ 1 & 4 & -1 \\ 1 & 7 & -1 \end{vmatrix} = 6$,

$|A_3| = \begin{vmatrix} 1 & 2 & 9 \\ 1 & 3 & 4 \\ 1 & 4 & 7 \end{vmatrix} = 8$, so $x_1 = \frac{-2}{2} = -1$, $x_2 = \frac{6}{2} = 3$, $x_3 = \frac{8}{2} = 4$.

(c) $|A| = \begin{vmatrix} 2 & 1 & 3 \\ 3 & -2 & 4 \\ 1 & 4 & -2 \end{vmatrix} = 28$, $|A_1| = \begin{vmatrix} 2 & 1 & 3 \\ 2 & -2 & 4 \\ 1 & 4 & -2 \end{vmatrix} = 14$, $|A_2| = \begin{vmatrix} 2 & 2 & 3 \\ 3 & 2 & 4 \\ 1 & 1 & -2 \end{vmatrix} = 7$,

$|A_3| = \begin{vmatrix} 2 & 1 & 2 \\ 3 & -2 & 2 \\ 1 & 4 & 1 \end{vmatrix} = 7$, so $x_1 = \frac{14}{28} = \frac{1}{2}$, $x_2 = \frac{7}{28} = \frac{1}{4}$, $x_3 = \frac{7}{28} = \frac{1}{4}$.

11. (a) $|A| = \begin{vmatrix} 1 & 4 & 2 \\ 1 & 4 & -1 \\ 2 & 6 & 1 \end{vmatrix} = -6$, $|A_1| = \begin{vmatrix} 5 & 4 & 2 \\ 2 & 4 & -1 \\ 7 & 6 & 1 \end{vmatrix} = -18$, $|A_2| = \begin{vmatrix} 1 & 5 & 2 \\ 1 & 2 & -1 \\ 2 & 7 & 1 \end{vmatrix} = 0$,

$|A_3| = \begin{vmatrix} 1 & 4 & 5 \\ 1 & 4 & 2 \\ 2 & 6 & 7 \end{vmatrix} = -6$, so $x_1 = \frac{-18}{-6} = 3$, $x_2 = \frac{0}{-6} = 0$, $x_3 = \frac{-6}{-6} = 1$.

(b) $|A| = \begin{vmatrix} 2 & -1 & 3 \\ 1 & 4 & 2 \\ 3 & 2 & 1 \end{vmatrix} = -35$, $|A_1| = \begin{vmatrix} 7 & -1 & 3 \\ 10 & 4 & 2 \\ 0 & 2 & 1 \end{vmatrix} = 70$, $|A_2| = \begin{vmatrix} 2 & 7 & 3 \\ 1 & 10 & 2 \\ 3 & 0 & 1 \end{vmatrix} = -35$,

$|A_3| = \begin{vmatrix} 2 & -1 & 7 \\ 1 & 4 & 10 \\ 3 & 2 & 0 \end{vmatrix} = -140$, so $x_1 = \frac{70}{-35} = -2$, $x_2 = \frac{-35}{-35} = 1$, $x_3 = \frac{-140}{-35} = 4$.

(c) $|A| = \begin{vmatrix} 8 & -2 & 1 \\ 2 & -1 & 6 \\ 6 & 1 & 4 \end{vmatrix} = -128$, $|A_1| = \begin{vmatrix} 1 & -2 & 1 \\ 3 & -1 & 6 \\ 3 & 1 & 4 \end{vmatrix} = -16$, $|A_2| = \begin{vmatrix} 8 & 1 & 1 \\ 2 & 3 & 6 \\ 6 & 3 & 4 \end{vmatrix} = -32$,

$|A_3| = \begin{vmatrix} 8 & -2 & 1 \\ 2 & -1 & 3 \\ 6 & 1 & 3 \end{vmatrix} = -64$, so $x_1 = \frac{-16}{-128} = \frac{1}{8}$, $x_2 = \frac{-32}{-128} = \frac{1}{4}$, $x_3 = \frac{-64}{-128} = \frac{1}{2}$.

12. (a) $|A| = 0$ so this system of equations cannot be solved using Cramer's rule.

(b) $|A| = \begin{vmatrix} 3 & 1 & -1 \\ 1 & 2 & 1 \\ 2 & 6 & 0 \end{vmatrix} = -18$, $|A_1| = \begin{vmatrix} 7 & 1 & -1 \\ 3 & 2 & 1 \\ -4 & 6 & 0 \end{vmatrix} = -72$, $|A_2| = \begin{vmatrix} 3 & 7 & -1 \\ 1 & 3 & 1 \\ 2 & -4 & 0 \end{vmatrix} = 36$,

$|A_3| = \begin{vmatrix} 3 & 1 & 7 \\ 1 & 2 & 3 \\ 2 & 6 & -4 \end{vmatrix} = -54$, so $x_1 = \frac{-72}{-18} = 4$, $x_2 = \frac{36}{-18} = -2$, $x_3 = \frac{-54}{-18} = 3$.

(c) $|A| = \begin{vmatrix} 3 & 6 & -1 \\ 1 & -2 & 3 \\ 4 & -2 & 5 \end{vmatrix} = 24$, $|A_1| = \begin{vmatrix} 3 & 6 & -1 \\ 2 & -2 & 3 \\ 5 & -2 & 5 \end{vmatrix} = 12$, $|A_2| = \begin{vmatrix} 3 & 3 & -1 \\ 1 & 2 & 3 \\ 4 & 5 & 5 \end{vmatrix} = 9$,

$|A_3| = \begin{vmatrix} 3 & 6 & 3 \\ 1 & -2 & 2 \\ 4 & -2 & 5 \end{vmatrix} = 18$, so $x_1 = \frac{12}{24} = \frac{1}{2}$, $x_2 = \frac{9}{24} = \frac{3}{8}$, $x_3 = \frac{18}{24} = \frac{3}{4}$.

13. (a) The determinant of the coefficient matrix is zero, so there is not a unique solution.

(b) The determinant of the coefficient matrix is -10, so there is a unique solution.

(c) The determinant of the coefficient matrix is zero, so there is not a unique solution.

14. (a) The determinant of the coefficient matrix is 42, so there is a unique solution.

(b) The determinant of the coefficient matrix is zero, so there is not a unique solution.

(c) The determinant of the coefficient matrix is zero, so there is not a unique solution.

15. The system of equations will have nontrivial solutions if the determinant of the coefficient matrix is zero, i.e., if

$$\begin{vmatrix} 1-\lambda & 6 \\ 5 & 2-\lambda \end{vmatrix} = 0.$$

$$\begin{vmatrix} 1-\lambda & 6 \\ 5 & 2-\lambda \end{vmatrix} = (1-\lambda)(2-\lambda) - 30 = 2 - 3\lambda + \lambda^2 - 30 = \lambda^2 - 3\lambda - 28 = 0, \text{ so}$$

$(\lambda-7)(\lambda+4) = 0$ and $\lambda = 7$ or $\lambda = -4$. Substituting $\lambda = 7$ in the given equations, one finds that the general solution is $x_1 = x_2 = r$. For $\lambda = -4$, the general solution is $x_1 = -6r/5$, $x_2 = r$.

16. The system of equations will have nontrivial solutions if the determinant of the coefficient matrix is zero, i.e., if

$$\begin{vmatrix} \lambda+4 & \lambda-2 \\ 4 & \lambda-3 \end{vmatrix} = 0.$$

$$\begin{vmatrix} \lambda+4 & \lambda-2 \\ 4 & \lambda-3 \end{vmatrix} = (\lambda+4)(\lambda-3) - (\lambda-2)4 = \lambda^2 + \lambda - 12 - 4\lambda + 8 = \lambda^2 - 3\lambda - 4 = 0,$$

so $(\lambda-4)(\lambda+1) = 0$ and $\lambda = 4$ or $\lambda = -1$. For $\lambda = 4$, the general solution is $x_1 = -r/4$, $x_2 = r$. For $\lambda = -1$, the general solution is $x_1 = x_2 = r$.

17. The system of equations will have nontrivial solutions if the determinant of the coefficient matrix is zero, i.e., if

$$\begin{vmatrix} 5-\lambda & 4 & 2 \\ 4 & 5-\lambda & 2 \\ 2 & 2 & 2-\lambda \end{vmatrix} = 0.$$

$$\begin{vmatrix} 5-\lambda & 4 & 2 \\ 4 & 5-\lambda & 2 \\ 2 & 2 & 2-\lambda \end{vmatrix} = (5-\lambda)(5-\lambda)(2-\lambda) + 16 + 16 - 4(5-\lambda) - 4(5-\lambda) - 16(2-\lambda)$$

$= 10 - 21\lambda + 12\lambda^2 - \lambda^3 = (1-\lambda)(1-\lambda)(10-\lambda) = 0$, so $\lambda = 1$ or $\lambda = 10$. For $\lambda = 1$, the general solution is $x_1 = -s - r/2$, $x_2 = s$, $x_3 = r$. For $\lambda = 10$, the general solution is $x_1 = x_2 = 2r$, $x_3 = r$.

18. $AX = \lambda X = \lambda I_n X$, so $AX - \lambda I_n X = 0$. Thus $(A - \lambda I_n)X = 0$, and this system of equations has a nontrivial solution if and only if $|A - \lambda I_n| = 0$.

19. If $A = A^t$, then the matrix obtained by deleting the ith row and jth column of A is the transpose of the matrix obtained by deleting the jth row and ith column. The determinant of the first of these matrices is C_{ij}, the (i,j)th element of $(\text{adj}(A))^t$, and the determinant of the second is C_{ji}, the (i,j)th element of adj(A). Since these determinants are equal, $\text{adj}(A) = (\text{adj}(A))^t$, and so adj(A) is symmetric.

20. If A is the zero matrix then adj(A) is the zero matrix also. If A is not the zero matrix and if A is not invertible then $|A| = 0$ and $A\,\text{adj}(A) = |A|\,I_n = O$. If adj(A) is invertible then $A = A\,I_n = A\,[\text{adj}(A)\,(\text{adj}(A))^{-1}] = [A\,\text{adj}(A)]\,(\text{adj}(A))^{-1} = O\,(\text{adj}(A))^{-1} = O$. Since A is not the zero matrix, this means that adj(A) has no inverse.

21. If A is invertible then $\frac{1}{|A|}\,\text{adj}(A) = A^{-1}$, so that $A\frac{1}{|A|}\,\text{adj}(A) = AA^{-1} = I_n$. Thus $\frac{1}{|A|}\,A = [\text{adj}(A)]^{-1}$.

22. $A = (A^{-1})^{-1} = \frac{1}{|A^{-1}|}\,\text{adj}(A^{-1})$, so $\text{adj}(A^{-1}) = |A^{-1}|\,A = \frac{1}{|A|}\,A = [\text{adj}(A)]^{-1}$ from Exercise 21

23. If AB is invertible then $|AB| \neq 0$. $|A||B| = |AB|$, so $|A| \neq 0$ and $|B| \neq 0$ and both A and B are invertible. The converse is also true. If A and B are invertible then $|AB| = |A||B| \neq 0$, so AB is invertible.

24. Let $B = A^{-1}$. The (i,j)th element of the product $AB = I_n$ is $a_{i1}b_{1j} + a_{i2}b_{2j} + \ldots + a_{in}b_{nj}$.
For i=n, $a_{n1}b_{1j} + a_{n2}b_{2j} + \ldots + a_{nn}b_{nj} = a_{nn}b_{nj}$, because $a_{n1} = a_{n2} = \ldots = a_{nn-1} = 0$.
This means that $a_{nn}b_{nj} = 0$ for $j < n$. But $a_{nn} \neq 0$ so $b_{nj} = 0$ for $j < n$.
For $i = n-1$, $a_{n-11}b_{1j} + a_{n-12}b_{2j} + \ldots + a_{n-1n-1}b_{n-1j} + a_{n-1n}b_{nj} = a_{n-1n-1}b_{n-1j}$, for $j < n$, because $a_{n-11} = a_{n-12} = \ldots = a_{n-2n-1} = b_{nj} = 0$. This means that $a_{n-1n-1}b_{n-1j} = 0$ for $j < n-1$. But $a_{n-1n-1} \neq 0$ so $b_{n-1j} = 0$ for $j < n-1$.
In the same manner each row of B is shown to have zero in all positions for which the column number is less than the row number. Thus B is upper triangular.

25. If $|A| = \pm 1$, then $A^{-1} = \frac{1}{|A|}$ adj(A) = ±adj(A), and since all elements of A are integers, all elements of adj(A) are integers (because adding and multiplying integers gives integer results).

26. If AX = 0 has only the trivial solution, then $|A| \neq 0$ so that $|A^k| = |A|^k \neq 0$. Thus $A^k X = 0$ has only the trivial solution.

27. $AX = B_2$ has a unique solution if and only if $|A| \neq 0$ if and only if $AX = B_1$ has a unique solution.

28. (a) True: $|A^2| = |AA| = |A||A| = (|A|)^2$.

(b) True: Determinant of diagonal matrix is the product of diagonal elements. If this product is zero at least one of the elements must be zero.

(c) True: $A^{-1} = \text{adj}(A)/|A|$. Thus if $|A| = 1$, $A^{-1} = \text{adj}(A)$.

(d) False: e.g., Let $A-B = \begin{bmatrix} 2 & 6 \\ 1 & 3 \end{bmatrix}$, so that $|A-B|=0$. Find two matrices A and B such that $A-B=\begin{bmatrix} 2 & 6 \\ 1 & 3 \end{bmatrix}$. Chances are they have distinct determinants. Say $A=\begin{bmatrix} 4 & 9 \\ 2 & 6 \end{bmatrix}$, $B=\begin{bmatrix} 2 & 3 \\ 1 & 3 \end{bmatrix}$. $|A|= 6$ and $|B|=3$.

(e) True: If A is nonsingular then it is invertible. Thus it is row equivalent to I (section 2.4).

<u>Exercise Set 3.4</u>

1. $\begin{vmatrix} 5-\lambda & 4 \\ 1 & 2-\lambda \end{vmatrix} = (5-\lambda)(2-\lambda) - 4 = \lambda^2 - 7\lambda + 6 = (\lambda-6)(\lambda-1)$, so the eigenvalues are $\lambda = 6$ and $\lambda = 1$. For $\lambda = 6$, the eigenvectors are the solutions of $\begin{bmatrix} -1 & 4 \\ 1 & -4 \end{bmatrix}\begin{bmatrix} x_1 \\ x_2 \end{bmatrix} = \mathbf{0}$, so the eigenvectors are vectors of the form $r\begin{bmatrix} 4 \\ 1 \end{bmatrix}$. For $\lambda = 1$, the eigenvectors are the solutions

of $\begin{bmatrix} 4 & 4 \\ 1 & 1 \end{bmatrix}\begin{bmatrix} x_1 \\ x_2 \end{bmatrix} = \mathbf{0}$, so the eigenvectors are vectors of the form $s\begin{bmatrix} -1 \\ 1 \end{bmatrix}$.

2. $\begin{vmatrix} 1-\lambda & -2 \\ 1 & 4-\lambda \end{vmatrix} = (1-\lambda)(4-\lambda) + 2 = \lambda^2 - 5\lambda + 6 = (\lambda-2)(\lambda-3)$, so the eigenvalues are $\lambda = 2$ and $\lambda = 3$. For $\lambda = 2$, the eigenvectors are the solutions of $\begin{bmatrix} -1 & -2 \\ 1 & 2 \end{bmatrix}\begin{bmatrix} x_1 \\ x_2 \end{bmatrix} = \mathbf{0}$, so the eigenvectors are vectors of the form $r\begin{bmatrix} -2 \\ 1 \end{bmatrix}$. For $\lambda = 3$, the eigenvectors are the solutions

of $\begin{bmatrix} -2 & -2 \\ 1 & 1 \end{bmatrix}\begin{bmatrix} x_1 \\ x_2 \end{bmatrix} = \mathbf{0}$, so the eigenvectors are vectors of the form $s\begin{bmatrix} -1 \\ 1 \end{bmatrix}$.

3. $\begin{vmatrix} 5-\lambda & 6 \\ -2 & -2-\lambda \end{vmatrix} = (5-\lambda)(-2-\lambda) + 12 = \lambda^2 - 3\lambda + 2 = (\lambda-2)(\lambda-1)$, so the eigenvalues are $\lambda = 2$ and $\lambda = 1$. For $\lambda = 2$, the eigenvectors are the solutions of $\begin{bmatrix} 3 & 6 \\ -2 & -4 \end{bmatrix}\begin{bmatrix} x_1 \\ x_2 \end{bmatrix} = \mathbf{0}$, so the eigenvectors are vectors of the form $r\begin{bmatrix} -2 \\ 1 \end{bmatrix}$. For $\lambda = 1$, the eigenvectors are the solutions of $\begin{bmatrix} 4 & 6 \\ -2 & -3 \end{bmatrix}\begin{bmatrix} x_1 \\ x_2 \end{bmatrix} = \mathbf{0}$, so the eigenvectors are vectors of the form $s\begin{bmatrix} -3 \\ 2 \end{bmatrix}$.

4. $\begin{vmatrix} 5-\lambda & 2 \\ -8 & -3-\lambda \end{vmatrix} = (5-\lambda)(-3-\lambda) + 16 = \lambda^2 - 2\lambda + 1 = (\lambda-1)(\lambda-1)$, so the only eigenvalue is

$\lambda = 1$. The eigenvectors are the solutions of $\begin{bmatrix} 4 & 2 \\ -8 & -4 \end{bmatrix}\begin{bmatrix} x_1 \\ x_2 \end{bmatrix} = \mathbf{0}$, so the eigenvectors are vectors of the form $r\begin{bmatrix} 1 \\ -2 \end{bmatrix}$.

5. $\begin{vmatrix} 1-\lambda & 2 \\ 2 & 1-\lambda \end{vmatrix} = (1-\lambda)(1-\lambda) - 4 = \lambda^2 - 2\lambda - 3 = (\lambda-3)(\lambda+1)$, so the eigenvalues are $\lambda = 3$ and $\lambda = -1$. For $\lambda = 3$, the eigenvectors are the solutions of $\begin{bmatrix} -2 & 2 \\ 2 & -2 \end{bmatrix}\begin{bmatrix} x_1 \\ x_2 \end{bmatrix} = \mathbf{0}$, so the eigenvectors are vectors of the form $r\begin{bmatrix} 1 \\ 1 \end{bmatrix}$. For $\lambda = -1$, the eigenvectors are the solutions

of $\begin{bmatrix} 2 & 2 \\ 2 & 2 \end{bmatrix}\begin{bmatrix} x_1 \\ x_2 \end{bmatrix} = \mathbf{0}$, so the eigenvectors are vectors of the form $s\begin{bmatrix} -1 \\ 1 \end{bmatrix}$.

6. $\begin{vmatrix} 2-\lambda & 1 \\ -1 & 4-\lambda \end{vmatrix} = (2-\lambda)(4-\lambda) + 1 = \lambda^2 - 6\lambda + 9 = (\lambda-3)(\lambda-3)$, so the only eigenvalue is $\lambda = 3$. The eigenvectors are the solutions of $\begin{bmatrix} -1 & 1 \\ -1 & 1 \end{bmatrix}\begin{bmatrix} x_1 \\ x_2 \end{bmatrix} = \mathbf{0}$, so the eigenvectors are vectors of the form $r\begin{bmatrix} 1 \\ 1 \end{bmatrix}$.

7. $\begin{vmatrix} 3-\lambda & -1 \\ 2 & -\lambda \end{vmatrix} = (3-\lambda)(-\lambda) + 2 = \lambda^2 - 3\lambda + 2 = (\lambda-2)(\lambda-1)$, so the eigenvalues are $\lambda = 2$ and

$\lambda = 1$. For $\lambda = 2$, the eigenvectors are the solutions of $\begin{bmatrix} 1 & -1 \\ 2 & -2 \end{bmatrix}\begin{bmatrix} x_1 \\ x_2 \end{bmatrix} = \mathbf{0}$, so the eigenvectors are vectors of the form $r\begin{bmatrix} 1 \\ 1 \end{bmatrix}$. For $\lambda = 1$, the eigenvectors are the solutions of $\begin{bmatrix} 2 & -1 \\ 2 & -1 \end{bmatrix}\begin{bmatrix} x_1 \\ x_2 \end{bmatrix} = \mathbf{0}$, so the eigenvectors are vectors of the form $s\begin{bmatrix} 1 \\ 2 \end{bmatrix}$.

8. $\begin{vmatrix} 2-\lambda & -4 \\ -1 & 2-\lambda \end{vmatrix} = (2-\lambda)(2-\lambda) - 4 = \lambda^2 - 4\lambda = \lambda(\lambda-4)$, so the eigenvalues are $\lambda = 0$ and $\lambda = 4$. For $\lambda = 0$, the eigenvectors are the solutions of $\begin{bmatrix} 2 & -4 \\ -1 & 2 \end{bmatrix}\begin{bmatrix} x_1 \\ x_2 \end{bmatrix} = \mathbf{0}$, so the eigenvectors are vectors of the form $r\begin{bmatrix} 2 \\ 1 \end{bmatrix}$. For $\lambda = 4$, the eigenvectors are the solutions of $\begin{bmatrix} -2 & -4 \\ -1 & -2 \end{bmatrix}\begin{bmatrix} x_1 \\ x_2 \end{bmatrix} = \mathbf{0}$, so the eigenvectors are vectors of the form $s\begin{bmatrix} -2 \\ 1 \end{bmatrix}$.

9. $\begin{vmatrix} 3-\lambda & 2 & -2 \\ -3 & -1-\lambda & 3 \\ 1 & 2 & -\lambda \end{vmatrix} = (3-\lambda)(-1-\lambda)(-\lambda) + 18 + 2(-1-\lambda) - 6(3-\lambda) - 6\lambda$

$= -\lambda^3 + 2\lambda^2 + \lambda - 2 = (1-\lambda^2)(\lambda-2)$, so the eigenvalues are $\lambda = 1$, $\lambda = -1$, and $\lambda = 2$.

For $\lambda = 1$, the eigenvectors are the solutions of $\begin{bmatrix} 2 & 2 & -2 \\ -3 & -2 & 3 \\ 1 & 2 & -1 \end{bmatrix}\begin{bmatrix} x_1 \\ x_2 \\ x_3 \end{bmatrix} = \mathbf{0}$, so the eigenvectors are vectors of the form $r\begin{bmatrix} 1 \\ 0 \\ 1 \end{bmatrix}$. For $\lambda = -1$,the eigenvectors are the solutions of $\begin{bmatrix} 4 & 2 & -2 \\ -3 & 0 & 3 \\ 1 & 2 & 1 \end{bmatrix}\begin{bmatrix} x_1 \\ x_2 \\ x_3 \end{bmatrix} = \mathbf{0}$, so the eigenvectors are vectors of the form $s\begin{bmatrix} 1 \\ -1 \\ 1 \end{bmatrix}$.

For $\lambda = 2$, the eigenvectors are the solutions of $\begin{bmatrix} 1 & 2 & -2 \\ -3 & -3 & 3 \\ 1 & 2 & -2 \end{bmatrix}\begin{bmatrix} x_1 \\ x_2 \\ x_3 \end{bmatrix} = \mathbf{0}$, so the

eigenvectors are vectors of the form $t\begin{bmatrix} 0 \\ 1 \\ 1 \end{bmatrix}$.

10. $\begin{vmatrix} 1-\lambda & -2 & 2 \\ -2 & 1-\lambda & 2 \\ -2 & 0 & 3-\lambda \end{vmatrix} = (1-\lambda)^2(3-\lambda)$, so the eigenvalues are $\lambda = 1$ and $\lambda = 3$. For $\lambda = 1$, the

eigenvectors are the solutions of $\begin{bmatrix} 0 & -2 & 2 \\ -2 & 0 & 2 \\ -2 & 0 & 2 \end{bmatrix}\begin{bmatrix} x_1 \\ x_2 \\ x_3 \end{bmatrix} = \mathbf{0}$, so the eigenvectors are

vectors of the form $r\begin{bmatrix} 1 \\ 1 \\ 1 \end{bmatrix}$. For $\lambda = 3$, the eigenvectors are the solutions of

$\begin{bmatrix} -2 & -2 & 2 \\ -2 & -2 & 2 \\ -2 & 0 & 0 \end{bmatrix}\begin{bmatrix} x_1 \\ x_2 \\ x_3 \end{bmatrix} = \mathbf{0}$, so the eigenvectors are vectors of the form $s\begin{bmatrix} 0 \\ 1 \\ 1 \end{bmatrix}$.

11. $\begin{vmatrix} 1-\lambda & 0 & 0 \\ -2 & 1-\lambda & 2 \\ -2 & 0 & 3-\lambda \end{vmatrix} = (1-\lambda)^2(3-\lambda)$, so the eigenvalues are $\lambda = 1$ and $\lambda = 3$. For $\lambda = 1$, the

eigenvectors are the solutions of $\begin{bmatrix} 0 & 0 & 0 \\ -2 & 0 & 2 \\ -2 & 0 & 2 \end{bmatrix}\begin{bmatrix} x_1 \\ x_2 \\ x_3 \end{bmatrix} = \mathbf{0}$, so the eigenvectors are vectors of the form $r\begin{bmatrix} 1 \\ 0 \\ 1 \end{bmatrix} + s\begin{bmatrix} 0 \\ 1 \\ 0 \end{bmatrix}$. For $\lambda = 3$, the eigenvectors are the solutions of

$\begin{bmatrix} -2 & 0 & 0 \\ -2 & -2 & 2 \\ -2 & 0 & 0 \end{bmatrix}\begin{bmatrix} x_1 \\ x_2 \\ x_3 \end{bmatrix} = \mathbf{0}$, so the eigenvectors are vectors of the form $t\begin{bmatrix} 0 \\ 1 \\ 1 \end{bmatrix}$.

12. $\begin{vmatrix} 1-\lambda & 0 & 0 \\ -2 & 5-\lambda & -2 \\ -2 & 4 & -1-\lambda \end{vmatrix} = (1-\lambda)(5-\lambda)(-1-\lambda) + 8(1-\lambda) = (1-\lambda)^2(3-\lambda)$, so the eigenvalues are

$\lambda = 1$ and $\lambda = 3$. For $\lambda = 1$, the eigenvectors are the solutions of $\begin{bmatrix} 0 & 0 & 0 \\ -2 & 4 & -2 \\ -2 & 4 & -2 \end{bmatrix}\begin{bmatrix} x_1 \\ x_2 \\ x_3 \end{bmatrix} = \mathbf{0}$,

so the eigenvectors are vectors of the form $r\begin{bmatrix} -1 \\ 0 \\ 1 \end{bmatrix} + s\begin{bmatrix} 1 \\ 1 \\ 1 \end{bmatrix}$. For $\lambda = 3$, the eigenvectors

are the solutions of $\begin{bmatrix} -2 & 0 & 0 \\ -2 & 2 & -2 \\ -2 & 4 & -4 \end{bmatrix}\begin{bmatrix} x_1 \\ x_2 \\ x_3 \end{bmatrix} = \mathbf{0}$, so the eigenvectors are vectors of the form

$t\begin{bmatrix} 0 \\ 1 \\ 1 \end{bmatrix}$.

13. $\begin{vmatrix} 15-\lambda & 7 & -7 \\ -1 & 1-\lambda & 1 \\ 13 & 7 & -5-\lambda \end{vmatrix} = (1-\lambda)(16-10\lambda+\lambda^2) = (1-\lambda)(2-\lambda)(8-\lambda)$, so the eigenvalues are

$\lambda = 1$, $\lambda = 2$, and $\lambda = 8$. For $\lambda = 1$, the eigenvectors are the solutions of

$\begin{bmatrix} 14 & 7 & -7 \\ -1 & 0 & 1 \\ 13 & 7 & -6 \end{bmatrix}\begin{bmatrix} x_1 \\ x_2 \\ x_3 \end{bmatrix} = \mathbf{0}$, so the eigenvectors are vectors of the form $r\begin{bmatrix} 1 \\ -1 \\ 1 \end{bmatrix}$. For $\lambda = 2$,

the eigenvectors are the solutions of $\begin{bmatrix} 13 & 7 & -7 \\ -1 & -1 & 1 \\ 13 & 7 & -7 \end{bmatrix}\begin{bmatrix} x_1 \\ x_2 \\ x_3 \end{bmatrix} = \mathbf{0}$, so the eigenvectors are

vectors of the form $s\begin{bmatrix} 0 \\ 1 \\ 1 \end{bmatrix}$. For $\lambda = 8$, the eigenvectors are the solutions of

$\begin{bmatrix} 7 & 7 & -7 \\ -1 & -7 & 1 \\ 13 & 7 & -13 \end{bmatrix}\begin{bmatrix} x_1 \\ x_2 \\ x_3 \end{bmatrix} = \mathbf{0}$, so the eigenvectors are vectors of the form $t\begin{bmatrix} 1 \\ 0 \\ 1 \end{bmatrix}$.

14. $\begin{vmatrix} 5-\lambda & -2 & 2 \\ 4 & -3-\lambda & 4 \\ 4 & -6 & 7-\lambda \end{vmatrix} = (5-\lambda)(3-\lambda)(1-\lambda)$, so the eigenvalues are $\lambda = 5$, $\lambda = 3$, and $\lambda = 1$.

For $\lambda = 5$, the eigenvectors are the solutions of $\begin{bmatrix} 0 & -2 & 2 \\ 4 & -8 & 4 \\ 4 & -6 & 2 \end{bmatrix}\begin{bmatrix} x_1 \\ x_2 \\ x_3 \end{bmatrix} = \mathbf{0}$, so the

eigenvectors are vectors of the form $r\begin{bmatrix} 1 \\ 1 \\ 1 \end{bmatrix}$. For $\lambda = 3$, the eigenvectors are the solutions

of $\begin{bmatrix} 2 & -2 & 2 \\ 4 & -6 & 4 \\ 4 & -6 & 4 \end{bmatrix}\begin{bmatrix} x_1 \\ x_2 \\ x_3 \end{bmatrix} = \mathbf{0}$, so the eigenvectors are vectors of the form $s\begin{bmatrix} 1 \\ 0 \\ -1 \end{bmatrix}$.

For $\lambda = 1$, the eigenvectors are the solutions of $\begin{bmatrix} 4 & -2 & 2 \\ 4 & -4 & 4 \\ 4 & -6 & 6 \end{bmatrix}\begin{bmatrix} x_1 \\ x_2 \\ x_3 \end{bmatrix} = \mathbf{0}$, so the eigenvectors are vectors of the form $t\begin{bmatrix}0\\1\\1\end{bmatrix}$.

15. $\begin{vmatrix}4-\lambda & 2 & -2 & 2\\ 1 & 3-\lambda & 1 & -1\\ 0 & 0 & 2-\lambda & 0\\ 1 & 1 & -3 & 5-\lambda\end{vmatrix} = (2-\lambda)\begin{vmatrix}4-\lambda & 2 & 2\\ 1 & 3-\lambda & -1\\ 1 & 1 & 5-\lambda\end{vmatrix} = (2-\lambda)(4-\lambda)(2-\lambda)(6-\lambda)$, so the

eigenvalues are $\lambda = 2$, $\lambda = 4$, and $\lambda = 6$. For $\lambda = 2$, the eigenvectors are the solutions of

$\begin{bmatrix}2 & 2 & -2 & 2\\ 1 & 1 & 1 & -1\\ 0 & 0 & 0 & 0\\ 1 & 1 & -3 & 3\end{bmatrix}\begin{bmatrix}x_1\\x_2\\x_3\\x_4\end{bmatrix} = \mathbf{0}$, so the eigenvectors are vectors of the form

$r\begin{bmatrix}1\\-1\\0\\0\end{bmatrix} + s\begin{bmatrix}0\\0\\1\\1\end{bmatrix}$. For $\lambda = 4$, the eigenvectors are the solutions of

$\begin{bmatrix}0 & 2 & -2 & 2\\ 1 & -1 & 1 & -1\\ 0 & 0 & -2 & 0\\ 1 & 1 & -3 & 1\end{bmatrix}\begin{bmatrix}x_1\\x_2\\x_3\\x_4\end{bmatrix} = \mathbf{0}$, so the eigenvectors are vectors of the form $t\begin{bmatrix}0\\1\\0\\-1\end{bmatrix}$.

For $\lambda = 6$, the eigenvectors are the solutions of $\begin{bmatrix}-2 & 2 & -2 & 2\\ 1 & -3 & 1 & -1\\ 0 & 0 & -4 & 0\\ 1 & 1 & -3 & -1\end{bmatrix}\begin{bmatrix}x_1\\x_2\\x_3\\x_4\end{bmatrix} = \mathbf{0}$, so the

eigenvectors are vectors of the form $p\begin{bmatrix}1\\0\\0\\1\end{bmatrix}$.

16. $\begin{vmatrix} 3-\lambda & 5 & -5 & 5 \\ 3 & 1-\lambda & 3 & -3 \\ -2 & 2 & -\lambda & 2 \\ 0 & 4 & -6 & 8-\lambda \end{vmatrix} = \begin{vmatrix} 3-\lambda & 5 & -5 & 8-\lambda \\ 3 & 1-\lambda & 3 & 0 \\ -2 & 2 & -\lambda & 0 \\ 0 & 4 & -6 & 8-\lambda \end{vmatrix} = \begin{vmatrix} 3-\lambda & 1 & 1 & 0 \\ 3 & 1-\lambda & 3 & 0 \\ -2 & 2 & -\lambda & 0 \\ 0 & 4 & -6 & 8-\lambda \end{vmatrix}$

$= (8-\lambda)\begin{vmatrix} 3-\lambda & 1 & 1 \\ 3 & 1-\lambda & 3 \\ -2 & 2 & -\lambda \end{vmatrix} = (8-\lambda)(4-\lambda)(2-\lambda)(2+\lambda)$, so the eigenvalues are $\lambda = 8$, $\lambda = 4$, $\lambda = 2$, and $\lambda = -2$. For $\lambda = 2$, the eigenvectors are the solutions of

$\begin{bmatrix} 1 & 5 & -5 & 5 \\ 3 & -1 & 3 & -3 \\ -2 & 2 & -2 & 2 \\ 0 & 4 & -6 & 6 \end{bmatrix}\begin{bmatrix} x_1 \\ x_2 \\ x_3 \\ x_4 \end{bmatrix} = \mathbf{0}$, so the eigenvectors are vectors of the form $r\begin{bmatrix} 0 \\ 0 \\ 1 \\ 1 \end{bmatrix}$. For $\lambda = 4$, the eigenvectors are the solutions of $\begin{bmatrix} -1 & 5 & -5 & 5 \\ 3 & -3 & 3 & -3 \\ -2 & 2 & -4 & 2 \\ 0 & 4 & -6 & 4 \end{bmatrix}\begin{bmatrix} x_1 \\ x_2 \\ x_3 \\ x_4 \end{bmatrix} = \mathbf{0}$, so the eigenvectors are vectors of the form $s\begin{bmatrix} 0 \\ 1 \\ 0 \\ -1 \end{bmatrix}$. For $\lambda = 8$, the eigenvectors are the

solutions of $\begin{bmatrix} -5 & 5 & -5 & 5 \\ 3 & -7 & 3 & -3 \\ -2 & 2 & -8 & 2 \\ 0 & 4 & -6 & 0 \end{bmatrix}\begin{bmatrix} x_1 \\ x_2 \\ x_3 \\ x_4 \end{bmatrix} = \mathbf{0}$, so the eigenvectors are vectors of the form

$t\begin{bmatrix} 1 \\ 0 \\ 0 \\ 1 \end{bmatrix}$. For $\lambda = -2$, the eigenvectors are the solutions of $\begin{bmatrix} 5 & 5 & -5 & 5 \\ 3 & 3 & 3 & -3 \\ -2 & 2 & 2 & 2 \\ 0 & 4 & -6 & 10 \end{bmatrix}\begin{bmatrix} x_1 \\ x_2 \\ x_3 \\ x_4 \end{bmatrix} = \mathbf{0}$, so the eigenvectors are vectors of the form $p\begin{bmatrix} 1 \\ -1 \\ 1 \\ 1 \end{bmatrix}$.

17. $\begin{vmatrix} 1-\lambda & 0 \\ 0 & 1-\lambda \end{vmatrix} = (1-\lambda)^2$, so the only eigenvalue is $\lambda = 1$. The eigenvectors are the solutions of $\begin{bmatrix} 0 & 0 \\ 0 & 0 \end{bmatrix}\begin{bmatrix} x_1 \\ x_2 \end{bmatrix} = \mathbf{0}$, so the eigenvectors are all the vectors in $\mathbf{R}^2$. The transformation represented by the identity matrix is the identity transformation that maps each vector in $\mathbf{R}^2$ into itself.

18. $\begin{vmatrix} 3-\lambda & 0 \\ 0 & 3-\lambda \end{vmatrix} = (3-\lambda)^2$, so the only eigenvalue is $\lambda = 3$. The eigenvectors are the solutions of $\begin{bmatrix} 0 & 0 \\ 0 & 0 \end{bmatrix}\begin{bmatrix} x_1 \\ x_2 \end{bmatrix} = \mathbf{0}$, so the eigenvectors are all the vectors in $\mathbf{R}^2$. The transformation represented by the given matrix is an expansion transformation that maps each vector **v** in $\mathbf{R}^2$ into the vector 3**v**. Thus each image in $\mathbf{R}^2$ has the same direction as the original vector.

19. $\begin{vmatrix} -2-\lambda & 0 \\ 0 & -2-\lambda \end{vmatrix} = (-2-\lambda)^2$, so the only eigenvalue is $\lambda = -2$. The eigenvectors are the solutions of $\begin{bmatrix} 0 & 0 \\ 0 & 0 \end{bmatrix}\begin{bmatrix} x_1 \\ x_2 \end{bmatrix} = \mathbf{0}$, so the eigenvectors are all the vectors in $\mathbf{R}^2$. The transformation represented by the given matrix maps each vector **v** in $\mathbf{R}^2$ into the vector −2**v**. Thus each image in $\mathbf{R}^2$ has the direction opposite the original vector.

20. $\begin{vmatrix} -\lambda & -1 \\ 1 & -\lambda \end{vmatrix} = \lambda^2 + 1 \neq 0$ for any real value of λ, so there are no real eigenvalues.

The given matrix is a rotation matrix that rotates each vector in $\mathbf{R}^2$ through a 90° angle. Thus no vector has the same or opposite direction as its image.

21. $\begin{vmatrix} 1-\lambda & 1 \\ -2 & -1-\lambda \end{vmatrix} = (1-\lambda)(-1-\lambda) + 2 = \lambda^2 + 1 \neq 0$ for any real value of λ, so there are no

real eigenvalues. Thus no eigenvectors. The image of every vector lies on a line different from the line on which the original vector lies.

22. $$|I_n - \lambda I_n| = \left| \begin{bmatrix} 1 & 0 & \dots & 0 \\ 0 & 1 & \dots & 0 \\ \vdots & \vdots & \vdots & \vdots \\ 0 & 0 & \dots & 1 \end{bmatrix} - \begin{bmatrix} \lambda & 0 & \dots & 0 \\ 0 & \lambda & \dots & 0 \\ \vdots & \vdots & \vdots & \vdots \\ 0 & 0 & \dots & \lambda \end{bmatrix} \right| = \begin{vmatrix} 1-\lambda & 0 & \dots & 0 \\ 0 & 1-\lambda & \dots & 0 \\ \vdots & \vdots & \vdots & \vdots \\ 0 & 0 & \dots & 1-\lambda \end{vmatrix} = (1-\lambda)^n|I_n| = (1-\lambda)^n$$

Eigenvalues of I_n are given by $|I_n - lI_n| = 0$. Thus eigenvalue $\lambda=1$, of multiplicity n.
$I_n\mathbf{x} = 1\mathbf{x}$ is true for all x in $\mathbf{R}^n$. Thus the eigenspace is $\mathbf{R}^n$. Note that if we multiply any vector in $\mathbf{R}^n$ by I_n it remains unchanged in direction and magnitude, confirming this result.
Every vector is an eigenvector. Geometrically, the matrix maps every vector into itself.

23. $$|A - \lambda I_n| = \left| \begin{bmatrix} 1 & 1 & \dots & 1 \\ 1 & 1 & \dots & 1 \\ \vdots & \vdots & \vdots & \vdots \\ 1 & 1 & \dots & 1 \end{bmatrix} - \begin{bmatrix} \lambda & 0 & \dots & 0 \\ 0 & \lambda & \dots & 0 \\ \vdots & \vdots & \vdots & \vdots \\ 0 & 0 & \dots & \lambda \end{bmatrix} \right| = \begin{vmatrix} 1-\lambda & 1 & \dots & 1 \\ 1 & 1-\lambda & \dots & 1 \\ \vdots & \vdots & \vdots & \vdots \\ 1 & 1 & \dots & 1-\lambda \end{vmatrix} =$$

$$\begin{vmatrix} n-\lambda & n-\lambda & \dots & n-\lambda \\ 1 & 1-\lambda & \dots & 1 \\ \vdots & \vdots & \vdots & \vdots \\ 1 & 1 & \dots & 1-\lambda \end{vmatrix} = (n-\lambda)\begin{vmatrix} 1 & 1 & \dots & 1 \\ 1 & 1-\lambda & \dots & 1 \\ \vdots & \vdots & \vdots & \vdots \\ 1 & 1 & \dots & 1-\lambda \end{vmatrix} = (n-\lambda)\begin{vmatrix} 1 & 1 & \dots & 1 \\ 0 & \lambda & \dots & 0 \\ \vdots & \vdots & \vdots & \vdots \\ 0 & 0 & \dots & \lambda \end{vmatrix} =$$

(on adding each of the following rows to the first)

(on subtracting the first row from each of the following rows)

$$(n-\lambda)\begin{vmatrix} \lambda & 0 & \dots & 0 \\ 0 & \lambda & \dots & 0 \\ \vdots & \vdots & & \vdots \\ 0 & 0 & \dots & \lambda \end{vmatrix} = (n-\lambda)\lambda^{n-1} = 0.$$ Thus $\lambda = n$ or 0.

$\lambda = n$: Eigenvectors given by $A\mathbf{x} = n\mathbf{x}$. Let $\mathbf{x} = (a, b, \dots, z)$. Then $a+b+\dots+z = na$; $a+b+\dots +z = nb$; ... ; $a+b+\dots +z = nz$. Thus $na=nb=,\dots,=nz$. $a=b=\dots=z$. Eigenvectors are of the form $k(1,1,\dots,1)$.
$\lambda = 0$: Eigenvectors given by $A\mathbf{x}=0\mathbf{x}$. $A\mathbf{x} = 0$. Let $\mathbf{x} = (a, b, \dots, z)$. Thus $a+b+\dots+z = 0$. $z=-(a+b+\dots)$. Eigenvectors are of the form $(a, b, \dots, -(a+b+\dots))$.

24. If A is a diagonal matrix with diagonal elements a_{ii}, then $A - \lambda I_n$ is also a diagonal matrix
with diagonal elements $a_{ii} - \lambda$. Thus $|A - \lambda I_n|$ is the product of the terms $a_{ii} - \lambda$, and the solutions of the equation $|A - \lambda I_n| = 0$ are the values $\lambda = a_{ii}$, the diagonal elements of A.

25. If A is an upper triangular matrix with diagonal elements a_{ii}, then $A - \lambda I_n$ is also an upper
triangular matrix with diagonal elements $a_{ii} - \lambda$. $|A - \lambda I_n|$ is the product of the terms $a_{ii} - \lambda$, and the solutions of the equation $|A - \lambda I_n| = 0$ are the values $\lambda = a_{ii}$, the diagonal elements of A.

26. $(A - \lambda I_n)^t = A^t - (\lambda I_n)^t = A^t - \lambda I_n$, so $|A - \lambda I_n| = |(A - \lambda I_n)^t| = |A^t - \lambda I_n|$, that is, A and A^t have the same characteristic polynomial and therefore the same eigenvalues.

27. $|A - 0I_n| = |A|$, so $|A - 0I_n| = 0$ if and only if $|A| = 0$.

28. If λ is an eigenvalue of A and $\mathbf{x}$ is an eigenvector corresponding to λ, then $A\mathbf{x} = \lambda\mathbf{x}$,

$A^2\mathbf{x} = A(A\mathbf{x}) = A(\lambda\mathbf{x}) = \lambda A\mathbf{x} = \lambda(\lambda\mathbf{x}) = \lambda^2\mathbf{x}$ (i.e., λ^2 is an eigenvalue of A^2 with

corresponding eigenvector $\mathbf{x}$), and $A^m\mathbf{x} = A^{m-1}(A\mathbf{x}) = A^{m-1}(\lambda\mathbf{x}) = \lambda A^{m-1}\mathbf{x} = \lambda A^{m-2}(A\mathbf{x})$

$= \lambda A^{m-2}(\lambda\mathbf{x}) = \lambda^2 A^{m-2}\mathbf{x} = \lambda^2 A^{m-3}(A\mathbf{x}) = \dots = \lambda^{m-1}A(A\mathbf{x}) = \lambda^{m-1}A(\lambda\mathbf{x}) = \lambda^m\mathbf{x}$, so that λ^m is an eigenvalue of A^m with corresponding eigenvector $\mathbf{x}$.

29. Have that A**x** = λ**x**. Thus $A^{-1}(A\mathbf{x}) = A^{-1}(\lambda\mathbf{x})$. $(A^{-1}A)\mathbf{x} = \lambda A^{-1}\mathbf{x}$. $(I)\mathbf{x} = \lambda A^{-1}\mathbf{x}$.

$\mathbf{x} = \lambda A^{-1}\mathbf{x}$. $\lambda^{-1}\mathbf{x} = A^{-1}\mathbf{x}$.

30. Have that A**x** = λ**x**. A**x** - c**x** = λ**x** - c**x**. (A - cI)**x** = (λ - c)**x**.
Thus λ - c is an eigenvalue of A - cI with corresponding eigenvector **x**.

31. The characteristic polynomial for the nxn zero matrix O_n is λ^n. Thus the only eigenvalue of O_n is λ = 0. However if λ is an eigenvalue of A then λ^k is an eigenvalue of A^k = O_n
(see Exercise 26). Thus $\lambda^k = 0$ so λ = 0.

32. The characteristic polynomial of A is $|A - \lambda I_n| = \lambda^n + c_{n-1}\lambda^{n-1} + \ldots + c_1\lambda + c_0$.
Substituting λ = 0, this equation becomes $|A| = c_0$.

33. $\begin{vmatrix} 1-\lambda & 2 \\ 1 & -\lambda \end{vmatrix} = (1-\lambda)(-\lambda) - 2 = \lambda^2 - \lambda - 2 = (\lambda-2)(\lambda+1)$, so the eigenvalues are λ = 2 and

λ = −1. The corresponding eigenvectors (written as row vectors) are [2 1] and [1 −1].

$\begin{vmatrix} 2-\lambda & 3 \\ 1 & -\lambda \end{vmatrix} = (2-\lambda)(-\lambda) - 3 = \lambda^2 - 2\lambda - 3 = (\lambda-3)(\lambda+1)$, so the eigenvalues are λ = 3

and λ = −1. The corresponding eigenvectors are [3 1] and [1 −1].

$\begin{vmatrix} 3-\lambda & 4 \\ 1 & -\lambda \end{vmatrix} = (3-\lambda)(-\lambda) - 4 = \lambda^2 - 3\lambda - 4 = (\lambda-4)(\lambda+1)$, so the eigenvalues are λ = 4

and λ = −1. The corresponding eigenvectors are [4 1] and [1 −1].

$\begin{vmatrix} 4-\lambda & 5 \\ 1 & -\lambda \end{vmatrix} = (4-\lambda)(-\lambda) - 5 = \lambda^2 - 4\lambda - 5 = (\lambda-5)(\lambda+1)$, so the eigenvalues are λ = 5

and λ = −1. The corresponding eigenvectors are [5 1] and [1 −1].

Conjecture: The matrix $\begin{bmatrix} a & a+1 \\ 1 & 0 \end{bmatrix}$ has eigenvalues λ = a+1 and λ = −1 and corresponding eigenvectors [a+1 1] and [1 −1].

Proof: $\begin{bmatrix} a & a+1 \\ 1 & 0 \end{bmatrix}\begin{bmatrix} a+1 \\ 1 \end{bmatrix} = (a+1)\begin{bmatrix} a+1 \\ 1 \end{bmatrix}$ and $\begin{bmatrix} a & a+1 \\ 1 & 0 \end{bmatrix}\begin{bmatrix} 1 \\ -1 \end{bmatrix} = \begin{bmatrix} -1 \\ 1 \end{bmatrix} = (-1)\begin{bmatrix} 1 \\ -1 \end{bmatrix}$.

34. (a) $\begin{vmatrix} -\lambda & 2 \\ -1 & 3-\lambda \end{vmatrix} = (-\lambda)(3-\lambda) + 2 = \lambda^2 - 3\lambda + 2.$

$$\begin{bmatrix} 0 & 2 \\ -1 & 3 \end{bmatrix}^2 - 3\begin{bmatrix} 0 & 2 \\ -1 & 3 \end{bmatrix} + 2\begin{bmatrix} 1 & 0 \\ 0 & 1 \end{bmatrix} = \begin{bmatrix} -2 & 6 \\ -3 & 7 \end{bmatrix} + \begin{bmatrix} 2 & -6 \\ 3 & -7 \end{bmatrix} = \begin{bmatrix} 0 & 0 \\ 0 & 0 \end{bmatrix}.$$

(b) $\begin{vmatrix} 8-\lambda & -10 \\ 5 & -7-\lambda \end{vmatrix} = (8-\lambda)(-7-\lambda) + 50 = \lambda^2 - \lambda - 6.$

$$\begin{bmatrix} 8 & -10 \\ 5 & -7 \end{bmatrix}^2 - \begin{bmatrix} 8 & -10 \\ 5 & -7 \end{bmatrix} - 6\begin{bmatrix} 1 & 0 \\ 0 & 1 \end{bmatrix} = \begin{bmatrix} 14 & -10 \\ 5 & -1 \end{bmatrix} + \begin{bmatrix} -14 & 10 \\ -5 & 1 \end{bmatrix} = \begin{bmatrix} 0 & 0 \\ 0 & 0 \end{bmatrix}.$$

(c) $\begin{vmatrix} 6-\lambda & -8 \\ 4 & -6-\lambda \end{vmatrix} = (6-\lambda)(-6-\lambda) + 32 = \lambda^2 - 4.$

$$\begin{bmatrix} 6 & -8 \\ 4 & -6 \end{bmatrix}^2 - 4\begin{bmatrix} 1 & 0 \\ 0 & 1 \end{bmatrix} = \begin{bmatrix} 4 & 0 \\ 0 & 4 \end{bmatrix} - \begin{bmatrix} 4 & 0 \\ 0 & 4 \end{bmatrix} = \begin{bmatrix} 0 & 0 \\ 0 & 0 \end{bmatrix}.$$

(d) $\begin{vmatrix} -1-\lambda & 5 \\ -10 & 14-\lambda \end{vmatrix} = (-1-\lambda)(14-\lambda) + 50 = \lambda^2 - 13\lambda + 36.$

$$\begin{bmatrix} -1 & 5 \\ -10 & 14 \end{bmatrix}^2 - 13\begin{bmatrix} -1 & 5 \\ -10 & 14 \end{bmatrix} + 36\begin{bmatrix} 1 & 0 \\ 0 & 1 \end{bmatrix} = \begin{bmatrix} -49 & 65 \\ -130 & 146 \end{bmatrix} + \begin{bmatrix} 49 & -65 \\ 130 & -146 \end{bmatrix}$$

$$= \begin{bmatrix} 0 & 0 \\ 0 & 0 \end{bmatrix}.$$

35. (a) False: Let A be a 3x3 matrix. The characteristic equation of A is $|A - \lambda I_n| = 0$. This will be a polynomial of degree 3 in λ. Thus 3, 2, or 1 distinct roots. 3, 2, or 1 distinct eigenvalues. In general nxn matrix has n, n-1, ..., 2, or 1 distinct eigenvalues.

(b) True: Suppose $A\mathbf{x} = \lambda_1\mathbf{x}$ and $A\mathbf{x} = \lambda_2\mathbf{x}$. Then $\lambda_1\mathbf{x} = \lambda_2\mathbf{x}$, $\lambda_1\mathbf{x} - \lambda_2\mathbf{x} = \mathbf{0}$, $(\lambda_1 - \lambda_2)\mathbf{x} = \mathbf{0}$. Since eigenvector $\mathbf{x}$ is nonzero, $(\lambda_1 - \lambda_2) = 0$, $\lambda_1 = \lambda_2$.

(c) False: Set of all eigenvectors for a given eigenvalue λ lie in a subspace. Thus the sum
of any two of these is an eigenvector in that subspace. However sum of two eigenvectors from different eigenspaces is not an eigenvector - Let $A\mathbf{x}_1 = \lambda_1\mathbf{x}_1$ and $A\mathbf{x}_2 = \lambda_2\mathbf{x}_2$. Add, $A\mathbf{x}_1+A\mathbf{x}_2 = \lambda_1\mathbf{x}_1+\lambda_2\mathbf{x}_2$, $A(\mathbf{x}_1+\mathbf{x}_2) = \lambda_1\mathbf{x}_1+\lambda_2\mathbf{x}_2 \neq \lambda(\mathbf{x}_1+\mathbf{x}_2)$ for any value of λ since $\lambda_1 \neq \lambda_2$.Thus in general, the sum of two eigenvectors is not an eigenevector.

(d) True: e.g. Let $A = \begin{bmatrix} 2 & 0 \\ 0 & -2 \end{bmatrix}$, $B = \begin{bmatrix} -2 & 0 \\ 0 & 2 \end{bmatrix}$. Then $|A-\lambda I|= (2-\lambda)(-2-\lambda)=0$, $\lambda=2$ or -2. $|B-\lambda I|= (-2-\lambda)(2-\lambda)=0$, $\lambda=-2$ or 2. A and B have the same eigenvalues.

<u>Exercise Set 3.5</u>

1. The eigenvectors of $\lambda = 1$ are vectors of the form $r\begin{bmatrix} 2 \\ 1 \end{bmatrix}$. If there is no change in total population $2r + r = 245 + 52 = 297$, so $r = 297/3$. Thus the long-term prediction is that population in metropolitan areas will be $2r=198$ million and population in nonmetropolitan areas will be $r=99$ million.

2. The eigenvectors of $\lambda = 1$ are vectors of the form $r\begin{bmatrix} .7 \\ 1.3 \\ 1 \end{bmatrix}$. If there is no change in total population $.7r + 1.3r + r = 82 + 163 + 52 = 297$, so $r = 297/3$. Thus the long-term prediction is that population in the cities will be $.7r = 60.3$ million, population in the suburbs will be $1.3r = 128.7$ million, and population in nonmetropolitan areas will be $r=99$ million.

3. $P^2 = \begin{bmatrix} .375 & .25 & .125 \\ .5 & .5 & .5 \\ .125 & .25 & .375 \end{bmatrix}$, and since all terms are positive, P is regular. The eigenvectors of $\lambda = 1$ are vectors of the form $r\begin{bmatrix} 1 \\ 2 \\ 1 \end{bmatrix}$. The powers of P approach the stochastic matrix $Q = \begin{bmatrix} s & s & s \\ 2s & 2s & 2s \\ s & s & s \end{bmatrix}$, so $s = .25$ and $Q = \begin{bmatrix} .25 & .25 & .25 \\ .5 & .5 & .5 \\ .25 & .25 & .25 \end{bmatrix}$.

The columns of Q indicate that when guinea pigs are bred with hybrids, only the long-term distribution of types AA, Aa, and aa will be 1:2:1. That is, the long-term probabilities of Types AA, Aa, and aa are .25, .5, and .25.

4.

213/326=.65, 117/511=.23 (to 2 dec places). $P = \begin{array}{c} \text{wet} \quad \text{dry} \\ \begin{bmatrix} .65 & .23 \\ .35 & .77 \end{bmatrix} \end{array} \begin{array}{l} \text{wet} \\ \text{dry} \end{array}$

(a) $P^2 = \begin{bmatrix} .5 & .33 \\ .5 & .67 \end{bmatrix}$. If Thursday is dry, the probability that Saturday will also be dry is .67, the (2,2) term in P^2.

(b) The eigenvectors of P corresponding to $\lambda = 1$ are vectors of the form $r\begin{bmatrix} 23 \\ 35 \end{bmatrix}$.

Thus the powers of P approach the stochastic matrix $Q = \begin{bmatrix} 23s & 23s \\ 35s & 35s \end{bmatrix}$, so

$(23+35)s = 1$ and $s = \frac{1}{58}$. $\frac{23}{58} = .4$, $\frac{35}{58} = .6$, and $Q = \begin{bmatrix} .4 & .4 \\ .6 & .6 \end{bmatrix}$, so the long-term probability for a wet day in December is .4 and for a dry day is .6.

5. P is regular since all terms of P^2 are positive. The eigenvectors of P corresponding to $\lambda = 1$ are vectors of the form $r\begin{bmatrix} 3 \\ 3 \\ 2 \end{bmatrix}$; thus the distribution of rats in rooms 1, 2, and 3 is 3:3:2. The powers of P approach the stochastic matrix $Q = \begin{bmatrix} 3s & 3s & 3s \\ 3s & 3s & 3s \\ 2s & 2s & 2s \end{bmatrix}$, so $3s + 3s + 2s = 1$ and $s = 1/8$. The long-term probability that a given rat will be in room 2 is therefore 3/8.

6.

$$P = \begin{bmatrix} 0 & 1/3 & 0 & 1/4 \\ 1/2 & 0 & 1/3 & 1/4 \\ 0 & 1/3 & 0 & 1/2 \\ 1/2 & 1/3 & 2/3 & 0 \end{bmatrix}$$

(columns: room 1, 2, 3, 4; rows: 1, 2, 3, 4)

P is regular since every term in P^2 is positive.

The eigenvectors of P corresponding to $\lambda = 1$ are vectors of the form

$r\begin{bmatrix} 2 \\ 3 \\ 3 \\ 4 \end{bmatrix}$; thus the distribution of rats in rooms 1, 2, 3, and 4 is 2:3:3:4. The powers of P

approach the stochastic matrix $Q = \begin{bmatrix} 2s & 2s & 2s & 2s \\ 3s & 3s & 3s & 3s \\ 3s & 3s & 3s & 3s \\ 4s & 4s & 4s & 4s \end{bmatrix}$, so $2s + 3s + 3s + 4s = 1$

and $s = 1/12$. The long-term probability that a given rat will be in room 4 is thus $4/12 = 1/3$.

7. $P = \begin{bmatrix} .75 & .20 \\ .25 & .80 \end{bmatrix}$. The eigenvectors of $\lambda = 1$ are vectors of the form $r\begin{bmatrix} 4 \\ 5 \end{bmatrix}$. The powers of P approach $Q = \begin{bmatrix} 4s & 4s \\ 5s & 5s \end{bmatrix}$, so $4s + 5s = 1$ and $s = 1/9$. If current trends continue, the eventual distribution will be 4/9 = 44.4% using company A and 5/9 = 55.6% using company B.

8. $P = \begin{bmatrix} .8 & .2 & .05 \\ .05 & .75 & .05 \\ .15 & .05 & .9 \end{bmatrix}$. The eigenvectors of $\lambda = 1$ are vectors of the form $r\begin{bmatrix} 9 \\ 5 \\ 16 \end{bmatrix}$. The powers of P approach $Q = \begin{bmatrix} 9s & 9s & 9s \\ 5s & 5s & 5s \\ 16s & 16s & 16s \end{bmatrix}$, so $9s + 5s + 16s = 1$ and $s = \frac{1}{30}$. If current buying patterns continue, the eventual distribution will be 9/30 = 30% using product I, $5/30 = 16\frac{2}{3}$ % using product II, and $16/30 = 53\frac{1}{3}$ % using product III.

9. The sum of the terms in each column of a stochastic matrix A is 1, so the sum of the terms in each column of A – I is zero. It has previously been proved (Exercise 18, Section 3.2) that if the sum of the terms in each column of a matrix is zero, the determinant of the matrix is zero. Thus |A – 1I| = |A – I| = 0, and 1 is an eigenvalue of A.

Chapter 3 Review Exercises

1. (a) 3x1 – 2x5 = –7. (b) –3x6 – 0x1 = –18. (c) 9x4 – 7x1 = 29.

2. (a) $M_{12} = \begin{vmatrix} -3 & 1 \\ 7 & 2 \end{vmatrix} = -3\text{x}2 - 1\text{x}7 = -13.$ $C_{12} = (-1)^{1+2} M_{12} = 13.$

(b) $M_{31} = \begin{vmatrix} 1 & 0 \\ 4 & 1 \end{vmatrix} = 1\text{x}1 - 0\text{x}4 = 1.$ $C_{31} = (-1)^{3+1} M_{31} = 1.$

(c) $M_{22} = \begin{vmatrix} 2 & 0 \\ 7 & 2 \end{vmatrix} = 2\text{x}2 - 0\text{x}4 = 4.$ $C_{22} = (-1)^{2+2} M_{22} = 4.$

3. (a) using row 1:

$$\begin{vmatrix} 1 & 2 & -3 \\ 0 & 2 & 5 \\ 4 & 1 & 2 \end{vmatrix} = 1\begin{vmatrix} 2 & 5 \\ 1 & 2 \end{vmatrix} - 2\begin{vmatrix} 0 & 5 \\ 4 & 2 \end{vmatrix} + (-3)\begin{vmatrix} 0 & 2 \\ 4 & 1 \end{vmatrix} = -1 + 40 + 24 = 63.$$

using column 1:

$$= 1\begin{vmatrix} 2 & 5 \\ 1 & 2 \end{vmatrix} - 0\begin{vmatrix} 2 & -3 \\ 1 & 2 \end{vmatrix} + 4\begin{vmatrix} 2 & -3 \\ 2 & 5 \end{vmatrix} = -1 + 0 + 64 = 63.$$

(b) using row 3:

$$\begin{vmatrix} 0 & 5 & 3 \\ 2 & -3 & 1 \\ 2 & 7 & 3 \end{vmatrix} = 2\begin{vmatrix} 5 & 3 \\ -3 & 1 \end{vmatrix} - 7\begin{vmatrix} 0 & 3 \\ 2 & 1 \end{vmatrix} + 3\begin{vmatrix} 0 & 5 \\ 2 & -3 \end{vmatrix} = 28 + 42 - 30 = 40.$$

using column 2:

$$= -5\begin{vmatrix} 2 & 1 \\ 2 & 3 \end{vmatrix} + (-3)\begin{vmatrix} 0 & 3 \\ 2 & 3 \end{vmatrix} - 7\begin{vmatrix} 0 & 3 \\ 2 & 1 \end{vmatrix} = -20 + 18 + 42 = 40.$$

4. $\begin{vmatrix} x & x \\ 2 & x-3 \end{vmatrix} = x(x-3) - 2x = x^2 - 5x = -6$, $x^2 - 5x + 6 = 0$. Thus $(x-3)(x-2) = 0$,

so $x = 3$ or $x = 2$.

5. (a) $$\begin{vmatrix} 1 & 2 & -1 \\ 3 & 1 & 1 \\ 2 & 4 & 1 \end{vmatrix} \underset{R3+(-2)R1}{=} \begin{vmatrix} 1 & 2 & -1 \\ 3 & 1 & 1 \\ 0 & 0 & 3 \end{vmatrix} = 3\begin{vmatrix} 1 & 2 \\ 3 & 1 \end{vmatrix} = -15.$$

(b) $$\begin{vmatrix} 5 & 3 & 4 \\ 4 & 6 & 1 \\ 2 & -3 & 7 \end{vmatrix} \underset{\substack{R2+(-2)R1 \\ R3+R1}}{=} \begin{vmatrix} 5 & 3 & 4 \\ -6 & 0 & -7 \\ 7 & 0 & 11 \end{vmatrix} = -3\begin{vmatrix} -6 & -7 \\ 7 & 11 \end{vmatrix} = 51.$$

(c) $$\begin{vmatrix} 1 & 4 & -2 \\ 2 & 3 & 1 \\ -1 & 5 & 6 \end{vmatrix} \underset{\substack{R2+(-2)R1 \\ R3+R1}}{=} \begin{vmatrix} 1 & 4 & -2 \\ 0 & -5 & 5 \\ 0 & 9 & 4 \end{vmatrix} = \begin{vmatrix} -5 & 5 \\ 9 & 4 \end{vmatrix} = -65.$$

6. (a) This matrix can be obtained from A by multiplying row 2 by 3, so its determinant is $3|A| = 6$.

(b) This matrix can be obtained from A by adding -2 times row 1 to row 2, so its determinant is $|A| = 2$.

(c) This matrix can be obtained from A by multiplying row 1 by 2, row 2 by -1, and row 3 by 3, so its determinant is $2\times -1 \times 3 \times |A| = -12$.

7. (a) $$\begin{vmatrix} 1 & 2 & 4 \\ -1 & 4 & 3 \\ 2 & 0 & 5 \end{vmatrix} \underset{R2+(-2)R1}{=} \begin{vmatrix} 1 & 2 & 4 \\ -3 & 0 & -5 \\ 2 & 0 & 5 \end{vmatrix} = -2\begin{vmatrix} -3 & -5 \\ 2 & 5 \end{vmatrix} = 10.$$

(b) $\begin{vmatrix} -1 & 3 & 2 \\ 0 & 5 & 2 \\ 1 & 7 & 6 \end{vmatrix} \underset{R3+R1}{=} \begin{vmatrix} -1 & 3 & 2 \\ 0 & 5 & 2 \\ 0 & 10 & 8 \end{vmatrix} = -1\begin{vmatrix} 5 & 2 \\ 10 & 8 \end{vmatrix} = -20.$

(c) $\begin{vmatrix} 2 & -3 & 5 \\ 4 & 0 & 6 \\ 1 & 2 & 7 \end{vmatrix} \underset{R3+(2/3)R1}{=} \begin{vmatrix} 2 & -3 & 5 \\ 4 & 0 & 6 \\ 7/3 & 0 & 31/3 \end{vmatrix} = -(-3)\begin{vmatrix} 4 & 6 \\ 7/3 & 31/3 \end{vmatrix} = 82.$

8. (a) $|3A| = 3^3\,|A| = 27x-2 = -54.$ (b) $|2AA^t| = 2^3\,|A||A^t| = 8|A||A| = 32.$

(c) $|A^3| = |A|^3 = (-2)^3 = -8.$

(d) $|(A^tA)^2| = (|A^tA|)^2 = (|A^t||A|)^2 = (|A||A|)^2 = |A|^4 = 16.$

(e) $|(A^t)^3| = (|A^t|)^3 = (|A|)^3 = (-2)^3 = -8.$

(f) $|2A^t(A^{-1})^2| = (2)^3\,|A^t||(A^{-1})^2| = 8|A^t||A^{-1}|^2 = 8|A|(1/|A|)^2 = 8/|A| = -4.$

9. $|B| \neq 0$ so B^{-1} exists, and A and B^{-1} can be multiplied. Let $C = AB^{-1}$. Then $CB = AB^{-1}B = A$.

10. $|C^{-1}AC| = |C^{-1}||AC| = |C^{-1}||A||C| = |C^{-1}||C||A| = |C^{-1}C||A| = |A|$

11. If A is upper triangular, then any element in A is zero if its row number is greater than its column number. If $i > j$, then for every k with $1 \le k \le n$, $a_{ik}a_{kj} = 0$ because either $i \ge k$ so that $a_{ik} = 0$, or $k \ge i > j$ so that $a_{kj} = 0$. Thus if $i > j$, the (i,j)th term $a_{i1}a_{1j} + a_{i2}a_{2j} + \ldots + a_{in}a_{nj}$ of A^2 is zero because each summand is zero. So A^2 is upper triangular. The proof is similar for lower triangular matrices.

12. If $A^2 = A$, then $|A||A| = |A|$ so $|A| = 1$ or zero. If A is also invertible, then $|A| \neq 0$ so $|A| = 1$.

13. (a) $\begin{vmatrix} 3 & 5 \\ 1 & 2 \end{vmatrix} = 1$, so $\begin{bmatrix} 3 & 5 \\ 1 & 2 \end{bmatrix}^{-1} = \begin{bmatrix} 2 & -5 \\ -1 & 3 \end{bmatrix}.$

(b) $\begin{vmatrix} 3 & 2 \\ -1 & 5 \end{vmatrix} = 17$, so $\begin{bmatrix} 3 & 2 \\ -1 & 5 \end{bmatrix}^{-1} = \frac{1}{17}\begin{bmatrix} 5 & -2 \\ 1 & 3 \end{bmatrix}.$

(c) $\begin{vmatrix} 1 & 4 & -1 \\ 0 & 2 & 0 \\ 1 & 6 & -1 \end{vmatrix} = 0$, so the inverse does not exist.

(d) $\begin{vmatrix} 2 & 1 & 3 \\ 0 & 2 & 9 \\ 4 & 2 & 11 \end{vmatrix} = 20$, so $\begin{bmatrix} 2 & 1 & 3 \\ 0 & 2 & 9 \\ 4 & 2 & 11 \end{bmatrix}^{-1}$

$$= \frac{1}{20}\begin{bmatrix} \begin{vmatrix} 2 & 9 \\ 2 & 11 \end{vmatrix} & -\begin{vmatrix} 1 & 3 \\ 2 & 11 \end{vmatrix} & \begin{vmatrix} 1 & 3 \\ 2 & 9 \end{vmatrix} \\ -\begin{vmatrix} 0 & 9 \\ 4 & 11 \end{vmatrix} & \begin{vmatrix} 2 & 3 \\ 4 & 11 \end{vmatrix} & -\begin{vmatrix} 2 & 3 \\ 0 & 9 \end{vmatrix} \\ \begin{vmatrix} 0 & 2 \\ 4 & 2 \end{vmatrix} & -\begin{vmatrix} 2 & 1 \\ 4 & 2 \end{vmatrix} & \begin{vmatrix} 2 & 1 \\ 0 & 2 \end{vmatrix} \end{bmatrix} = \frac{1}{20}\begin{bmatrix} 4 & -5 & 3 \\ 36 & 10 & -18 \\ -8 & 0 & 4 \end{bmatrix}.$$

14. (a) $x_1 = \dfrac{\begin{vmatrix} -1 & 1 \\ 18 & -5 \end{vmatrix}}{\begin{vmatrix} 2 & 1 \\ 3 & -5 \end{vmatrix}} = \dfrac{-13}{-13} = 1$, $x_2 = \dfrac{\begin{vmatrix} 2 & -1 \\ 3 & 18 \end{vmatrix}}{\begin{vmatrix} 2 & 1 \\ 3 & -5 \end{vmatrix}} = \dfrac{39}{-13} = -3$.

(b) $|A| = \begin{vmatrix} 1 & 1 & 1 \\ 2 & -1 & 3 \\ 4 & 5 & 1 \end{vmatrix} = 8$, $|A_1| = \begin{vmatrix} 1 & 1 & 1 \\ 5 & -1 & 3 \\ 3 & 5 & 1 \end{vmatrix} = 16$, $|A_2| = \begin{vmatrix} 1 & 1 & 1 \\ 2 & 5 & 3 \\ 4 & 3 & 1 \end{vmatrix} = -8$,

$|A_3| = \begin{vmatrix} 1 & 1 & 1 \\ 2 & -1 & 5 \\ 4 & 5 & 3 \end{vmatrix} = 0$, so $x_1 = \dfrac{16}{8} = 2$, $x_2 = \dfrac{-8}{8} = -1$, and $x_3 = \dfrac{0}{8} = 0$.

15. If A is not invertible, $|A|=0$. The (i,j)th term of A[adj(A)] is $a_{i1}C_{j1} + a_{i2}C_{j2} + \ldots + a_{in}C_{jn}$. If $i = j$, this term is $|A| = 0$ and if $i \neq j$ it is the determinant of the matrix obtained from A by replacing row j with row i; that is, it is the determinant of a matrix having two equal rows, and so it is zero. Thus A[adj(A)] is the zero matrix.

16. $|A|$ is the product of the diagonal elements, and since $|A| \neq 0$ all diagonal elements must be nonzero.

17. If $|A| = \pm 1$ then $A^{-1} = \pm\ \text{adj}(A)$, so $X = A^{-1}AX = A^{-1}B = \pm\ \text{adj}(A)B$. If all the elements of A and of B are integers, then all the elements of adj(A) are integers and all the elements of the product adj(A)B are integers, so X has all integer components.

18. $\begin{vmatrix} 5-\lambda & -7 & 7 \\ 4 & -3-\lambda & 4 \\ 4 & -1 & 2-\lambda \end{vmatrix} = (5-\lambda)(1-\lambda)(-2-\lambda)$, so the eigenvalues are $\lambda = 5$, $\lambda = 1$, and $\lambda = -2$. For $\lambda = 5$, the eigenvectors are the solutions of $\begin{bmatrix} 0 & -7 & 7 \\ 4 & -8 & 4 \\ 4 & -1 & -3 \end{bmatrix}\begin{bmatrix} x_1 \\ x_2 \\ x_3 \end{bmatrix} = \mathbf{0}$, so the eigenvectors are vectors of the form $r\begin{bmatrix} 1 \\ 1 \\ 1 \end{bmatrix}$. For $\lambda = 1$, the eigenvectors are the solutions of $\begin{bmatrix} 4 & -7 & 7 \\ 4 & -4 & 4 \\ 4 & -1 & 1 \end{bmatrix}\begin{bmatrix} x_1 \\ x_2 \\ x_3 \end{bmatrix} = \mathbf{0}$, so the eigenvectors are vectors of the form $s\begin{bmatrix} 0 \\ 1 \\ 1 \end{bmatrix}$.

For $\lambda = -2$, the eigenvectors are the solutions of $\begin{bmatrix} 7 & -7 & 7 \\ 4 & -1 & 4 \\ 4 & -1 & 4 \end{bmatrix}\begin{bmatrix} x_1 \\ x_2 \\ x_3 \end{bmatrix} = \mathbf{0}$, so the eigenvectors are vectors of the form $t\begin{bmatrix} 1 \\ 0 \\ -1 \end{bmatrix}$.

19. Let λ be an eigenvalue of A with eigenvector $\mathbf{x}$. A is invertible so $\lambda \neq 0$. $A\mathbf{x} = \lambda\mathbf{x}$, so $\mathbf{x} = A^{-1}A\mathbf{x} = A^{-1}\lambda\mathbf{x} = \lambda A^{-1}\mathbf{x}$. Thus $\frac{1}{\lambda}\ \mathbf{x} = A^{-1}\mathbf{x}$, so the eigenvalues for A^{-1} are the inverses of the eigenvalues for A and the corresponding eigenvectors are the same.

20. $A\mathbf{x} = \lambda\mathbf{x}$, so $A\mathbf{x} - kI\mathbf{x} = \lambda\mathbf{x} - kI\mathbf{x} = \lambda\mathbf{x} - k\mathbf{x}$, so $(A - kI)\mathbf{x} = (\lambda - k)\mathbf{x}$. Thus $\lambda - k$ is an eigenvalue for $A - kI$ with corresponding eigenvector $\mathbf{x}$.

Chapter 4

Exercise Set 4.1

1. $\mathbf{u} = \begin{bmatrix} a & b \\ c & d \end{bmatrix}$ and $\mathbf{v} = \begin{bmatrix} e & f \\ g & h \end{bmatrix}$ in M_{22} ; k and l are scalars.

axiom 2 $\quad k\mathbf{u} = k\begin{bmatrix} a & b \\ c & d \end{bmatrix} = \begin{bmatrix} ka & kb \\ kc & kd \end{bmatrix}$ is in M_{22} .

axiom 7 $\quad k(\mathbf{u} + \mathbf{v}) = k\begin{bmatrix} a+e & b+f \\ c+g & d+h \end{bmatrix} = \begin{bmatrix} k(a+e) & k(b+f) \\ k(c+g) & k(d+h) \end{bmatrix} = \begin{bmatrix} ka+ke & kb+kf \\ kc+kg & kd+kh \end{bmatrix}$

$= \begin{bmatrix} ka & kb \\ kc & kd \end{bmatrix} + \begin{bmatrix} ke & kf \\ kg & kh \end{bmatrix} = k\mathbf{u} + k\mathbf{v}.$

axiom 8 $\quad (k + l)\mathbf{u} = (k + l)\begin{bmatrix} a & b \\ c & d \end{bmatrix} = \begin{bmatrix} (k+l)a & (k+l)b \\ (k+l)c & (k+l)d \end{bmatrix} = \begin{bmatrix} ka+la & kb+lb \\ kc+lc & kd+ld \end{bmatrix}$

$= \begin{bmatrix} ka & kb \\ kc & kd \end{bmatrix} + \begin{bmatrix} la & lb \\ lc & ld \end{bmatrix} = k\mathbf{u} + l\mathbf{u}.$

axiom 9 $\quad k(l\mathbf{u}) = k\begin{bmatrix} la & lb \\ lc & ld \end{bmatrix} = \begin{bmatrix} kla & klb \\ klc & kld \end{bmatrix} = \begin{bmatrix} (kl)a & (kl)b \\ (kl)c & (kl)d \end{bmatrix} = (kl)\begin{bmatrix} a & b \\ c & d \end{bmatrix} = (kl)\mathbf{u}.$

axiom 10 $\quad 1\mathbf{u} = 1\begin{bmatrix} a & b \\ c & d \end{bmatrix} = \begin{bmatrix} a & b \\ c & d \end{bmatrix} = \mathbf{u}.$

2. (a) $(f + g)(x) = f(x) + g(x) = x + 2 + x^2 - 1 = x^2 + x + 1,$

$(2f)(x) = 2(f(x)) = 2(x + 2) = 2x + 4$, and $(3g)(x) = 3(g(x)) = 3(x^2 - 1) = 3x^2 - 3.$

(b) $(f + g)(x) = f(x) + g(x) = 2x + 4 - 2x = 4$, $(3f)(x) = 3(f(x)) = 3(2x) = 6x$, and

$-g(x) = -(4 - 2x) = -4 + 2x.$

3. f, g, and h are functions and c and d are scalars.

axiom 3 $\quad (f + g)(x) = f(x) + g(x) = g(x) + f(x) = (g + f)(x)$, so $f + g = g + f$.

axiom 4 $((f + g) + h)(x) = (f + g)(x) + h(x) = (f(x) + g(x)) + h(x) = f(x) + (g(x) + h(x))$
$= f(x) + (g + h)(x) = (f + (g + h))(x)$, so $(f + g) + h = f + (g + h)$.

axiom 7 $(c(f + g))(x) = c(f(x) + g(x)) = c(f(x)) + c(g(x)) = (cf)(x) + (cg)(x) = (cf + cg)(x)$, so $c(f + g) = cf + cg$.

axiom 8 $((c + d)f)(x) = (c + d)(f(x)) = c(f(x)) + d(f(x)) = (cf)(x) + (df)(x) = (cf + df)(x)$, so $(c + d)f = cf + df$.

axiom 9 $(c(df))(x) = c((df)(x)) = c(d(f(x))) = (cd)f(x) = ((cd)f)(x)$, so $(cd)f = c(df)$.

axiom 10 $(1f)(x) = 1(f(x)) = f(x)$, so $1f = f$.

4. Let $\mathbf{u} = (u_1, \ldots, u_n)$, $\mathbf{v} = (v_1, \ldots, v_n)$ and $\mathbf{w} = (w_1, \ldots, w_n)$ be vectors in $\mathbf{C}^n$ and c and d be (complex) scalars.

axiom 1 $(u_1, \ldots, u_n) + (v_1, \ldots, v_n) = (u_1 + v_1, \ldots, u_n + v_n)$ is in $\mathbf{C}^n$, so the set is closed under vector addition.

axiom 2 $c(u_1, \ldots, u_n) = (cu_1, \ldots, cu_n)$ is in $\mathbf{C}^n$, so the set is closed under scalar multiplication.

axiom 3 $\mathbf{u}+\mathbf{v}=(u_1, \ldots, u_n) + (v_1, \ldots, v_n) = (u_1 + v_1, \ldots, u_n + v_n) =$
$(v_1 + u_1, \ldots, v_n + u_n) = (v_1, \ldots, v_n) + (u_1, \ldots, u_n) = \mathbf{v} + \mathbf{u}$.

axiom 4 $(\mathbf{u} + \mathbf{v}) + \mathbf{w} = ((u_1, \ldots, u_n) + (v_1, \ldots, v_n)) + (w_1, \ldots, w_n)$

$= (u_1 + v_1, \ldots, u_n + v_n) + (w_1, \ldots, w_n) = ((u_1 + v_1) + w_1, \ldots, (u_n + v_n) + w_n)$

$= (u_1 + (v_1 + w_1), \ldots, u_n + (v_n + w_n)) = (u_1, \ldots, u_n) + (v_1 + w_1, \ldots, v_n + w_n)$

$= (u_1, \ldots, u_n) + ((v_1, \ldots, v_n) + (w_1, \ldots, w_n)) = \mathbf{u} + (\mathbf{v} + \mathbf{w})$.

axiom 5 $\mathbf{u} + \mathbf{0} = (u_1, \ldots, u_n) + (0, \ldots, 0) = (u_1, \ldots, u_n) = \mathbf{u}$.

axiom 6 $\mathbf{u} + (-\mathbf{u}) = (u_1, \ldots, u_n) + (-u_1, \ldots, -u_n) = (u_1 - u_1, \ldots, u_n - u_n) = (0, \ldots, 0) = \mathbf{0}$.

axiom 7 $c(\mathbf{u} + \mathbf{v}) = c(u_1 + v_1, \ldots, u_n + v_n) = (c(u_1 + v_1), \ldots, c(u_n + v_n))$

$= (cu_1 + cv_1, \ldots, cu_n + cv_n) = (cu_1, \ldots, cu_n) + (cv_1, \ldots, cv_n) = c\mathbf{u} + c\mathbf{v}$.

axiom 8 $(c + d)\mathbf{u} = (c + d)(u_1, \ldots, u_n) = ((c+d)u_1, \ldots, (c+d)u_n)$
$= (cu_1 + du_1, \ldots, cu_n + du_n) = (cu_1, \ldots, cu_n) + (du_1, \ldots, du_n) = c\mathbf{u} + d\mathbf{u}$.

axiom 9 $c(d\mathbf{u}) = c(du_1, \ldots, du_n) = (cdu_1, \ldots, cdu_n) = ((cd)u_1, \ldots, (cd)u_n) =$
$= (cd)(u_1, \ldots, u_n) = (cd)\mathbf{u}$.

axiom 10 $1\mathbf{u} = 1(u_1, \ldots, u_n) = (1u_1, \ldots, 1u_n) = (u_1, \ldots, u_n) = \mathbf{u}$.

(a) $\mathbf{u} + \mathbf{v} = (2 - i, 3 + 4i) + (5, 1 + 3i) = (7 - i, 4 + 7i)$ and $c\mathbf{u} = (3 - 2i)(2 - i, 3 + 4i) = (6 - 2 - 3i - 4i, 9 + 8 + 12i - 6i) = (4 - 7i, 17 + 6i)$.

(b) $\mathbf{u} + \mathbf{v} = (1 + 5i, -2 - 3i) + (2i, 3 - 2i) = (1 + 7i, 1 - 5i)$ and $c\mathbf{u} = (4 + i)(1 + 5i, -2 - 3i) = (4 - 5 + 20i + i, -8 + 3 - 12i - 2i) = (-1 + 21i, -5 - 14i)$.

5. W = {a(1,2,3)}. Axiom 1: Let a(1,2,3) and b(1,2,3) be elements of W. a(1,2,3)+b(1,2,3) = (a+b)(1,2,3). This is an element of W since it is a scalar multiple of (1,2,3). W is closed under addition. Axiom 2: Let c be a scalar. c(a(1,2,3)) = (ca)(1,2,3). This is an element of W since it is a scalar multiple of (1,2,3). W is closed under scalar multiplication. W inherits all other vector space properties from $\mathbf{R}^3$ - e.g., for **u** and **v** in W, **u**+**v**=**v**+**u** since **u** and **v** are in vector space $\mathbf{R}^3$. Similarly associative property is true. The zero vector of $\mathbf{R}^3$ is (0,0,0). This is also in W, and is the zero vector of W. The inverse of **u**=a(1,2,3) in $\mathbf{R}^3$ is a(-1,-2,-3). This is also the inverse of **u** In W. The scalar multiplication properties hold from $\mathbf{R}^3$.

6. Axiom 1: Let **v** and **w** be vectors in U. Thus **v**·**u**=0 and **w**·**u**=0. By the properties of dot product, **v**·**u**+**w**·**u**=0, (**v**+**w**)·**u**=0. The vector (**v**+**w**) is perpendicular to **u**; it is in U. U is closed under addition. Axiom 2: Let c be a scalar. by properties of dot product, (c**v**)·**u**=c(**v**·**u**) = 0. c**v** is perpendicular to **u**; it is in U. U is closed under scalar multiplication. W inherits all other vector space properties from $\mathbf{R}^3$ - e.g., for **u** and **v** in W, **u**+**v**=**v**+**u** since **u** and **v** are in vector space $\mathbf{R}^3$. Similarly associative property is true. (0,0,0) is in U since (0.0.0)·**u**=**0**, and is the zero vector of U. (-**v**) is in U since (-**v**)·**u** =-(**u**·**v**)=**0**, and is the negative of **v**. The scalar multiplication properties hold from $\mathbf{R}^3$.

7. Let A and B be 2x2 matrices with all elements positive. A+B will have all positive elements. Thus closed under addition. But (-1)A will have all negative elements. Thus (-1)A is not in W. W is not closed under scalar multiplication, thus not a vector space.

8. Let A and B be 3x3 symmetric matrices. Thus $A=A^t$ and $B=B^t$. $(A+B)^t = A^t + B^t = A+B$.

Therefore A+B is symmetric. Closure under addition. Let c be a scalar. $(cA)^t=cA^t=cA$. cA is symmetric. Closure under scalar multiplication. Let **0** be the zero 3x3 matrix. $\mathbf{0}^t=\mathbf{0}$. Thus the zero matrix is symmetric. All other properties are satisfied as for 3x3 matrices. The set of 3x3 matrices is a vector space.

9. Consider the stochastic matrices $A = \begin{bmatrix} .8 & .3 \\ .2 & .7 \end{bmatrix}$ and $B = \begin{bmatrix} .5 & 0 \\ .5 & 1 \end{bmatrix}$. Get $A+B=\begin{bmatrix} 1.3 & .3 \\ .7 & 1.7 \end{bmatrix}$
A+B is not stochastic. Has elements greater than 1, and columns do not add to 1. Thus set not closed under addition. Not a vector space. It is not closed under scalar multiplication either. Multiplication by a negative number gives all negative elements.

10. (a) Let $f(x)=k_1$ and $g(x)=k_2$ be constant functions on $(-\infty,\infty)$. Then $(f+g)(x)=k_1+k_2$, a constant function on $(-\infty,\infty)$. $(cf)(x)=ck_1$, a constant function on $(-\infty,\infty)$. U is closed under addition and scalar multiplication. It is a subset of the vector space of all functions on $(-\infty,\infty)$. It inherits all the other vector space properties from this larger space. Thus U is a vector space.

(b) V is not closed under addition. For example $f(x)=x$ is non constant on $(-\infty,\infty)$. $g(x) = -x$ is non-constant on $(-\infty,\infty)$. But $(f+g)(x)=0$, constant on $(-\infty,\infty)$. Thus not a vector space.

11. (a) The set of all continuous functions on [0,1] is closed under addition and scalar multiplication – sum of two continuous functions is continuous, and so is the scalar multiple, on [0,1]. It is a subset of the set of all functions on [0,1], which is a vector space by the same reasoning as that for the set of all functions with domain the real numbers. It inherits all the other vector space properties from this larger space. Thus the set of all continuous functions on [0,1] is a vector space.

(b) This set is not closed under addition. For example, if $f(x) = 2$ for $0 \le x \le 1/2$ and $f(x) = 3$ for $1/2 < x \le 1$, and $g(x) = 3$ for $0 \le x \le 1/2$, and $g(x) = 2$ for $1/2 < x \le 1$ then both f and g are discontinuous functions on [0,1], but f + g is the constant function $(f + g)(x) = 5$, which is continuous. Thus not a vector space.

12. (a) Let f and g be even functions. Thus $f(-x)=f(x)$ and $g(-x)=g(x)$. Get $(f+g)(-x)= f(-x)+g(-x)=f(x)+g(x) = (f+g)(x)$. Therefore f+g is even. Closed under addition. Let c be a scalar. $(cf)(-x)=c(f(-x))=c(f(x))=(cf)(x)$. cf is even; closed under scalar multiplication. $\mathbf{0}(-x)=0=\mathbf{0}(x)$. Thus zero function is even. All the other properties of a vector space are satisfied as for a space of functions. Thus it is a vector space.

(b) Let f and g be odd functions. Thus $f(-x)=-f(x)$ and $g(-x)=-g(x)$. Get $(f+g)(-x)= f(-x)+g(-x)=-f(x)-g(x) = -(f+g)(x)$. Therefore f+g is odd. Closed under addition. Let c be a scalar. $(cf)(-x)=c(f(-x))=c(-f(x))=-(cf)(x)$. cf is odd; closed under scalar multiplication. $\mathbf{0}(-x)=0=-\mathbf{0}(x)$. Thus zero function is odd. All other properties of a vector space are satisfied as for a space of functions. It is a vector space.

13. $f(x) = x^2 + 2x + 3$ and $g(x) = -x^2 + x + 1$ are polynomials of degree 2. Neither $f(x) + g(x) = 3x + 4$ nor $0f(x) = 0$ is a polynomial of degree 2, so the set is not closed under addition or under scalar multiplication. Thus not a vector space.

14. (a) The set {...,-3,-2,-1,0,1,2,3,...}. Let a and b be integers. Then a+b is an integer. Closed under addition. Consider (1/2)3. This is not an integer. Thus not closed under scalar multiplication, not a vector space.

(b) Let a and b be positive numbers. Then a+b is a positive number. Closed under addition. Consider (-3)4. This is not a positive number. Thus not closed under scalar multiplication, not a vector space.

15. (a) $c\mathbf{0} \underset{\text{axiom5}}{=} c\mathbf{0} + \mathbf{0} \underset{\text{axiom6}}{=} c\mathbf{0} + (c\mathbf{0} + -(c\mathbf{0})) \underset{\text{axiom4}}{=} (c\mathbf{0} + c\mathbf{0}) + -(c\mathbf{0})$

$\underset{\text{axiom7}}{=} c(\mathbf{0} + \mathbf{0}) + -(c\mathbf{0}) \underset{\text{axiom5}}{=} c\mathbf{0} + -(c\mathbf{0}) \underset{\text{axiom6}}{=} \mathbf{0}.$

(b) $c\mathbf{v} = \mathbf{0}$. Suppose $c \neq 0$. Then $\mathbf{v} \underset{\text{axiom10}}{=} 1\mathbf{v} = (\frac{1}{c}c)\mathbf{v} \underset{\text{axiom9}}{=} \frac{1}{c}(c\mathbf{v}) = \frac{1}{c}(\mathbf{0}) \underset{\text{(a)}}{=} \mathbf{0}.$

Now suppose $\mathbf{v} \neq \mathbf{0}$. It was just shown that if $c \neq 0$ then $\mathbf{v} = \mathbf{0}$. Since $\mathbf{v} \neq \mathbf{0}$, it must be that $c = 0$.

(c) By axiom 6 there is a vector $-(-\mathbf{v})$ such that $(-\mathbf{v}) + (-(-\mathbf{v})) = \mathbf{0}$.

However $(-\mathbf{v}) + \mathbf{v} \underset{\text{axiom3}}{=} \mathbf{v} + (-\mathbf{v}) \underset{\text{axiom6}}{=} \mathbf{0}$. Thus the vector $-(-\mathbf{v})$ is $\mathbf{v}$.

(d) $\mathbf{u} \underset{\text{axiom5}}{=} \mathbf{u} + \mathbf{0} \underset{\text{axiom6}}{=} \mathbf{u} + (\mathbf{w} + (-\mathbf{w})) \underset{\text{axiom4}}{=} (\mathbf{u} + \mathbf{w}) + (-\mathbf{w}) \underset{\text{given}}{=} (\mathbf{v} + \mathbf{w}) + (-\mathbf{w})$

$\underset{\text{axiom4}}{=} \mathbf{v} + (\mathbf{w} + (-\mathbf{w})) \underset{\text{axiom6}}{=} \mathbf{v} + \mathbf{0} \underset{\text{axiom5}}{=} \mathbf{v}.$

(e) $\mathbf{0} \underset{\text{axiom6}}{=} a\mathbf{u} + (-a\mathbf{u}) \underset{\text{given}}{=} b\mathbf{u} + (-a\mathbf{u}) \underset{\text{axiom8}}{=} (b + -a)\,\mathbf{u}$, so from part (b), $b + -a = 0$ and therefore $b = a$.

16. Axioms 1 and 2: Let L_1 and L_2 be lines through the origin, elements of W. Then by definition of addition and scalar multiplication L_1+L_2 and cL_1 are lines through the origin. Thus W is closed under addition and scalar multiplication. Axiom 3: Let L_1 have slope m_1 and L_2 have slope m_2. Then L_1+L_2 has slope m_1+m_2. L_2+L_1 has

slope m_2+m_1, which is equal to m_1+m_2. Thus L_1+L_2 and L_2+L_1 are both lines through the origin with the same slope. $L_1+L_2 = L_2+L_1$. Axiom 4: Can be shown similarly that $L_1+(L_2+L_3) = (L_1+L_2)+L_3$, each being the line through the origin of slope $m_1+m_2+m_3$. Axiom 5: Consider the line L_0 of slope 0, y=0x, (the x-axis). Then L_1+L_0 has slope $m_1+0=m_1$. $L_1+L_0 = L_1$. Thus L_0 is the zero vector. Axiom 6: Let $-L_1$ be the line of slope $-m_1$. Then $L_1+(-L_1)$ has slope of zero. $L_1+(-L_1) = L_0$. Thus $-L_1$ is the negative of L_1. Axiom 7: $c(L_1+L_2)$ is the line with slope $c(m_1+m_2)=cm_1+cm_2$. cL_1+cL_2 has slope cm_1+cm_2. Thus $c(L_1+L_2) = cL_1+cL_2$. Axiom 8: $(c+d)L_1$ and cL_1+dL_1 are the line with slope $(c+d)m_1$. $(c+d)L_1 = cL_1+dL_1$. Axiom 9: $c(dL_1)$ and $(cd)L_1$ are the line with slope cdm_1. $c(dL_1) = (cd)L_1$. Axiom 10: $1L_1$ is the line with slope $1m_1$, i.e., m_1. Thus $1L_1=L_1$. All axioms hold, W is a vector space.

17. This is an interesting example because the closure axioms hold - some of the others break down. We suggest that kC_1 be the circle radius $|k|r_1$ since this is a positive number for all real values of k.
Axiom 1: C_1+C_2 is the circle center the origin and radius r_1+r_2. Thus U is closed under addition. Axiom 2: kC_1 is the circle center the origin and radius $|k|r_1$. Thus U is closed under scalar multiplication. Axiom 3: C_1+C_2 is the circle center the origin and radius r_1+r_2. C_2+C_1 is the circle center the origin and radius r_2+r_1; that is radius r_1+r_2. Thus $C_1+C_2 =C_2+C_1$. Axiom 4: $C_1+(C_2+C_3)$ and $(C_1+C_2)+C_3$ are both the circle center the origin, radius $r_1+r_2+r_3$. $C_1+(C_2+C_3) = (C_1+C_2)+C_3$. Axiom 5: Let **0** be the circle center the origin, radius zero (the origin). Then $C_1+\mathbf{0}=C_1$. Axiom 6: There is no circle we can call $-C_1$, such that $C_1+(-C_1)=\mathbf{0}$. This axiom breaks down. Axiom 7: $k(C_1+C_2)$ is the circle through the origin with radius $|k|(r_1+r_2)$. That is, radius $|k|r_1+|k|r_2$. This is also the circle kC_1+kC_2. Thus $k(C_1+C_2)=kC_1+kC_2$. Axiom 8: $(k+m)C_1$ is the circle through the origin with radius $|k+m|r_1$. kC_1+mC_1 is the circle through the origin with radius $|k|r_1+|m|r_1$. Since $|k+m|r_1 \neq (|k|+|m|)r_1$ in general, $(k+m)C_1 \neq kC_1+mC_1$. Axiom breaks down. Axiom 9: $k(mC_1)$ is the circle through the origin with radius $|k||m|r_1$. $(km)C_1$ is the circle with radius $|km|r_1$. Since $|k||m|r_1 = |km|r_1$, $k(mC_1) =(km)C_1$. Axiom 10: $1C_1$ is the circle through the origin with radius $1r_1$; that is, radius r_1. Thus $1C_1 = C_1$.
All axioms hold except 6 and 8.

18. (a) (a,3a,5a) + (b,3b,5b) = (a+b,3(a+b),5(a+b)) and c(a,3a,5a) = (ca,3ca,5ca); thus the sum and scalar product of vectors in the set are also in the set, and so the set is a subspace of $\mathbf{R}^2$. The set is a line defined by the vector (1,3,5).

(b) (a,-a,2a) + (b,-b,2b) = (a+b,-(a+b),2(a+b)) and c(a,-a,2a) = (ca,-ca,2ca); thus the sum and scalar product of vectors in the set are also in the set, and so the set is a subspace of $\mathbf{R}^2$. The set is the line defined by the vector (1,-1,2).

(c) (a,b,a+2b) + (c,d,c+2d) = (a+c,b+d,(a+c)+2(b+d)) and k(a,b,a+2b) = (ka,kb,ka+2kb); thus the sum and scalar product of vectors in the set are also in the set, and so the set is a subspace of $\mathbf{R}^2$. The set is the plane z=x+2y.

(d) (a,b,a-b) + (c,d,c-d) = (a+c,b+d,(a+c)-(b+d)) and k(a,b,a-b) = (ka,kb,ka-kb); thus the sum and scalar product of vectors in the set are also in the set. This set is all of $\mathbf{R}^2$.

19. (a) (a,0) + (b,0) = (a+b,0) and c(a,0) = (ca,0); the sum and scalar product of vectors in the set are also in the set, and so the set is a subspace of $\mathbf{R}^2$. The set is a line, the x axis.

(b) (a,2a) + (b,2b) = (a+b,2(a+b)) and c(a,2a) = (ca,2ca); the sum and scalar product of vectors in the set are also in the set, and so the set is a subspace of $\mathbf{R}^2$. The set is the line defined by the vector (1,2).

(c) (a,1) + (b,1) = (a+b, 2). Last component is 2. (a+b,2) is not in the set. Set is not closed under addition. Thus not a subspace. Let us check scalar mult. k(a, 1)=(ka, k). (ka,k) is not in subset unless k=1. Thus not closed under scalar multiplication either.

(d) (a,a+3)+(b,b+3)=(a+b,(a+b)+6). Not closed under addition. Not a subspace.

(e) (a,b,0) + (d,e,0) = (a+d,b+e,0) and c(a,b,0) = (ca,cb,0); the sum and scalar product of vectors in the set are also in the set, and so the set is a subspace of $\mathbf{R}^3$ .The set is the xy plane.

(f) (a,b,2) + (c,d,2) = (a+c,b+d,4). Last component is 4. (a+c,b+d,4) is not in the set. Not closed under addition. Thus not a subspace. Let us check scalar mult. k(a,b,2) = (ka,kb,2k). Not in subset unless k=1. Thus not closed under scalar multiplication either.

(g) (a,b,2a+3b)+(c,d,2c+3d)=(a+c,b+d,2(a+c)+3(b+d)). k(a,b,2a+3d)=(ka,kb,2ka+3kb). Sum and scalar product are in the set. It is a subspace. The set is the plane z=2x+3y.

20. (a) Yes, this set is a subspace of $\mathbf{R}^3$. (a,b,c)+(d,e,f) = (a+d,b+e,c+f) and (a+d)+(b+e)+(c+f) = (a+b+c)+(d+e+f) = 0. Also k(a,b,c) = (ka,kb,kc) and ka+kb+kc = k(a+b+c) = 0.

(b) No, this set is not a subspace of $\mathbf{R}^3$. Neither the sum nor the scalar product of these
vectors is in the set. (a+b+c)+(d+e+f) = 1+1 = 2 and k(a+b+c) = k.

(c) No, this set is not a subspace of $\mathbf{R}^3$. The sum of two such vectors is not necessarily
such a vector. (0,1,1)+(1,0,1) = (1,1,2), which is not in the set.

(d) No, this set is not a subspace of $\mathbf{R}^3$. Neither the sum nor the scalar product of these
vectors is necessarily in the set. (5,1,1)+(5,1,1) = 2(5,1,1) = (10,2,2), which is not in the set.

(e) No, this set is not a subspace of $\mathbf{R}^3$. The sum of two such vectors is not necessarily in the set. (0,1,2)+(2,3,3) = (2,4,5), which is not in the set.

(f) Yes, this set is a subspace of $\mathbf{R}^3$. (a,b,c)+(d,e,f) = (a+d,b+e,c+f) and k(a,b,c) = (ka,kb,kc). If a = b+c and d = e+f then a+d = (b+c)+(e+f) = (b+e)+(c+f) and ka = kb+kc.

21. (a) No, this set is not a subspace of $\mathbf{R}^3$. If k = 1/2, then k(1,b,c)=(1/2,b/2,c/2); the first component is not an integer.

(b) No, this set is not a subspace of $\mathbf{R}^3$. If k is negative, then k(1,b,c)=(k,kb,kc); the first component is negative.

(c) No, this set is not a subspace of $\mathbf{R}^3$. If k is irrational, then k(1,b,c)=(k,kb,kc); the first component is irrational.

22. (a) No, this set is not a subspace of $\mathbf{R}^2$. If $b \neq 0$ and k is negative, then $k(a,b^2) = (ka,kb^2)$, and kb^2 is negative and so cannot be the square of a real number.

(b) Yes, this set is a subspace of $\mathbf{R}^2$. $(a,b^3)+(d,e^3) = (a+d,b^3+e^3)$ and $c(a,b^3) = (ca,cb^3)$. b^3+e^3 is the cube of some real number and so is cb^3.

(c) No, this set is not a subspace of $\mathbf{R}^2$. If k is negative, then k(a,b) = (ka,kb) and ka is negative.

(d) No, this set is not a subspace of $\mathbf{R}^2$. If k = 0, then k(a,b) = (0,0), which is not in the set.

(e) No, this set is not a subspace of $\mathbf{R}^2$. If k is negative, then k(a,b) = (ka,kb) and ka is positive if a is negative and kb is negative if b is positive.

23. (a) The set of vectors (a,0,0), where a is nonnegative, is closed under addition but not under scalar multiplication. If k is negative, then k(a,0,0) = (ka,0,0) and ka is negative if a is positive.

(b) The set of vectors (a,b,0), where $a \neq b$ unless a = b = 0, is closed under scalar multiplication but not under addition. (1,2,0)+(2,1,0) = (3,3,0).

24. (a) If (a,a+1,b) = (0,0,0), then a = a+1 = b = 0. a = a+1 is impossible. Therefore (0,0,0) is not in the set.

(b) If (a,3,2a) = (0,0,0), then 3 = 0, so (0,0,0) is not in the set.

(c) If (a,b,a+b−4) = (0,0,0), then a = b = a+b−4 = 0, but if a = b = 0 then a+b−4 = −4, not zero, so (0,0,0) is not in the set.

(d) If a > 0, then $a \neq 0$ so the first component of a vector in this set cannot be zero. Thus (0,0,0) is not in the set.

25. (a) The set of vectors of the form (a,a^2) contains the zero vector but is not a subspace of $\mathbf{R}^2$.

(b) The set of vectors of the form (a,a^2,a) contains the zero vector but is not a subspace of $\mathbf{R}^3$.

26. If U is a subspace of $\mathbf{R}^3$, then the sum of every pair of vectors in U is in U and cu is in U for every scalar c and every vector u in U. If u_1 and u_2 are vectors in U and a and b are scalars, then au_1 is in U and bu_2 is in U and their sum $au_1 + bu_2$ is in U.

Let U be a subset of $\mathbf{R}^3$ such that for every pair of vectors u_1 and u_2 in U and every pair of scalars a and b, the vector $au_1 + bu_2$ is in U. Let a=b=1. Then $au_1 + bu_2 = u_1 + u_2$ is in U for every pair of vectors u_1 and u_2 in U. Next let a=c, any scalar, and b = 0. Then $au_1 + bu_2 = cu_1$ is in U for any scalar c and any vector u_1 in U. Thus U is a subspace of $\mathbf{R}^3$.

27. (a) This set is a subspace. $\begin{bmatrix} 0 & a \\ b & 0 \end{bmatrix} + \begin{bmatrix} 0 & c \\ d & 0 \end{bmatrix} = \begin{bmatrix} 0 & a+c \\ b+d & 0 \end{bmatrix}$ and

$k\begin{bmatrix} 0 & a \\ b & 0 \end{bmatrix} = \begin{bmatrix} 0 & ka \\ kb & 0 \end{bmatrix}$.

(b) $2\begin{bmatrix} 2 & -1 \\ 0 & 5 \end{bmatrix} = \begin{bmatrix} 4 & -2 \\ 0 & 10 \end{bmatrix}$, which is not in the set, so the set is not a subspace.

(c) For $k \neq 0$ or 1, $k\begin{bmatrix} a & a^2 \\ b & b^2 \end{bmatrix} = \begin{bmatrix} ka & ka^2 \\ kb & kb^2 \end{bmatrix} \neq \begin{bmatrix} ka & (ka)^2 \\ kb & (kb)^2 \end{bmatrix}$, so the set is not a subspace.

(d) $2\begin{bmatrix} a & a+2 \\ b & c \end{bmatrix} = \begin{bmatrix} 2a & 2a+4 \\ 2b & 2c \end{bmatrix}$, and $2a + 4 \neq 2a + 2$, so the set is not a subspace.

28. (a) This set is a subspace: Let A and B be nxn symmetric matrices. Therefore $A = A^t$ and $B = B^t$. By theorems on transpose, $(A + B)^t = A^t + B^t = A + B$. Thus A + B is symmetric. $(cA)^t = cA^t = cA$. Therefore cA is symmetric. The set is closed under addition and scalar multiplication. It is a subspace.

(b) This set is not a subspace: e.g., one counterexample suffices.

Let $A = \begin{bmatrix} 1 & 2 \\ 3 & 4 \end{bmatrix}$ and $B = \begin{bmatrix} 0 & 5 \\ 4 & 1 \end{bmatrix}$. Neither A or B are symmetric. But

$A + B = \begin{bmatrix} 1 & 7 \\ 7 & 5 \end{bmatrix}$, symmetric. Thus not closed under addition. Not a subspace.

Note that if $A \neq A^t$, then $(cA)^t = cA^t \neq cA$. The set is closed under scalar multiplication.

(c) Set is a subspace: Let A and B be antisymmetric. Thus $A = -A^t$ and $B = -B^t$.
$A+B = -A^t - B^t = -(A^t+B^t) = -(A+B)^t$. Thus A+B is antisymmetric.
$cA = c(-A^t) = -(cA^t) = -(cA)^t$. Thus cA is antisymmetric. It is a subspace.

(d) This set is not a subspace: e.g., one counterexample suffices.
The matrices $\begin{bmatrix} 1 & 1 \\ 0 & 1 \end{bmatrix}$ and $\begin{bmatrix} -1 & 0 \\ 0 & 2 \end{bmatrix}$ are invertible but their sum $\begin{bmatrix} 0 & 1 \\ 0 & 3 \end{bmatrix}$ is not, so the set is not a subspace.
Or, more general: Let A be invertible. Then -A is invertible with inverse $-A^{-1}$, since $(-A)(-A^{-1}) = (-A^{-1})(A) = I$. But $A + (-A) = 0$, which is not invertible.

29. (a) $\begin{bmatrix} a & b & 0 \\ c & d & 0 \end{bmatrix} + \begin{bmatrix} e & f & 0 \\ g & h & 0 \end{bmatrix} = \begin{bmatrix} a+e & b+f & 0 \\ c+g & d+h & 0 \end{bmatrix}$ and $k\begin{bmatrix} a & b & 0 \\ c & d & 0 \end{bmatrix} = \begin{bmatrix} ka & kb & 0 \\ kc & kd & 0 \end{bmatrix}$,

so the set is closed under addition and scalar multiplication and is therefore a subspace.

(b) $\begin{bmatrix} a & 2a & 3a \\ b & 2b & 3b \end{bmatrix} + \begin{bmatrix} c & 2c & 3c \\ d & 2d & 3d \end{bmatrix} = \begin{bmatrix} a+c & 2a+2c & 3a+3c \\ b+d & 2b+2d & 3b+3d \end{bmatrix} = \begin{bmatrix} a+c & 2(a+c) & 3(a+c) \\ b+d & 2(b+d) & 3(b+d) \end{bmatrix}$

and $k\begin{bmatrix} a & 2a & 3a \\ b & 2b & 3b \end{bmatrix} = \begin{bmatrix} ka & 2ka & 3ka \\ kb & 2kb & 3kb \end{bmatrix}$, so the set is closed under addition and scalar multiplication and is therefore a subspace.

(c) $\begin{bmatrix} a & 1 & b \\ c & d & e \end{bmatrix} + \begin{bmatrix} p & 1 & q \\ r & s & t \end{bmatrix} = \begin{bmatrix} a+p & 2 & b+q \\ c+r & d+s & e+t \end{bmatrix}$, not in the set.

$k\begin{bmatrix} a & 1 & b \\ c & d & e \end{bmatrix} = \begin{bmatrix} ka & k & kb \\ kc & kd & ke \end{bmatrix}$, not in the set.

Set not closed under addition or scalar multiplication. Not a subspace.

30. Every element of P_2 is an element of P_3. Both P_2 and P_3 are vector spaces with the same operations and the same set of scalars, so P_2 is a subspace of P_3.

31. If $f(x) = ax^2 + bx + 3$ then $2f(x) = 2ax^2 + 2bx + 6$, which is not in S, so S is not closed under scalar multiplication and therefore not a subspace of P_2.

32. $\mathbf{R}^n$ is a subset of $\mathbf{C}^n$, but the scalars for $\mathbf{C}^n$ are the complex numbers. If a nonzero

element of $\mathbf{R}^n$ is multiplied by a complex number such as 1 + 2i, the result is an element
of $\mathbf{C}^n$ that is not an element of $\mathbf{R}^n$. Thus $\mathbf{R}^n$ is not closed under multiplication by the scalars in $\mathbf{C}^n$.

33. (a) $\begin{bmatrix} a & 1 \\ b & c \end{bmatrix} \neq \begin{bmatrix} 0 & 0 \\ 0 & 0 \end{bmatrix}$ for any a,b,c, so the zero vector $\begin{bmatrix} 0 & 0 \\ 0 & 0 \end{bmatrix}$ is not in the subset.

(b) ax + 2 ≠ 0 for any a and all x, so the zero vector is not in the subset.

34. (a) (f+g)(x) = f(x) + g(x) so (f+g)(0) = 0 + 0 = 0, and (cf)(x) = c(f(x)) so (cf)(0) = c0 = 0. Thus this subset is a subspace.

(b) (f+g)(0) = f(0) + g(0) = 3 + 3 = 6, so the subset is not closed under addition, and the subset is not a subspace.

(c) If f(x) = a and g(x) = b then (f+g)(x) = a + b, a constant, and (cf)(x) = ca, a constant, so the subset is closed under addition and scalar multiplication and is therefore a subspace.

35. If **u** and **v** are both in V but not in U, then it is possible for **u** + **v** to be in U. For example, let U be the subspace of V = $\mathbf{R}^2$ with basis (1,1). Then (1,0) and (0,1) are in $\mathbf{R}^2$ but not in U. However (1,0) + (0,1) = (1,1) is in U.
If **u** is in U then −**u** is in U and (**u** + **v**) + −**u** = **v**, so if **u** + **v** and **u** are in U then so is **v**. Therefore if **u** is in U and **v** is not, then their sum is not in U.
If c**u** is in U then $\frac{1}{c}$ c**u** = **u** is in U, so if **u** is not in U, then c**u** is also not in U for all c ≠ 0.

36. If a**u** + b**v** is in U for all vectors **u** and **v** in U and all scalars a and b, then in particular if a = b = 1, **u** + **v** is in U and if a = c and b = 0, c**u** is in U, so U is a subspace of V.
If U is a subspace of V, then **u** + **v** and c**u** are in U for all vectors **u** and **v** in U and all scalars c. Therefore if **u** is in U and a is a scalar, then a**u** is in U. Likewise b**v** is in U, so the sum a**u** + b**v** is in U.

37. (a) Let U be the subspace of $\mathbf{R}^2$ with basis (1,0) and let V be the subspace of $\mathbf{R}^2$ with basis (0,1). (1,0) and (0,1) are in the union of U and V but their sum, (1,1), is not.

(b) If **u** and **v** are two vectors in the intersection of two subspaces U and V, then **u** + **v** is in U and in V, so **u** + **v** is in the intersection. Likewise if c is any scalar, then c**u** is in U and in V, so c**u** is in the intersection of U and V. Thus the intersection of U and V is a subspace.

38. (a) Zero vector is unique: Let **u**+**w**=**0** for every vector **u** in V. Then **w**+**u**=**0** by axiom 3. Let **u**=**0**. Then **w**+**0**=**0**. **w**=**0** by axiom 5.

(b) Inverse vector is unique: Suppose **u**+**w**=**0**. Add (-**u**) to both sides. (**u**)+[**u**+**w**]=(- **u**)+**0**. [(-**u**)+**u**]+**w**=(-**u**) by axioms 4 and 5. [**u**+(-**u**)]+**w**=(-**u**) by axiom 3. **0**+**w**=(-**u**) by axiom 6. **w**+**0**=(-**u**) by axiom 3. **w**=(-**u**) by axiom 5.

Exercise Set 4.2

1. (a) Consider $a(1,-1) + b(2,4)=(-1,7)$. Get $a+2b=-1$, $-a+4b=7$. Unique solution, $a = -3$ and $b = 1$, so that $(-1,7) = -3(1,-1) + (2,4)$. Thus $(-1,7)$ is a linear combination of $(1,-1)$ and $(2,4)$.

   (b) $a(1, 2)+b(2, 3)=(8,13)$ gives $a+2b=8$, $2a+3b=13$. $a=2$, $b=3$. $(8,13) = 2(1,2) + 3(2,3)$. Is a linear combination.

   (c) $a(-1,4) + b(2,-8)=(-1,15)$ gives $-a + 2b=-1$ and $4a - 8b=15$. This system of equations has no solution, so $(-1,15)$ is not a linear combination of$(-1,4)$ and $(2,-8)$.

   (d) $a(1,3)+b(4,1)=(13,6)$ gives $a+4b=13$ and $3a+b=6$. $a=1$, $b=3$. $(13,6)= (1,3) + 3(4,1)$. Is a linear combination.

2. (a) $a(1,-1, 2)+b(2,1,0)+c(-1,2,1)=(-3,3,7)$ gives $a+2b-c=-3$, $-a+b+2c=3$, $2a+c=7$. Unique solution $a=2,b=-1,c=3$. Thus $(-3,3,7) = 2(1,-1,2) - (2,1,0) + 3(-1,2,1)$. Is a linear combination.

   (b) $a(1,-1,0)+b(2,1,4)+c(-2,4,1)=(-2,11,7)$ gives $a+2b-2c=-2$, $-a+b+4c=11$, $4b+c=7$. Unique solution $a=2,b=1,c=3$. Thus $(-2,11,7) = 2(1,-1,0) + (2,1,4) + 3(-2,4,1)$. Is a linear combination.

   (c) $a(1,2,3)+b(-1,2,4)+c(1,6,10)=(2,7,13)$. Then $a-b+c=2$, $2a+2b+6c=7$, and $3a+4b+10c=13$. This system of equations has no solution, so $(2,7,13)$ is not a linear combination of $(1,2,3)$, $(-1,2,4)$, and $(1,6,10)$.

3. (a) $a(-1,2,3)+b(1,3,1)+c(1,8,5)=(0,10,8)$ gives $-a+b+c=0$, $2a+3b+8c=10$, $3a +b+5c=8$. This system has many solutions. The general solution is $a = 2-c$, $b = 2-2c$, $c =$ any real number. Thus many linear combinations. $(0,10,8) = (2-c)(-1,2,3) + (2-2c)(1,3,1) + c(1,8,5)$, where c is any real number. Vector is in the space.

   (b) $a((1,0,1)+b(1,1,0)+c(3,1,2) = (1,4,-3)$ gives $a+b+3c=1$, $b+c=4$, $a+2c=-3$ has many solutions. The general solution is $a=-3-2c$, $b=4-c$. Thus many linear combinations. $(1, 4, -3)=(3-2c)(1,0,1)+(4-c)(1,1,0)+c(3,1,2)$. Vector is in the space.

   (c) $a(0,1,0)+b(3,5,6)+c(1,2,1)=(1,1,2)$ gives $3b+c=1$, $a+5b+2c=1$, $6b+c=2$ has no solution. $(1,1,2)$ is not a linear combination of $(0, 1, 0)$, $(3, 5, 6)$, $(1, 2, 1)$. Vector is not in the space.

4. (a) Any vector of the form $a(1, 2) + b(3, -5)$. e.g., $1(1, 2) + 2(3, -5) = (7, -8)$. $3(1, 2)-2(3, -5) = (-3, 16)$.

   (b) Any vector of the form $a(-1, 0)+b(3, 1)+c(2, 4)$. e.g., $1(-1, 0)+2(3, 1)+3(2, 4) = (11, 14)$. $0(-1, 0)+3(3, 1)-2(2, 4) = (5, -5)$.

   (c) Any vector of the form $a(1, -3, 5)+b(0, 1, 2)$ e.g., $1(1, -3, 5)+1(0, 1, 2) = (1, -2, 7)$. $2(1, -3, 5)-3(0, 1, 2) = (2, -9, 4)$.

   (d) Any vector of the form $a(1, 2, 3)+b(1, 1, 1)+c(0, 7, 2)+d(4, 3, -2)$ e.g. $1(1, 2, 3)+$

2(1, 1, 1)+3(0, 7, 2)+4(4, 3, -2) = (19, 37, 3). 0(1, 2, 3)+1(1, 1, 1)-2(0, 7, 2)-3(4, 3, -2) = (-11, -22, 3)

5. (a) e.g. (1,2,3) + (1,2,0) = (2,4,3), (1,2,3) − (1,2,0) = (0,0,3), 2(1,2,3) = (2,4,6).

(b) e.g. (1,2,1) + (2,1,4) = (3,3,5), (1,2,1) − (2,1,4) = (−1,1,−3), 2(1,2,1) = (2,4,2).

(c) e.g. −(1,2,3) = (−1,−2,−3), 2(1,2,3) = (2,4,6), (1/2)(1,2,3) = (1/2,1,3/2).

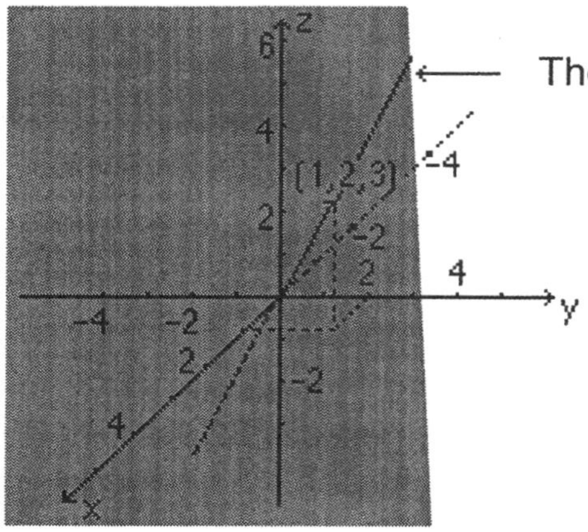

The subspace is all vectors on this line.

6. (a) e.g. 0(4, -1, 3) = (0, 0, 0), -2(4, -1, 3) = (-8, 2, -6), 3(4, -1, 3) = (12, -3, 9)

(b) e.g., −(1,2) = (−1,−2), 2(1,2) = (2,4), 5(1,2) = (5,10).

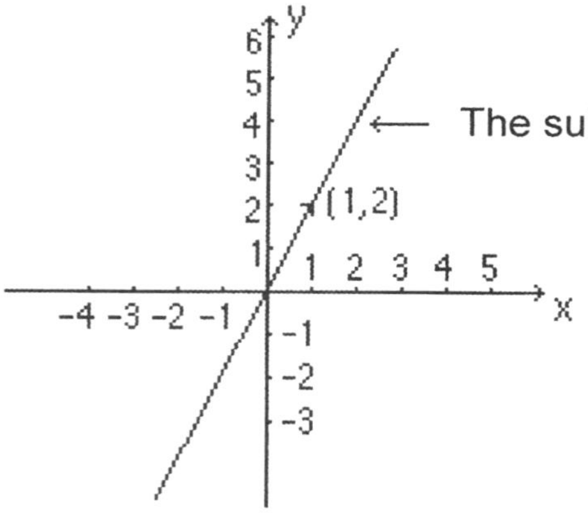

The subspace is all vectors on this line.

(c) e.g., −(1,2,−1,3) = (−1,−2,1,−3), 2(1,2,−1,3) = (2,4,−2,6), (.1)(1,2,−1,3) = (.1,.2,−.1,.3).

7. e.g., −(2,1,−3,4) = (−2,−1,3,−4), (2,1,−3,4) + (−3,0,1,5) = (−1,1,−2,9),

2(2, 1, -3, 4) + 3(-3, 0, 1, 5) – (4, 1, 2, 0) = (-9, 1, -5, 23).

8. (a) $a\begin{bmatrix} 1 & 2 \\ 3 & -4 \end{bmatrix} + b\begin{bmatrix} 0 & 3 \\ 1 & 2 \end{bmatrix} + c\begin{bmatrix} 1 & 2 \\ 0 & 0 \end{bmatrix} = \begin{bmatrix} 5 & 7 \\ 5 & -10 \end{bmatrix}$. Have a+c=5, 2a+3b+2c=7, 3a+b=5, -4a+2b=-10. Unique solution, a=2, b=-1, c=3.

Thus linear combination $2\begin{bmatrix} 1 & 2 \\ 3 & -4 \end{bmatrix} - \begin{bmatrix} 0 & 3 \\ 1 & 2 \end{bmatrix} + 3\begin{bmatrix} 1 & 2 \\ 0 & 0 \end{bmatrix} = \begin{bmatrix} 5 & 7 \\ 5 & -10 \end{bmatrix}$.

(b) $a\begin{bmatrix} 3 & 0 \\ 1 & 1 \end{bmatrix} + b\begin{bmatrix} 0 & 1 \\ 3 & 4 \end{bmatrix} + c\begin{bmatrix} 1 & 2 \\ 0 & 1 \end{bmatrix} = \begin{bmatrix} 7 & 6 \\ -5 & -3 \end{bmatrix}$. Have 3a+c=7, b+2c=6, a+3b=-5, a+4b+c=-3. Unique solution a=1, b=-2, c=4.

Thus linear combination $\begin{bmatrix} 3 & 0 \\ 1 & 1 \end{bmatrix} - 2\begin{bmatrix} 0 & 1 \\ 3 & 4 \end{bmatrix} + 4\begin{bmatrix} 1 & 2 \\ 0 & 1 \end{bmatrix} = \begin{bmatrix} 7 & 6 \\ -5 & -3 \end{bmatrix}$.

(c) $a\begin{bmatrix} 1 & 1 \\ 1 & 1 \end{bmatrix} + b\begin{bmatrix} 3 & 1 \\ 0 & 0 \end{bmatrix} + c\begin{bmatrix} -1 & -1 \\ 2 & 3 \end{bmatrix} = \begin{bmatrix} 4 & 1 \\ 7 & 10 \end{bmatrix}$ gives 4 = a + 3b − c, 1 = a + b − c, 7 = a + 2c, and 10 = a + 3c. This system of equations has no solution, so the first matrix is not a linear combination of the others.

9. (a) $a(x^2 + 1) + b(x + 3) = 3x^2 + 2x + 9$, gives $ax^2 + bx + a+3b = 3x^2 + 2x + 9$. Get system a=3, b=2, a+3b=9. Unique solution, a=3, b=2. Is unique linear combination, $3x^2 + 2x + 9 = 3(x^2 + 1) + 2(x + 3)$.

(b) $a(x^2 - x + 1) + b(x^2 + 2x - 2) = 2x^2 + x - 3$ gives a + b=2, −a + 2b=1, a − 2b = -3. This system of equations has no solution, so the first function is not a linear combination of the others.

(c) $a(x^2 + x - 1) + b(x^2 + 2x + 1) = x^2 + 4x + 5$, gives $(a+b)x^2 + (a+2b)x + (-a+b) = x^2 + 4x + 5$. Get system a+b=1, a+2b=4, -a+b=5. Unique solution, a=-2, b=3. Is unique linear combination, $x^2 + 4x + 5 = -2(x^2 + x - 1) + 3(x^2 + 2x + 1)$.

10. For example:

(a) h(x) = f(x) + g(x) = x + (x+3) = 2x + 3; h(x) = 2f(x) + 3g(x) = 2x + 3(x+3) = 5x + 9; h(x) = 0f(x) – 3g(x) = -3(x+3) = -3x – 9.

(b) $h(x) = f(x) - g(x) = (2x+1) - (3x^2 + x - 3) = -3x^2 + x + 4$; h(x) = 2f(x) + 0g(x) = 4x + 2; $h(x) = 2f(x) + 4g(x) = 2(2x+1) + 4(3x^2 + x - 3) = 12x^2 + 8x - 10$.

(c) $k(x) = 3f(x) + g(x) - 2h(x) = 3(2x^3 - 5x + 3) + (7x + 2) - 2(x^2 + x) = 6x^3 - 2x^2 - 10x + 11$;
k(x) = 3g(x) = 3(7x + 2) = 21x + 6;
$k(x) = f(x) - 2g(x) = (2x^3 - 5x + 3) - 2(7x + 2) = 2x^3 -- 19x - 1$

(d) $k(x) = f(x) + g(x) + h(x) = x^2 + x + 1$; $k(x) = 3f(x) + 2g(x) + h(x) = 3x^2 + 2x + 1$;
$k(x) = = 0f(x) + 0g(x) + 0(x) = 0$

11. (a) $a(x+1) + b(x+3) = x+5$. $(a+b)x + (a+3b) = x+5$. Get system $a+b=1$, $a+3b=5$.
Unique solution $a=-1$, $b=2$. It is in the space. $f(x)=2h(x)-g(x)$.

(b) $a(2x^2 +3) + b(x^2 +3x-1) = 3x^2+5x+1$. Get system $2a + b = 3$, $3b = 5$, $3a - b = 1$.
This system of equations has no solution, so f(x) is not in the subspace generated by g(x) and h(x).

(c) e.g. $g(x) + h(x) = (2x^2 + 3) + (x^2 + 3x - 1) = 3x^2 + 3x + 2$;
$g(x) - h(x) = (2x^2 + 3) - (x^2 + 3x - 1) = x^2 - 3x + 4$; and
$2g(x) = 2(2x^2 + 3) = 4x^2 + 6$.

12. Let $\mathbf{v} = a\mathbf{v}_1 + b\mathbf{v}_2$ and $\mathbf{u} = c\mathbf{u}_1 + d\mathbf{u}_2$. Then $\mathbf{v} + \mathbf{u} = a\mathbf{v}_1 + b\mathbf{v}_2 + c\mathbf{u}_1 + d\mathbf{u}_2$.

13. Let $\mathbf{v} = c_1\mathbf{v}_1 + \ldots + c_m\mathbf{v}_m$. Then we can write $\mathbf{v} = c_1\mathbf{v}_1 + \ldots + c_m\mathbf{v}_m + 0\mathbf{v}_{m+1}$.

14. Let $\mathbf{v} = a\mathbf{v}_1 + b\mathbf{v}_2$. Since both c_1 and c_2 are nonzero we can write
$\mathbf{v} = \frac{a}{c_1}(c_1\mathbf{v}_1) + \frac{b}{c_2}(c_2\mathbf{v}_2)$. Thus v is a linear combination of $c_1\mathbf{v}_1$ and $c_2\mathbf{v}$.

15. Suppose $\mathbf{v}$ is not a linear combination of $\mathbf{v}_1$ and $\mathbf{v}_2$ but $\mathbf{v} = a(c_1\mathbf{v}_1) + b(c_2\mathbf{v}_2)$.
Then $\mathbf{v} = (ac_1)\mathbf{v}_1 + (bc_2)\mathbf{v}_2$; i.e., $\mathbf{v}$ is a linear combination of $\mathbf{v}_1$ and $\mathbf{v}_2$ – canot be. Thus $\mathbf{v}$ must not be a linear combination of $c_1\mathbf{v}_1$ and $c_2\mathbf{v}_2$.

16. If $\mathbf{v}_1$and $\mathbf{v}_2$ span V, then any vector $\mathbf{v}$ in V can be written as a linear combination of $\mathbf{v}_1$ and $\mathbf{v}_2$: $\mathbf{v} = a_1\mathbf{v}_1 + a_2\mathbf{v}_2$. In particular $\mathbf{v}_3 = b_1\mathbf{v}_1 + b_2\mathbf{v}_2$. Thus
$\mathbf{v} = a_1\mathbf{v}_1 + a_2\mathbf{v}_2 - \mathbf{v}_3 + \mathbf{v}_3 = a_1\mathbf{v}_1 + a_2\mathbf{v}_2 - (b_1\mathbf{v}_1 + b_2\mathbf{v}_2) + \mathbf{v}_3 =$
$(a_1 - b_1)\mathbf{v}_1 + (a_2 - b_1)\mathbf{v}_2 + \mathbf{v}_3$, so that $\mathbf{v}_1$, $\mathbf{v}_2$, and $\mathbf{v}_3$ span V.
(It is also trivially true that $\mathbf{v} = a_1\mathbf{v}_1 + a_2\mathbf{v}_2 + 0\mathbf{v}_3$.)

17. Let $\mathbf{v}$ be an arbitrary vector in V. Since $\mathbf{v}_1$, $\mathbf{v}_2$, $\mathbf{v}_3$ span V there exist scalars a_1, a_2, a_3

such

that $\mathbf{v} = a_1\mathbf{v}_1 + a_2\mathbf{v}_2 + a_3\mathbf{v}_3$. Gradually rewrite this expression as follows.

$$\mathbf{v} = a_1\mathbf{v}_1 + a_2\mathbf{v}_2 + a_3\mathbf{v}_3$$

$$= a_1\mathbf{v}_1 + a_2\mathbf{v}_2 + a_3(\mathbf{v}_1 + \mathbf{v}_2 + \mathbf{v}_3) - a_3\mathbf{v}_1 - a_3\mathbf{v}_2$$

$= a_1\mathbf{v}_1 - a_3\mathbf{v}_1 + a_2\mathbf{v}_2 - a_3\mathbf{v}_2 + a_3(\mathbf{v}_1 + \mathbf{v}_2 + \mathbf{v}_3)$

$= a_1\mathbf{v}_1 - a_3\mathbf{v}_1 + (a_2 - a_3)\mathbf{v}_2 + a_3(\mathbf{v}_1 + \mathbf{v}_2 + \mathbf{v}_3)$

$= a_1\mathbf{v}_1 - a_2\mathbf{v}_1 + a_2\,\mathbf{v}_1 - a_3\mathbf{v}_1 + (a_2 - a_3)\mathbf{v}_2 + a_3(\mathbf{v}_1 + \mathbf{v}_2 + \mathbf{v}_3)$

$= a_1\mathbf{v}_1 - a_2\,\mathbf{v}_1 + (a_2 - a_3)\,\mathbf{v}_1 + (a_2 - a_3)\mathbf{v}_2 + a_3(\mathbf{v}_1 + \mathbf{v}_2 + \mathbf{v}_3)$

$= (a_1 - a_2)\mathbf{v}_1 + (a_2 - a_3)(\,\mathbf{v}_1 + \mathbf{v}_2) + a_3(\mathbf{v}_1 + \mathbf{v}_2 + \mathbf{v}_3)$

Thus **v** can be written as a linear combination of $\mathbf{v}_1$, $\mathbf{v}_1 + \mathbf{v}_2$, $\mathbf{v}_1 + \mathbf{v}_2 + \mathbf{v}_3$. These vectors span V.

18. U is subspace generated by $\mathbf{v}_1$, $\mathbf{v}_2$ and $\mathbf{v}_3$. Let **v** be an arbitrary vector in U. Thus we can write $\mathbf{v} = a_1\mathbf{v}_1 + a_2\mathbf{v}_2 + a_3\mathbf{v}_3$.
 (a) Suppose v_3 is a linear combination of v_1 and v_2. Let $\mathbf{v}_3 = b_1\mathbf{v}_1 + b_2\mathbf{v}_2$.
 Thus $\mathbf{v} = a_1\mathbf{v}_1 + a_2\mathbf{v}_2 + a_3(b_1\mathbf{v}_1 + b_2\mathbf{v}_2) = (a_1 + a_3b_1)\mathbf{v}_1 + (a_2 + a_3b_2)\mathbf{v}_2$.
 Therefore $\mathbf{v}_1$ and $\mathbf{v}_2$ generate U.

 (b) Suppose $\mathbf{v}_1$ and $\mathbf{v}_2$ generate U. Thus we can write $\mathbf{v}_3 = c_1\mathbf{v}_1 + c_2\mathbf{v}_2$.
 Therefore v_3 is a linear combination of v_1 and v_2.

19. (a) Let $\mathbf{w}_1$ and $\mathbf{w}_2$ be vectors in U + V. Thus there exist vectors $\mathbf{u}_1$ and $\mathbf{u}_2$ in U and $\mathbf{v}_1$ and $\mathbf{v}_2$ in V such that $\mathbf{w}_1 = \mathbf{u}_1 + \mathbf{v}_1$ and $\mathbf{w}_2 = \mathbf{u}_2 + \mathbf{v}_2$. Adding, $\mathbf{w}_1 + \mathbf{w}_2 = (\mathbf{u}_1 + \mathbf{v}_1) + (\mathbf{u}_2 + \mathbf{v}_2)$. This can be written $\mathbf{w}_1 + \mathbf{w}_2 = (\mathbf{u}_1 + \mathbf{u}_2) + (\mathbf{v}_1 + \mathbf{v}_2)$; the sum of an element of U and an element of V. Therefore $\mathbf{w}_1 + \mathbf{w}_2$ is in U + V. Let k be a scalar. Then $k\mathbf{w}_1 = k(\mathbf{u}_1 + \mathbf{v}_1) = k\mathbf{u}_1 + k\mathbf{v}_1$, an element of U + V. Thus U + V is closed under addition and scalar multiplcation. It is a subspace.

 (b) Given U = Span$\{\mathbf{u}_1, \mathbf{u}_2\}$ and V = Span$\{\mathbf{v}_1, \mathbf{v}_2\}$. Let **w** be a vector in U + V. Thus there exist scalars a, b, c, d such that $\mathbf{w} = (a\mathbf{u}_1 + b\mathbf{u}_2) + (c\mathbf{v}_1 + d\mathbf{v}_2)$. We can write $\mathbf{w} = a\mathbf{u}_1 + b\mathbf{u}_2 + c\mathbf{v}_1 + d\mathbf{v}_2$. Thus **w** is an element of Span$(\mathbf{u}_1, \mathbf{u}_2, \mathbf{v}_1, \mathbf{v}_2\}$. Conversely, if **w'** is an element of Span$(\mathbf{u}_1, \mathbf{u}_2, \mathbf{v}_1, \mathbf{v}_2\}$, then there are scalars p, q, r, s such that $\mathbf{w}' = p\mathbf{u}_1 + q\mathbf{u}_2 + r\mathbf{v}_1 + s\mathbf{v}_2$. We can write $\mathbf{w}' = (p\mathbf{u}_1 + q\mathbf{u}_2\,) + (r\mathbf{v}_1 + s\mathbf{v}_2)$. Thus is **w'** is an element of U + V. Thus U + V = Span$(\mathbf{u}_1, \mathbf{u}_2, \mathbf{v}_1, \mathbf{v}_2\}$.

<u>Exercise Set 4.3</u>

1. (a) $a(-1,2) + b(2,-4) = \mathbf{0}$ gives $-a+2b=0$, $2a-4b=0$. Many solutions, a=2r, b=r. Let r=1 say.
 Get solution a=2, b=1. Thus $2(-1,2) + (2,-4) = \mathbf{0}$. Vectors are linearly dependent.
 Geometry: (2, -4) = -2(-1, 2). (-1, 2) and (2, -4) are collinear, thus are linearly dependent.

(b) $a(-1,3) + b(2,5) = (0, 0)$ gives $-a+2b=0$, $3a+5b=0$. Unique solution $a=0$, $b=0$.
Thus vectors are linearly independent.
Geometry: There is no constant k such that $(2, 5) = k(-1, 3)$. $(-1, 3)$ and $(2, 5)$ are not collinear. Thus are linearly independent.

(c) $a(1,-2,3) + b(-2,4,1) + c(-4,8,9) = \mathbf{0}$ gives $a-2b-4c=0$, $-2a+4b+8c=0$, $3a+b+9c=0$.
Many solutions, $a=-2r$, $b=-3r$, $c=r$. Let $r=1$ say. Get solution $a=-2$, $b=-3$, $c=1$.
Thus $-2(1,-2,3) - 3(-2,4,1) + (-4,8,9) = \mathbf{0}$. The vectors are linearly dependent.
This means that the vectors lie in a plane.

(d) $a(1,0,2) + b(2,6,4) + c(1,12,2) = \mathbf{0}$ gives $a+2b+c=0$, $6b+12c=0$, $2a+4b+2c=0$.
Many solutions, $a=3r$, $b=-2r$, $c=r$. Let $r=1$ say. Get solution $a=3$, $b=-2$, $c=1$.
$3(1,0,2) - 2(2,6,4) + (1,12,2) = \mathbf{0}$. The vectors are linearly dependent.
This means that the vectors lie in a plane.

(e) $a(1,2,5) + b(1,-2,1) + c(2,1,4) = \mathbf{0}$ gives $a+b+2c=0$, $2a-2b+c=0$, $5a+b+4c=0$.
Unique solution $a=0$, $b=0$, $c=0$. Thus the vectors are linearly independent.
This means that the vectors do not lie in a plane.

(f) $a(1,1,1) + b(-4,3,2) + c(4,1,2) = \mathbf{0}$ gives $a - 4b+4c = 0$, $a+3b+c = 0$, $a+2b+2c = 0$.
Unique solution $a=0$, $b=0$, $c=0$. Thus the vectors are linearly independent.
This means that the vectors do not lie in a plane.

2. (a) $2(2,-1,3) + (-4,2,-6) = \mathbf{0}$, so the vectors $(2,-1,3)$ and $(-4,2,-6)$ are linearly dependent and any set of vectors containing these vectors is linearly dependent.

   (b) $-3(1,-2,3) + (3,-6,9) = \mathbf{0}$, so the vectors $(1,-2,3)$ and $(3,-6,9)$ are linearly dependent and any set of vectors containing these vectors is linearly dependent.

   (c) $(3, 0, 4) + (-3, 0, -4) = \mathbf{0}$, so the vectors $(3, 0, 4)$ and $(-3, 0, -4)$ are linearly dependent and any set of vectors containing these vectors is linearly dependent.

   (d) $-2(1, 1, 1) + (2, 2, 2) = 0$, so the vectors $(1, 1, 1)$ and $(2, 2, 2)$ are linearly dependent and any set of vectors containing these vectors is linearly dependent.

3. (a) e.g., of linear dependence of $\{(-1,2), (t,-4)\}$ is when $(t,-4)=k(-1,2)$. Vectors are collinear.
   [This also amounts to examining the identity $k(-1,2)+1(t,-4)=\mathbf{0}$]
   Then $t=-k$ and $-4=2k$. Thus $k=-2$, giving $t=2$.

   (b) e.g. of linear dependence of $\{(3,t), (6,t-1)\}$ is when $(6,t-1)=k(3,t)$.
   Then $6=3k$ and $t-1=kt$. Thus $k=2$, giving $t=-1$.

(c) e.g. of linear dependence of {(2,-t), (2t+6,4t)} is when (2t+6,4t)=k(2,-t).
Then 2t+6=2k, 4t=-tk. Thus k=-4, giving t= –7.

4. (a) If (0,0) = a(1,1) + b(0,2) = (a,a+2b), then a = 0 and a+2b = 0+2b = 0, so b = 0. Thus the vectors are linearly independent.

(b) If (0,0,0) = a(1,1,2) + b(0,−1,3) + c(0,0,5) = (a, a−b, 2a+3b+5c), then a = 0, a−b = 0−b = 0, so b = 0 and 2a+3b+5c = 0+0+5c = 0, so c = 0. Thus the vectors are linearly independent.

(c) If (0,0,0,0) = a(3,−2,4,5) + b(0,2,3,−4) + c(0,0,2,7) + d(0,0,0,4) = (3a, −2a+2b, 4a+3b+2c, 5a−4b+7c+4d), then 3a = 0 so a = 0, −2a+2b = 0 so b = 0, 4a+3b+2c = 0 so c = 0, and 5a−4b+7c+4d = 0 so d = 0. Thus the vectors are linearly independent.

(d) The first vector has all nonzero components, the second vector has its first component zero and all other components nonzero, the third vector has its first and second components zero and all other components nonzero. In general, the kth vector has its first k−1 components zero and all others nonzero. The set {(1,2,4,−1,3), (0,1,−1,3,3), (0,0,2,4,10), (0,0,0,−1,−1), (0,0,0,0,1)} is linearly independent in $\mathbf{R}^5$.

5. If (0,0,0,0) = a(1,0,0,7) + b(0,1,0,4) + c(0,0,1,3) = (a, b, c, 7a+4b+3c), then equating the first three components of (a, b, c, 7a+4b+3c) and (0,0,0,0) we see that a = 0, b = 0, and c = 0. Thus the vectors are linearly independent.
The set of nonzero row vectors of any matrix in reduced echelon form is linearly independent using the same reasoning as above.

6. (a) Observe that (2, 3, 4) = (1, 2, 3) + (1, 1, 1). Thus 1(1, 2, 3) + 1(1, 1, 1) – 1(2, 3, 4) = **0**.
(b) (3, 6, 15) = 3(1, 2, 5). Thus 3(1, 2, 5) – (3, 6, 15) + 0(-7, 3, 2) = **0**.
(c) (5, 6, 7) = (3, 4, 5) + 2(1, 1, 1). Thus 1(3, 4, 5) + 2(1, 1, 1) –1(5, 6, 7) = **0**.
(d) 0(1, 1, 1) + 0(0, 2, -3) + a(0, 0, 0) = **0**, for any non-zero value of a. Thus linearly dependent. Similarly any set that contains the zero vector is linearly dependent.

7. (a) Add any vector of the form **v**=a(1,-1,0)+b(2,1,3). Since then a(1,-1,0)+b(2,1,3)-**v**=**0**. e.g., 2(1,-1,0) + 3(2,1,3) = (8,1,9). Set {(1,-1,0), (2,1,3), (8,1,9)} is linearly dependent.
(b) Add any vector of the form a(3, -5, 1) + b(1, 4, 3). e.g., 1(3, -5, 1) + 2(1, 4, 3)=(5, 3, 7). Set {(3, -5, 1), (1, 4, 3), (5, 3, 7)} is linearly dependent.

(c) Add any vector of the form a(1, 2, 4) + b(0, 2, 5). e.g., 2(1, 2, 4) - 1(0, 2, 5) = (2, 2, 3). Set {(1, 2, 4), (0, 2, 5), (2, 2, 3)} is linearly dependent.

(d) Add any vector of the form a(3, 2, -4) + b(2, 5, 7). e.g., 4(3, 2, -4) + 1(2, 5, 7) = (14, 13, -9). Set {(3, 2, -4), (2, 5, 7), (14, 13, -9} is linearly dependent.

8. (a) If $a\begin{bmatrix} 1 & 0 \\ 0 & 0 \end{bmatrix} + b\begin{bmatrix} 0 & 2 \\ 0 & 0 \end{bmatrix} + c\begin{bmatrix} 0 & 0 \\ 3 & 0 \end{bmatrix} + d\begin{bmatrix} 0 & 0 \\ 0 & 4 \end{bmatrix} = \begin{bmatrix} 0 & 0 \\ 0 & 0 \end{bmatrix}$,

then $\begin{bmatrix} a & 2b \\ 3c & 4d \end{bmatrix} = \begin{bmatrix} 0 & 0 \\ 0 & 0 \end{bmatrix}$. a=b=c=d=0. Matrices are linearly independent.

(b) If $a\begin{bmatrix} 1 & 2 \\ 3 & 1 \end{bmatrix} + b\begin{bmatrix} 1 & 1 \\ 1 & 1 \end{bmatrix} + c\begin{bmatrix} 2 & 1 \\ 4 & 2 \end{bmatrix} = \begin{bmatrix} 0 & 0 \\ 0 & 0 \end{bmatrix}$, then a + b + 2c = 0,

2a + b + c = 0, 3a + b + 4c = 0, and a + b + 2c = 0. This system of equations has the unique solution a = b = c = 0, so the set is linearly independent.

(c) $2\begin{bmatrix} 1 & 2 \\ -1 & 0 \end{bmatrix} + (-3)\begin{bmatrix} 1 & 2 \\ 1 & 1 \end{bmatrix} + \begin{bmatrix} 1 & 2 \\ 5 & 3 \end{bmatrix} = \begin{bmatrix} 0 & 0 \\ 0 & 0 \end{bmatrix}$. Linearly dependent.

(d) $3\begin{bmatrix} 2 & 4 \\ 0 & 1 \end{bmatrix} - \begin{bmatrix} 0 & -2 \\ 8 & 3 \end{bmatrix} + 2\begin{bmatrix} -3 & -7 \\ 4 & 0 \end{bmatrix} = \begin{bmatrix} 0 & 0 \\ 0 & 0 \end{bmatrix}$. Linearly dependent.

9. (a) $1(2x^2 + 1) + (-1)(x^2 + 4x) + (-1)(x^2 - 4x + 1) = 0$, so the set is linearly dependent.

(b) $-2(x^2 + 3) + 3(x + 1) + (2x^2 - 3x + 3) = 0$, so the set is linearly dependent.

(c) If $a(x^2 + 3x - 1) + b(x + 3) + c(2x^2 - x + 1) = 0$, then a + 2c = 0, 3a + b − c = 0, and −a + 3b + c = 0. This system of homogeneous equations has the unique solution a = b = c = 0, so the functions are linearly independent.

(d) If $a(-x^2 + 2x - 5) + b(5x - 1) + c(7) = 0$, then -a = 0, 2a + 5b = 0, and −5a - b + 7c = 0. This system of homogeneous equations has the unique solution a = b = c = 0, so the functions are linearly independent.

10. Have $a\mathbf{v_1} + b\mathbf{v_2} + (-1)(a\mathbf{v_1} + b\mathbf{v_2}) = \mathbf{0}$ for all values of a and b.

11. Consider $c_1\mathbf{v}_1 + c_2\mathbf{v}_2 + c_3\mathbf{v}_3 = \mathbf{0}$. If this is true for some $c_3 \neq 0$ we can write

$\mathbf{v}_3 = -\frac{c_1}{c_3}\mathbf{v}_1 - \frac{c_2}{c_3}\mathbf{v}_2$ (in the form $\mathbf{v}_3 = a\mathbf{v}_1 + b\mathbf{v}_2$). If $\mathbf{v}_3$ cannot be written in this form must have $c_3 = 0$. Then $c_1\mathbf{v}_1 + c_2\mathbf{v}_2 = \mathbf{0}$, and since $\mathbf{v}_1$ and $\mathbf{v}_2$ are linearly independent, $c_1 = c_2 = 0$. Thus $\mathbf{v}_1$, $\mathbf{v}_2$, $\mathbf{v}_3$ are linearly independent.

12. Let $c_1\mathbf{v}_1 + c_2\mathbf{v}_2 = \mathbf{0}$, where c_1 and c_2 are not both zero. Any linear combination of

$\mathbf{v}_1 + \mathbf{v}_2$ and $\mathbf{v}_1 - \mathbf{v}_2$ is a linear combination of $\mathbf{v}_1$ and $\mathbf{v}_2$: $a_1(\mathbf{v}_1 + \mathbf{v}_2) + a_2(\mathbf{v}_1 - \mathbf{v}_2)$ $= (a_1 + a_2)\mathbf{v}_1 + (a_1 - a_2)\mathbf{v}_2$. Thus if $a_1 + a_2 = c_1$ and $a_1 - a_2 = c_2$,
i.e., $a_1 = \frac{c_1+c_2}{2}$ and $a_2 = \frac{c_1-c_2}{2}$,then at least one of $a_1$and a_2 is not zero and $a_1(\mathbf{v}_1 + \mathbf{v}_2) + a_2(\mathbf{v}_1 - \mathbf{v}_2) = \mathbf{0}$.

13. Let V = {$\mathbf{v}_1, \mathbf{v}_2, \ldots, \mathbf{v}_n$} be linearly independent. Let {$\mathbf{v}_1, \mathbf{v}_2, \ldots, \mathbf{v}_m$} be a subset. Consider the identity $a_1\mathbf{v}_1 + a_2\mathbf{v}_2 + \ldots + a_m\mathbf{v}_m = \mathbf{0}$. This means that $a_1\mathbf{v}_1 + a_2\mathbf{v}_2 + \ldots + a_m\mathbf{v}_m + 0\mathbf{v}_{m+1} + \ldots + 0\mathbf{v}_n = \mathbf{0}$. Since {$\mathbf{v}_1, \mathbf{v}_2, \ldots, \mathbf{v}_n$} is linearly independent, a_1=0, ..., a_m=0. Thus {$\mathbf{v}_1, \mathbf{v}_2, \ldots, \mathbf{v}_m$} is linearly independent.

A linearly dependent set can have a linearly independent subset. For example, the set {(1,−2,3), (−1,2,−3), (1,0,0)} is linearly dependent since (1,−2,3) + (−1,2,−3) + 0(1,0,0) = **0**, but the subset {(−1,2,−3), (1,0,0)} is linearly independent.

14. Given $\mathbf{u}\cdot\mathbf{v} = \mathbf{0}$. Consider $a\mathbf{u} + b\mathbf{v} = \mathbf{0}$. Take the dot product with $\mathbf{u}$ and simplify. $(a\mathbf{u} + b\mathbf{v})\cdot\mathbf{u} = \mathbf{0}\cdot\mathbf{u}$, $a\mathbf{u}\cdot\mathbf{u} + b\mathbf{v}\cdot\mathbf{u} = \mathbf{0}$, $a\mathbf{u}\cdot\mathbf{u} + b\mathbf{u}\cdot\mathbf{v} = \mathbf{0}$, $a\mathbf{u}\cdot\mathbf{u} + \mathbf{0} = \mathbf{0}$, $a\mathbf{u}\cdot\mathbf{u} = \mathbf{0}$. Since $\mathbf{u}$ is a nonzero vector this means that a=0. Similarly on taking the dot product with $\mathbf{v}$, b=0. Thus $\mathbf{u}$ and $\mathbf{v}$ are linearly independent.

15. (a) $\mathbf{v}_1$, $\mathbf{v}_2$, and $\mathbf{v}_3$ span $\mathbf{R}^3$. They do not lie on a line or in a plane. $\mathbf{v}_1$, $\mathbf{v}_2$, and $\mathbf{v}_3$ are linearly independent.

(b) $\mathbf{v}_3$ lies in the space spanned by $\mathbf{v}_1$ and $\mathbf{v}_2$. This space may be a plane or a line, depending on whether $\mathbf{v}_1$ and $\mathbf{v}_2$ are independent or not. $\mathbf{v}_1$, $\mathbf{v}_2$, and $\mathbf{v}_3$ are linearly dependent.

(c) $\mathbf{v}_2$ and $\mathbf{v}_3$ lie in the space spanned by $\mathbf{v}_1$. They all lie on the line defined by $\mathbf{v}_1$. $\mathbf{v}_1$, $\mathbf{v}_2$, and $\mathbf{v}_3$ are linearly dependent.

16. (a) False: Consider (1, 4, 2) = a(1, 0, 0) + b(0, 1, 0). Then (1, 4, 2) = (a, b, 0). System 1=a, 4=b, 2=0 has no solution. Geometry: (1, 0, 0) and (0, 1, 0) have no component in the z direction while (1, 4, 2) does. Thus (1, 4, 2) cannot be a linear combination of (1, 0, 0) and (0, 1, 0).

(b) False: A vector can be expressed as a linear combination of (1, 0, 0), (2, 1, 0), (-1, 3, 0) only if the last component is zero. For example, (0, 0, 1) cannot be expressed as a linear combination of these vectors.

(c) True: We can write (a, a, b) = a(1, 1, 0) + (-b)(0, 0, -1)

(d) False: a(1, 2, 3) + b(1, 2, 0) + c(1, 0, 0) = 0, a+b+c=0, 2a+2b=0, 3a=0, means that a=b=c=0. Thus linearly independent. Geometry: (1, 0, 0) is on x-axis, (1, 2, 0) lies in xy-plane, not collinear with (1, 0, 0). Thus the two vectors are linearly independent. (1, 2, 3) lies outside xy-plane thus linearly independent of the other two vectors. The three vectors are linearly independent.

17. (a) True: $\mathbf{v} = a\mathbf{v}_1 + b\mathbf{v}_2 + 0\mathbf{v}_3$.

(b) True: One vector spans the line defined by it. Two vectors span the plane defined by them. Need at least three vectors that do not lie in a plane to span $\mathbf{R}^3$.

(c) False: e.g., $\mathbf{R}^2$ is spanned by (1, 0), (0, 1) since (a, b) = a(1, 0) + b(0, 1). It is also spanned by (k, 0), (0, k) for any nonzero k, since (a, b) = $\frac{a}{k}$(k, 0) + $\frac{b}{k}$ (0, k).

(d) True: $\mathbf{v} = a\mathbf{v}_1 + b\mathbf{v}_2$, an element of Span{$\mathbf{v}_1$, $\mathbf{v}_2$}, can be written as $\mathbf{v} = a\mathbf{v}_1 + \frac{b}{2}2\mathbf{v}_2$, an element of Span{$\mathbf{v}_1$, $2\mathbf{v}_2$}, and vice-versa. Spaces consist of the same vectors.

(e) True: Three vectors that lie in a plane and are linearly dependent. (See discussion in this section.)

18. It is more likely that the vectors will be independent. For them to be dependent,one has to be a multiple of the other. For them to be dependent they would also have to lie in a line.

19. It is more likely that the vectors will be independent. For them to be dependent, they must all lie in the same plane.

20. The computer is more likely to state that the vectors are linearly independent when they are linearly dependent, because round-off error causes zero to look like a nonzero number close to zero to the computer. Thinking geometrically, a small move by one or more of the vectors (round-off error) is not likely to place them in the same plane if they are not already in the same plane, but is more likely to move one of the vectors out of the plane of the other two if they are in fact in the same plane.

<u>Exercise Set 4.4</u>

1. For two vectors to be linearly dependent one must be a multiple of the other. In each case below, neither vector is a multiple of the other, so the vectors are linearly independent. We show that each set spans $\mathbf{R}^2$.

(a) $(x_1, x_2) = \frac{3x_2 - x_1}{5}(1,2) + \frac{2x_1 - x_2}{5}(3,1)$.

(b) $(x_1, x_2) = \frac{-5x_1+2x_2}{13}(-1,4) + \frac{4x_1+x_2}{13}(2,5)$.

(c) $(x_1, x_2) = \frac{x_1+x_2}{2}(1,1) + \frac{-x_1+x_2}{2}(-1,1)$.

(d) $(x_1, x_2) = (x_1 - x_2)(1,0) + x_2(1,1)$.

2. It is necessary to show either that the set of vectors is linearly independent or that the set spans $\mathbf{R}^2$. For two vectors to be linearly dependent, one must be a multiple of the other. In each case neither vector is a multiple of the other, so the vectors are linearly independent and therefore a basis for $\mathbf{R}^2$. (Theorem 4.11)

3. Use Theorem 4.11. Two vectors in each case, thus examining dependence will suffice.

(a) These two vectors are linearly independent since neither is a multiple of the other. Thus they are a basis for $\mathbf{R}^2$.

(b) $(-2,6) = -2(1,-3)$, so these vectors are not linearly independent and therefore not a basis for $\mathbf{R}^2$.

(c) These two vectors are linearly independent since neither is a multiple of the other. Thus they are a basis for $\mathbf{R}^2$.

(d) $(3,-6) = -3(-1,2)$, so these vectors are not linearly independent and therefore not a basis for $\mathbf{R}^2$.

4. Use Theorem 4.11. Two vectors in each case, thus examining dependence will suffice.

(a) $a(1,1,1) + b(0,1,2) + c(3,0,1) = (0,0,0)$ if and only if $a+3c = 0$, $a+b = 0$, and $a+2b+c = 0$. System has the unique solution $a = b = c = 0$, so the three vectors are linearly independent. Thus they are a basis for $\mathbf{R}^3$.

(b) $a(1,2,3) + b(2,4,1) + c(3,0,0) = (0,0,0)$ if and only if $a+2b+3c = 0$, $2a+4b = 0$, and $3a+b = 0$. System has the unique solution $a = b = c = 0$, so the three vectors are linearly independent. Thus they are a basis for $\mathbf{R}^3$.

(c) $a(0,0,1) + b(2,3,1) + c(4,1,2) = (0,0,0)$ if and only if $2b+4c = 0$, $3b+c = 0$, and $a+b+2c = 0$. The system has the unique solution $a = b = c = 0$, so the three vectors are linearly independent. Thus they are a basis for $\mathbf{R}^3$.

(d) $a(1,1,4) + b(2,1,3) + c(0,1,6) = (0,0,0)$ if and only if $a+2b = 0$, $a+b+c = 0$, and

4a+3b+6c = 0. The system has the unique solution a = b = c = 0, so the three vectors are linearly independent. Thus they are a basis for $\mathbf{R}^3$.

5. (a) a(1,−1,2) + b(2,0,1) +c(3,0,0) = (0,0,0) if and only if a+2b+3c = 0, −a = 0, and 2a+b = 0. The system has the unique solution a = b = c = 0, so the three vectors are linearly independent. Thus they are a basis for $\mathbf{R}^3$.

(b) 2(2,1,0) + (−1,1,1) = (3,3,1), so the vectors are linearly dependent and therefore not a basis for $\mathbf{R}^3$.

(c) $2\begin{bmatrix}3\\1\\-1\end{bmatrix} + 2\begin{bmatrix}-1\\-1\\0\end{bmatrix} = \begin{bmatrix}4\\0\\-2\end{bmatrix}$, so the vectors are linearly dependent and therefore not a basis for $\mathbf{R}^3$.

(d) $a\begin{bmatrix}1\\2\\2\end{bmatrix} + b\begin{bmatrix}-1\\0\\1\end{bmatrix} + c\begin{bmatrix}-3\\1\\-1\end{bmatrix} = \begin{bmatrix}0\\0\\0\end{bmatrix}$ if and only if a−b−3c = 0, 2a+c = 0, and 2a+b−c = 0. The system has the unique solution a = b = c = 0, so the three vectors are linearly independent. Thus they are a basis for $\mathbf{R}^3$

6. (a) These vectors are linearly dependent.

(b) A basis for $\mathbf{R}^2$ can contain only two vectors. Also, any set of three vectors in $\mathbf{R}^2$ is l.d.

(c) These vectors are linearly dependent.

(d) The third vector is a multiple of the second, so the set is not linearly independent.

(e) A basis for $\mathbf{R}^3$ can contain only three vectors. Also, any set of four vectors in $\mathbf{R}^3$ is l.d.

(f) (1,4) is not in $\mathbf{R}^3$.

(g) None of these vectors is in $\mathbf{R}^4$.

7. (1,4,3) = 3(−1,2,1) + 2(2,−1,0), and (−1,2,1) and (2,−1,0) are linearly independent, so the dimension is 2 and (−1,2,1) and (2,−1,0) are a basis for the subspace.

8. (1,2,−1) = 2(1,3,1) − (1,4,3).

9. (2,1,4) = (1,0,2) + (1,1,2).

10. $(-3,3,-6) = -\frac{3}{2}(2,-2,4)$.

11. No, the three vectors (1,2,−1), (1,−1,0) and (3,−1,2) are linearly independent.

12. Basis for $\mathbf{R}^2$ has two linearly independent (non-colinear) vectors. {(1,2),(0,1)} is a basis.

13. Need three linearly independent vectors. The set {(1,1,1),(1,0,−2),(1,0,0)} is a basis for $\mathbf{R}^3$.

14. The set {(−1,0,2),(0,1,1),(1,0,0)} is a basis for $\mathbf{R}^3$.

15. (a) (a,a,b) = a(1,1,0) + b(0,0,1), so the linearly independent set {(1,1,0),(0,0,1)} spans the subspace of vectors of the form (a,a,b) and is therefore a basis. The dimension of the space is 2 since there are 2 vectors in the basis.

(b) (a,a,2a) = a(1,1,2), so the single vector (1,1,2) is a basis for the subspace of vectors of the form (a,a,2a). The dimension of the space is 1 since there is 1 vector in the basis.

(c) (a, b, a+b) = a(1,0,1) + b(0,1,1), so the linearly independent set {(1,0,1),(0,1,1)} spans the subspace of vectors of the form (a,b,a+b) and is therefore a basis. The dimension of the space is 2 since there are 2 vectors in the basis.

(d) (a, 2b, a+3b) = a(1,0,1) + b(0,2,3), so the linearly independent set {(1,0,1),(0,2,3)} spans the subspace of vectors of the form (a,2b,a+3b) and is therefore a basis. The dimension of the space is 2 since there are 2 vectors in the basis.

(e) (a,b,c) = (a, b, −a−b) = a(1,0,−1) + b(0,1,−1), so the linearly independent set {(1,0,−1),(0,1,−1)} spans the subspace of vectors of the form (a,b,c) where a+b+c = 0. Thus the set {(1,0,−1),(0,1,−1)} is a basis for the subspace. The dimension of the space is 2 since there are 2 vectors in the basis.

16. (a) (a, b, a+b, a−b) = a(1,0,1,1) + b(0,1,1,−1), so the linearly independent set {(1,0,1,1),(0,1,1,−1)} spans the subspace of vectors of the form (a,b,a+b,a−b) and is therefore a basis. The dimension of the space is 2 since there are 2 vectors in the basis.

(b) (a,2a,b,0) = a(1,2,0,0) + b(0,0,1,0), so the linearly independent set {(1,2,0,0),(0,0,1,0)} spans the subspace of vectors of the form (a,2a,b,0) and is therefore a basis. The dimension of the space is 2 since there are 2 vectors in the basis.

(c) (2a, b, a+3b, c) = a(2,0,1,0) + b(0,1,3,0)+c(0,0,0,1), so the linearly independent set {(2,0,1,0), (0,1,3,0),(0,0,0,1)} spans the subspace of vectors of the form (2a, b, a+3b, 0) and is therefore a basis. The dimension of the space is 3 since there are 3 vectors in the basis.

(d) (a,a,a,a) = a(1,1,1,1), so the single vector (1,1,1,1) is a basis for the subspace of vectors of the form (a,a,a,a). The dimension of the space is 1 since there is 1 vector in the basis.

17. (a) The set $\{x^3, x^2, x, 1\}$ is a basis. The dimension is 4.

(b) $\begin{bmatrix} 1 & 0 & 0 \\ 0 & 0 & 0 \\ 0 & 0 & 0 \end{bmatrix}, \begin{bmatrix} 0 & 1 & 0 \\ 0 & 0 & 0 \\ 0 & 0 & 0 \end{bmatrix}, \begin{bmatrix} 0 & 0 & 1 \\ 0 & 0 & 0 \\ 0 & 0 & 0 \end{bmatrix}, \begin{bmatrix} 0 & 0 & 0 \\ 1 & 0 & 0 \\ 0 & 0 & 0 \end{bmatrix}, \begin{bmatrix} 0 & 0 & 0 \\ 0 & 1 & 0 \\ 0 & 0 & 0 \end{bmatrix}, \begin{bmatrix} 0 & 0 & 0 \\ 0 & 0 & 1 \\ 0 & 0 & 0 \end{bmatrix},$
$\begin{bmatrix} 0 & 0 & 0 \\ 0 & 0 & 0 \\ 1 & 0 & 0 \end{bmatrix}, \begin{bmatrix} 0 & 0 & 0 \\ 0 & 0 & 0 \\ 0 & 1 & 0 \end{bmatrix}$, and $\begin{bmatrix} 0 & 0 & 0 \\ 0 & 0 & 0 \\ 0 & 0 & 1 \end{bmatrix}$ are a basis. The dimension is 9.

(c) $\begin{bmatrix} 1 & 0 & 0 \\ 0 & 0 & 0 \end{bmatrix}, \begin{bmatrix} 0 & 1 & 0 \\ 0 & 0 & 0 \end{bmatrix}, \begin{bmatrix} 0 & 0 & 1 \\ 0 & 0 & 0 \end{bmatrix}, \begin{bmatrix} 0 & 0 & 0 \\ 1 & 0 & 0 \end{bmatrix}, \begin{bmatrix} 0 & 0 & 0 \\ 0 & 1 & 0 \end{bmatrix}$, and $\begin{bmatrix} 0 & 0 & 0 \\ 0 & 0 & 1 \end{bmatrix}$

are a basis. The dimension is 6.

(d) $\begin{bmatrix} 1 & 0 \\ 0 & 0 \end{bmatrix}$ and $\begin{bmatrix} 0 & 0 \\ 0 & 1 \end{bmatrix}$ are a basis. The dimension is 2.

(e) $\begin{bmatrix} 1 & 0 \\ 0 & 0 \end{bmatrix}, \begin{bmatrix} 0 & 0 \\ 0 & 1 \end{bmatrix}$, and $\begin{bmatrix} 0 & 1 \\ 1 & 0 \end{bmatrix}$ are a basis. The dimension is 3.

18. $V_1 = \{\begin{bmatrix} a & b \\ -a & c \end{bmatrix}\}$, $V_2 = \{\begin{bmatrix} p & -p \\ q & r \end{bmatrix}\}$.

(a) Let $u = \begin{bmatrix} a_1 & b_1 \\ -a_1 & c_1 \end{bmatrix}$ and $w = \begin{bmatrix} a_2 & b_2 \\ -a_2 & c_2 \end{bmatrix}$ be elements of V_1. u + w =
$\begin{bmatrix} a_1+a_2 & b_1+b_2 \\ -(a_1+a_2) & c_1+c_2 \end{bmatrix}$, in V_1. $ku = \begin{bmatrix} ka_1 & kb_1 \\ -(ka_1) & kc_1 \end{bmatrix}$, in V_1. V_1 is closed under addition and scalar multiplication. It is a subspace.

Let $u = \begin{bmatrix} p_1 & -p_1 \\ q_1 & r_1 \end{bmatrix}$ and $w = \begin{bmatrix} p_2 & -p_2 \\ q_2 & r_2 \end{bmatrix}$ be elements of V_2. $u + w = \begin{bmatrix} p_1+p_2 & -(p_1+p_2) \\ q_1+q_2 & r_1+r_2 \end{bmatrix}$, in V_1. $ku = \begin{bmatrix} kp_1 & -(kp_1) \\ kq_1 & kr_1 \end{bmatrix}$, in V_1. V_2 is closed under addition and scalar multiplication. It is a subspace.

$V_1 \cap V_2 = \{ \begin{bmatrix} a & -a \\ -a & b \end{bmatrix} \}$. Let $u = \begin{bmatrix} a_1 & -a_1 \\ -a_1 & b_1 \end{bmatrix}$ and $w = \begin{bmatrix} a_2 & -a_2 \\ -a_2 & b_2 \end{bmatrix}$ be elements of $V_1 \cap V_2$.
$u+w = \begin{bmatrix} a_1+a_2 & -(a_1+a_2) \\ -(a_1+a_2) & b_1+b_2 \end{bmatrix}$, in $V_1 \cap V_2$. $ku = \begin{bmatrix} ka_1 & -(ka_1) \\ -(ka_1) & kb_1 \end{bmatrix}$, in $V_1 \cap V_2$.
$V_1 \cap V_2$ is closed under addition and scalar multiplication. It is a subspace.

(b) In V_1: $\begin{bmatrix} a & b \\ -a & c \end{bmatrix} = a\begin{bmatrix} 1 & 0 \\ -1 & 0 \end{bmatrix} + b\begin{bmatrix} 0 & 1 \\ 0 & 0 \end{bmatrix} + c\begin{bmatrix} 0 & 0 \\ 0 & 1 \end{bmatrix}$.
Matrices $\begin{bmatrix} 1 & 0 \\ -1 & 0 \end{bmatrix}, \begin{bmatrix} 0 & 1 \\ 0 & 0 \end{bmatrix}, \begin{bmatrix} 0 & 0 \\ 0 & 1 \end{bmatrix}$ span V_1 and are linearly independent.
V_1 is a subspace of dimension 3.
In V_2: $\begin{bmatrix} p & -p \\ q & r \end{bmatrix} = p\begin{bmatrix} 1 & -1 \\ 0 & 0 \end{bmatrix} + q\begin{bmatrix} 0 & 0 \\ 1 & 0 \end{bmatrix} + r\begin{bmatrix} 0 & 0 \\ 0 & 1 \end{bmatrix}$.

Matrices $\begin{bmatrix} 1 & -1 \\ 0 & 0 \end{bmatrix}, \begin{bmatrix} 0 & 0 \\ 1 & 0 \end{bmatrix}, \begin{bmatrix} 0 & 0 \\ 0 & 1 \end{bmatrix}$ span V_2 and are linearly independent.
V_2 is a subspace of dimension 3.
In $V_1 \cap V_2$: $\begin{bmatrix} a & -a \\ -a & b \end{bmatrix} = a\begin{bmatrix} 1 & -1 \\ -1 & 0 \end{bmatrix} + b\begin{bmatrix} 0 & 0 \\ 0 & 1 \end{bmatrix}$.
Matrices $\begin{bmatrix} 1 & -1 \\ -1 & 0 \end{bmatrix}$ and $\begin{bmatrix} 0 & 0 \\ 0 & 1 \end{bmatrix}$ span $V_1 \cap V_2$ and are linearly independent.
$V_1 \cap V_2$ is a subspace of dimension 2.

19. (a) Yes; $f(x) = 2h(x) - g(x)$.

(b) If $f(x) = ag(x) + bh(x)$, then $3x^2 + 5x + 1 = (2a+b)x^2 + 3bx + 3a - b$, so that $3 = 2a + b$, $5 = 3b$, and $1 = 3a - b$. This system of equations has no solution, so $f(x)$ is not in the subspace spanned by $g(x)$ and $h(x)$.

(c) $g(x) + h(x) = (2x^2 + 3) + (x^2 + 3x - 1) = 3x^2 + 3x + 2$,
$g(x) - h(x) = (2x^2 + 3) - (x^2 + 3x - 1) = x^2 - 3x + 4$, and
$2g(x) = 2(2x^2 + 3) = 4x^2 + 6$ are functions in the space spanned by $g(x)$ and $h(x)$.

(d) $f(x)$ and $g(x)$ are linearly independent and $h(x) = g(x) - f(x)$, so the set $\{f(x), g(x)\}$ is a basis.

20. (a) A basis for P_2 must consist of three linearly independent elements of P_2.

$f(x) + g(x) - h(x) = 0$ so the three given functions are not linearly independent and therefore not a basis for P_2 .

(b) These three functions are linearly independent, so they are a basis for P_2 .

(c) $\begin{bmatrix} 1 & 2 \\ 0 & 1 \end{bmatrix} - \begin{bmatrix} 3 & 4 \\ 1 & 1 \end{bmatrix} + 2\begin{bmatrix} 1 & 2 \\ 1 & 1 \end{bmatrix} - \begin{bmatrix} 0 & 2 \\ 1 & 2 \end{bmatrix} = \begin{bmatrix} 0 & 0 \\ 0 & 0 \end{bmatrix}$, so these matrices are linearly dependent and therefore not a basis for M_{22} .

(d) These four matrices are linearly independent, so they are a basis for M_{22} .

(e) The set {(1,0), (0,1)} is a basis for $\mathbf{C}^2$, so the dimension of $\mathbf{C}^2$ is 2, and any set of two linearly independent vectors in $\mathbf{C}^2$ is a basis. The given vectors are linearly independent since neither is a multiple of the other, so they are a basis.

(f) $2(1+2i, 3-i, 1) - (4+i, 3i, 1+i) = (-2+3i, 6-5i, 1-i)$, so the vectors are linearly dependent and therefore not a basis.

21. Since dim(V) = 2, any basis for V must contain 2 vectors. In Example 3 in Section 4.3, it was shown that the vectors $\mathbf{u}_1 = \mathbf{v}_1 + \mathbf{v}_2$ and $\mathbf{u}_2 = \mathbf{v}_1 - \mathbf{v}_2$ are linearly independent. Thus by Theorem 4.11 the set $\{\mathbf{u}_1, \mathbf{u}_2\}$ is a basis for V.

22. Since V is three-dimensional, any basis for V must contain three vectors. We need only show that the vectors $\mathbf{u}_1 = \mathbf{v}_1$, $\mathbf{u}_2 = \mathbf{v}_1 + \mathbf{v}_2$, and $\mathbf{u}_3 = \mathbf{v}_1 + \mathbf{v}_2 + \mathbf{v}_3$ are linearly independent. If $a\mathbf{v}_1 + b(\mathbf{v}_1 + \mathbf{v}_2) + c(\mathbf{v}_1 + \mathbf{v}_2 + \mathbf{v}_3) = (a+b+c)\mathbf{v}_1 + (b+c)\mathbf{v}_2 + c\mathbf{v}_3 = \mathbf{0}$, then $+b+c = b+c = c = 0$ so that $a = b = c = 0$. Thus $\mathbf{u}_1$, $\mathbf{u}_2$, and $\mathbf{u}_3$ are linearly independent and a basis for V.

23. Since V is n-dimensional any basis for V must contain n vectors. We need only show that the vectors $c\mathbf{v}_1, c\mathbf{v}_2, \ldots, c\mathbf{v}_n$ are linearly independent.
If $a_1 c\mathbf{v}_1 + a_2 c\mathbf{v}_2 + \ldots + a_n c\mathbf{v}_n = \mathbf{0}$, then $a_1 c = a_2 c = \ldots = a_n c = 0$ and since $c \neq 0$, $a_1 = a_2 = \ldots = a_n = 0$. Thus the vectors $c\mathbf{v}_1, c\mathbf{v}_2, \ldots, c\mathbf{v}_n$ are linearly independent and a basis for V.

24. If a set of $n-1$ vectors spans V, then either that set is linearly independent and is a basis for V, so that $\dim(V) = n-1$, or there is a linearly independent subset of those vectors that is a basis for V and $\dim(v) < n-1$. Both possibilities contradict $\dim(V) = n$.

25. Let $\{\mathbf{w}_1, \ldots, \mathbf{w}_m\}$ be a basis for W. If $m>n$ these vectors would have to be linearly dependent by Theorem 4.8. Being base vectors they are linearly independent. Thus $m \leq n$.

26. Suppose (a,b) is orthogonal to (u_1, u_2). Then $(a,b)\cdot(u_1, u_2) = au_1 + bu_2 = 0$, so $au_1 = -bu_2$

Thus if $u_1 \neq 0$ (a,b) is orthogonal to (u_1, u_2) if and only if (a,b) is of the form $\left(\frac{-bu_2}{u_1}, b\right) = \frac{b}{u_1}(-u_2, u_1)$; that is, (a,b) is orthogonal to (u_1, u_2) if and only if (a,b) is in the vector space with basis $(-u_2, u_1)$. If $u_1 = 0$, then b = 0 and (a,b) is orthogonal to (u_1, u_2) if and only if (a,b) is of the form a(1,0); that is, (a,b) is in the vector space with basis (1,0).

27. Suppose (a,b,c) is orthogonal to (u_1, u_2, u_3). Then $(a,b,c)\cdot(u_1, u_2, u_3) = au_1 + bu_2 + cu_3 = 0$, so $au_1 = -bu_2 - cu_3$. Thus if $u_1 \neq 0$, (a,b,c) is orthogonal to (u_1, u_2, u_3) if and only if (a,b,c) is of the form $\left(\frac{-bu_2 - cu_3}{u_1}, b, c\right) = \frac{b}{u_1}(-u_2, u_1, 0) + \frac{c}{u_1}(-u_3, 0, u_1)$; that is, (a,b,c) is orthogonal to (u_1, u_2, u_3) if and only if (a,b,c) is in the vector space with basis vectors $(-u_2, u_1, 0)$ and $(-u_3, 0, u_1)$. If $u_1 = 0$ then $-bu_2 = cu_3$ and (a,b,c) is orthogonal to (u_1, u_2, u_3) if and only if (a,b,c) is of the form $\left(a, b, \frac{-bu_2}{u_3}\right) = a(1,0,0) - \frac{b}{u_3}(0, -u_3, u_2)$ or $\left(a, \frac{-cu_3}{u_2}, c\right) = a(1,0,0) + \frac{c}{u_2}(0, -u_3, u_2)$; that is, (a,b,c) is in the vector space with basis vectors (1,0,0) and $(0, -u_3, u_2)$.

28. Suppose (a,b,c) is orthogonal to **u**$=(u_1, u_2, u_3)$ and **v**$=(v_1, v_2, v_3)$. Then $(a,b,c)\cdot(u_1, u_2, u_3) = au_1 + bu_2 + cu_3 = 0$ and $(a,b,c)\cdot(v_1, v_2, v_3) = av_1 + bv_2 + cv_3 = 0$. At least one of $u_2v_1 - u_1v_2$, $u_3v_1 - u_1v_3$, and $u_3v_2 - u_2v_3$ is nonzero. (Otherwise **v** = k**u**.) We assume $u_2v_1 - u_1v_2$ is nonzero and solve for a and b in terms of c. $a = \frac{c(u_3v_2 - u_2v_3)}{u_2v_1 - u_1v_2}$ and $b = \frac{c(u_1v_3 - u_3v_1)}{u_2v_1 - u_1v_2}$, so $(a,b,c) = c\left(\frac{u_3v_2 - u_2v_3}{u_2v_1 - u_1v_2}, \frac{u_1v_3 - u_3v_1}{u_2v_1 - u_1v_2}, 1\right)$.

So (a,b,c) is orthogonal to both **u** and **v** if and only if (a,b,c) is in the one-dimensional vector space with basis vector $\left(\frac{u_3v_2 - u_2v_3}{u_2v_1 - u_1v_2}, \frac{u_1v_3 - u_3v_1}{u_2v_1 - u_1v_2}, 1\right)$.

29. The two-dimensional subspace with basis (1,0,0) and (0,1,0) is orthogonal to the one-dimensional subspace with basis (0,0,1).

30. If m = n then S is a basis for V. If m > n then S is linearly dependent (Theorem 4.8). Let S' be the set found by deleting one vector from S that is a linear combination of the

remaining vectors. Let it be $\mathbf{v}_m$. Then any vector $\mathbf{v} = a_1\mathbf{v}_1 + \ldots + a_{m-1}\mathbf{v}_{m-1} + a_m\mathbf{v}_m$ can be written $\mathbf{v} = a_1\mathbf{v}_1 + \ldots + a_{m-1}\mathbf{v}_{m-1} + (b_1\mathbf{v}_1 + \ldots + b_{m-1}\mathbf{v}_{m-1}) = c_1\mathbf{v}_1 + \ldots + c_{m-1}\mathbf{v}_{m-1}$. Thus S' spans V, but has m-1 vectors. Continue deleting vectors in this manner until a set of linearly independent vectors is found. This set will span V, is linearly independent, and is thus a basis.

31. (a) False: The vectors are linearly dependent.

 (b) False: Any set that spans $\mathbf{R}^3$ must contain at least three vectors.

 (c) True: (1,2,3) + (0,1,4) = (1,3,7), and (1,2,3) and (0,1,4) are linearly independent, so the dimension is 2.

 (d) True: In fact any set of two linearly independent vectors in $\mathbf{R}^2$ spans $\mathbf{R}^2$ and any set of two vectors that spans $\mathbf{R}^2$ is linearly independent.

 (e) True: The number of vectors in any set of linearly independent vectors is less than or equal to the dimension of the vector space.

32. (a) False: The three vectors lie in the 2D subspace of $\mathbf{R}^3$ with basis {(1,0,0), (0,1,0)}. Thus they are linearly dependent.

 (b) False: If the two vectors are linearly dependent you can't get a basis no matter what vector you add to them.

 (c) False: In fact any set of more than two vectors in $\mathbf{R}^2$ is a linearly dependent set.

 (d) True: (1,2) and (2,1) span $\mathbf{R}^2$. (1,2), (2,1), (1,2)+(2,1) span but are linearly dependent in $\mathbf{R}^2$. i.e., (1,2), (2,1), (3,3) span $\mathbf{R}^2$ and are not linearly independent.

 (e) True: e.g., (1,2,3), (4,5,6) are linearly independent but do not span $\mathbf{R}^3$. Dimension of $\mathbf{R}^3$ is 3, thus need 3 linearly independent vectors to span it.

Exercise Set 4.5

1. (a) The row vectors are linearly independent. Dim row space = 2. Rank = 2.

 (b) R2 = -2R1. The rows are linearly dependent. Dim row space = 1. Rank = 1.

 (c) The row vectors are linearly independent. Dim row space = 2. Rank = 2.

(d) R2 = (3/2) R1. The rows are linearly dependent. Dim row space = 1. Rank = 1.

2. (a) R3=R2-R1. The third row is in the space spanned by the other two linearly independent rows. Dim row space = 2. Rank = 2.

(b) The rows are the standard basis for $\mathbf{R}^3$, so the rank of the matrix is 3.

(c) Rows 2 and 3 are multiples of row 1. Dim row space = 1. Rank = 1.

(d) The rows are linearly independent. Dim row space = 3. Rank = 3.

(e) R3=R2+R1. The third row is in the space spanned by the other two linearly independent rows. Dim row space = 2. Rank = 2.

(f) R3=2R2-R1. The third row is in the space spanned by the other two linearly independent rows. Dim row space = 2. Rank = 2.

3. (a) $\begin{bmatrix} 1 & 2 & -1 \\ 2 & 5 & 2 \\ 0 & 2 & 9 \end{bmatrix} \approx \begin{bmatrix} 1 & 2 & -1 \\ 0 & 1 & 4 \\ 0 & 2 & 9 \end{bmatrix} \approx \begin{bmatrix} 1 & 0 & -9 \\ 0 & 1 & 4 \\ 0 & 0 & 1 \end{bmatrix} \approx \begin{bmatrix} 1 & 0 & 0 \\ 0 & 1 & 0 \\ 0 & 0 & 1 \end{bmatrix}$, so the vectors (1,0,0),

(0,1,0), and (0,0,1) are a basis for the row space and the rank of the matrix is 3.

(b) $\begin{bmatrix} 1 & 1 & 8 \\ 0 & 1 & 3 \\ -1 & 1 & -2 \end{bmatrix} \approx \begin{bmatrix} 1 & 1 & 8 \\ 0 & 1 & 3 \\ 0 & 2 & 6 \end{bmatrix} \approx \begin{bmatrix} 1 & 0 & 5 \\ 0 & 1 & 3 \\ 0 & 0 & 0 \end{bmatrix}$, so the vectors (1,0,5) and (0,1,3) are

a basis for the row space and the rank of the matrix is 2.

(c) $\begin{bmatrix} 1 & -3 & 2 \\ -2 & 6 & -4 \\ -1 & 3 & -2 \end{bmatrix} \approx \begin{bmatrix} 1 & -3 & 2 \\ 0 & 0 & 0 \\ 0 & 0 & 0 \end{bmatrix}$, so the vector (1,−3,2) is a basis for the row space

and the rank of the matrix is 1.

4. (a) $\begin{bmatrix} 1 & 4 & 0 \\ -1 & -3 & 3 \\ 2 & 9 & 5 \end{bmatrix} \approx \begin{bmatrix} 1 & 4 & 0 \\ 0 & 1 & 3 \\ 0 & 1 & 5 \end{bmatrix} \approx \begin{bmatrix} 1 & 0 & -12 \\ 0 & 1 & 3 \\ 0 & 0 & 2 \end{bmatrix} \approx \begin{bmatrix} 1 & 0 & 0 \\ 0 & 1 & 0 \\ 0 & 0 & 1 \end{bmatrix}$, so the vectors (1,0,0),

(0,1,0), and (0,0,1) are a basis for the row space and the rank of the matrix is 3.

(b) $\begin{bmatrix} 1 & 2 & 0 \\ 0 & 1 & 1 \\ -1 & 2 & 3 \end{bmatrix} \approx \begin{bmatrix} 1 & 2 & 0 \\ 0 & 1 & 1 \\ 0 & 4 & 3 \end{bmatrix} \approx \begin{bmatrix} 1 & 0 & -2 \\ 0 & 1 & 1 \\ 0 & 0 & -1 \end{bmatrix} \approx \begin{bmatrix} 1 & 0 & 0 \\ 0 & 1 & 0 \\ 0 & 0 & 1 \end{bmatrix}$, so the vectors (1,0,0),

(0,1,0), and (0,0,1) are a basis for the row space and the rank of the matrix is 3.

(c) $\begin{bmatrix} 1 & 2 & 3 \\ 0 & -1 & -1 \\ 3 & 4 & 7 \end{bmatrix} \approx \begin{bmatrix} 1 & 2 & 3 \\ 0 & 1 & 1 \\ 0 & -2 & -2 \end{bmatrix} \approx \begin{bmatrix} 1 & 0 & 1 \\ 0 & 1 & 1 \\ 0 & 0 & 0 \end{bmatrix}$, so the vectors (1,0,1) and (0,1,1)

are a basis for the row space and the rank of the matrix is 2.

5. (a) $\begin{bmatrix} 1 & 2 & 3 & 4 \\ -1 & 2 & 0 & 1 \\ 0 & 1 & 0 & 2 \end{bmatrix} \approx \begin{bmatrix} 1 & 2 & 3 & 4 \\ 0 & 4 & 3 & 5 \\ 0 & 1 & 0 & 2 \end{bmatrix} \approx \begin{bmatrix} 1 & 2 & 3 & 4 \\ 0 & 1 & .75 & 1.25 \\ 0 & 1 & 0 & 2 \end{bmatrix} \approx \begin{bmatrix} 1 & 0 & 1.5 & 1.5 \\ 0 & 1 & .75 & 1.25 \\ 0 & 0 & -.75 & .75 \end{bmatrix} \approx$

$\begin{bmatrix} 1 & 0 & 1.5 & 1.5 \\ 0 & 1 & .75 & 1.25 \\ 0 & 0 & 1 & -1 \end{bmatrix} \approx \begin{bmatrix} 1 & 0 & 0 & 3 \\ 0 & 1 & 0 & 2 \\ 0 & 0 & 1 & -1 \end{bmatrix}$, so the vectors (1,0,0,3), (0,1,0,2), and

(0,0,1,−1) are a basis for the row space and the rank of the matrix is 3.

(b) $\begin{bmatrix} 1 & 2 & -1 & 4 \\ 0 & 1 & -2 & 3 \\ -1 & 0 & -3 & 2 \end{bmatrix} \approx \begin{bmatrix} 1 & 2 & -1 & 4 \\ 0 & 1 & -2 & 3 \\ 0 & 2 & -4 & 6 \end{bmatrix} \approx \begin{bmatrix} 1 & 0 & 3 & -2 \\ 0 & 1 & -2 & 3 \\ 0 & 0 & 0 & 0 \end{bmatrix}$, so the vectors (1,0,3,-2)

and (0,1,−2,3) are a basis for the row space and the rank of the matrix is 2.

(c) $\begin{bmatrix} 1 & 1 & 0 & -1 \\ 2 & 1 & 0 & 0 \\ 3 & 2 & 0 & -1 \\ -1 & 0 & 1 & 1 \end{bmatrix} \approx \begin{bmatrix} 1 & 1 & 0 & -1 \\ 0 & -1 & 0 & 2 \\ 0 & -1 & 0 & 2 \\ 0 & 1 & 1 & 0 \end{bmatrix} \approx \begin{bmatrix} 1 & 1 & 0 & -1 \\ 0 & 1 & 0 & -2 \\ 0 & -1 & 0 & 2 \\ 0 & 1 & 1 & 0 \end{bmatrix} \approx \begin{bmatrix} 1 & 0 & 0 & 1 \\ 0 & 1 & 0 & -2 \\ 0 & 0 & 0 & 0 \\ 0 & 0 & 1 & 2 \end{bmatrix} \approx$

$\begin{bmatrix} 1 & 0 & 0 & 1 \\ 0 & 1 & 0 & -2 \\ 0 & 0 & 1 & 2 \\ 0 & 0 & 0 & 0 \end{bmatrix}$, so the vectors (1,0,0,1), (0,1,0,−2), and (0,0,1,2) are a basis for the row space and the rank of the matrix is 3.

6. (a) The given vectors are linearly independent so they are a basis for the space they span, which is $\mathbf{R}^3$. Also $\begin{bmatrix} 1 & 3 & 2 \\ 0 & 1 & 4 \\ 1 & 4 & 9 \end{bmatrix} \approx \begin{bmatrix} 1 & 3 & 2 \\ 0 & 1 & 4 \\ 0 & 1 & 7 \end{bmatrix} \approx \begin{bmatrix} 1 & 0 & -10 \\ 0 & 1 & 4 \\ 0 & 0 & 3 \end{bmatrix} \approx \begin{bmatrix} 1 & 0 & 0 \\ 0 & 1 & 0 \\ 0 & 0 & 1 \end{bmatrix}$, so the row vectors of any of these matrices are a basis for the same vector space.

(b) The given vectors are linearly independent so they are a basis for the space they span, which is $\mathbf{R}^3$. Also $\begin{bmatrix} 1 & -2 & 5 \\ 1 & -1 & 4 \\ 2 & -5 & 14 \end{bmatrix} \approx \begin{bmatrix} 1 & -2 & 5 \\ 0 & 1 & -1 \\ 0 & -1 & 4 \end{bmatrix} \approx \begin{bmatrix} 1 & 0 & 3 \\ 0 & 1 & -1 \\ 0 & 0 & 1 \end{bmatrix} \approx$

$\begin{bmatrix} 1 & 0 & 0 \\ 0 & 1 & 0 \\ 0 & 0 & 1 \end{bmatrix}$

so the row vectors of any of these matrices are a basis for the same vector space.

(c) $\begin{bmatrix} 1 & -1 & 3 \\ 1 & 0 & 1 \\ -2 & 1 & -4 \end{bmatrix} \approx \begin{bmatrix} 1 & -1 & 3 \\ 0 & 1 & -2 \\ 0 & -1 & 2 \end{bmatrix} \approx \begin{bmatrix} 1 & 0 & 1 \\ 0 & 1 & -2 \\ 0 & 0 & 0 \end{bmatrix}$, so the vectors (1,0,1) and (0,1,−2) are a basis for the space spanned by the given vectors.

(d) All three of the given vectors are multiples of the vector (1,−3,2), so it is a basis for the vector space.

7. (a) $\begin{bmatrix} 1 & 3 & -1 & 4 \\ 1 & 3 & 0 & 6 \\ -1 & -3 & 0 & -8 \end{bmatrix} \approx \begin{bmatrix} 1 & 3 & -1 & 4 \\ 0 & 0 & 1 & 2 \\ 0 & 0 & -1 & -4 \end{bmatrix} \approx \begin{bmatrix} 1 & 3 & 0 & 6 \\ 0 & 0 & 1 & 2 \\ 0 & 0 & 0 & 0 \end{bmatrix} \approx \begin{bmatrix} 1 & 3 & 0 & 0 \\ 0 & 0 & 1 & 0 \\ 0 & 0 & 0 & 1 \end{bmatrix}$, so the vectors (1,3,0,0), (0,0,1,0), and (0,0,0,1) are a basis for the subspace. The given vectors are linearly independent so they are a basis also.

(b) $\begin{bmatrix} 1 & 2 & 0 & 1 \\ -1 & -1 & 3 & 1 \\ 2 & 3 & -2 & 4 \end{bmatrix} \approx \begin{bmatrix} 1 & 2 & 0 & 1 \\ 0 & 1 & 3 & 2 \\ 0 & -1 & -2 & 2 \end{bmatrix} \approx \begin{bmatrix} 1 & 0 & -6 & -3 \\ 0 & 1 & 3 & 2 \\ 0 & 0 & 1 & 4 \end{bmatrix} \approx \begin{bmatrix} 1 & 0 & 0 & 21 \\ 0 & 1 & 0 & -10 \\ 0 & 0 & 1 & 4 \end{bmatrix}$, so the vectors (1,0,0,21), (0,1,0,-10), and (0,0,1,4) are a basis for the subspace. The given vectors are linearly independent so they are a basis also.

(c) $\begin{bmatrix} 1 & 2 & 3 & 4 \\ 0 & -1 & 2 & 3 \\ 2 & 3 & 8 & 11 \\ 2 & 3 & 6 & 8 \end{bmatrix} \approx \begin{bmatrix} 1 & 2 & 3 & 4 \\ 0 & 1 & -2 & -3 \\ 0 & -1 & 2 & 3 \\ 0 & -1 & 0 & 0 \end{bmatrix} \approx \begin{bmatrix} 1 & 0 & 7 & 10 \\ 0 & 1 & -2 & -3 \\ 0 & 0 & 0 & 0 \\ 0 & 0 & -2 & -3 \end{bmatrix} \approx \begin{bmatrix} 1 & 0 & 7 & 10 \\ 0 & 1 & -2 & -3 \\ 0 & 0 & 1 & 1.5 \\ 0 & 0 & 0 & 0 \end{bmatrix}$

$\approx \begin{bmatrix} 1 & 0 & 0 & -.5 \\ 0 & 1 & 0 & 0 \\ 0 & 0 & 1 & 1.5 \\ 0 & 0 & 0 & 0 \end{bmatrix}$, so the vectors (1,0,0,-.5), (0,1,0,0), and (0,0,1,1.5) are a basis for the subspace.

(d) $\begin{bmatrix} 1 & -3 & -1 & 2 \\ 0 & 1 & -4 & 1 \\ 1 & -4 & 5 & 1 \\ 2 & -5 & -6 & 5 \end{bmatrix} \approx \begin{bmatrix} 1 & -3 & -1 & 2 \\ 0 & 1 & -4 & 1 \\ 0 & -1 & 6 & -1 \\ 0 & 1 & -4 & 1 \end{bmatrix} \approx \begin{bmatrix} 1 & 0 & -13 & 5 \\ 0 & 1 & -4 & 1 \\ 0 & 0 & 2 & 0 \\ 0 & 0 & 0 & 0 \end{bmatrix} \approx \begin{bmatrix} 1 & 0 & 0 & 5 \\ 0 & 1 & 0 & 1 \\ 0 & 0 & 1 & 0 \\ 0 & 0 & 0 & 0 \end{bmatrix}$, so the vectors (1,0,0,5), (0,1,0,1), and (0,0,1,0) are a basis for the subspace.

8. $A = \begin{bmatrix} 1 & 2 & -1 \\ 0 & 1 & 3 \\ 1 & 4 & 6 \end{bmatrix} \approx \begin{bmatrix} 1 & 2 & -1 \\ 0 & 1 & 3 \\ 0 & 2 & 7 \end{bmatrix} \approx \begin{bmatrix} 1 & 0 & -7 \\ 0 & 1 & 3 \\ 0 & 0 & 1 \end{bmatrix} \approx \begin{bmatrix} 1 & 0 & 0 \\ 0 & 1 & 0 \\ 0 & 0 & 1 \end{bmatrix}$, so the vectors (1,0,0), (0,1,0), and (0,0,1) are a basis for the row space of A.

$A^t = \begin{bmatrix} 1 & 0 & 1 \\ 2 & 1 & 4 \\ -1 & 3 & 6 \end{bmatrix} \approx \begin{bmatrix} 1 & 0 & 1 \\ 0 & 1 & 2 \\ 0 & 3 & 7 \end{bmatrix} \approx \begin{bmatrix} 1 & 0 & 1 \\ 0 & 1 & 2 \\ 0 & 0 & 1 \end{bmatrix} \approx \begin{bmatrix} 1 & 0 & 0 \\ 0 & 1 & 0 \\ 0 & 0 & 1 \end{bmatrix}$, so the vectors (1,0,0), (0,1,0), and (0,0,1) are a basis for the row space of A^t. Therefore the column vectors $\begin{bmatrix} 1 \\ 0 \\ 0 \end{bmatrix}$, $\begin{bmatrix} 0 \\ 1 \\ 0 \end{bmatrix}$, and $\begin{bmatrix} 0 \\ 0 \\ 1 \end{bmatrix}$ are a basis for the column space of A. Both the row space and the column space of A have dimension 3.

9. $A = \begin{bmatrix} 1 & 3 & 2 \\ 1 & 4 & 1 \\ 2 & 5 & 5 \end{bmatrix} \approx \begin{bmatrix} 1 & 3 & 2 \\ 0 & 1 & -1 \\ 0 & -1 & 1 \end{bmatrix} \approx \begin{bmatrix} 1 & 0 & 5 \\ 0 & 1 & -1 \\ 0 & 0 & 0 \end{bmatrix}$, so the vectors (1,0,5) and (0,1,−1)

are a basis for the row space of A.

$A^t = \begin{bmatrix} 1 & 1 & 2 \\ 3 & 4 & 5 \\ 2 & 1 & 5 \end{bmatrix} \approx \begin{bmatrix} 1 & 1 & 2 \\ 0 & 1 & -1 \\ 0 & -1 & 1 \end{bmatrix} \approx \begin{bmatrix} 1 & 0 & 3 \\ 0 & 1 & -1 \\ 0 & 0 & 0 \end{bmatrix}$, so the vectors (1,0,3) and (0,1,−1)

are a basis for the row space of A^t. Therefore the column vectors $\begin{bmatrix} 1 \\ 0 \\ 3 \end{bmatrix}$ and $\begin{bmatrix} 0 \\ 1 \\ -1 \end{bmatrix}$ are a basis for the column space of A. Both the row space and the column space of A have dimension 2.

10. (a) (i) $\begin{bmatrix} 1 & 0 & 0 \\ 0 & 1 & 0 \\ 0 & 0 & 1 \end{bmatrix}$. (ii) Ranks 3 and 3. Thus unique solution.

(iii) $x_1=2$, $x_2=1$, $x_3=3$. (iv) $\begin{bmatrix} 12 \\ 18 \\ -8 \end{bmatrix} = 2\begin{bmatrix} 1 \\ 2 \\ -1 \end{bmatrix} + 1\begin{bmatrix} -2 \\ -1 \\ 3 \end{bmatrix} + 3\begin{bmatrix} 4 \\ 5 \\ -3 \end{bmatrix}$.

(b) (i) $\begin{bmatrix} 1 & 0 & 3 \\ 0 & 1 & 2 \\ 0 & 0 & 0 \end{bmatrix}$. (ii) Ranks 2 and 2. Many solutions. (iii) $x_1 = -3r + 4$, $x_2 = -2r + 1$, $x_3 = r$.

(iv) $\begin{bmatrix} 9 \\ 7 \\ 7 \end{bmatrix} = (-3r + 4)\begin{bmatrix} 3 \\ 2 \\ 3 \end{bmatrix} + (-2r + 1)\begin{bmatrix} -3 \\ -1 \\ -5 \end{bmatrix} + r\begin{bmatrix} 3 \\ 4 \\ -1 \end{bmatrix}$.

(c) (i) $\begin{bmatrix} 1 & 0 & -3 \\ 0 & 1 & 1 \\ 0 & 0 & 0 \end{bmatrix}$. (ii) Ranks 2 and 3. No solution. (iii) No solution.

(iv) $\begin{bmatrix} 2 \\ 0 \\ -3 \end{bmatrix}$ is not a linear combination of $\begin{bmatrix} 1 \\ 1 \\ 2 \end{bmatrix}, \begin{bmatrix} 4 \\ 2 \\ 6 \end{bmatrix}, \begin{bmatrix} 1 \\ -1 \\ 0 \end{bmatrix}$.

(d) (i) $\begin{bmatrix} 1 & 0 & 0 \\ 0 & 1 & 0 \\ 0 & 0 & 1 \end{bmatrix}$. (ii) Ranks 3 and 3. Unique solution. (iii) $x_1=-1$, $x_2=1$, $x_3=2$.

(iv) $\begin{bmatrix} 6 \\ 9 \\ 11 \end{bmatrix} = -1\begin{bmatrix} 1 \\ 1 \\ 2 \end{bmatrix} + 1\begin{bmatrix} 1 \\ 2 \\ 1 \end{bmatrix} + 2\begin{bmatrix} 3 \\ 4 \\ 6 \end{bmatrix}$.

11. Use Theorem 4.17. (a) Unique solution. (b) No solution. (c) Many solutions. (d) No solution. (e) Many solutions. (f) Unique solution.

12. (a) There are three rows, so the row space must have dimension ≤ 3.
Five columns, column space must have dimension ≤ 5.
Since dim row space = dim col space, this dim ≤3. Rank(A) ≤ 3.

(b) The largest possible rank is the smaller of the two numbers m and n.

13. dim(column space of A) = dim(row space of A) ≤ number of rows of A = m < n

14. (a) The rank of A cannot be greater than 3, so a basis for the column space of A can contain no more than three vectors. The four column vectors in A must therefore be linearly dependent by Theorem 4.8.

(b) The rank of A cannot be greater than 3, so a basis for the row space of A can contain no more than three vectors. The seven row vectors in A must therefore be linearly dependent by Theorem 4.8.

(c) The rank of A cannot be greater than m, so a basis for the column space of A can contain no more than m vectors. Since n>m the n column vectors in A must therefore be linearly dependent by Theorem 4.8.

15. (a) If |A| ≠ 0, then rank (A)=n (Theorem 4.18). Thus the n row vectors are linearly independent. They therefore form a basis for $\mathbf{R}^n$ (Theorem 4.11).

(b) A is invertible if and only if |A| ≠ 0, so the proof is the same as in part (a).

16. If the n columns of A are linearly independent, they are a basis for the column space of A (Theorem 4.11) and dim(column space of A) is n, so rank(A) = n.

If rank(A) = n, then the n column vectors of A are basis for the column space of A because they span the column space of A (Theorem 4.11), but this means they are linearly independent.

17. If the rows of A are linearly independent, then dim(row space of A) = rank(A) = n. Thus the n columns of A are a basis for the column space of A. But the columns are linearly independent vectors in $\mathbf{R}^n$, which has dimension n. Any set of n linearly independent vectors in a space of dimension n is a basis. Thus they are a basis for $\mathbf{R}^n$.

If the columns of A span $\mathbf{R}^n$, then they are linearly independent and rank(A) = n = dim(row space of A) so the n rows of A are linearly independent.

18. Each row in A + B is a sum of a row in A and a row in B. Each row in A is a linear

combination of rank(A) basis vectors and each row in B is a linear combination of rank(B) basis vectors. Thus the row space of A + B is spanned by rank(A) + rank(B) vectors. A basis for the row space of A + B therefore consists of no more than rank(A) + rank(B) vectors, so rank(A + B) ≤ rank(A) + rank(B).

19. Use Theorem 4.18 and an awareness of Theorem 4.11.
(a)⇔(b) by Thm 4.18 (a) & (f). (a)⇔(f) by Thm 4.18 (a) & (c) One set {(a),(b),(f)}.
(c)⇔(d) by Thm 4.18 (d) & (b). (c)⇔(e) by Thm 4.18 (d) & (e). Other set {(c),(d),(e)}.

20. (a) False. Let A be mxn with m>n. If columns of A are linearly independent its rank is n.

(b) False. Consider $A = \begin{bmatrix} 1 & 0 & 0 \\ 0 & 1 & 0 \\ 0 & 0 & 1 \end{bmatrix}$ and $B = \begin{bmatrix} 1 & 0 & 0 \\ 0 & 1 & 0 \\ 0 & 0 & -1 \end{bmatrix}$. Rank(A)=3, rank(B)=3.

$A+B = \begin{bmatrix} 2 & 0 & 0 \\ 0 & 2 & 0 \\ 0 & 0 & 0 \end{bmatrix}$. Rank(A)+Rank(B) = 6, but Rank(A+B)=2.

$3A = \begin{bmatrix} 3 & 0 & 0 \\ 0 & 3 & 0 \\ 0 & 0 & 3 \end{bmatrix}$. Rank(3A)=3, but 3Rank(A) = 9.

(c) True. Let A be the augmented matrix and the system be A**X**=**B**. The rank of A is n, thus
the columns of A are linearly independent. They form a basis for $\mathbf{R}^n$. **B** is a unique linear
combination of these vectors.

(d) False. Suppose **b** is linearly independent of each of $\mathbf{a}_1, \mathbf{a}_2, \ldots, \mathbf{a}_n$. Then the system $[\mathbf{a}_1\ \mathbf{a}_2 \ldots \mathbf{a}_n]\mathbf{x} = \mathbf{b}$ does not have any solutions.

(e) True. If $[\mathbf{a}_1\ \mathbf{a}_2 \ldots \mathbf{a}_n]\mathbf{x} = \mathbf{b}$ has no solution then **b** is not a linear of $\mathbf{a}_1, \mathbf{a}_2, \ldots, \mathbf{a}_n$. Thus if $\text{rank}([\mathbf{a}_1\ \mathbf{a}_2 \ldots \mathbf{a}_n]) = m$, $\text{rank}([\mathbf{a}_1\ \mathbf{a}_2 \ldots \mathbf{a}_n\ \mathbf{b}]) = m+1$.

Exercise Set 4.6

1. (a) (1,2)·(2,−1) = 0, so the vectors are orthogonal.

(b) (3,−1)·(0,5) = −5, so the vectors are not orthogonal.

(c) (0,−2)·(3,0) = 0, so the vectors are orthogonal.

(d) (4,1)·(2,−3) = 5, so the vectors are not orthogonal.

(e) (−3,2)·(2,3) = 0, so the vectors are orthogonal.

2. (a) (1,2,1)·(4,−2,0) = 0, (1,2,1)·(2,4,−10) = 0, and (4,−2,0)·(2,4,−10) = 0, so the set of vectors is orthogonal.

(b) (3,−1,1)·(2,0,1) = 7, so the set of vectors is not orthogonal.

(c) $(-2,-1,3)\cdot(6,-1,-1) = -14$, so the set of vectors is not orthogonal.

(d) $(1,2,-1,1)\cdot(3,1,4,-1) = 0$, $(1,2,-1,1)\cdot(0,1,-1,-3) = 0$, and $(3,1,4,-1)\cdot(0,1,-1,-3) = 0$, so the set of vectors is orthogonal.

3. (a) $(\frac{1}{3}, \frac{2}{3}, \frac{2}{3})\cdot(\frac{2}{3}, -\frac{2}{3}, \frac{1}{3}) = 0$, $(\frac{1}{3}, \frac{2}{3}, \frac{2}{3})\cdot(\frac{2}{3}, \frac{1}{3}, -\frac{2}{3}) = 0$, and

$(\frac{2}{3}, -\frac{2}{3}, \frac{1}{3})\cdot(\frac{2}{3}, \frac{1}{3}, -\frac{2}{3}) = 0$, so the set of vectors is orthogonal.

It is easiest here to use the form $||\mathbf{u}||^2 = \mathbf{u}\cdot\mathbf{u}$ (sec 1.6) to check magnitude. Then

$(\frac{1}{3}, \frac{2}{3}, \frac{2}{3})\cdot(\frac{1}{3}, \frac{2}{3}, \frac{2}{3}) = 1$, $(\frac{2}{3}, -\frac{2}{3}, \frac{1}{3})\cdot(\frac{2}{3}, -\frac{2}{3}, \frac{1}{3}) = 1$, and $(\frac{2}{3}, \frac{1}{3}, -\frac{2}{3})\cdot(\frac{2}{3}, \frac{1}{3}, -\frac{2}{3}) = 1$,

so all the vectors are unit vectors. Thus the vectors are an orthonormal set.

(b) $(\frac{1}{\sqrt{10}}, \frac{3}{\sqrt{10}})\cdot(\frac{-3}{\sqrt{10}}, \frac{1}{\sqrt{10}}) = 0$, so the vectors are orthogonal.

$(\frac{1}{\sqrt{10}}, \frac{3}{\sqrt{10}})\cdot(\frac{1}{\sqrt{10}}, \frac{3}{\sqrt{10}}) = 1$ and $(\frac{-3}{\sqrt{10}}, \frac{1}{\sqrt{10}})\cdot(\frac{-3}{\sqrt{10}}, \frac{1}{\sqrt{10}}) = 1$,

so the vectors are unit vectors. Thus the vectors are an orthonormal set.

(c) $(\frac{1}{\sqrt{2}}, 0, \frac{1}{\sqrt{2}})\cdot(\frac{1}{\sqrt{2}}, 0, \frac{-1}{\sqrt{2}}) = 0$, $(\frac{1}{\sqrt{2}}, 0, \frac{1}{\sqrt{2}})\cdot(0,1,0) = 0$, and

$(\frac{1}{\sqrt{2}}, 0, \frac{-1}{\sqrt{2}})\cdot(0,1,0) = 0$, so the set of vectors is orthogonal.

$(\frac{1}{\sqrt{2}}, 0, \frac{1}{\sqrt{2}})\cdot(\frac{1}{\sqrt{2}}, 0, \frac{1}{\sqrt{2}}) = 1$, $(\frac{1}{\sqrt{2}}, 0, \frac{-1}{\sqrt{2}})\cdot(\frac{1}{\sqrt{2}}, 0, \frac{-1}{\sqrt{2}}) = 1$, and

$(0,1,0)\cdot(0,1,0) = 1$, so the vectors are unit vectors. Thus the vectors are an orthonormal set.

(d) $(\frac{1}{\sqrt{6}}, \frac{-1}{\sqrt{6}}, \frac{2}{\sqrt{6}}) \cdot (0, \frac{2}{\sqrt{5}}, \frac{1}{\sqrt{5}}) = 0$, $(\frac{1}{\sqrt{6}}, \frac{-1}{\sqrt{6}}, \frac{2}{\sqrt{6}}) \cdot (\frac{5}{\sqrt{30}}, \frac{1}{\sqrt{30}}, \frac{-2}{\sqrt{30}}) = 0$, and

$(0, \frac{2}{\sqrt{5}}, \frac{1}{\sqrt{5}}) \cdot (\frac{5}{\sqrt{30}}, \frac{1}{\sqrt{30}}, \frac{-2}{\sqrt{30}}) = 0$, so the set of vectors is orthogonal.

$(\frac{1}{\sqrt{6}}, \frac{-1}{\sqrt{6}}, \frac{2}{\sqrt{6}}) \cdot (\frac{1}{\sqrt{6}}, \frac{-1}{\sqrt{6}}, \frac{2}{\sqrt{6}}) = 1$, $(0, \frac{2}{\sqrt{5}}, \frac{1}{\sqrt{5}}) \cdot (0, \frac{2}{\sqrt{5}}, \frac{1}{\sqrt{5}}) = 1$, and

$(\frac{5}{\sqrt{30}}, \frac{1}{\sqrt{30}}, \frac{-2}{\sqrt{30}}) \cdot (\frac{5}{\sqrt{30}}, \frac{1}{\sqrt{30}}, \frac{-2}{\sqrt{30}}) = 1$, so the vectors are unit vectors. Thus the set of vectors is an orthonormal set.

(e) $(\frac{1}{\sqrt{32}}, \frac{-2}{\sqrt{32}}, \frac{5}{\sqrt{32}}) \cdot (\frac{1}{\sqrt{32}}, \frac{-2}{\sqrt{32}}, \frac{5}{\sqrt{32}}) = \frac{30}{32} \neq 1$, so $(\frac{1}{\sqrt{32}}, \frac{-2}{\sqrt{32}}, \frac{5}{\sqrt{32}})$ is not a unit vector. Therefore the set is not orthonormal.

4. $\mathbf{v} = (\mathbf{v}\cdot\mathbf{u}_1)\mathbf{u}_1 + (\mathbf{v}\cdot\mathbf{u}_2)\mathbf{u}_2 + (\mathbf{v}\cdot\mathbf{u}_3)\mathbf{u}_3$.
$\mathbf{v}\cdot\mathbf{u}_1 = (2,-3,1) \cdot (0,-1,0) = 3$, $\mathbf{v}\cdot\mathbf{u}_2 = (2,-3,1) \cdot (\frac{3}{5}, 0, \frac{-4}{5}) = \frac{2}{5}$, and
$\mathbf{v}\cdot\mathbf{u}_3 = (2,-3,1) \cdot (\frac{4}{5}, 0, \frac{3}{5}) = \frac{11}{5}$, so $\mathbf{v} = 3\mathbf{u}_1 + \frac{2}{5}\mathbf{u}_2 + \frac{11}{5}\mathbf{u}_3$.

5. $\mathbf{v} = (\mathbf{v}\cdot\mathbf{u}_1)\mathbf{u}_1 + (\mathbf{v}\cdot\mathbf{u}_2)\mathbf{u}_2 + (\mathbf{v}\cdot\mathbf{u}_3)\mathbf{u}_3$.
$\mathbf{v}\cdot\mathbf{u}_1 = (7,5,-1) \cdot (1,0,0) = 7$, $\mathbf{v}\cdot\mathbf{u}_2 = (7,5,-1) \cdot (0, \frac{1}{\sqrt{2}}, \frac{1}{\sqrt{2}}) = \frac{4}{\sqrt{2}} = 2\sqrt{2}$, and
$\mathbf{v}\cdot\mathbf{u}_3 = (7,5,-1) \cdot (0, \frac{1}{\sqrt{2}}, \frac{-1}{\sqrt{2}}) = \frac{6}{\sqrt{2}} = 3\sqrt{2}$, so $\mathbf{v} = 7\mathbf{u}_1 + 2\sqrt{2}\,\mathbf{u}_2 + 3\sqrt{2}\,\mathbf{u}_3$.

6. We show that the column vectors are unit vectors and that they are mutually orthogonal.

(a) $\mathbf{a}_1 = \begin{bmatrix} 1 \\ 0 \end{bmatrix}$ and $\mathbf{a}_2 = \begin{bmatrix} 0 \\ 1 \end{bmatrix}$. $\|\mathbf{a}_1\| = \|\mathbf{a}_2\| = 1$ and $\mathbf{a}_1 \cdot \mathbf{a}_2 = 0$.

(b) $\mathbf{a}_1 = \begin{bmatrix} 0 \\ 1 \end{bmatrix}$ and $\mathbf{a}_2 = \begin{bmatrix} -1 \\ 0 \end{bmatrix}$. $\|\mathbf{a}_1\| = \|\mathbf{a}_2\| = 1$ and $\mathbf{a}_1 \cdot \mathbf{a}_2 = 0$.

(c) $\mathbf{a}_1 = \begin{bmatrix} \frac{\sqrt{3}}{2} \\ \frac{-1}{2} \end{bmatrix}$ and $\mathbf{a}_2 = \begin{bmatrix} \frac{1}{2} \\ \frac{\sqrt{3}}{2} \end{bmatrix}$. $||\mathbf{a}_1||^2 = \frac{3}{4} + \frac{1}{4} = 1$ and $||\mathbf{a}_2||^2 = \frac{1}{4} + \frac{3}{4} = 1$,

so $||\mathbf{a}_1|| = ||\mathbf{a}_2|| = 1$, and $\mathbf{a}_1 \cdot \mathbf{a}_2 = \frac{\sqrt{3}}{2} \times \frac{1}{2} - \frac{1}{2} \times \frac{\sqrt{3}}{2} = 0$.

(d) $\mathbf{a}_1 = \begin{bmatrix} 1 \\ 0 \\ 0 \end{bmatrix}$, $\mathbf{a}_2 = \begin{bmatrix} 0 \\ 0 \\ 1 \end{bmatrix}$, and $\mathbf{a}_3 = \begin{bmatrix} 0 \\ -1 \\ 0 \end{bmatrix}$.

$||\mathbf{a}_1|| = ||\mathbf{a}_2|| = ||\mathbf{a}_3|| = 1$ and $\mathbf{a}_1 \cdot \mathbf{a}_2 = \mathbf{a}_1 \cdot \mathbf{a}_3 = \mathbf{a}_2 \cdot \mathbf{a}_3 = 0$.

(e) $\mathbf{a}_1 = \begin{bmatrix} 2/3 \\ -2/3 \\ 1/3 \end{bmatrix}$, $\mathbf{a}_2 = \begin{bmatrix} 2/3 \\ 1/3 \\ -2/3 \end{bmatrix}$, and $\mathbf{a}_3 = \begin{bmatrix} 1/3 \\ 2/3 \\ 2/3 \end{bmatrix}$. $||\mathbf{a}_1||^2 = \frac{4}{9} + \frac{4}{9} + \frac{1}{9} = 1$,

$||\mathbf{a}_2||^2 = \frac{4}{9} + \frac{1}{9} + \frac{4}{9} = 1$, and $||\mathbf{a}_3||^2 = \frac{1}{9} + \frac{4}{9} + \frac{4}{9} = 1$, so $||\mathbf{a}_1|| = ||\mathbf{a}_2|| = ||\mathbf{a}_3|| = 1$,

and $\mathbf{a}_1 \cdot \mathbf{a}_2 = \frac{4}{9} - \frac{2}{9} - \frac{2}{9} = 0$, $\mathbf{a}_1 \cdot \mathbf{a}_3 = \frac{2}{9} - \frac{4}{9} + \frac{2}{9} = 0$, $\mathbf{a}_2 \cdot \mathbf{a}_3 = \frac{2}{9} + \frac{2}{9} - \frac{4}{9} = 0$.

7. (a) The row vectors are $\mathbf{r}_1 = \left[\frac{1}{\sqrt{2}}\ \frac{1}{\sqrt{2}}\right]$ and $\mathbf{r}_2 = \left[\frac{-1}{\sqrt{2}}\ \frac{1}{\sqrt{2}}\right]$.

$||\mathbf{r}_1||^2 = ||\mathbf{r}_2||^2 = \frac{1}{2} + \frac{1}{2} = 1$, so $||\mathbf{r}_1|| = ||\mathbf{r}_2|| = 1$, and $\mathbf{r}_1 \cdot \mathbf{r}_2 = \frac{1}{\sqrt{2}} \times \frac{-1}{\sqrt{2}} + \frac{1}{\sqrt{2}} \times \frac{1}{\sqrt{2}} = 0$. Thus the rows are orthonormal vectors. Similarly for columns.

$$AA^t = \begin{bmatrix} \frac{1}{\sqrt{2}} & \frac{1}{\sqrt{2}} \\ \frac{-1}{\sqrt{2}} & \frac{1}{\sqrt{2}} \end{bmatrix} \begin{bmatrix} \frac{1}{\sqrt{2}} & \frac{-1}{\sqrt{2}} \\ \frac{1}{\sqrt{2}} & \frac{1}{\sqrt{2}} \end{bmatrix} = \begin{bmatrix} 1 & 0 \\ 0 & 1 \end{bmatrix}, \text{ so } A^{-1} = A^t.$$

The rows of A, which have been shown to be orthonormal, are the columns of $A^{-1} = A^t$, so A^{-1} is orthogonal. $|A| = \frac{1}{\sqrt{2}} \times \frac{1}{\sqrt{2}} - \frac{1}{\sqrt{2}} \times \frac{-1}{\sqrt{2}} = 1$.

(b) The row vectors are $\mathbf{r}_1 = \begin{bmatrix} \frac{\sqrt{3}}{2} & \frac{1}{2} \end{bmatrix}$ and $\mathbf{r}_2 = \begin{bmatrix} \frac{1}{2} & \frac{-\sqrt{3}}{2} \end{bmatrix}$. $||\mathbf{r}_1||^2 = \frac{3}{4} + \frac{1}{4} = 1$ and $||\mathbf{r}_2||^2 = \frac{1}{4} + \frac{3}{4} = 1$, so $||\mathbf{r}_1|| = ||\mathbf{r}_2|| = 1$, and $\mathbf{r}_1 \cdot \mathbf{r}_2 = \frac{\sqrt{3}}{2} \times \frac{1}{2} - \frac{1}{2} \times \frac{\sqrt{3}}{2} = 0$. Thus the rows are orthonormal vectors. Similarly for columns.

$$AA^t = \begin{bmatrix} \frac{\sqrt{3}}{2} & \frac{1}{2} \\ \frac{1}{2} & \frac{-\sqrt{3}}{2} \end{bmatrix} \begin{bmatrix} \frac{\sqrt{3}}{2} & \frac{1}{2} \\ \frac{1}{2} & \frac{-\sqrt{3}}{2} \end{bmatrix} = \begin{bmatrix} 1 & 0 \\ 0 & 1 \end{bmatrix}, \text{ so } A^{-1} = A^t.$$

The rows of A, which have been shown to be orthonormal, are the columns of $A^{-1} = A^t$, so A^{-1} is orthogonal. $|A| = \frac{\sqrt{3}}{2} \times \frac{-\sqrt{3}}{2} - \frac{1}{2} \times \frac{1}{2} = -1$.

8. In matrices of (a), (b), (c) column vectors are orthonormal. Thus each matrix is orthogonal. Inverse is the transpose. Inverses are:

(a) $\begin{bmatrix} 0 & -1 \\ 1 & 0 \end{bmatrix}$ (b) $\begin{bmatrix} \frac{2}{\sqrt{5}} & -\frac{1}{\sqrt{5}} \\ \frac{1}{\sqrt{5}} & \frac{1}{\sqrt{5}} \end{bmatrix}$ (c) $\begin{bmatrix} 1/3 & 2/3 & -2/3 \\ 2/3 & -2/3 & -1/3 \\ -2/3 & 1/3 & 2/3 \end{bmatrix}$

9. Let C = AB. Then $C^t = (AB)^t = B^t A^t = B^{-1}A^{-1} = (AB)^{-1} = C^{-1}$. Thus C is orthogonal.

10. $AA^t = (I - 2uu^t)(I - 2uu^t)^t = (I - 2uu^t)(I - 2(u^t)^t u^t) = (I - 2uu^t)(I - 2uu^t)$
$= I - 4uu^t + 4uu^t uu^t = I - 4uu^t + 4u(u^t u)u^t = I - 4uu^t + 4uIu^t = I$.
Thus $AA^t = I$. Multiply both sides by A^{-1}. $A^{-1}AA^t = A^{-1}I$, $A^t = A^{-1}$. Thus A is orthogonal.

11. To show that A^{-1} is unitary it is necessary to show that $(A^{-1})^{-1} = (\overline{A^{-1}})^t$.

Since A is unitary $A^{-1} = \overline{A}^t$. Thus $(A^{-1})^t = (\overline{A}^t)^t = \overline{A}$, so that $(\overline{A^{-1}})^t = A = (A^{-1})^{-1}$.

12. The (i,j)th element of $A\overline{A}^t$ is $a_{i1}\overline{a_{j1}} + a_{i2}\overline{a_{j2}} + \ldots + a_{in}\overline{a_{jn}}$. If $A\overline{A}^t = I$,

then $a_{i1}\overline{a_{j1}} + a_{i2}\overline{a_{j2}} + \ldots + a_{in}\overline{a_{jn}} = 1$ if i = j and zero if i ≠ j, so that the columns of A are a set of mutually orthogonal unit vectors in $\mathbf{C}^n$. Conversely, if the columns of A are a set of mutually orthogonal unit vectors in $\mathbf{C}^n$, then $a_{i1}\overline{a_{j1}} + a_{i2}\overline{a_{j2}} + \ldots + a_{in}\overline{a_{jn}}$ = 1 if i = j and 0 if i ≠ j, so that $A\overline{A}^t = I$.

13. $\text{proj}_{\mathbf{u}}\,\mathbf{v} = \frac{\mathbf{v}\cdot\mathbf{u}}{\mathbf{u}\cdot\mathbf{u}}\,\mathbf{u}$.

(a) $\text{proj}_{\mathbf{u}}\,\mathbf{v} = \frac{(7,4)\cdot(1,2)}{(1,2)\cdot(1,2)}(1,2) = 3(1,2) = (3,6)$.

(b) $\text{proj}_{\mathbf{u}}\,\mathbf{v} = \frac{(-1,5)\cdot(3,-2)}{(3,-2)\cdot(3,-2)}(3,-2) = -(3,-2) = (-3,2)$.

(c) $\text{proj}_{\mathbf{u}}\,\mathbf{v} = \frac{(4,6,4)\cdot(1,2,3)}{(1,2,3)\cdot(1,2,3)}(1,2,3) = 2(1,2,3) = (2,4,6)$.

(d) $\text{proj}_{\mathbf{u}}\,\mathbf{v} = \frac{(6,-8,7)\cdot(-1,3,0)}{(-1,3,0)\cdot(-1,3,0)}(-1,3,0) = -3(-1,3,0) = (3,-9,0)$.

(e) $\text{proj}_{\mathbf{u}}\,\mathbf{v} = \frac{(1,2,3,0)\cdot(1,-1,2,3)}{(1,-1,2,3)\cdot(1,-1,2,3)}(1,-1,2,3) = \frac{1}{3}(1,-1,2,3) = \left(\frac{1}{3}, \frac{-1}{3}, \frac{2}{3}, 1\right)$.

14. (a) $\text{proj}_{\mathbf{u}}\,\mathbf{v} = \frac{(1,2)\cdot(2,5)}{(2,5)\cdot(2,5)}(2,5) = \frac{12}{29}(2,5) = \left(\frac{24}{29}, \frac{60}{29}\right)$.

(b) $\text{proj}_{\mathbf{u}}\,\mathbf{v} = \frac{(-1,3)\cdot(2,4)}{(2,4)\cdot(2,4)}(2,4) = \frac{1}{2}(2,4) = (1,2)$.

(c) $\text{proj}_{\mathbf{u}}\,\mathbf{v} = \frac{(1,2,3)\cdot(1,2,0)}{(1,2,0)\cdot(1,2,0)}(1,2,0) = (1,2,0)$.

(d) $\text{proj}_{\mathbf{u}}\,\mathbf{v} = \frac{(2,1,4)\cdot(-1,-3,2)}{(-1,-3,2)\cdot(-1,-3,2)}(-1,-3,2) = \frac{3}{14}(-1,-3,2) = \left(\frac{-3}{14}, \frac{-9}{14}, \frac{6}{14}\right)$.

(e) $\text{proj}_{\mathbf{u}}\,\mathbf{v} = \frac{(2,-1,3,1)\cdot(-1,2,1,3)}{(-1,2,1,3)\cdot(-1,2,1,3)}(-1,2,1,3) = \frac{2}{15}(-1,2,1,3) = \left(\frac{-2}{15}, \frac{4}{15}, \frac{2}{15}, \frac{2}{5}\right)$.

15. (a) $\mathbf{u}_1 = (1,2)$, $\mathbf{u}_2 = (-1,3) - \frac{(-1,3)\cdot(1,2)}{(1,2)\cdot(1,2)}(1,2) = (-1,3) - (1,2) = (-2,1)$, is an orthogonal basis for $\mathbf{R}^3$.

$\|\mathbf{u}_1\| = \sqrt{5}$ and $\|\mathbf{u}_2\| = \sqrt{5}$, so the set $\left\{\left(\frac{1}{\sqrt{5}}, \frac{2}{\sqrt{5}}\right), \left(\frac{-2}{\sqrt{5}}, \frac{1}{\sqrt{5}}\right)\right\}$ is an orthonormal basis for $\mathbf{R}^2$.

(b) $\mathbf{u}_1 = (1,1)$, $\mathbf{u}_2 = (6,2) - \frac{(6,2)\cdot(1,1)}{(1,1)\cdot(1,1)}(1,1) = (6,2) - \frac{8}{2}(1,1) = (2,-2)$, is an orthogonal basis for $\mathbf{R}^3$.

$||\mathbf{u}_1|| = \sqrt{2}$ and $||\mathbf{u}_2|| = \sqrt{8} = 2\sqrt{2}$, so the set $\left\{(\frac{1}{\sqrt{2}},\frac{1}{\sqrt{2}}),(\frac{1}{\sqrt{2}},\frac{-1}{\sqrt{2}})\right\}$ is an orthonormal basis for $\mathbf{R}^2$.

(c) $\mathbf{u}_1 = (1,-1)$, $\mathbf{u}_2 = (4,-2) - \frac{(4,-2)\cdot(1,-1)}{(1,-1)\cdot(1,-1)}(1,-1) = (1,1)$, is an orthogonal basis for $\mathbf{R}^3$. $||\mathbf{u}_1|| = \sqrt{2}$ and $||\mathbf{u}_2|| = \sqrt{2}$, so the set $\left\{(\frac{1}{\sqrt{2}},\frac{-1}{\sqrt{2}}),(\frac{1}{\sqrt{2}},\frac{1}{\sqrt{2}})\right\}$ is an orthonormal basis for $\mathbf{R}^2$.

16. (a) $\mathbf{u}_1 = (1,1,1)$, $\mathbf{u}_2 = (2,0,1) - \frac{(2,0,1)\cdot(1,1,1)}{(1,1,1)\cdot(1,1,1)}(1,1,1) = (1,-1,0)$,

$$\mathbf{u}_3 = (2,4,5) - \frac{(2,4,5)\cdot(1,1,1)}{(1,1,1)\cdot(1,1,1)}(1,1,1) - \frac{(2,4,5)\cdot(1,-1,0)}{(1,-1,0)\cdot(1,-1,0)}(1,-1,0)$$

$= (2,4,5) - \frac{11}{3}(1,1,1) - (-1)(1,-1,0) = (\frac{-2}{3},\frac{-2}{3},\frac{4}{3})$, is an orthogonal basis for $\mathbf{R}^3$. $||\mathbf{u}_1|| = \sqrt{3}$, $||\mathbf{u}_2|| = \sqrt{2}$, and $||\mathbf{u}_3|| = \frac{\sqrt{24}}{3} = \frac{2\sqrt{6}}{3}$, so the set

$\{(\frac{1}{\sqrt{3}},\frac{1}{\sqrt{3}},\frac{1}{\sqrt{3}}),(\frac{1}{\sqrt{2}},\frac{-1}{2},0)(\frac{-1}{\sqrt{6}},\frac{-1}{\sqrt{6}},\frac{2}{\sqrt{6}})\}$ is an orthonormal basis for $\mathbf{R}^3$.

(b) $\mathbf{u}_1 = (3,2,0)$, $\mathbf{u}_2 = (1,5,-1) - \frac{(1,5,-1)\cdot(3,2,0)}{(3,2,0)\cdot(3,2,0)}(3,2,0) = (-2,3,-1)$,

$$\mathbf{u}_3 = (5,-1,2) - \frac{(5,-1,2)\cdot(3,2,0)}{(3,2,0)\cdot(3,2,0)}(3,2,0) - \frac{(5,-1,2)\cdot(-2,3,-1)}{(-2,3,-1)\cdot(-2,3,-1)}(-2,3,-1)$$

$= (5,-1,2) - (3,2,0) - \frac{-15}{14}(-2,3,-1) = (\frac{-2}{14},\frac{3}{14},\frac{13}{14})$, is an orthogonal basis for $\mathbf{R}^3$. $||\mathbf{u}_1|| = \sqrt{13}$, $||\mathbf{u}_2|| = \sqrt{14}$, and $||\mathbf{u}_3|| = \frac{\sqrt{182}}{14}$, so the set

$$\{(\frac{3}{\sqrt{13}},\frac{2}{\sqrt{13}},0),(\frac{-2}{\sqrt{14}},\frac{3}{\sqrt{14}},\frac{-1}{\sqrt{14}}),(\frac{-2}{\sqrt{182}},\frac{3}{\sqrt{182}},\frac{13}{\sqrt{182}})\}$$

is an orthonormal basis for $\mathbf{R}^3$.

17. (a) $\mathbf{u}_1 = (1,0,2)$, $\mathbf{u}_2 = (-1,0,1) - \dfrac{(-1,0,1)\cdot(1,0,2)}{(1,0,2)\cdot(1,0,2)}(1,0,2) = (-1,0,1) - \dfrac{1}{5}(1,0,2)$

$= \left(\dfrac{-6}{5}, 0, \dfrac{3}{5}\right)$, is an orthogonal basis for the subspace of $\mathbf{R}^3$. $\|\mathbf{u}_1\| = \sqrt{5}$ and

$\|\mathbf{u}_2\| = \dfrac{3}{5}\sqrt{5}$, so the set $\left\{\left(\dfrac{1}{\sqrt{5}}, 0, \dfrac{2}{\sqrt{5}}\right), \left(\dfrac{-2}{\sqrt{5}}, 0, \dfrac{1}{\sqrt{5}}\right)\right\}$ is an orthonormal basis

for the subspace.

(b) $\mathbf{u}_1 = (1,-1,1)$, $\mathbf{u}_2 = (1,2,-1) - \dfrac{(1,2,-1)\cdot(1,-1,1)}{(1,-1,1)\cdot(1,-1,1)}(1,-1,1) = (1,2,-1) - \dfrac{-2}{3}(1,-1,1)$

$= \left(\dfrac{5}{3}, \dfrac{4}{3}, \dfrac{-1}{3}\right)$, is an orthogonal basis for the subspace of $\mathbf{R}^3$. $\|\mathbf{u}_1\| = \sqrt{3}$ and

$\|\mathbf{u}_2\| = \dfrac{1}{3}\sqrt{42}$, so the set $\left\{\left(\dfrac{1}{\sqrt{3}}, \dfrac{-1}{\sqrt{3}}, \dfrac{1}{\sqrt{3}}\right), \left(\dfrac{5}{\sqrt{42}}, \dfrac{4}{\sqrt{42}}, \dfrac{-1}{\sqrt{42}}\right)\right\}$ is an

orthonormal basis for the subspace.

18. (a) $\mathbf{u}_1 = (1,2,3,4)$, $\mathbf{u}_2 = (-1,1,0,1) - \dfrac{(-1,1,0,1)\cdot(1,2,3,4)}{(1,2,3,4)\cdot(1,2,3,4)}(1,2,3,4)$

$= (-1,1,0,1) - \dfrac{1}{6}(1,2,3,4) = \left(\dfrac{-7}{6}, \dfrac{4}{6}, \dfrac{-3}{6}, \dfrac{2}{6}\right)$, is an orthogonal basis for the

subspace of $\mathbf{R}^4$. $\|\mathbf{u}_1\| = \sqrt{30}$ and $\|\mathbf{u}_2\| = \dfrac{1}{6}\sqrt{78}$, so the set

$\{(\dfrac{1}{\sqrt{30}}, \dfrac{2}{\sqrt{30}}, \dfrac{3}{\sqrt{30}}, \dfrac{4}{\sqrt{30}}), (\dfrac{-7}{\sqrt{78}}, \dfrac{4}{\sqrt{78}}, \dfrac{-3}{\sqrt{78}}, \dfrac{2}{\sqrt{78}})\}$ is an orthonormal basis for the subspace.

(b) $\mathbf{u}_1 = (3,0,0,0)$, $\mathbf{u}_2 = (0,1,2,1)$ (because $\mathbf{u}_1 \cdot \mathbf{u}_2 = 0$),

$\mathbf{u}_3 = (0,-1,3,2) - 0(3,0,0,0) - \dfrac{(0,-1,3,2)\cdot(0,1,2,1)}{(0,1,2,1)\cdot(0,1,2,1)}(0,1,2,1) = \left(0, \dfrac{-13}{6}, \dfrac{4}{6}, \dfrac{5}{6}\right)$, is an

orthogonal basis for the subspace of $\mathbf{R}^4$. $\|\mathbf{u}_1\| = 3$, $\|\mathbf{u}_2\| = \sqrt{6}$, and $\|\mathbf{u}_3\| = \dfrac{1}{6}\sqrt{210}$, so the set $\{(1,0,0),(0,\frac{1}{\sqrt{6}},\frac{2}{\sqrt{6}},\frac{1}{\sqrt{6}}),(0,\frac{-13}{\sqrt{210}},\frac{4}{\sqrt{210}},\frac{5}{\sqrt{210}})\}$ is an orthonormal basis for the subspace.

19. To construct such a vector, start with any vector that is not a multiple of the given vector (1,2,−1,−1) and use the Gram-Schmidt process to find a vector orthogonal to (1,2,−1,−1). The vector (−2,1,0,0) is orthogonal to (1,2,−1,−1).

20. To construct such a vector, start with any vector that is not a multiple of the given vector (2,0,1,1) and use the Gram-Schmidt process to find a vector orthogonal to (2,0,1,1). The vector (1,1,−2,0) is orthogonal to (2,0,1,1).

21. First find an orthonormal basis for W. $\mathbf{u}_1 = (0,-1,3)$,

$\mathbf{u}_2 = (1,1,2) - \frac{(1,1,2)\cdot(0,-1,3)}{(0,-1,3)\cdot(0,-1,3)}(0,-1,3) = (1,1,2) - \frac{1}{2}(0,-1,3) = (1,\frac{3}{2},\frac{1}{2})$, is an

orthogonal basis for the subspace of $\mathbf{R}^3$. $\|\mathbf{u}_1\| = \sqrt{10}$ and $\|\mathbf{u}_2\| = \frac{1}{2}\sqrt{14}$, so the set $\left\{(0,\frac{-1}{\sqrt{10}},\frac{3}{\sqrt{10}}),(\frac{2}{\sqrt{14}},\frac{3}{\sqrt{14}},\frac{1}{\sqrt{14}})\right\}$ is an orthonormal basis for W.

(a) $\text{proj}_W(3,-1,2) = \left((3,-1,2)\cdot(0,\frac{-1}{\sqrt{10}},\frac{3}{\sqrt{10}})\right)(0,\frac{-1}{\sqrt{10}},\frac{3}{\sqrt{10}})$

$+\left((3,-1,2)\cdot(\frac{2}{\sqrt{14}},\frac{3}{\sqrt{14}},\frac{1}{\sqrt{14}})\right)(\frac{2}{\sqrt{14}},\frac{3}{\sqrt{14}},\frac{1}{\sqrt{14}}) = \frac{7}{10}(0,-1,3) + \frac{5}{14}(2,3,1)$

$= (\frac{5}{7},\frac{13}{35},\frac{86}{35})$

(b) $\text{proj}_W(1,1,1) = \left((1,1,1)\cdot(0,\frac{-1}{\sqrt{10}},\frac{3}{\sqrt{10}})\right)(0,\frac{-1}{\sqrt{10}},\frac{3}{\sqrt{10}})$

$+\left((1,1,1)\cdot(\frac{2}{\sqrt{14}},\frac{3}{\sqrt{14}},\frac{1}{\sqrt{14}})\right)(\frac{2}{\sqrt{14}},\frac{3}{\sqrt{14}},\frac{1}{\sqrt{14}}) = \frac{1}{5}(0,-1,3) + \frac{3}{7}(2,3,1)$

$= (\frac{6}{7},\frac{38}{35},\frac{36}{35})$

(c) $\text{proj}_W(4,2,1) = \left((4,2,1)\cdot(0, \frac{-1}{\sqrt{10}}, \frac{3}{\sqrt{10}})\right)(0, \frac{-1}{\sqrt{10}}, \frac{3}{\sqrt{10}})$

$+ \left((4,2,1)\cdot(\frac{2}{\sqrt{14}}, \frac{3}{\sqrt{14}}, \frac{1}{\sqrt{14}})\right)(\frac{2}{\sqrt{14}}, \frac{3}{\sqrt{14}}, \frac{1}{\sqrt{14}}) = \frac{1}{10}(0,-1,3) + \frac{15}{14}(2,3,1)$

$= (\frac{15}{7}, \frac{109}{35}, \frac{48}{35})$

22. First find an orthonormal basis for V. $\mathbf{u}_1 = (-1,0,2,1)$,

$\mathbf{u}_2 = (1,-1,0,3) - \frac{(1,-1,0,3)\cdot(-1,0,2,1)}{(-1,0,2,1)\cdot(-1,0,2,1)}(-1,0,2,1) = (1,-1,0,3) - \frac{1}{3}(-1,0,2,1)$

$= (\frac{4}{3}, -1, \frac{-2}{3}, \frac{8}{3})$, is an orthogonal basis for the subspace of $\mathbf{R}^3$. $\|\mathbf{u}_1\| = \sqrt{6}$ and $\|\mathbf{u}_2\| = \frac{1}{3}\sqrt{93}$, so the set $\left\{(\frac{-1}{\sqrt{6}}, 0, \frac{2}{\sqrt{6}}, \frac{1}{\sqrt{6}}), (\frac{4}{\sqrt{93}}, \frac{-3}{\sqrt{93}}, \frac{-2}{\sqrt{93}}, \frac{8}{\sqrt{93}})\right\}$ is an orthonormal basis for V.

(a) $\text{proj}_V(1,-1,1,-1) = \left((1,-1,1,-1)\cdot(\frac{-1}{\sqrt{6}}, 0, \frac{2}{\sqrt{6}}, \frac{1}{\sqrt{6}})\right)(\frac{-1}{\sqrt{6}}, 0, \frac{2}{\sqrt{6}}, \frac{1}{\sqrt{6}})$

$+ \left((1,-1,1,-1)\cdot(\frac{4}{\sqrt{93}}, \frac{-3}{\sqrt{93}}, \frac{-2}{\sqrt{93}}, \frac{8}{\sqrt{93}})\right)(\frac{4}{\sqrt{93}}, \frac{-3}{\sqrt{93}}, \frac{-2}{\sqrt{93}}, \frac{8}{\sqrt{93}})$

$= 0(-1,0,2,1) + \frac{-1}{31}(4,-3,-2,8) = (\frac{-4}{31}, \frac{3}{31}, \frac{2}{31}, \frac{-8}{31})$

(b) $\text{proj}_V(2,0,1,-1) = \left((2,0,1,-1)\cdot(\frac{-1}{\sqrt{6}}, 0, \frac{2}{\sqrt{6}}, \frac{1}{\sqrt{6}})\right)(\frac{-1}{\sqrt{6}}, 0, \frac{2}{\sqrt{6}}, \frac{1}{\sqrt{6}})$

$+ \left((2,0,1,-1)\cdot(\frac{4}{\sqrt{93}}, \frac{-3}{\sqrt{93}}, \frac{-2}{\sqrt{93}}, \frac{8}{\sqrt{93}})\right)(\frac{4}{\sqrt{93}}, \frac{-3}{\sqrt{93}}, \frac{-2}{\sqrt{93}}, \frac{8}{\sqrt{93}})$

$= \frac{-1}{6}(-1,0,2,1) + \frac{-2}{93}(4,-3,-2,8) = (\frac{5}{62}, \frac{2}{31}, \frac{-9}{31}, \frac{-21}{62})$

(c) $\text{proj}_V(3,2,1,0) = \left((3,2,1,0)\cdot(\frac{-1}{\sqrt{6}}, 0, \frac{2}{\sqrt{6}}, \frac{1}{\sqrt{6}})\right)(\frac{-1}{\sqrt{6}}, 0, \frac{2}{\sqrt{6}}, \frac{1}{\sqrt{6}})$

$+ \left((3,2,1,0)\cdot(\frac{4}{\sqrt{93}}, \frac{-3}{\sqrt{93}}, \frac{-2}{\sqrt{93}}, \frac{8}{\sqrt{93}})\right)(\frac{4}{\sqrt{93}}, \frac{-3}{\sqrt{93}}, \frac{-2}{\sqrt{93}}, \frac{8}{\sqrt{93}})$

$= \frac{-1}{6}(-1,0,2,1) + \frac{4}{93}(4,-3,-2,8) = (\frac{21}{62}, \frac{-4}{31}, \frac{-13}{31}, \frac{11}{62})$

23. $V = \{a(1,1,0) + b(0,0,1)\}$. The set $\{(1,1,0), (0,0,1)\}$ is an orthogonal basis. The vectors $(\frac{1}{\sqrt{2}}, \frac{1}{\sqrt{2}}, 0)$ and $(0,0,1)$ are therefore an orthonormal basis. $\mathbf{v} = \mathbf{w} + \mathbf{w}_\perp$, where

$\mathbf{w} = \text{proj}_V \mathbf{v} = \left((1,2,-1)\cdot(\frac{1}{\sqrt{2}}, \frac{1}{\sqrt{2}}, 0)\right)(\frac{1}{\sqrt{2}}, \frac{1}{\sqrt{2}}, 0) + \left((1,2,-1)\cdot(0,0,1)\right)(0,0,1)$

$= \frac{3}{2}(1,1,0) + (-1)(0,0,1) = (\frac{3}{2}, \frac{3}{2}, -1)$ and

$\mathbf{w}_\perp = \mathbf{v} - \text{proj}_V \mathbf{v} = (1,2,-1) - (\frac{3}{2}, \frac{3}{2}, -1) = (\frac{-1}{2}, \frac{1}{2}, 0)$

24. $V = \{a(1,2,0) + b(0,0,1)\}$. The set $\{(1,2,0), (0,0,1)\}$ is an orthogonal basis. The vectors $(\frac{1}{\sqrt{5}}, \frac{2}{\sqrt{5}}, 0)$ and $(0,0,1)$ are therefore an orthonormal basis. $\mathbf{v} = \mathbf{w} + \mathbf{w}_\perp$, where

$\mathbf{w} = \text{proj}_V \mathbf{v} = \left((4,1,-2)\cdot(\frac{1}{\sqrt{5}}, \frac{2}{\sqrt{5}}, 0)\right)(\frac{1}{\sqrt{5}}, \frac{2}{\sqrt{5}}, 0) + \left((4,1,-2)\cdot(0,0,1)\right)(0,0,1)$

$= \frac{6}{5}(1,2,0) + (-2)(0,0,1) = (\frac{6}{5}, \frac{12}{5}, -2)$ and

$\mathbf{w}_\perp = \mathbf{v} - \text{proj}_V \mathbf{v} = (4,1,-2) - (\frac{6}{5}, \frac{12}{5}, -2) = (\frac{14}{25}, \frac{-7}{25}, 0)$

25. $W = \{a(1,0,1) + b(0,1,1)\}$. We must find an orthogonal basis. $\mathbf{u}_1 = (1,0,1)$ and

$\mathbf{u}_2 = (0,1,1) - \frac{(0,1,1)\cdot(1,0,1)}{(1,0,1)\cdot(1,0,1)}(1,0,1) = (0,1,1) - \frac{1}{2}(1,0,1) = (\frac{-1}{2}, 1, \frac{1}{2})$ are an orthogonal basis. The vectors $(\frac{1}{\sqrt{2}}, 0, \frac{1}{\sqrt{2}})$ and $(\frac{-1}{\sqrt{6}}, \frac{2}{\sqrt{6}}, \frac{1}{\sqrt{6}})$ are therefore an orthonormal basis. $\mathbf{v} = \mathbf{w} + \mathbf{w}_\perp$, where

$\mathbf{w} = \text{proj}_W\, \mathbf{v} = \left((3,2,1)\cdot(\frac{1}{\sqrt{2}}, 0, \frac{1}{\sqrt{2}})\right)(\frac{1}{\sqrt{2}}, 0, \frac{1}{\sqrt{2}})$

$\quad + \left((3,2,1)\cdot(\frac{-1}{\sqrt{6}}, \frac{2}{\sqrt{6}}, \frac{1}{\sqrt{6}})\right)(\frac{-1}{\sqrt{6}}, \frac{2}{\sqrt{6}}, \frac{1}{\sqrt{6}}) = 2(1,0,1) + \frac{1}{3}(-1,2,1) = (\frac{5}{3},\frac{2}{3},\frac{7}{3})$

$\mathbf{w}_\perp = \mathbf{v} - \text{proj}_V\, \mathbf{v} = (3,2,1) - (\frac{5}{3},\frac{2}{3},\frac{7}{3}) = (\frac{4}{3}, \frac{4}{3}, \frac{-4}{3})$

26. $W = \{a(1,1,0) + b(0,0,1)\}$. The set $\{(1,1,0), (0,0,1)\}$ is an orthogonal basis. The vectors $(\frac{1}{\sqrt{2}}, \frac{1}{\sqrt{2}}, 0)$ and $(0,0,1)$ are therefore an orthonormal basis.

$\text{proj}_W\, \mathbf{x} = \left((1,3,-2)\cdot(\frac{1}{\sqrt{2}}, \frac{1}{\sqrt{2}}, 0)\right)(\frac{1}{\sqrt{2}}, \frac{1}{\sqrt{2}}, 0) + \left((1,3,-2)\cdot(0,0,1)\right)(0,0,1)$

$\quad = 2(1,1,0) + (-2)(0,0,1) = (2,2,-2)$, so

$d(\mathbf{x},W) = \|\mathbf{x} - \text{proj}_W\, \mathbf{x}\| = \|(1,3,-2) - (2,2,-2)\| = \|(-1,1,0)\| = \sqrt{2}$.

27. $W = \{a(1,-2,0) + b(0,0,1)\}$. The set $\{(1,-2,0), (0,0,1)\}$ is an orthogonal basis. The vectors $(\frac{1}{\sqrt{5}}, \frac{-2}{\sqrt{5}}, 0)$ and $(0,0,1)$ are therefore an orthonormal basis.

$\text{proj}_W\, \mathbf{x} = \left((2,4,-1)\cdot(\frac{1}{\sqrt{5}}, \frac{-2}{\sqrt{5}}, 0)\right)(\frac{1}{\sqrt{5}}, \frac{-2}{\sqrt{5}}, 0) + \left((2,4,-1)\cdot(0,0,1)\right)(0,0,1)$

$\quad = \frac{-6}{5}(1,-2,0) + (-1)(0,0,1) = (\frac{-6}{5}, \frac{12}{5}, -1)$, so

$d(\mathbf{x},W) = \|\mathbf{x} - \text{proj}_W\, \mathbf{x}\| = \|(2,4,-1) - (\frac{-6}{5}, \frac{12}{5}, -1)\| = \|(\frac{16}{5}, \frac{8}{5}, 0)\| = \frac{8}{5}\sqrt{5}$.

28. $W = \{a(1,2,3)\}$. The vector $(\frac{1}{\sqrt{14}}, \frac{2}{\sqrt{14}}, \frac{3}{\sqrt{14}})$ is an orthonormal basis for W.

$\text{proj}_W\, \mathbf{x} = \left((1,3,-2)\cdot(\frac{1}{\sqrt{14}}, \frac{2}{\sqrt{14}}, \frac{3}{\sqrt{14}})\right)(\frac{1}{\sqrt{14}}, \frac{2}{\sqrt{14}}, \frac{3}{\sqrt{14}}) = (\frac{1}{14}, \frac{2}{14}, \frac{3}{14})$, so

$d(\mathbf{x},W) = \|\mathbf{x} - \text{proj}_W\, \mathbf{x}\| = \|(1,3,-2) - (\frac{1}{14}, \frac{2}{14}, \frac{3}{14})\| = \|(\frac{13}{14}, \frac{40}{14}, \frac{-31}{14})\| = \frac{\sqrt{2730}}{14}$.

29. (1,2,−2) and (6,1,4) are orthogonal. If the vector (a,b,c) is orthogonal to both, then (1,2,−2)·(a,b,c) = a + 2b − 2c = 0 and (6,1,4)·(a,b,c) = 6a + b + 4c = 0. One solution to this system of equations is a = −10, b = 16, and c = 11. The set {(1,2,−2), (6,1,4), (−10,16,11)} is therefore an orthogonal basis for $\mathbf{R}^3$.

30. $\|\mathbf{v}\| = \sqrt{\mathbf{v}\cdot\mathbf{v}}$ so we need to show that $\mathbf{v}\cdot\mathbf{v} = (\mathbf{v}\cdot\mathbf{u}_1)^2 + (\mathbf{v}\cdot\mathbf{u}_2)^2 + \ldots + (\mathbf{v}\cdot\mathbf{u}_n)^2$.
$\mathbf{v} = (\mathbf{v}\cdot\mathbf{u}_1)\mathbf{u}_1 + (\mathbf{v}\cdot\mathbf{u}_2)\mathbf{u}_2 + \ldots + (\mathbf{v}\cdot\mathbf{u}_n)\mathbf{u}_n$, so
$\mathbf{v}\cdot\mathbf{v} = ((\mathbf{v}\cdot\mathbf{u}_1)\mathbf{u}_1 + (\mathbf{v}\cdot\mathbf{u}_2)\mathbf{u}_2 + \ldots + (\mathbf{v}\cdot\mathbf{u}_n)\mathbf{u}_n)\cdot((\mathbf{v}\cdot\mathbf{u}_1)\mathbf{u}_1 + (\mathbf{v}\cdot\mathbf{u}_2)\mathbf{u}_2 + \ldots + (\mathbf{v}\cdot\mathbf{u}_n)\mathbf{u}_n)$
$= (\mathbf{v}\cdot\mathbf{u}_1)\mathbf{u}_1\cdot(\mathbf{v}\cdot\mathbf{u}_1)\mathbf{u}_1 + (\mathbf{v}\cdot\mathbf{u}_2)\mathbf{u}_2\cdot(\mathbf{v}\cdot\mathbf{u}_2)\mathbf{u}_2 + \ldots + (\mathbf{v}\cdot\mathbf{u}_n)\mathbf{u}_n\cdot(\mathbf{v}\cdot\mathbf{u}_n)\mathbf{u}_n$. All the other terms are zero because $\mathbf{u}_i\cdot\mathbf{u}_j = 0$ for $i \neq j$, and since $\mathbf{u}_i\cdot\mathbf{u}_i = 1$ we have
$\mathbf{v}\cdot\mathbf{v} = (\mathbf{v}\cdot\mathbf{u}_1)^2 + (\mathbf{v}\cdot\mathbf{u}_2)^2 + \ldots + (\mathbf{v}\cdot\mathbf{u}_n)^2$.

31. $\{\mathbf{u}_1, \mathbf{u}_2, \ldots, \mathbf{u}_n\}$ is a set of n linearly independent vectors in a vector space with dimension n. Therefore $\{\mathbf{u}_1, \mathbf{u}_2, \ldots, \mathbf{u}_n\}$ is a basis for the vector space.

32. $\text{proj}_{\mathbf{u}}\,\mathbf{v} = \frac{\mathbf{v}\cdot\mathbf{u}}{\mathbf{u}\cdot\mathbf{u}}\,\mathbf{u}$, so $(\mathbf{v} - \text{proj}_{\mathbf{u}}\,\mathbf{v})\cdot\mathbf{u} = (\mathbf{v} - \frac{\mathbf{v}\cdot\mathbf{u}}{\mathbf{u}\cdot\mathbf{u}}\,\mathbf{u})\cdot\mathbf{u} = \mathbf{v}\cdot\mathbf{u} - \frac{\mathbf{v}\cdot\mathbf{u}}{\mathbf{u}\cdot\mathbf{u}}\,\mathbf{u}\cdot\mathbf{u} = 0$.

<u>Exercise Set 4.7</u>

1. $T((x_1,y_1)+(x_2,y_2)) = T(x_1+x_2, y_1+y_2) = (3(x_1+x_2), x_1+x_2+y_1+y_2)$
$= (3x_1+3x_2, x_1+y_1+x_2+y_2) = (3x_1, x_1+y_1) + (3x_2, x_2+y_2)$
$= T(x_1,y_1) + T(x_2,y_2)$ and
T(c(x,y)) = T(cx,cy) = (3cx,cx+cy) = c(3x,x+y) = cT(x,y). Thus T is linear.
T(1,3) = (3(1),1+3) = (3,4). T(-1,2) = (3(-1),-1+2)=(-3,1).

2. $T((x_1,y_1)+(x_2,y_2)) = T(x_1+x_2, y_1+y_2)$
$= (x_1+x_2+2(y_1+y_2),\ 3(y_1+y_2),\ x_1+x_2-(y_1+y_2))$
$= (x_1+x_2+2y_1+2y_2,\ 3y_1+3y_2,\ x_1+x_2-y_1-y_2))$
$=(x_1+2y_1,\ 3y_1,\ x_1-y_1)+(x_2+2y_2,\ 3y_2,\ x_2-y_2)$
$= T(x_1,y_1) + T(x_2,y_2)$.
T(c(x,y)) = T(cx,cy) = (cx+2cy, 3cy, cx-cy) = c(x+2y, 3y, x-y) = cT(x,y), so T is linear.
T(0,4) = (0+2(4), 3(4), 0-4) = (8,12,-4). T(1,1) = (1+2(1), 3(3), 1-1) = (3,9,0).

3. (a) Often easiest to check 2nd condition when we suspect mapping is nonlinear:
$T(c(x,y)) = T(cx,cy) = (cy)^2 = c^2y^2$. But $cT(x,y) = cy^2$. $T(c(x,y)) \neq cT(x,y)$. T is not linear.

(b) T(c(x,y)) = T(cx,cy) = cx-3. But cT(x,y) = c(x-3)=cx-3c. $T(c(x,y)) \neq cT(x,y)$. T is not linear.

4. $T((ax^2 + bx + c)+(px^2 + qx + r)) = T((a+p)x^2 + (b+q)x + c+ r) = (c+r)x^2 + (a+p)$

$= (cx^2 + a) + (rx^2 + p) = T(ax^2 + bx + c)+ T(px^2 + qx + r)$.

$T(k(ax^2 + bx + c)) = T(kax^2 + kbx + kc) = kcx^2 + ka = k(cx^2 + a) = kT(ax^2 + bx + c)$.

T is linear. $T(3x^2-x+2) = 2x^2 + 3$. $T(3x^2+bx+2)= 2x^2 + 3$ for any value of b.

5. $T((ax^2 + bx + c)+(px^2 + qx + r)) = T((a+p)x^2 + (b+q)x + c+ r) = (b+q)x + c + r$

$= (bx+c) + (qx+r) = T((ax^2 + bx + c) + T((px^2 + qx + r)$.

$T(k(ax^2 + bx + c)) = T(kax^2 + kbx + kc) = kbx + kc = k(bx+c) = kT(ax^2 + bx + c)$.

T is linear. $T(3x^2 - x + 2) = -x+2$. $T(ax^2 - x + 2) = -x+2$ for any value of a.

6. $T(k(ax + b)) = T(kax + kb) = x + ka$. But $kT(ax + b) = k(x+a) = kx + ka$.

$T(k(x,y)) \neq kT(x,y)$. Thus T is not linear.

7. $J(f) = \int_0^1 f(x)\ dx$. $J(f + g) = \int_0^1 (f(x)+g(x))\ dx = \int_0^1 f(x)\ dx + \int_0^1 g(x)\ dx = J(f) + J(g)$ and

$J(cf) = \int_0^1 cf(x)\ dx = c\int_0^1 f(x)\ dx = cJ(f)$, so J is a linear mapping of P_n to **R**.

8. (a) $\begin{bmatrix} 1 & 2 \\ 3 & 0 \end{bmatrix}\begin{bmatrix} x \\ y \end{bmatrix} = \begin{bmatrix} x+2y \\ 3x \end{bmatrix}$; thus (x,y) is in the kernel of T if $x+2y = 0$ and $3x = 0$.

The only solution of this system of homogeneous equations is $x = 0$, $y = 0$, so only the zero vector is in the kernel of T. The columns of the matrix are linearly independent, so the range is $\mathbf{R}^2$. Dim ker(T) = 0, dim range(T) = 2, and dim domain(T) = 2, so dim ker(T) + dim range(T) = dim domain(T).

(b) $\begin{bmatrix} 2 & 0 \\ 3 & 0 \end{bmatrix}\begin{bmatrix} x \\ y \end{bmatrix} = \begin{bmatrix} 2x \\ 3x \end{bmatrix}$; thus (x,y) is in the kernel of T if $2x = 0$ and $3x = 0$, i.e., if $x = 0$.

The kernel of T is the set {(0,r)} and the range is the set {(2r,3r)}. Dim ker(T) = 1, dim range(T) = 1, and dim domain(T) = 2, so dim ker(T) + dim range(T) = dim domain(T).

(c) $\begin{bmatrix} 2 & 4 \\ 4 & 8 \end{bmatrix}\begin{bmatrix} x \\ y \end{bmatrix} = \begin{bmatrix} 2x+4y \\ 4x+8y \end{bmatrix}$; thus (x,y) is in the kernel of T if 2x+4y = 0 and 4x+8y = 0,

i.e., if x = −2y. The kernel of T is the set {(−2r,r)} and the range is the set {(r,2r)}. Dim ker(T) = 1, dim range(T) = 1, and dim domain(T) = 2, so dim ker(T) + dim range(T) = dim domain(T).

(d) $\begin{bmatrix} 1 & 2 \\ -1 & 3 \end{bmatrix}\begin{bmatrix} x \\ y \end{bmatrix} = \begin{bmatrix} x+2y \\ -x+3y \end{bmatrix}$; thus (x,y) is in the kernel of T if x+2y = 0 and −x+3y = 0.

The only solution of this system of homogeneous equations is x = 0, y = 0, so only the zero vector is in the kernel of T. The range is $\mathbf{R}^2$. Dim ker(T) = 0, dim range(T) = 2, and dim domain(T) = 2, so dim ker(T) + dim range(T) = dim domain(T).

(e) $\begin{bmatrix} 1 & 2 & 3 \\ 0 & 1 & 2 \end{bmatrix}\begin{bmatrix} x \\ y \\ z \end{bmatrix} = \begin{bmatrix} x+2y+3z \\ y+2z \end{bmatrix}$; thus (x,y,z) is in the kernel of T if x+2y+3z = 0 and

y+2z = 0. The general solution to this system of equations is x = r, y = −2r, and z = r, so the kernel of T is the set {(r,−2r,r)}. The range is $\mathbf{R}^2$.
Dim ker(T) = 1, dim range(T) = 2, and dim domain(T) = 3, so dim ker(T) + dim range(T) = dim domain(T).

(f) $\begin{bmatrix} 1 & 0 & 0 \\ 0 & 2 & 0 \\ 0 & 0 & 3 \end{bmatrix}\begin{bmatrix} x \\ y \\ z \end{bmatrix} = \begin{bmatrix} x \\ 2y \\ 3z \end{bmatrix}$; thus (x,y,z) is in the kernel of T if x = 0, 2y = 0, and 3z = 0,

i.e., if x = y = z = 0. The kernel of T is the zero vector. The range is $\mathbf{R}^3$.
Dim ker(T) = 0, dim range(T) = 3, and dim domain(T) = 3, so dim ker(T) + dim range(T) = dim domain(T).

(g) $\begin{bmatrix} 0 & 1 & 0 \\ 0 & 2 & 0 \\ 0 & 0 & 4 \end{bmatrix}\begin{bmatrix} x \\ y \\ z \end{bmatrix} = \begin{bmatrix} y \\ 2y \\ 4z \end{bmatrix}$; thus (x,y,z) is in the kernel of T if y = 0, 2y = 0, and 4z = 0,

i.e., if y = z = 0. The kernel of T is the set {(r,0,0)}. The range is the set {(a,2a,b)}. Dim ker(T) = 1, dim range(T) = 2, and dim domain(T) = 3, so dim ker(T) + dim range(T) = dim domain(T).

(h) $\begin{bmatrix} 1 & 2 & 1 \\ -1 & -2 & 0 \\ 2 & 4 & 1 \end{bmatrix}\begin{bmatrix} x \\ y \\ z \end{bmatrix} = \begin{bmatrix} x+2y+z \\ -x-2y \\ 2x+4y+z \end{bmatrix}$; thus (x,y,z) is in the kernel of T if x+2y+z = 0,

−x−2y = 0 and 2x+4y+z = 0, i.e., if x = −2y and z = 0. The kernel of T is the set {(−2r,r,0)} and the range is the set {(a+b,−a,2a+b)}, using the first and third columns of the matrix as a basis. Dim ker(T) = 1, dim range(T) = 2, and dim domain(T) = 3, so dim ker(T) + dim range(T) = dim domain(T).

(i) $\begin{bmatrix} 1 & 1 & 5 \\ 0 & 1 & 3 \\ 2 & 1 & 7 \end{bmatrix}\begin{bmatrix} x \\ y \\ z \end{bmatrix} = \begin{bmatrix} x+y+5z \\ y+3z \\ 2x+y+7z \end{bmatrix}$; thus (x,y,z) is in the kernel of T if x+y+5z = 0,

y+3z = 0, and 2x+y+7z = 0, i.e., if x = −2z and y = −3z. The kernel of T is the set {(−2r,−3r,r)} and the range is the set {(a+b,b,2a+b)}, using the first two columns of the matrix as a basis. Dim ker(T) = 1, dim range(T) = 2, and dim domain(T) = 3, so dim ker(T) + dim range(T) = dim domain(T).

9. (a) The kernel is the set {(0,r,s)} and the range is the set {(a,0,0)}. Dim ker(T) = 2, dim range(T) = 1, and dim domain(T) = 3, so dim ker(T) + dim range(T) = dim domain(T).

(b) The kernel is the set {(r,−r,0)} and the range is $\mathbf{R}^2$. Dim ker(T) = 1, dim range(T) = 2, and dim domain(T) = 3, so dim ker(T) + dim range(T) = dim domain(T).

(c) The kernel is the set {(−r−s,r,s)} and the range is **R**. Dim ker(T) = 2, dim range(T)=1,
and dim domain(T) = 3, so dim ker(T) + dim range(T) = dim domain(T).

(d) The kernel is the set {(0,r)} and the range is the set {(a,2a,3a)}. Dim ker(T) = 1, dim range(T) = 1, and dim domain(T) = 2, so dim ker(T) + dim range(T) = dim domain(T).

(e) The kernel is the zero vector and the range is the set {(3a,a−b,b)}. Dim ker(T) = 0, dim range(T) = 2, and dim domain(T) = 2, so dim ker(T) + dim range(T) = dim domain(T).

10. (a) T(**u**+**w**) = 5(**u**+**w**) = 5**u** + 5**w** = T(**u**) + T(**w**) and T(c**u**) = 5(c**u**) = 5c**u** = c(5**u**) = cT(**u**), so T is linear. ker(T) is the zero vector and range(T) is U (since for any vector **u** in U, **u** = T(.2**u**)).

(b) T(c**u**) = 2(c**u**) + 3**v** ≠ c(2**u** + 3**v**) = cT(**u**), so T is not linear.

(c) T(**u**+**w**) = **u**+**w** = T(**u**) + T(**w**) and T(c**u**) = c**u** = cT(**u**), so T is linear. ker(T) is the zero vector and range(T) is U.

(d) T(**u**+**w**) = **0** = **0** + **0** = T(**u**) + T(**w**) and T(c**u**) = **0** = c**0** = cT(**u**), so T is linear. ker(T) is U and range(T) is **0**.

11. (a) Let $\mathbf{u}_1, \mathbf{u}_2, \ldots, \mathbf{u}_n$ be a basis for U.

The vectors $T(\mathbf{u}_1), T(\mathbf{u}_2), \ldots, T(\mathbf{u}_n)$ span range(T): If **v** is a vector in range(T), then **v** = T(**u**) for some vector $\mathbf{u} = a_1\mathbf{u}_1 + a_2\mathbf{u}_2 + \ldots + a_n\mathbf{u}_n$ in U. We get $\mathbf{v} = T(\mathbf{u}) = T(a_1\mathbf{u}_1 + a_2\mathbf{u}_2 + \ldots + a_n\mathbf{u}_n) = a_1T(\mathbf{u}_1) + a_2T(\mathbf{u}_2) + \ldots + a_nT(\mathbf{u}_n)$.

The vectors $T(\mathbf{u}_1), T(\mathbf{u}_2), \ldots, T(\mathbf{u}_n)$ are linearly independent:
Consider the identity $a_1 T(\mathbf{u}_1) + a_2 T(\mathbf{u}_2) + \ldots + a_n T(\mathbf{u}_n) = \mathbf{0}$. The linearity of T gives $T(a_1 \mathbf{u}_1 + a_2 \mathbf{u}_2 + \ldots + a_n \mathbf{u}_n) = \mathbf{0}$. Since ker(T) = **0** this implies that $a_1 \mathbf{u}_1 + a_2 \mathbf{u}_2 + \ldots + a_n \mathbf{u}_n = \mathbf{0}$. Since $\mathbf{u}_1, \mathbf{u}_2, \ldots, \mathbf{u}_n$ are linearly independent this means that $a_1 = a_2 = \ldots = a_n = 0$.
Thus $T(\mathbf{u}_1), T(\mathbf{u}_2), \ldots, T(\mathbf{u}_n)$ are a basis for range(T). So dim range(T) = dim domain(T), and since dim ker(T) = 0, the equality is proved.

(b) If ker(T) = U, then range(T) = **0**, since all vectors in U are mapped into **0**. dim ker(T) = dim U = dim domain(T), and since dim range(T) = 0, the equality is proved.

12. (a) If ker(T) = **0** and $\mathbf{u}_1, \mathbf{u}_2, \ldots, \mathbf{u}_n$ are a basis for U, then $T(\mathbf{u}_1), T(\mathbf{u}_2), \ldots, T(\mathbf{u}_n)$ are a basis for range(T) (see Exercise 11(a)). Thus $\{T(\mathbf{u}_1), T(\mathbf{u}_2), \ldots, T(\mathbf{u}_n)\}$ is a set of n linearly independent vectors in V. If dim(V) = dim(U) = n, this set of vectors is a basis for V. Since V and range(T) have the same basis vectors, they are the same vector space.
If $\mathbf{u}_1, \mathbf{u}_2, \ldots, \mathbf{u}_n$ are a basis for U, then $T(\mathbf{u}_1), T(\mathbf{u}_2), \ldots, T(\mathbf{u}_n)$ span range(T). If range(T) = V then dim range(T) = n and $T(\mathbf{u}_1), T(\mathbf{u}_2), \ldots, T(\mathbf{u}_n)$ are a basis for range(T) = V. If $\mathbf{0} = T(\mathbf{u}) = T(a_1 \mathbf{u}_1 + a_2 \mathbf{u}_2 + \ldots + a_n \mathbf{u}_n)$
$= a_1 T(\mathbf{u}_1) + a_2 T(\mathbf{u}_2) + \ldots + a_n T(\mathbf{u}_n)$ then $a_1 = a_2 = \ldots = a_n = 0$ since $T(\mathbf{u}_1), T(\mathbf{u}_2), \ldots, T(\mathbf{u}_n)$ are a basis. Thus $\mathbf{u} = \mathbf{0}$ and ker(T) = **0**.

(b) If range(T) = **0**, then T(**u**) = **0** for all vectors **u** in U. Thus ker(T) = U.

If ker(T) = U, then T(**u**) = **0** for all vectors **u** in U. Thus the only vector in range(T) is **0**.

13. dim ker(T) + dim range(T) = dim domain(T). dim ker(T) ≥ 0, so dim range(T) $\leq$ dim domain(T).

14. dim range(T) = rank A = rank A^t = dim range(T^t).

15. T(x,y,z) = (x,y,0) = (1,2,0) if x = 1, y = 2, z = r. Thus the set of vectors mapped by T into (1,2,0) is the set {(1,2,r)}.

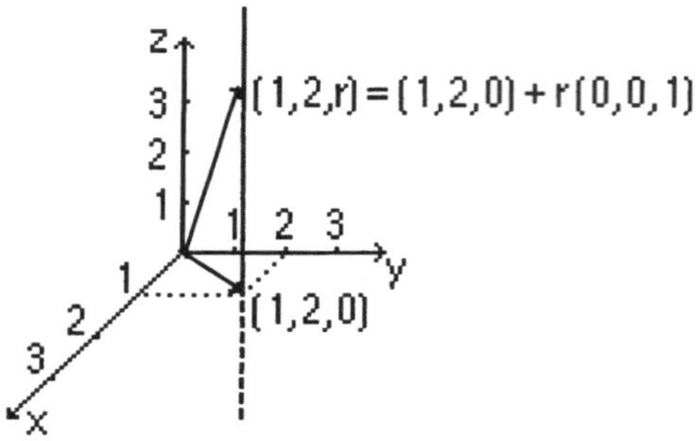

Figure for Exercise 15

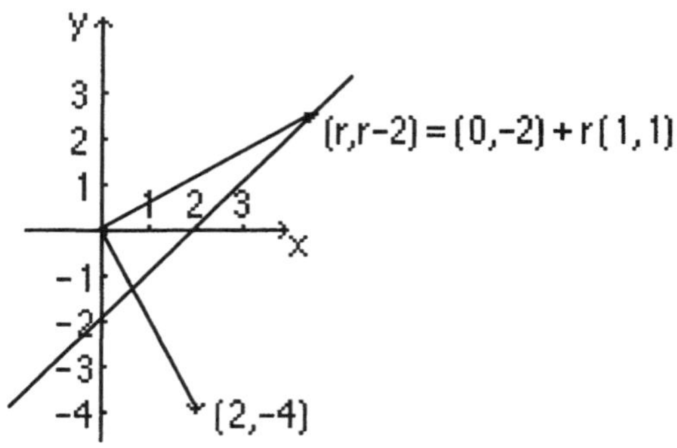

Figure for Exercise16

16. $T(x,y) = (x-y,2y-2x) = (2,-4)$ if $x-y = 2$ and $2y-2x = -4$, i.e., if $y = x-2$. Thus the set of vectors mapped by T into $(2,-4)$ is the set $\{(r, r-2)\}$.

17. $T(x,y) = (2x,3x) = (4,6)$ if $2x = 4$ and $3x = 6$, i.e., if $x = 2$. Thus the set of vectors mapped by T into $(4,6)$ is the set $\{(2, r)\}$. This set is not a subspace of $\mathbf{R}^2$. It does not contain the zero vector.

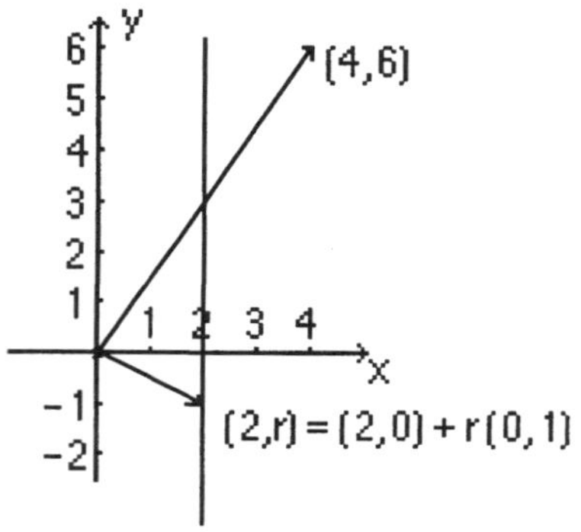

Figure for Exercise 17

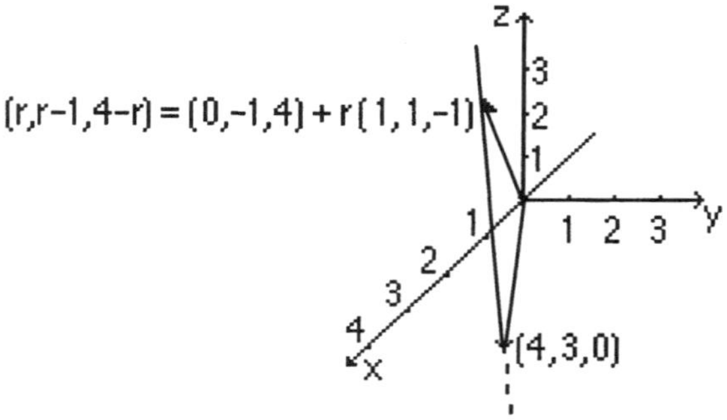

Figure for Exercise 18

18. $T(x,y,z) = (x-y,x+z) = (1,4)$ if $x-y = 1$ and $x+z = 4$. Thus the set of vectors mapped by T into $(1,4)$ is the set $\{(r,r-1,4-r)\}$. This set is not a subspace of $\mathbf{R}^3$. It does not contain the zero vector.

19. $T(a_2 x^2 + a_1 x + a_0) + T(b_2 x^2 + b_1 x + b_0)$
$= (a_2 + a_1)x^2 + a_1 x + 2a_0 + (b_2 + b_1)x^2 + b_1 x + 2b_0$
$= (a_2 + a_1 + b_2 + b_1)x^2 + (a_1 + b_1)x + 2(a_0 + b_0)$
$= T((a_2 + b_2)x^2 + (a_1 + b_1)x + a_0 + b_0)$
$= T(a_2 x^2 + a_1 x + a_0 + b_2 x^2 + b_1 x + b_0)$ and

$T(c(a_2x^2 + a_1x + a_0)) = T(ca_2x^2 + ca_1x + ca_0) = (ca_2 + ca_1)x^2 + ca_1x + 2ca_0$
$= c((a_2 + a_1)x^2 + a_1x + 2a_0) = cT(a_2x^2 + a_1x + a_0)$,
so T is linear.
$T(a_2x^2 + a_1x + a_0) = 0$ if $a_2 = a_1 = a_0 = 0$, so ker(T) is the zero polynomial and range(T) is P_2. A basis for P_2 is the set $\{1, x, x^2\}$.

20. $T(a_3x^3 + a_2x^2 + a_1x + a_0) + T(b_3x^3 + b_2x^2 + b_1x + b_0)$
$= a_3x^2 - a_0 + b_3x^2 - b_0 = (a_3 + b_3)x^2 - (a_0 + b_0)$
$= T((a_3 + b_3)x^3 + (a_2 + b_2)x^2 + (a_1 + b_1)x + a_0 + b_0)$
$= T(a_3x^3 + a_2x^2 + a_1x + a_0 + b_3x^3 + b_2x^2 + b_1x + b_0)$ and
$T(c(a_3x^3 + a_2x^2 + a_1x + a_0)) = T(ca_3x^3 + ca_2x^2 + ca_1x + ca_0)$
$= ca_3x^2 - ca_0 = c(a_3x^2 - a_0) = cT(a_3x^3 + a_2x^2 + a_1x + a_0)$,
so T is linear. $T(a_3x^3 + a_2x^2 + a_1x + a_0) = 0$ if $a_3 = a_0 = 0$, thus
ker(T) = $\{a_2x^2 + a_1x\}$ (basis = $\{x, x^2\}$) and range(T) = $\{a_3x^2 - a_0\}$ (basis = $\{1, x^2\}$).

21. $g(a_2x^2 + a_1x + a_0) + g(b_2x^2 + b_1x + b_0)$
$= 2a_2x^3 + a_1x + 3a_0 + 2b_2x^3 + b_1x + 3b_0$
$= 2(a_2 + b_2)x^3 + (a_1 + b_1)x + 3(a_0 + b_0) = g((a_2 + b_2)x^2 + (a_1 + b_1)x + a_0 + b_0)$
$= g(a_2x^2 + a_1x + a_0 + b_2x^2 + b_1x + b_0)$ and
$g(c(a_2x^2 + a_1x + a_0)) = g(ca_2x^2 + ca_1x + ca_0)$
$= 2ca_2x^3 + ca_1x + 3ca_0 = c(2a_2x^3 + a_1x + 3a_0)$
$= cg(a_2x^2 + a_1x + a_0)$, so g is linear.
$g(a_2x^2 + a_1x + a_0) = 0$ if $a_2 = a_1 = a_0 = 0$, so ker(g) is the zero polynomial and range(g) = $\{a_3x^3 + a_1x + a_0\}$. A basis for range(g) is the set $\{1, x, x^3\}$.

22. $D(x^3 - 3x^2 + 2x + 1) = 3x^2 - 6x + 2$.
$D(a_nx^n + \ldots + a_1x + a_0) = 0$ if $a_n = \ldots = a_1 = 0$, i.e., ker(D) = the set of constant polynomials. Range(D) = P_{n-1}, because every polynomial of degree less than or equal to $n-1$ is the derivative of a polynomial of degree one larger than its own degree, and no polynomial of degree n is the derivative of a polynomial of degree n or less.

23. (a) $D^2(2x^3 + 3x^2 - 5x + 4) = D(6x^2 + 6x - 5) = 12x + 6$.
$D^2(a_nx^n + \ldots + a_1x + a_0) = D(na_nx^{n-1} + \ldots + 2a_2x + a_1)$
$= n(n-1)a_nx^{n-2} + \ldots + 6a_3x + 2a_2$.

$D^2(a_n x^n + ... + a_1 x + a_0) + D^2(b_n x^n + ... + b_1 x + b_0)$
$= n(n-1)a_n x^{n-2} + ... + 6a_3 x + 2a_2 + n(n-1)b_n x^{n-2} + ... + 6b_3 x + 2b_2$
$= n(n-1)(a_n + b_n)x^{n-2} + ... + 6(a_3 + b_3)x + 2(a_2 + b_2)$
$= D^2((a_n + b_n)x^n + ... + (a_1 + b_1)x + a_0 + b_0)$ and
$D^2(c(a_n x^n + ... + a_1 x + a_0)) = D^2(ca_n x^n + ... + ca_1 x + ca_0)$
$= n(n-1)ca_n x^{n-2} + ... + 6ca_3 x + 2ca_2 = c(n(n-1)a_n x^{n-2} + ... + 6a_3 x + 2a_2)$
$= cD^2(a_n x^n + ... + a_1 x + a_0)$, so D^2 is linear.

$D^2(a_n x^n + ... + a_1 x + a_0) = n(n-1)a_n x^{n-2} + ... + 6a_3 x + 2a_2 = 0$ if $a_n = ... = a_3 = a_2 = 0$. Thus $\ker(D^2) = \{a_1 x + a_0\} = P_1$. Range$(D^2) = P_{n-2}$ because every polynomial of degree less than or equal to n−2 is the second derivative of a polynomial of degree two larger than its own degree, and no polynomial of degree n−1 or larger is the second derivative of a polynomial of degree n or less.

(b) $(D^2 + D + 3)(x^3 - 2x^2 + 6x + 1) = 6x - 4 + 3x^2 - 4x + 6 + 3(x^3 - 2x^2 + 6x + 1)$
$= 3x^3 - 3x^2 + 20x + 5$.

$(D^2 + D + 3)(a_n x^n + ... + a_1 x + a_0)$
$= n(n-1)a_n x^{n-2} + ... + 6a_3 x + 2a_2 + na_n x^{n-1} + ... + 2a_2 x + a_1 + 3(a_n x^n + ... + a_1 x + a_0)$.

$(D^2 + D + 3)(a_n x^n + ... + a_1 x + a_0) + (D^2 + D + 3)(b_n x^n + ... + b_1 x + b_0)$
$= n(n-1)a_n x^{n-2} + ... + 6a_3 x + 2a_2 + na_n x^{n-1} + ... + 2a_2 x + a_1$
$+ 3(a_n x^n + ... + a_1 x + a_0) + n(n-1)b_n x^{n-2} + ... + 6b_3 x + 2b_2$
$+ nb_n x^{n-1} + ... + 2b_2 x + b_1 + 3(b_n x^n + ... + b_1 x + b_0)$
$= n(n-1)(a_n + b_n)x^{n-2} + ... + 6(a_3 + b_3)x + 2(a_2 + b_2)$
$+ n(a_n + b_n)x^{n-1} + ... + 2(a_2 + b_2)x + a_1 + b_1$
$+ 3((a_n + b_n)x^n + ... + (a_1 + b_1)x + a_0 + b_0)$
$= (D^2 + D + 3)((a_n + b_n)x^n + ... + (a_1 + b_1)x + a_0 + b_0)$ and
$(D^2 + D + 3)(c(a_n x^n + ... + a_1 x + a_0)) = (D^2 + D + 3)(ca_n x^n + ... + ca_1 x + ca_0)$
$= n(n-1)ca_n x^{n-2} + ... + 6ca_3 x + 2ca_2 + nca_n x^{n-1} + ... + 2ca_2 x + ca_1$
$+ 3(ca_n x^n + ... + ca_1 x + ca_0) = c(n(n-1)a_n x^{n-2} + ... + 6a_3 x + 2a_2$
$+ na_n x^{n-1} + ... + 2a_2 x + a_1 + 3(a_n x^n + ... + a_1 x + a_0))$.

$= c(D^2 + D + 3)(a_n x^n + \ldots + a_1 x + a_0)$, so $(D^2 + D + 3)$ is linear.

$(D^2 + D + 3)(a_n x^n + \ldots + a_1 x + a_0) = 3a_n x^n + (3a_{n-1} + na_n)x^{n-1}$
$+(3a_{n-2} + (n-1)a_{n-1} + n(n-1)a_n)x^{n-2} + \ldots + (3a_1 + 2a_2 + 6a_3)x + (3a_0 + a_1 + 2a_2)$.

Thus if $(D^2 + D + 3)(a_n x^n + \ldots + a_1 x + a_0) = 0$, then

$a_n = a_{n-1} = \ldots = a_2 = a_1 = a_0 = 0$, so that $\ker(D^2 + D + 3)$ is the zero polynomial and range $(D^2 + D + 3) = P_n$.

24. $D(a_n x^n + \ldots + a_1 x + a_0) = na_n x^{n-1} + \ldots + 3a_3 x^2 + 2a_2 x + a_1 = 3x^2 - 4x + 7$ if $a_n = a_{n-1} = \ldots = a_4 = 0$, $a_3 = 1$, $a_2 = -2$, $a_1 = 7$, and $a_0 = r$, where r is any real number. Thus the set of polynomials mapped into $3x^2 - 4x + 7$ is the set $\{x^3 - 2x^2 + 7x + r\}$.

25. $D^2(a_n x^n + \ldots + a_1 x + a_0) = n(n-1)a_n x^{n-2} + \ldots + 20a_5 x^3 + 12a_4 x^2 + 6a_3 x + 2a_2$ $= 4x^3 + 6x^2 - 2x + 3$ if $a_n = a_{n-1} = \ldots = a_6 = 0$, $a_5 = 1/5$, $a_4 = 1/2$, $a_3 = -1/3$, $a_2 = 3/2$, $a_1 = r$, and $a_0 = s$, where r and s are any real numbers. Thus the set of polynomials mapped into $4x^3 + 6x^2 - 2x + 3$ is the set $\{x^5/5 + x^4/2 - x^3/3 + 3x^2/2 + rx + s\}$.

26. $\int_0^1 (8x^3 + 6x^2 + 4x + 1)\, dx = [2x^4 + 2x^3 + 2x^2 + x]_0^1 = 7$.

$$\int_0^1 (a_n x^n + a_{n-1}x^{n-1} + \ldots + a_1 x + a_0)\, dx = \frac{a_n}{n+1} + \frac{a_{n-1}}{n} + \ldots + \frac{a_1}{2} + a_0,$$

so ker(T) is all polynomials $a_n x^n + a_{n-1} x^{n-1} + \ldots + a_1 x + a_0$ for which

$\frac{a_n}{n+1} + \frac{a_{n-1}}{n} + \ldots + \frac{a_1}{2} - a_0 = 0$. For any real number c, let $p(x) = c$. Then

$\int_0^1 p(x)\, dx = \int_0^1 c\, dx = [cx]_0^1 = c$, so range(T) = **R**. The set of elements mapped into 2 is the set of all polynomials $a_n x^n + a_{n-1} x^{n-1} + \ldots + a_1 x + a_0$ for which

$\frac{a_n}{n+1} + \frac{a_{n-1}}{n} + \ldots + \frac{a_1}{2} + a_0 = 2$.

27. $T(\mathbf{x}_1 + \mathbf{x}_2) = (\mathbf{x}_1 + \mathbf{x}_2)\cdot\mathbf{y} = \mathbf{x}_1\cdot\mathbf{y} + \mathbf{x}_2\cdot\mathbf{y} = T(\mathbf{x}_1) + T(\mathbf{x}_2)$ and $T(c\mathbf{x}) = c\mathbf{x}\cdot\mathbf{y} = c(\mathbf{x}\cdot\mathbf{y}) = cT(\mathbf{x})$, so T

is a linear transformation from $\mathbf{R}^n$ to $\mathbf{R}$.

28. Easiest to check 2nd linearity condition: Let A be nxn matrix. Then det(cA) = |cA| (See Theorem 3.4) = $c^2|A| = c^2\det(A) \neq c\det(A)$. Thus det is not linear.

29. (a) $T\left(\begin{bmatrix} 2 & 0 \\ 1 & 3 \end{bmatrix}\right) = \begin{bmatrix} 1 & 2 \\ 3 & 4 \end{bmatrix}\begin{bmatrix} 2 & 0 \\ 1 & 3 \end{bmatrix} = \begin{bmatrix} 4 & 6 \\ 10 & 12 \end{bmatrix}$.

(b) $T(B+C) = \begin{bmatrix} 1 & 2 \\ 3 & 4 \end{bmatrix}(B+C) = \begin{bmatrix} 1 & 2 \\ 3 & 4 \end{bmatrix}B + \begin{bmatrix} 1 & 2 \\ 3 & 4 \end{bmatrix}C$.

$T(kB) = \begin{bmatrix} 1 & 2 \\ 3 & 4 \end{bmatrix}(kB) = k\begin{bmatrix} 1 & 2 \\ 3 & 4 \end{bmatrix}B = kT(B)$.

30. (a) $(T_1 + T_2)(\mathbf{u}+\mathbf{v}) = T_1(\mathbf{u}+\mathbf{v}) + T_2(\mathbf{u}+\mathbf{v}) = T_1(\mathbf{u}) + T_1(\mathbf{v}) + T_2(\mathbf{u}) + T_2(\mathbf{v})$

$= T_1(\mathbf{u}) + T_2(\mathbf{u}) + T_1(\mathbf{v}) + T_2(\mathbf{v}) = (T_1 + T_2)(\mathbf{u}) + (T_1 + T_2)(\mathbf{v})$ and $(T_1 + T_2)(c\mathbf{u})$

$= T_1(c\mathbf{u}) + T_2(c\mathbf{u}) = cT_1(\mathbf{u}) + cT_2(\mathbf{u}) = c(T_1(\mathbf{u}) + T_2(\mathbf{u})) = c(T_1 + T_2)(\mathbf{u})$,
so $T_1 + T_2$ is linear.

$(cT_1)(\mathbf{u}+\mathbf{v}) = cT_1(\mathbf{u}+\mathbf{v}) = c(T_1(\mathbf{u}) + T_1(\mathbf{v})) = cT_1(\mathbf{u}) + cT_1(\mathbf{v}) = (cT_1)(\mathbf{u}) + (cT_1)(\mathbf{v})$ and

$(cT_1)(k\mathbf{u}) = cT_1(k\mathbf{u}) = c(kT_1)(\mathbf{u}) = ckT_1(\mathbf{u}) = kcT_1(\mathbf{u}) = k(cT_1)(\mathbf{u})$, so cT_1 is linear.

(b) If $T_1(\mathbf{x}) = A_1\mathbf{x}$ and $T_2(\mathbf{x}) = A_2\mathbf{x}$, then $(T_1 + T_2)(\mathbf{x}) = T_1(\mathbf{x}) + T_2(\mathbf{x}) = A_1\mathbf{x} + A_2\mathbf{x}$

$= (A_1 + A_2)\mathbf{x}$ and $(cT_1)(\mathbf{x}) = cT_1(\mathbf{x}) = c(A_1\mathbf{x}) = cA_1\mathbf{x}$.

(c) $T_1 + T_2$ and cT_1 are both linear transformations, so the closure conditions (axioms 1 and 2) for a vector space are satisfied.

Axiom 3 $(T_1 + T_2)(\mathbf{u}) = T_1(\mathbf{u}) + T_2(\mathbf{u}) = T_2(\mathbf{u}) + T_1(\mathbf{u}) = (T_2 + T_1)(\mathbf{u})$, so
$T_1 + T_2 = T_2 + T_1$.

Axiom 4 $((T_1 + T_2) + T_3)(\mathbf{u}) = (T_1 + T_2)(\mathbf{u}) + T_3(\mathbf{u}) = (T_1(\mathbf{u}) + T_2(\mathbf{u})) + T_3(\mathbf{u})$

$= T_1(\mathbf{u}) + (T_2(\mathbf{u}) + T_3(\mathbf{u})) = T_1(\mathbf{u}) + (T_2 + T_3)(\mathbf{u}) = (T_1 + (T_2 + T_3))(\mathbf{u})$, so $(T_1 + T_2) + T_3 = T_1 + (T_2 + T_3)$.

Axiom 5 Let O be the linear transformation defined by $O(\mathbf{u}) = \mathbf{0}$. Then

$(T_1 + O)(\mathbf{u}) = T_1(\mathbf{u}) + O(\mathbf{u}) = T_1(\mathbf{u}) + \mathbf{0} = T_1(\mathbf{u})$, so $T_1 + O = T_1$.

Axiom 6 $-T_1$ is the linear transformation defined by $(-T_1)(\mathbf{u}) = -T_1(\mathbf{u})$.

$(T_1 + (-T_1))(\mathbf{u}) = T_1(\mathbf{u}) + (-T_1)(\mathbf{u}) = \mathbf{0}$, so $T_1 + (-T_1) = O$.

Axiom 7 $(c(T_1 + T_2))(\mathbf{u}) = c(T_1 + T_2)(\mathbf{u}) = c(T_1(\mathbf{u}) + T_2(\mathbf{u})) = cT_1(\mathbf{u}) + cT_2(\mathbf{u})$

$= (cT_1)(\mathbf{u}) + (cT_2)(\mathbf{u})$, so $c(T_1 + T_2) = cT_1 + cT_2$.

Axiom 8 $((c+d)T_1)(\mathbf{u}) = (c+d)T_1(\mathbf{u}) = cT_1(\mathbf{u}) + dT_1(\mathbf{u}) = (cT_1)(\mathbf{u}) + (dT_1)(\mathbf{u})$

$= (cT_1 + dT_1)(\mathbf{u})$, so $(c+d)T_1 = cT_1 + dT_1$.

Axiom 9 $((cd)T_1)(\mathbf{u}) = (cd)T_1(\mathbf{u}) = c(dT_1)(\mathbf{u})$, so $(cd)T_1 = c(dT_1)$.

Axiom 10 $(1T_1)(\mathbf{u}) = 1T_1(\mathbf{u}) = T_1(\mathbf{u})$, so $1T_1 = T_1$.

31. (a) $T(A+C) = (A+C)^t = A^t + C^t = T(A) + T(C)$ and $T(cA) = (cA)^t = cA^t = cT(A)$, so T is linear. $A^t = O$ if and only if $A = O$, so ker(T) = O and range(T) = U.

(b) $T(A+C) = |A+C|$ and $T(A) + T(C) = |A| + |C| \neq |A+C|$, so T is not linear. e.g. for $A = \begin{bmatrix} 2 & 0 \\ 0 & 1 \end{bmatrix}$ and $C = \begin{bmatrix} -2 & 0 \\ 0 & -1 \end{bmatrix} = |A| + |C| = 2 + 2 = 4$ and $|A+C| = 0$.

(c) $T(A+C) = tr(A+C)$ and $T(A) + T(C) = tr(A) + tr(C) = tr(A+C)$, so T is not linear.
Proof: $tr(A) + tr(C) = (a_{11}+a_{22})+(c_{11}+c_{22}) = (a_{11}+c_{11}+a_{22}+c_{22}) = tr(A+C)$,
and $tr(cA) = ca_{11}+ca_{22} = c(a_{11}+a_{22}) = ctr(A)$. T is linear. Ker(T)= $\begin{bmatrix} a & b \\ c & -a \end{bmatrix}$, range (T)=**R**.

(d) $T(A+C) = (A+C)^2 = A^2 + AC + CA + C^2 \neq A^2 + C^2 = T(A) + T(C)$, so T is not linear.

(e) $T(A+C) = A+C+B \neq (A+B) + (C+B) = T(A) + T(C)$, so T is not linear.

(f) $T(A+C) = a_{11} + c_{11} = T(A) + T(C)$ and $T(cA) = ca_{11} = cT(A)$, so T is linear.

Ker(T) = $\left\{\begin{bmatrix} 0 & a \\ b & c \end{bmatrix}\right\}$ and range(T) = **R**.

(g) T(A+C) = 0 = 0 + 0 = T(A) + T(C) and T(cA) = 0 = cx0 = cT(A), so T is linear. Ker(T) = U and range(T) = 0.

(h) T(A+C) = $I_2 \neq I_2 + I_2$ = T(A) + T(C), so T is not linear.

(i) T(A+C) = A+C + $(A+C)^t$ = A+C + A^t + C^t = A + A^t + C + C^t = T(A) + T(C) and T(cA) = cA + $(cA)^t$ = cA + cA^t = c(A + A^t) = cT(A), so T is linear.

$\begin{bmatrix} a & b \\ c & d \end{bmatrix} + \begin{bmatrix} a & c \\ b & d \end{bmatrix} = \begin{bmatrix} 2a & b+c \\ b+c & 2d \end{bmatrix}$, so ker(T) = $\left\{\begin{bmatrix} 0 & b \\ -b & 0 \end{bmatrix}\right\}$ and range(T) is the set of all symmetric matrices.

(j) If A = $\begin{bmatrix} 1 & 0 \\ 0 & 1 \end{bmatrix}$ and C = $\begin{bmatrix} 1 & 0 \\ 0 & 1 \end{bmatrix}$ then A+C = $\begin{bmatrix} 2 & 0 \\ 0 & 2 \end{bmatrix}$ and

T(A+C) = $I_2 \neq I_2 + I_2$ = T(A) + T(C), so T is not linear.

32. This set is not a subspace because it does not contain the zero vector.

33. (a) Kernel (T_A) is given by A**x** = **0**; Kernel (T_B) is given by B**x** = **0**. Since A$\approx$B the solutions to A**x** = **0** and B**x** = **0** are the same. Thus T_A and T_B have the same kernel. (b) Range (T_A) is spanned by the column vectors of A. To find a basis for these vectors we use the reduced echelon form of A^t. Let E be the reduced echelon form of A^t. The nonzero row vectors of E form a basis for the range of T_A. Since A $\approx$ B then $A^t \approx B^t$.

Thus the reduced echelon form of B^t is also E, and the nonzero rows of E also form a basis for the range of T_B. Thus range(T_A) = range (T_B).

Exercise Set 4.8

1. (a) The dimension of the range of the transformation is the rank of the matrix = 2. The domain is $\mathbf{R}^3$, which has dimension 3, so the dimension of the kernel is 1. This transformation is not one-to-one.

(b) The dimension of the range of the transformation is the rank of the matrix = 3. The domain is $\mathbf{R}^3$, which has dimension 3, so the dimension of the kernel is 0. This transformation is one-to-one.

(c) The dimension of the range of the transformation is the rank of the matrix = 3. The domain is $\mathbf{R}^4$, which has dimension 4, so the dimension of the kernel is 1. This transformation is not one-to-one.

(d) The dimension of the range of the transformation is the rank of the matrix = 2. The domain is $\mathbf{R}^3$, which has dimension 3, so the dimension of the kernel is 1. This transformation is not one-to-one.

(e) The dimension of the range of the transformation is the rank of the matrix = 2. The domain is $\mathbf{R}^4$, which has dimension 4, so the dimension of the kernel is 2. This transformation is not one-to-one.

(f) The dimension of the range of the transformation is the rank of the matrix = 3. The domain is $\mathbf{R}^3$, which has dimension 3, so the dimension of the kernel is 0. This transformation is one-to-one.

2. (a) $|A| = 12 \neq 0$, so the transformation is nonsingular and therefore one-to-one.
(b) $|B| = 0$, so the transformation is not one-to-one.
(c) $|C| = 0$, so the transformation is not one-to-one.
(d) $|D| = 0$, so the transformation is not one-to-one.
(e) $|E| = -24$, so the transformation is one-to-one.
(f) $|F| = 212$, so the transformation is one-to-one.

3. (a) The set of fixed points is the set of all points for which (x,y) = (x,3y). This is the set {(r,0)}.

(b) The set of fixed points is the set of all points for which (x,y) = (x,2). This is the set {(r,2)}.

(c) This transformation has no fixed points since there are no solutions to the equation y = y + 1.

(d) All points in U are fixed points, since T maps each point to itself.
(e) The set of fixed points is the set of all points for which (x,y) = (y,x). This is the set {(r,r)}.
(f) The set of fixed points is the set of all points for which (x,y) = (x+y,x−y). This set contains only the zero vector.
(g) If **u** and **v** are fixed points of a linear transformation T: U $\rightarrow$ U, then T(**u**) = **u** and T(**v**) = **v**, so that T(**u**+**v**) = T(**u**) + T(**v**) = **u** + **v** and T(c**u**) = cT(**u**) = c**u**. Thus both **u**+**v** and c**u** are fixed points of T, and the set of fixed points is therefore a subspace of U.

4. (a) $T\left(\begin{bmatrix} x \\ y \end{bmatrix}\right) = \begin{bmatrix} 7x-3y \\ 5x-2y \end{bmatrix}$. $T\left(\begin{bmatrix} 1 \\ 0 \end{bmatrix}\right) = \begin{bmatrix} 7 \\ 5 \end{bmatrix}$, $T\left(\begin{bmatrix} 0 \\ 1 \end{bmatrix}\right) = \begin{bmatrix} -3 \\ -2 \end{bmatrix}$. $A = \begin{bmatrix} 7 & -3 \\ 5 & -2 \end{bmatrix}$, $A^{-1} = \begin{bmatrix} -2 & 3 \\ -5 & 7 \end{bmatrix}$.

T is invertible. $T^{-1}\left(\begin{bmatrix} x \\ y \end{bmatrix}\right) = \begin{bmatrix} -2 & 3 \\ -5 & 7 \end{bmatrix}\begin{bmatrix} x \\ y \end{bmatrix} = \begin{bmatrix} -2x+3y \\ -5x+7y \end{bmatrix}$. T^{-1}(x, y) = (-2x+3y,-5x+7y).

T^{-1}(2, 3) = (5, 11).

(b) $T\left(\begin{bmatrix} x \\ y \end{bmatrix}\right) = \begin{bmatrix} 7x+6y \\ 8x+7y \end{bmatrix}$. $T\left(\begin{bmatrix} 1 \\ 0 \end{bmatrix}\right) = \begin{bmatrix} 7 \\ 8 \end{bmatrix}$, $T\left(\begin{bmatrix} 0 \\ 1 \end{bmatrix}\right) = \begin{bmatrix} 6 \\ 7 \end{bmatrix}$. $A = \begin{bmatrix} 7 & 6 \\ 8 & 7 \end{bmatrix}$, $A^{-1} = \begin{bmatrix} 7 & -6 \\ -8 & 7 \end{bmatrix}$.

T is invertible. $T^{-1}\left(\begin{bmatrix} x \\ y \end{bmatrix}\right) = \begin{bmatrix} 7 & -6 \\ -8 & 7 \end{bmatrix}\begin{bmatrix} x \\ y \end{bmatrix} = \begin{bmatrix} 7x-6y \\ -8x+7y \end{bmatrix}$. $T^{-1}(x, y) = (7x\text{-}6y,\text{-}8x\text{+}7y)$.

$T^{-1}(2, 3) = (-4,5)$.

(c) $T\left(\begin{bmatrix} x \\ y \end{bmatrix}\right) = \begin{bmatrix} 2x-y \\ -4x+2y \end{bmatrix}$. $T\left(\begin{bmatrix} 1 \\ 0 \end{bmatrix}\right) = \begin{bmatrix} 2 \\ -4 \end{bmatrix}$, $T\left(\begin{bmatrix} 0 \\ 1 \end{bmatrix}\right) = \begin{bmatrix} -1 \\ 2 \end{bmatrix}$. $A = \begin{bmatrix} 2 & -1 \\ -4 & 2 \end{bmatrix}$, $|A|=0$.

T is singular, T^{-1} does not exist.

(d) $T\left(\begin{bmatrix} x \\ y \end{bmatrix}\right) = \begin{bmatrix} x+4y \\ 0 \end{bmatrix}$. $T\left(\begin{bmatrix} 1 \\ 0 \end{bmatrix}\right) = \begin{bmatrix} 1 \\ 0 \end{bmatrix}$, $T\left(\begin{bmatrix} 0 \\ 1 \end{bmatrix}\right) = \begin{bmatrix} 4 \\ 0 \end{bmatrix}$. $A = \begin{bmatrix} 1 & 4 \\ 0 & 0 \end{bmatrix}$, $|A|=0$.

T is singular, T^{-1} does not exist.

5. (a) $T\left(\begin{bmatrix} x \\ y \\ z \end{bmatrix}\right) = \begin{bmatrix} x+y-z \\ -3x+2y-z \\ 3x-3y+2z \end{bmatrix}$. $T\left(\begin{bmatrix} 1 \\ 0 \\ 0 \end{bmatrix}\right) = \begin{bmatrix} 1 \\ -3 \\ 3 \end{bmatrix}$, $T\left(\begin{bmatrix} 0 \\ 1 \\ 0 \end{bmatrix}\right) = \begin{bmatrix} 1 \\ 2 \\ -3 \end{bmatrix}$, $T\left(\begin{bmatrix} 0 \\ 0 \\ 1 \end{bmatrix}\right) = \begin{bmatrix} -1 \\ -1 \\ 2 \end{bmatrix}$. $A = \begin{bmatrix} 1 & 1 & -1 \\ -3 & 2 & -1 \\ 3 & -3 & 2 \end{bmatrix}$.

$A^{-1} = \begin{bmatrix} 1 & 1 & 1 \\ 3 & 5 & 4 \\ 3 & 6 & 5 \end{bmatrix}$. T is invertible. $T^{-1}\left(\begin{bmatrix} x \\ y \\ z \end{bmatrix}\right) = \begin{bmatrix} 1 & 1 & 1 \\ 3 & 5 & 4 \\ 3 & 6 & 5 \end{bmatrix}\begin{bmatrix} x \\ y \\ z \end{bmatrix} = \begin{bmatrix} x+y+z \\ 3x+5y+4z \\ 3x+6y+5z \end{bmatrix}$.

$T^{-1}(x, y,z) = (x\text{+}y\text{+}z,3x\text{+}5y\text{+}4z,3x\text{+}6y\text{+}5z)$. $T^{-1}(1,-1,2) = (2,6,7)$.

(b) $T\left(\begin{bmatrix} x \\ y \\ z \end{bmatrix}\right) = \begin{bmatrix} x+z \\ 4x+4y+3z \\ -4x-3y-3z \end{bmatrix}$. $T\left(\begin{bmatrix} 1 \\ 0 \\ 0 \end{bmatrix}\right) = \begin{bmatrix} 1 \\ 4 \\ -4 \end{bmatrix}$, $T\left(\begin{bmatrix} 0 \\ 1 \\ 0 \end{bmatrix}\right) = \begin{bmatrix} 0 \\ 4 \\ -3 \end{bmatrix}$, $T\left(\begin{bmatrix} 0 \\ 0 \\ 1 \end{bmatrix}\right) = \begin{bmatrix} 1 \\ 3 \\ -3 \end{bmatrix}$. $A = \begin{bmatrix} 1 & 0 & 1 \\ 4 & 4 & 3 \\ -4 & -3 & -3 \end{bmatrix}$.

$A^{-1} = \begin{bmatrix} -3 & -3 & -4 \\ 0 & 1 & 1 \\ 4 & 3 & 4 \end{bmatrix}$. T is invertible. $T^{-1}\left(\begin{bmatrix} x \\ y \\ z \end{bmatrix}\right) = \begin{bmatrix} -3 & -3 & -4 \\ 0 & 1 & 1 \\ 4 & 3 & 4 \end{bmatrix}\begin{bmatrix} x \\ y \\ z \end{bmatrix} = \begin{bmatrix} -3x-3y-4z \\ y+z \\ 4x+3y+4z \end{bmatrix}$.

$T^{-1}(x, y,z) = (\text{-}3x\text{-}3y\text{-}4z,y\text{+}z,4x\text{+}3y\text{+}4z)$. $T^{-1}(1,-1,2) = (-8,1,9)$.

(c) $T\left(\begin{bmatrix} x \\ y \\ z \end{bmatrix}\right) = \begin{bmatrix} x+2y+3z \\ y-z \\ x+3y+2z \end{bmatrix}$. $T\left(\begin{bmatrix} 1 \\ 0 \\ 0 \end{bmatrix}\right) = \begin{bmatrix} 1 \\ 0 \\ 1 \end{bmatrix}$, $T\left(\begin{bmatrix} 0 \\ 1 \\ 0 \end{bmatrix}\right) = \begin{bmatrix} 2 \\ 1 \\ 3 \end{bmatrix}$, $T\left(\begin{bmatrix} 0 \\ 0 \\ 1 \end{bmatrix}\right) = \begin{bmatrix} 3 \\ -1 \\ 2 \end{bmatrix}$. $A = \begin{bmatrix} 1 & 2 & 3 \\ 0 & 1 & -1 \\ 1 & 3 & 2 \end{bmatrix}$, $|A|=0$.

T is singular, T^{-1} does not exist.

(d) $T\left(\begin{bmatrix} x \\ y \\ z \end{bmatrix}\right) = \begin{bmatrix} y \\ z \\ x \end{bmatrix}$. $T\left(\begin{bmatrix} 1 \\ 0 \\ 0 \end{bmatrix}\right) = \begin{bmatrix} 0 \\ 0 \\ 1 \end{bmatrix}$, $T\left(\begin{bmatrix} 0 \\ 1 \\ 0 \end{bmatrix}\right) = \begin{bmatrix} 1 \\ 0 \\ 0 \end{bmatrix}$, $T\left(\begin{bmatrix} 0 \\ 0 \\ 1 \end{bmatrix}\right) = \begin{bmatrix} 0 \\ 1 \\ 0 \end{bmatrix}$. $A = \begin{bmatrix} 0 & 1 & 0 \\ 0 & 0 & 1 \\ 1 & 0 & 0 \end{bmatrix}$.

$A^{-1} = \begin{bmatrix} 0 & 0 & 1 \\ 1 & 0 & 0 \\ 0 & 1 & 0 \end{bmatrix}$. T is invertible. $T^{-1}\left(\begin{bmatrix} x \\ y \\ z \end{bmatrix}\right) = \begin{bmatrix} 0 & 0 & 1 \\ 1 & 0 & 0 \\ 0 & 1 & 0 \end{bmatrix}\begin{bmatrix} x \\ y \\ z \end{bmatrix} = \begin{bmatrix} z \\ x \\ y \end{bmatrix}$.

$T^{-1}(x, y,z) = (z, x, y)$. $T^{-1}(1,-1,2) = (2, 1, -1)$.

6. T is one-to-one if and only if ker(T) is the zero vector if and only if dim ker(T) = 0 if and only if dim range(T) = dim domain(T).

7. Let T be a linear transformation from $\mathbf{R}^3$ to $\mathbf{R}^2$. Dim domain(T) = 3 and dim range(T) ≤ 2, so dim ker(T) = dim domain(T) − dim range(T) ≥ 1. Thus ker(T) is not the zero vector and T is not one-to-one.
In general, if dim(U) > dim(V) and T: U → V is a linear transformation, then dim ker(T) = dim domain(T) − dim range(T) ≥ dim(U) − dim(V) ≥ 1. Thus ker(T) is not the zero vector and T is not one-to-one.

8. Converse Theorem 4.31: Let T: U → V be a linear transformation. If T preserves linear independence then it is one-to-one.
Proof: Suppose T(**u**) = T(**w**). Then **0** = T(**u**) − T(**w**) = T(**u** − **w**).
Let **u** − **w** = $a_1 \mathbf{u}_1 + a_2 \mathbf{u}_2 + \ldots + a_n \mathbf{u}_n$, where $\{\mathbf{u}_1, \mathbf{u}_2, \ldots, \mathbf{u}_n\}$ is a basis for U.
Thus **0** = T(**u** − **w**) = $T(a_1 \mathbf{u}_1 + a_2 \mathbf{u}_2 + \ldots + a_n \mathbf{u}_n) = a_1 T(\mathbf{u}_1) + a_2 T(\mathbf{u}_2) + \ldots + a_n T(\mathbf{u}_n)$. But $\{T(\mathbf{u}_1), T(\mathbf{u}_2), \ldots, T(\mathbf{u}_n)\}$ is a linearly independent set in V so $a_1 = a_2 = \ldots = a_n = 0$. Therefore **u** − **w** = **0**, **u** = **w**, and T is one-to-one.

9. (a) $T(\mathbf{u}_1)=\mathbf{v}_1$ and $T(\mathbf{u}_2)=\mathbf{v}_2$. $T(\mathbf{u}_1)+T(\mathbf{u}_2)=\mathbf{v}_1+\mathbf{v}_2$. $T(\mathbf{u}_1+\mathbf{u}_2)=\mathbf{v}_1+\mathbf{v}_2$ since T is linear.
Thus $T^{-1}(\mathbf{v}_1+\mathbf{v}_2)=\mathbf{u}_1+\mathbf{u}_2$.
(b) T(**u**)= **v**. cT(**u**)=c**v**. T(c**u**)=c**v**, since T is linear. T^{-1}(c**v**)=c**u**.

10. T_1 and T_2 are invertible. Let standard matrices be A and B. Thus T_1(**u**)=A**u**, and T_2(**u**)=B**u**.

$T_2 \circ T_1(\mathbf{u}) = T_2(T_1(\mathbf{u})) = T_2(A(\mathbf{u})) = BA(\mathbf{u})$. Standard matrix of $T_2 \circ T_1$ is BA. BA is invertible. $(BA)^{-1} = A^{-1}B^{-1}$. Thus $(T_2 \circ T_1)^{-1}(\mathbf{u}) = A^{-1}B^{-1}\mathbf{u}$; $(T_2 \circ T_1)^{-1}(\mathbf{u}) = T_1^{-1} \circ T_2^{-1}(\mathbf{u})$. $(T_2 \circ T_1)^{-1} = T_1^{-1} \circ T_2^{-1}$.

Exercise Set 4.9

1. (r, r, 1) = r(1,1,0) + (0,0,1).
2. (0,1,2) = r(0,0,0) + (0,1,2).
3. (r+1, 2r, r) = r(1,2,1) + (1,0,0).
4. (r−s+3, r, s) = r(1,1,0) + s(−1,0,1) + (3,0,0).
5. (−4r+3s−2, −5r+5s−6, r, s) = r(−4,−5,1,0) + s(3,5,0,1) + (−2,−6,0,0).

6. (−32r−23s−19, 13r+9s+8, −5r−3s−3, r, s)
   = r(−32,13, −5,1,0) + s(−23,9,−3, 0,1) + (−19,8,−3,0,0).

7. The system A**x** = **y** has solutions if the vector **y** is in range(T), i.e., if the coordinates of **y** = (y_1, y_2, y_3) satisfy the condition $y_3 = y_1 + y_2$.

   (a) 2 = 1 + 1, so solutions exist. (b) 3 ≠ −1 + 2, so there are no solutions.

   (c) 5 = 3 + 2, so solutions exist. (d) 5 ≠ 2 + 4, so there are no solutions.

8. Any particular solution of the system will do in place of $\mathbf{x}_1$. (2,3,1) is another particular solution (obtained from the general solution by letting r = 1).
9. (0,−2,−1,0) is another particular solution (obtained from the general solution by letting r = −1 and s = 0).

10. The coefficient matrix is A. Since $\mathbf{x}_1 = (1,-1,4)$ is a solution, substitute $x_1 = 1$, $x_2 = -1$, $x_3 = 4$ in the left side of each equation to find the number on the right.

$$\begin{aligned} x_1 + 2x_2 + x_3 &= 3 \\ x_2 + 2x_3 &= 7 \\ x_1 + x_2 - x_3 &= -4 \end{aligned}$$

11. The coefficient matrix is A. Since $\mathbf{x}_1 = (2,0,3)$ is a solution, substitute $x_1 = 2$, $x_2 = 0$, $x_3 = 3$ in the left side of each equation to find the number on the right.

$$\begin{aligned} 2x_1 + x_2 &= 4 \\ 3x_1 + 3x_2 + x_3 &= 9 \\ x_2 + x_3 &= 3 \end{aligned}$$

12. Let T(**x**) = A**x**. First suppose ker(T) = **0**. If A$\mathbf{x}_1$ = **y** and A$\mathbf{x}_2$ = **y** then A($\mathbf{x}_1$ − $\mathbf{x}_2$) = **0**. Thus $\mathbf{x}_1$ − $\mathbf{x}_2$ is in ker(T), so $\mathbf{x}_1$ − $\mathbf{x}_2$ = **0**; that is, $\mathbf{x}_1$ = $\mathbf{x}_2$, so there is exactly one solution to A**x** = **y**. Now suppose A**x** = **y** has $\mathbf{x}_1$ as its unique solution. If **z** is in ker(T), then A**z** = **0**, so A($\mathbf{x}_1$ + **z**) = **y** + **0** = **y**. Thus $\mathbf{x}_1$ + **z** is a solution to A**x** = **y**. That means $\mathbf{x}_1$ + **z** = $\mathbf{x}_1$, so **z** = **0**, and ker(T) is the zero vector.

13. (a) $(D^2 + D-2)e^{mx} = 0$, so $(m^2 + m-2)e^{mx} = 0$, which gives $(m+2)(m-1) = 0$, so that $m = -2$ or $m = 1$. Thus a basis for the kernel is the set $\{e^{-2x}, e^x\}$. $\ker(D^2 + D-2) = \{ae^{-2x} + be^x\}$.

(b) $(D^2 + 4D+3)e^{mx} = 0$, so $(m^2 + 4m+3)e^{mx} = 0$, which gives $(m+3)(m+1) = 0$, so that $m = -3$ or $m = -1$. Thus a basis for the kernel is the set $\{e^{-3x}, e^{-x}\}$. $\ker(D^2 + 4D+3) = \{ae^{-3x} + be^{-x}\}$.

(c) $(D^2 + 2D-8)e^{mx} = 0$, so $(m^2 + 2m-8)e^{mx} = 0$, which gives $(m+4)(m-2) = 0$, so that $m = -4$ or $m = 2$. Thus a basis for the kernel is the set $\{e^{-4x}, e^{2x}\}$. $\ker(D^2 + 2D-8) = \{ae^{-4x} + be^{2x}\}$.

14. (a) $(D^2 + 5D+6)e^{mx} = 0$ gives $(m^2 + 5m+6)e^{mx} = 0$, which gives $(m+3)(m+2) = 0$, so that $m = -3$ or $m = -2$. Thus a basis for the kernel is the set $\{e^{-3x}, e^{-2x}\}$.

A particular solution is given by $6y = 8$ or $y = 4/3$, so the general solution is $y = re^{-3x} + se^{-2x} + 4/3$.

(b) $(D^2 - 3D)e^{mx} = 0$ gives $(m^2 - 3m)e^{mx} = 0$, which gives $(m-3)m = 0$, so that $m = 3$ or $m = 0$. Thus a basis for the kernel is the set $\{e^{3x}, 1\}$.

A particular solution is given by $-3dy/dx = 8$ or $y = -8x/3$, so the general solution is $y = re^{3x} + s - 8x/3$.

(c) $(D^2 - 7D+12)e^{mx} = 0$ gives $(m^2 - 7m+12)e^{mx} = 0$, which gives $(m-3)(m-4) = 0$, so that $m = 3$ or $m = 4$. Thus a basis for the kernel is the set $\{e^{3x}, e^{4x}\}$. A particular solution is given by $12y = 24$ or $y = 2$, so the general solution is $y = re^{3x} + se^{4x} + 2$.

(d) $(D^2 + 8D)e^{mx} = 0$ gives $(m^2 + 8m)e^{mx} = 0$, which gives $(m+8)m = 0$, so that $m = -8$ or $m = 0$. Thus a basis for the kernel is the set $\{e^{-8x}, 1\}$.

A particular solution is of the form $y = ke^{2x}$. We determine k. $dy/dx = 2ke^{2x}$ and $d^2y/dx^2 = 4ke^{2x}$, so that $d^2y/dx^2 + 8dy/dx = 20ke^{2x} = 3e^{2x}$. Thus $k = 3/20$, and the general solution is $y = re^{-8x} + s + 3e^{2x}/20$.

Chapter 4 Review Exercises

1. W is the set of vectors of the form a(1, 3, 7). Let a(1, 3, 7) and b(1, 3, 7) be elements of W. Then a(1, 3, 7) + b(1, 3, 7) = (a+b)(1, 3, 7), an element of V. Let k be a scalar. Then k(a(1, 3, 7)) = (ka)(1, 3, 7), an element of V. V is closed under addition and scalar multiplication. V is a subset of the vector space $\mathbf{R}^3$. V inherits all the other algebraic properties of a vector space from $\mathbf{R}^3$ - it is a subspace of $\mathbf{R}^3$. Thus V is a vector space. It is the line defined by the vector (1, 3, 7).

2. U is the set of 2x2 matrices where all the elements are nonnegative. Let A be an element of U, and k be a negative number. Then kA has all negative elements. kA is not in U. U is not closed under scalar multiplication. U is not a vector space.

3. W is the set of 2x2 matrices whose elements add up to zero. Let A and B be in W and let
k be a scalar. Then $a_{11}+ a_{12} + a_{21}+ a_{22} = 0$, and $b_{11}+ b_{12} + b_{21}+ b_{22} = 0$. Thus $a_{11}+ a_{12} + a_{21}+ a_{22} + b_{11}+ b_{12} + b_{21}+ b_{22} = 0$, implying that $(a_{11}+b_{11})+(a_{12} +b_{12})+(a_{21} +b_{21})+(a_{22} +b_{22}) = 0$. A+B is in W.
$k(a_{11}+ a_{12} + a_{21}+ a_{22}) = 0$, implying that $ka_{11}+ ka_{12} + ka_{21}+ ka_{22} = 0$. kA is in W. W is closed under addition and scalar multiplication. W is a subset of the vector space of 2x2 matrices. It inherits all the other vector space properties from this space - it is a subspace of M_{22}. Thus W is a vector space.

4. $(f + g)(x) = 3x - 1 + 2x^2 + 3 = 2x^2 + 3x + 2$, $3f(x) = 3(3x - 1) = 9x - 3$, and $(2f - 3g)(x) = 2(3x - 1) - 3(2x^2 + 3) = -6x^2 + 6x - 11$.

5. (a) Let f and g be in V. Then f(2) = 0 and g(2) = 0. (f+g)(2) = f(2)+g(2) = 0. Thus f+g is in V. Let k be a scalar. Then kf(2) = k(f(2)) = k(0) = 0. kf is in V. V is closed under addition and
scalar multiplication. V is a subset of the vector space U of functions having the real numbers as domain. V inherits all the other vector space properties from this space. V is a vector space - a subspace of this vector space U.
(b) Let f and g be in W. Then f(2) = 1 and g(2) = 1. (f+g)(2) = f(2)+g(2) = 2. Thus f+g is not in W. W is not closed under addition. It is not a vector space.

6. (a) (a, b, a−2) + (c, d, c−2) = (a+c, b+d, a+c−4), so the sum of two such vectors is not in the set. Thus the set is not a subspace of $\mathbf{R}^3$.

   (b) (a,−2a,3a) + (b,−2b,3b) = (a+b, −2(a+b), 3(a+b)) and c(a, −2a, 3a) = (ca, −2ca, 3ca), so the sum and scalar product of vectors in the set is in the set. Thus the set is a subspace of $\mathbf{R}^3$.

   (c) (a, b, 2a−3b) + (e, f, 2e−3f) = (a+e, b+f, 2(a+e)−3(b+f)) and c(a, b, 2a−3b) = (ca, cb, 2ca−3cb), so the sum and scalar product of vectors in the set is in the set. Thus the set is a subspace of $\mathbf{R}^3$.

(d) (a,2,b) + (c,2,d) = (a+c, 4, b+d), so the sum of vectors in the set is not in the set. Thus the set is not a subspace of $\mathbf{R}^3$.

7. Only the subset (a) is a subspace of $\mathbf{R}^3$. None of the other subsets is closed under scalar multiplication.

8. (a) Not a subspace: $\begin{bmatrix} 1 & 2 \\ 3 & 4 \end{bmatrix}$ is in the subset. $0\begin{bmatrix} 1 & 2 \\ 3 & 4 \end{bmatrix} = \begin{bmatrix} 0 & 0 \\ 0 & 0 \end{bmatrix}$ is not in the subset. Not closed under scalar multiplication.

(b) A subspace: Let A = $\begin{bmatrix} a & 0 \\ b & c \end{bmatrix}$ and B = $\begin{bmatrix} p & 0 \\ q & r \end{bmatrix}$; (2, 2) elements are zero.

Then A+B = $\begin{bmatrix} a & 0 \\ b & c \end{bmatrix} + \begin{bmatrix} p & 0 \\ q & r \end{bmatrix} = \begin{bmatrix} a+p & 0 \\ b+q & c+r \end{bmatrix}$. Thus closed under addition.

Let k be a scalar. Then $k\begin{bmatrix} a & 0 \\ b & c \end{bmatrix} = \begin{bmatrix} ka & 0 \\ kb & kc \end{bmatrix}$. Closed under scalar multiplication.

(c) A subspace: $\begin{bmatrix} a & b \\ -b & c \end{bmatrix} + \begin{bmatrix} p & q \\ -q & r \end{bmatrix} = \begin{bmatrix} a+p & b+q \\ -(b+q) & c+r \end{bmatrix}$.

$k\begin{bmatrix} a & b \\ -b & c \end{bmatrix} = \begin{bmatrix} ka & kb \\ -kb & kc \end{bmatrix}$. Closed under addition and under scalar multiplication.

(d) Not a subspace: Consider A=$\begin{bmatrix} 1 & 1 \\ 1 & 1 \end{bmatrix}$ and B=$\begin{bmatrix} 1 & 2 \\ 3 & 6 \end{bmatrix}$. |A|=0 and |B|=0.

A+B=$\begin{bmatrix} 2 & 3 \\ 4 & 7 \end{bmatrix}$. |A+B|=2≠0. Not closed under addition.

9. $3x^2 + ax - b + 3x^2 + cx - d = 6x^2 + (a+c)x - (b+d)$ is not in S, so the set is not closed under addition and therefore is not a subspace of P_2.

10. Necessary: Let **v** be in the subspace and 0 be the zero scalar. Then 0**v** = **0**, the zero vector (Theorem 4.1(a)). The subspace is closed under scalar multiplication. Thus **0** is in the subspace.
Not sufficient: The subset of $\mathbf{R}^2$ consisting of vectors of the form (a, a^2) contains the zero vector. It is not closed under addition, thus not a subspace.

11. (a) (3,15,−4) = 2(1,2,−1) + 3(2,4,0) − (5,1,2).

(b) Not a linear combination. The three vectors lie in the xz plane. (-3, -4, 7) does not lie in this plane.

12. No: $2A+3B-C = 2\begin{bmatrix} 1 & 2 \\ 4 & 3 \end{bmatrix} +3\begin{bmatrix} 0 & -2 \\ 1 & 5 \end{bmatrix} - \begin{bmatrix} 1 & 1 \\ 1 & 1 \end{bmatrix} = \begin{bmatrix} 1 & -3 \\ 10 & 20 \end{bmatrix}$.

13. (1, -2, 3) and (-2, 4, -6) are linearly dependent. Need at least three linearly independent vectors to span $\mathbf{R}^3$. Thus vectors do not span $\mathbf{R}^3$.
Consider (x_1, x_2, x_3) = a(1, -2, 3) + b(-2, 4, -6) + c(0, 6, 4) = (a-2b)(1, -2, 3) + c(0, 6, 4).
Vectors (1, -2, 3), (0, 6, 4) are linearly independent (they are not collinear).
Thus vectors span a 2D subspace of $\mathbf{R}^3$ having basis {(1, -2, 3), (0, 6, 4)}.

14. (10,9,8) = 2(-1,3,1) + 3(4,1,2).
15. $13x^2 + 8x - 21 = 2(2x^2 + x - 3) - 3(-3x^2 - 2x + 5)$.

16. (a) a(1,−2,0) + b(0,1,3) + c(2,0,12) = (0,0,0) if and only if a + 2c = 0, −2a + b = 0, and 3b + 12c = 0. System has many solutions a=-2r, b=-4r, c=r.
Get -2r(1,-2,0)-4r(0,1,3)+r(2,0,12)=(0,0,0).
e.g., let r=1, -2(1,-2,0)-4(0,1,3)+(2,0,12)=(0,0,0). Vectors are linearly dependent.

(b) a(−1,18,7) + b(−1,4,1) + c(1,3,2) = (0,0,0) if and only if −a − b + c = 0, 18a + 4b + 3c = 0, and 7a + b + 2c = 0. Many solutions, a=(-1/2)r, b=(3/2)r, c=r.
Get (-1/2)r(-1,18,7)+(3/2)r(-1,4,1)+r(1,3,2)=(0,0,0). e.g., let r=2, get
-1(-1,18,7)+3(-1,4,1)+2(1,3,2)=(0,0,0). Vectors are linearly dependent.

(c) a(5,−1,3) + b(2,1,0) + c(3,−2,2) = (0,0,0) if and only if 5a + 2b + 3c = 0, −a + b − 2c = 0, and 3a + 2c = 0. System has unique solution a = b = c = 0. The vectors are therefore linearly independent.

17. (a) and (b) In each case the set consists of two linearly independent vectors. By Theorem 4.11 they are therefore a basis for $\mathbf{R}^2$.

(c) and (d) In each case the set consists of three linearly independent vectors. By Theorem 4.11 they are therefore a basis for $\mathbf{R}^3$.

18. $ax^2 + bx + c = -a(-x^2) + (b/3)(3x) +(c/2)2$.

19. Any of the sets {(1,−2,3), (4,1,−1), (1,0,0)}, {(1,−2,3), (4,1,−1), (0,1,0)}, {(1,−2,3), (4,1,−1), (0,0,1)} is a basis for $\mathbf{R}^3$.

20. (a,b,c,a−2b+3c) = a(1,0,0,1) + b(0,1,0,−2) + c(0,0,1,3). The linearly independent set {(1,0,0,1), (0,1,0,−2), (0,0,1,3)} is a basis for the subspace.

21. $\begin{bmatrix} 1 & 0 & 0 \\ 0 & 0 & 0 \\ 0 & 0 & 0 \end{bmatrix}, \begin{bmatrix} 0 & 1 & 0 \\ 0 & 0 & 0 \\ 0 & 0 & 0 \end{bmatrix}, \begin{bmatrix} 0 & 0 & 1 \\ 0 & 0 & 0 \\ 0 & 0 & 0 \end{bmatrix}, \begin{bmatrix} 0 & 0 & 0 \\ 0 & 1 & 0 \\ 0 & 0 & 0 \end{bmatrix}, \begin{bmatrix} 0 & 0 & 0 \\ 0 & 0 & 1 \\ 0 & 0 & 0 \end{bmatrix}$, and $\begin{bmatrix} 0 & 0 & 0 \\ 0 & 0 & 0 \\ 0 & 0 & 1 \end{bmatrix}$ are a basis for the vector space of upper triangular 3x3 matrices.

22. $a(x^2 + 2x - 3) + b(3x^2 + x - 1) + c(4x^2 + 3x - 3) = 0$ if and only if $a + 3b + 4c = 0$, $2a + b + 3c = 0$, and $-3a - b - 3c = 0$. This system of homogeneous equations has the unique solution $a = b = c = 0$. Thus the given functions are linearly independent. The dimension of P_2 is 3, so the three functions are a basis.

23. (a) $\begin{bmatrix} 1 & 2 & -1 \\ -1 & 3 & 4 \\ 0 & 5 & 3 \end{bmatrix} \approx \begin{bmatrix} 1 & 2 & -1 \\ 0 & 5 & 3 \\ 0 & 5 & 3 \end{bmatrix} \approx \begin{bmatrix} 1 & 2 & -1 \\ 0 & 1 & 3/5 \\ 0 & 0 & 0 \end{bmatrix}$, so the rank of the matrix is 2.

(b) $\begin{bmatrix} 2 & 1 & 4 \\ -2 & 0 & -1 \\ 3 & 2 & 7 \end{bmatrix} \approx \begin{bmatrix} 1 & 1/2 & 2 \\ 0 & 1 & 3 \\ 0 & 1/2 & 1 \end{bmatrix} \approx \begin{bmatrix} 1 & 1/2 & 2 \\ 0 & 1 & 3 \\ 0 & 0 & 1 \end{bmatrix}$, so the rank of the matrix is 3

(c) $\begin{bmatrix} -2 & 4 & 8 \\ 1 & -2 & 4 \\ 4 & -8 & 16 \end{bmatrix} \approx \begin{bmatrix} 1 & -2 & -4 \\ 0 & 0 & 1 \\ 0 & 0 & 0 \end{bmatrix}$, so the rank of the matrix is 2.

24. $\begin{bmatrix} 1 & -2 & 3 & 4 \\ -1 & 3 & 1 & -2 \\ 2 & -3 & 10 & 10 \end{bmatrix} \approx \begin{bmatrix} 1 & -2 & 3 & 4 \\ 0 & 1 & 4 & 2 \\ 0 & 1 & 4 & 2 \end{bmatrix} \approx \begin{bmatrix} 1 & -2 & 3 & 4 \\ 0 & 1 & 4 & 2 \\ 0 & 0 & 0 & 0 \end{bmatrix}$, so the vectors

(1,−2, 3, 4) and (0, 1, 4, 2) are a basis for the subspace.

25. $\mathbf{v} = a\mathbf{v}_1 + b\mathbf{v}_2 = a\mathbf{v}_1 + b\mathbf{v}_2 + 0\mathbf{v}_3$.

26. If $a\mathbf{v}_1 + b\mathbf{v}_2 = \mathbf{0}$ then $a\mathbf{v}_1 + b\mathbf{v}_2 + 0\mathbf{v}_3 = \mathbf{0}$ and since the set $\{\mathbf{v}_1, \mathbf{v}_2, \mathbf{v}_3\}$ is linearly independent this means $a = b = 0$, so $\{\mathbf{v}_1, \mathbf{v}_2\}$ must be linearly independent.

If $a\mathbf{v}_1 + c\mathbf{v}_3 = \mathbf{0}$ then $a\mathbf{v}_1 + 0\mathbf{v}_2 + c\mathbf{v}_3 = \mathbf{0}$ and since the set $\{\mathbf{v}_1, \mathbf{v}_2, \mathbf{v}_3\}$ is linearly independent this means $a = c = 0$, so $\{\mathbf{v}_1, \mathbf{v}_3\}$ must be linearly independent.
If $b\mathbf{v}_2 + c\mathbf{v}_3 = \mathbf{0}$ then $0\mathbf{v}_1 + b\mathbf{v}_2 + c\mathbf{v}_3 = \mathbf{0}$ and since the set $\{\mathbf{v}_1, \mathbf{v}_2, \mathbf{v}_3\}$ is linearly independent this means $b = c = 0$, so $\{\mathbf{v}_2, \mathbf{v}_3\}$ must be linearly independent.
Since $\{\mathbf{v}_1, \mathbf{v}_2, \mathbf{v}_3\}$ is linearly independent, none of the three vectors can be the zero vector. Thus $a\mathbf{v}_1 = \mathbf{0}$ (or $b\mathbf{v}_2 = \mathbf{0}$ or $c\mathbf{v}_3 = \mathbf{0}$) only if $a = 0$ (or $b = 0$ or $c = 0$).

27. $a(\mathbf{v}_1 + 2\mathbf{v}_2) + b(3\mathbf{v}_1 - \mathbf{v}_2) = (a + 3b)\mathbf{v}_1 + (2a - b)\mathbf{v}_2$. There are scalars c and d, not both zero, with $c\mathbf{v}_1 + d\mathbf{v}_2 = \mathbf{0}$. Solve the system $c = a + 3b$, $d = 2a - b$: $a = \frac{3d+c}{7}$, $b = \frac{2c-d}{7}$. $3d+c$ and $2c-d$ cannot both be zero unless both c and d are zero, so at least one of a and b is nonzero and $a(\mathbf{v}_1 + 2\mathbf{v}_2) + b(3\mathbf{v}_1 - \mathbf{v}_2) = c\mathbf{v}_1 + d\mathbf{v}_2 = \mathbf{0}$. So $\mathbf{v}_1 + 2\mathbf{v}_2$ and $3\mathbf{v}_1 - \mathbf{v}_2$ are linearly dependent.

28. If rank(A) = n, the row space of A is $\mathbf{R}^n$, so the reduced echelon form of A must have n linearly independent rows; i.e., it must be I_n. If A is row equivalent to I_n, the rows of I_n are linear combinations of the rows of A. But the rows of I_n span $\mathbf{R}^n$, so the rows of A span $\mathbf{R}^n$ and rank(A) = n.

29. $\text{proj}_{\mathbf{u}}\, \mathbf{v} = \frac{\mathbf{v}\cdot\mathbf{u}}{\mathbf{u}\cdot\mathbf{u}}\, \mathbf{u}$.

(a) $\text{proj}_{\mathbf{u}}\, \mathbf{v} = \frac{(1,3)\cdot(2,4)}{(2,4)\cdot(2,4)}(2,4) = \frac{7}{10}(2,4) = (\frac{7}{5}, \frac{14}{5})$

(b) $\text{proj}_{\mathbf{u}}\, \mathbf{v} = \frac{(-1,3,4)\cdot(-1,2,4)}{(-1,2,4)\cdot(-1,2,4)}(-1,2,4) = \frac{23}{21}(-1,2,4) = \left(-\frac{23}{21}, \frac{46}{21}, \frac{92}{21}\right)$.

30. $\mathbf{u}_1 = (1,2,3,-1)$, $\mathbf{u}_2 = (2,0,-1,1) - \frac{(2,0,-1,1)\cdot(1,2,3,-1)}{(1,2,3,-1)\cdot(1,2,3,-1)}(1,2,3,-1)$

$= (2,0,-1,1) - \frac{-2}{15}(1,2,3,-1) = (\frac{32}{15}, \frac{4}{15}, \frac{-9}{15}, \frac{13}{15})$, and

$\mathbf{u}_3 = (3,2,0,1) - \frac{(3,2,0,1)\cdot(1,2,3,-1)}{(1,2,3,-1)\cdot(1,2,3,-1)}(1,2,3,-1) - \frac{(3,2,0,1)\cdot(32,4,-9,13)}{(32,4,-9,13)\cdot(32,4,-9,13)}(32,4,-9,13)$

$= (3,2,0,1) - \frac{2}{5}(1,2,3,-1) - \frac{39}{430}(32,4,-9,13)$

$= (\frac{13}{5}, \frac{6}{5}, \frac{-6}{5}, \frac{7}{5}) - \frac{39}{86}(\frac{32}{5}, \frac{4}{5}, \frac{-9}{5}, \frac{13}{5}) = (\frac{-26}{86}, \frac{72}{86}, \frac{-33}{86}, \frac{19}{86})$,

are an orthogonal basis for the subspace of $\mathbf{R}^4$. $||\mathbf{u}_1|| = \sqrt{15}$, $||\mathbf{u}_2|| = \frac{1}{15}\sqrt{1290}$, and $||\mathbf{u}_3|| = \frac{1}{86}\sqrt{7310}$, so the set $\{(\frac{1}{\sqrt{15}}, \frac{2}{\sqrt{15}}, \frac{3}{\sqrt{15}}, \frac{-1}{\sqrt{15}})$,

$(\frac{32}{\sqrt{1290}}, \frac{4}{\sqrt{1290}}, \frac{-9}{\sqrt{1290}}, \frac{13}{\sqrt{1290}}), (\frac{-26}{\sqrt{7310}}, \frac{72}{\sqrt{7310}}, \frac{-33}{\sqrt{7310}}, \frac{19}{\sqrt{7310}})\}$

is an orthonormal basis for the subspace.

31. (x,y,x+2y) = x(1,0,1) + y(0,1,2). The vectors (1,0,1) and (0,1,2) span the subspace and are linearly independent so they are a basis. $\mathbf{u}_1$ = (1,0,1) and

$\mathbf{u}_2 = (0,1,2) - \frac{(0,1,2)\cdot(1,0,1)}{(1,0,1)\cdot(1,0,1)}(1,0,1) = (0,1,2) - (1,0,1) = (-1,1,1)$, are an orthogonal basis, and $\|\mathbf{u}_1\| = \sqrt{2}$ and $\|\mathbf{u}_2\| = \sqrt{3}$, so the set $\left\{(\frac{1}{\sqrt{2}},0,\frac{1}{\sqrt{2}}),(\frac{-1}{\sqrt{3}},\frac{1}{\sqrt{3}},\frac{1}{\sqrt{3}})\right\}$

is an orthonormal basis for the subspace.

32. $\mathbf{u}_1 = (2,1,1)$ and $\mathbf{u}_2 = (1,-1,3) - \frac{(1,-1,3)\cdot(2,1,1)}{(2,1,1)\cdot(2,1,1)}(2,1,1) = (1,-1,3) - \frac{2}{3}(2,1,1)$

$= (\frac{-1}{3}, \frac{-5}{3}, \frac{7}{3})$, are an orthogonal basis for W. $\|\mathbf{u}_1\| = \sqrt{6}$ and $\|\mathbf{u}_2\| = \frac{5}{3}\sqrt{3}$,

so $(\frac{2}{\sqrt{6}}, \frac{1}{\sqrt{6}}, \frac{1}{\sqrt{6}})$ and $(\frac{-1}{5\sqrt{3}}, \frac{-5}{5\sqrt{3}}, \frac{7}{5\sqrt{3}})$ are an orthonormal basis.

$$\text{proj}_W(3,1,-2) = \left((3,1,-2)\cdot(\frac{2}{\sqrt{6}}, \frac{1}{\sqrt{6}}, \frac{1}{\sqrt{6}})\right)(\frac{2}{\sqrt{6}}, \frac{1}{\sqrt{6}}, \frac{1}{\sqrt{6}})$$

$$+ \left((3,1,-2)\cdot(\frac{-1}{5\sqrt{3}}, \frac{-5}{5\sqrt{3}}, \frac{7}{5\sqrt{3}})\right)(\frac{-1}{5\sqrt{3}}, \frac{-5}{5\sqrt{3}}, \frac{7}{5\sqrt{3}})$$

$$= \frac{5}{6}(2,1,1) + \frac{-22}{75}(-1,-5,7) = (\frac{294}{150}, \frac{345}{150}, \frac{-183}{150})$$

33. W = {a(1,3,0) + b(0,0,1)}. (1,3,0) and (0,0,1) are an orthogonal basis for the subspace, so

$(\frac{1}{\sqrt{10}}, \frac{3}{\sqrt{10}}, 0)$ and (0,0,1) are an orthonormal basis. Let $\mathbf{x}$ = (1,2,−4). $\text{proj}_W\,\mathbf{x} =$

$$\left((1,2,-4)\cdot(\frac{1}{\sqrt{10}}, \frac{3}{\sqrt{10}}, 0)\right)(\frac{1}{\sqrt{10}}, \frac{3}{\sqrt{10}}, 0) + ((1,2,-4)\cdot(0,0,1))(0,0,1)$$

$= \frac{7}{10}(1,3,0) - 4(0,0,1) = (\frac{7}{10}, \frac{21}{10}, -4)$ Thus

$d(\mathbf{x},W) = \|\mathbf{x} - \text{proj}_W\,\mathbf{x}\| = \|(1,2,-4) - (\frac{7}{10}, \frac{21}{10}, -4)\| = \|(\frac{3}{10}, \frac{-1}{10}, 0)\| = \frac{1}{10}\sqrt{10}$.

34. If A is orthogonal, the rows of A form an orthonormal set. The rows of A are the columns of A^t, so from the definition of orthogonal matrix, A^t is orthogonal. Interchange A and

A^t in the argument above to show that if A^t is orthogonal then A is orthogonal.

35. $W = \{a(1,0,1) + b(0,1,-2)\}$. $\mathbf{u}_1 = (1,0,1)$ and $\mathbf{u}_2 = (0,1,-2) - \dfrac{(0,1,-2)\cdot(1,0,1)}{(1,0,1)\cdot(1,0,1)}(1,0,1)$

$= (0,1,-2) - (-1)(1,0,1) = (1,1,-1)$ are an orthogonal basis. The vectors $\left(\frac{1}{\sqrt{2}}, 0, \frac{1}{\sqrt{2}}\right)$ and $\left(\frac{1}{\sqrt{3}}, \frac{1}{\sqrt{3}}, \frac{-1}{\sqrt{3}}\right)$ are therefore an orthonormal basis.

$\mathbf{v} = (1,3,-1) = \mathbf{w} + \mathbf{w}\perp$, where $\mathbf{w} = \text{proj}_W \mathbf{v} = \left((1,3,-1)\cdot\left(\frac{1}{\sqrt{2}}, 0, \frac{1}{\sqrt{2}}\right)\right)\left(\frac{1}{\sqrt{2}}, 0, \frac{1}{\sqrt{2}}\right)$

$+ \left((1,3,-1)\cdot\left(\frac{1}{\sqrt{3}}, \frac{1}{\sqrt{3}}, \frac{-1}{\sqrt{3}}\right)\right)\left(\frac{1}{\sqrt{3}}, \frac{1}{\sqrt{3}}, \frac{-1}{\sqrt{3}}\right) = \frac{5}{3}(1,1,-1) = \left(\frac{5}{3}, \frac{5}{3}, \frac{-5}{3}\right)$ and

$\mathbf{w}_\perp = \mathbf{v} - \text{proj}_V \mathbf{v} = (1,3,-1) - \left(\frac{5}{3}, \frac{5}{3}, \frac{-5}{3}\right) = \left(\frac{-2}{3}, \frac{4}{3}, \frac{2}{3}\right)$

36. Suppose **u** and **v** are orthogonal and that a**u** + b**v** = 0, so that a**u** = −b**v**. Then a**u**·a**u** = a**u**·(−b**v**) = (−ab)**u**·**v** = 0, so a**u** = **0**. Thus a = 0 since **u** ≠ **0**. In the same way b = 0, and so
u and **v** are linearly independent.

37. If **u**·**v** = 0 and **u**·**w** = 0 then **u**·(a**v** + b**w**) = **u**·(a**v**) + **u**·(b**w**) = a(**u**·**v**) + b(**u**·**w**) = 0.

38. (a) True: The vectors are both in the subspace that has dimension 2, and they are linearly independent.
(b) False: The dimension of $\mathbf{R}^2$ is 2. No set of more than two vectors can be linearly independent.
(c) True: By Theorem 4.11 they form a basis for R^3. Thus they span R^3.
(d) True: If two vectors in $\mathbf{R}^2$ are not a basis, they are linearly dependent and therefore collinear.
(e) False: For the vectors to be linearly dependent all three would have to lie on the same line or in the same plane. It is much more likely that the first two vectors would not lie on the same line and that the third vector would not lie in the plane of the first two.

39. (a) False: Can have any number of linearly dependent vectors. e.g., **v**, 2**v**, 3**v**, 4**v**, ...

(b) True: Space is of dim n. Thus basis consists of n linearly independent vectors. Any other vector is l.d. on these.
(c) True: Space is of dim n. Thus need at least n vectors to span the space.
(d) False: Let $\{\mathbf{v}_1, \ldots, \mathbf{v}_n\}$ be a basis. Then add any vector v. $\{\mathbf{v}_1, \ldots, \mathbf{v}_n, \mathbf{v}\}$ spans V and is linearly dependent.

40. $T((x_1, y_1)+(x_2, y_2)) = T(x_1+x_2, y_1+y_2) = (2(x_1+x_2), x_1+x_2+ 3(y_1+y_2))$
$= (2x_1+2x_2, x_1+3y_1+x_2+3y_2) = (2x_1, x_1+3y_1) + (2x_2, x_2+3y_2)$
$= T(x_1, y_1) + T(x_2, y_2)$ and
$T(c(x,y)) = T(cx,cy) = (2cx,cx+3cy) = c(2x,x+3y) = cT(x,y)$. Thus T is linear.
$T(1,2) = (2(1),1+6) = (2,7)$.

41. $T((ax^2 + bx + c)+(px^2 + qx + r)) = T((a+p)x^2 + (b+q)x + c+ r) = 2(a+p)x + (b+q)$

$= (2ax +b) +(2px+q) = T(ax^2 + bx + c)+ T(px^2 + qx + r)$.
$T(k(ax^2 + bx + c)) = T(kax^2 + kbx + kc) = 2kax + kb = k(2x^2 + b) = kT(ax^2 + bx + c)$.
T is linear. $T(3x^2-2x+1) = 6x - 2$. $T(3x^2-2x+c)= 6x -2$ for any value of c.

42. $\begin{bmatrix} 6 & 4 \\ 3 & 2 \end{bmatrix}\begin{bmatrix} x \\ y \end{bmatrix} = \begin{bmatrix} 6x+4y \\ 3x+2y \end{bmatrix}$, thus (x,y) is in the kernel of T if $3x+2y = 0$, i.e., if $y = -3x/2$.

The kernel of T is the set $\{(r,-3r/2)\}$ and the range is the set $\{(2r,r)\}$. Dim ker(T) = 1,

dim range(T) = 1, and dim domain(T) = 2, so dim ker(T) + dim range(T) = dim domain(T).

43. $\begin{bmatrix} 1 & 1 & 1 \\ 0 & 1 & -1 \\ 2 & 3 & 1 \end{bmatrix}\begin{bmatrix} x \\ y \\ z \end{bmatrix} = \begin{bmatrix} x+y+z \\ y-z \\ 2x+3y+z \end{bmatrix}$; thus (x,y,z) is in the kernel of T if $x+y+z = 0$,

$y-z = 0$, and $2x+3y+z = 0$, i.e., if $x = -2y$ and $z = y$. The kernel of T is the set $\{(-2r,r,r)\}$ with basis $\{(-2,1,1)\}$, and a basis for the range is the set $\{(1,0,2), (1,1,3)\}$.

44. (a) The dimension of the range of the transformation is the rank of the matrix is 2.

The domain is $\mathbf{R}^4$, which has dimension 4, so the dimension of the kernel is 2.

This transformation is not one-to-one.

(b) $|B| \neq 0$, so the transformation is one-to-one.

45. The kernel is the set $\{(0,r,r)\}$ with basis $\{(0,1,1)\}$ and the range is the set $\{(a,2a,b)\}$ with basis $\{(1,2,0), (0,0,1)\}$.

46. $g(a_2x^2 + a_1x + a_0) + g(b_2x^2 + b_1x + b_0)$

$= (a_2 - a_1)x^3 - a_1x + 2a_0 + (b_2 - b_1)x^3 - b_1x + 2b_0$

$= (a_2 + b_2 - a_1 - b_1)x^3 - (a_1 + b_1)x + 2(a_0 + b_0)$

$= g((a_2 + b_2)x^2 + (a_1 + b_1)x + a_0 + b_0)$

$= g(a_2 x^2 + a_1 x + a_0 + b_2 x^2 + b_1 x + b_0)$.

Addition is preserved.

$g(c(a_2 x^2 + a_1 x + a_0)) = g(ca_2 x^2 + ca_1 x + ca_0) = (ca_2 - ca_1)x^3 - ca_1 x + 2ca_0$

$= c((a_2 - a_1)x^3 - a_1 x + 2a_0) = cg(a_2 x^2 + a_1 x + a_0)$.

Scalar multiplication is preserved. Thus g is linear.

Ker(g) is the set of all polynomials $a_2 x^2 + a_1 x + a_0$ with $a_2 - a_1 = 0$, $a_1 = 0$, and $a_0 = 0$, i.e., $a_2 = a_1 = a_0 = 0$, so ker(g) is the zero polynomial. Range(g) is the set of all polynomials $ax^3 + bx + c$. The set $\{x^3, x, 1\}$ is a basis for range(g).

47. $(D^2 - 2D + 1)(a_n x^n + \ldots + a_1 x + a_0) =$

$n(n-1)a_n x^{n-2} + \ldots + 6a_3 x + 2a_2 - 2(na_n x^{n-1} + \ldots + 2a_2 x + a_1)$

$+ (a_n x^n + \ldots + a_1 x + a_0)$. Thus,

$(D^2 - 2D + 1)(a_n x^n + \ldots + a_1 x + a_0) + (D^2 - 2D + 1)(b_n x^n + \ldots + b_1 x + b_0)$

$= n(n-1)a_n x^{n-2} + \ldots + 6a_3 x + 2a_2 - 2(na_n x^{n-1} + \ldots + 2a_2 x + a_1)$

$+ (a_n x^n + \ldots + a_1 x + a_0) + n(n-1)b_n x^{n-2} + \ldots + 6b_3 x + 2b_2$

$- 2(nb_n x^{n-1} + \ldots + 2b_2 x + b_1) + (b_n x^n + \ldots + b_1 x + b_0)$

$-$

$= n(n-1)(a_n + b_n)x^{n-2} + \ldots + 6(a_3 + b_3)x + 2(a_2 + b_2)$

$- 2(n(a_n + b_n)x^{n-1} + \ldots + 2(a_2 + b_2)x + a_1 + b_1) + ((a_n + b_n)x^n + \ldots$

$+ (a_1 + b_1)x + a_0 + b_0)$

$= (D^2 - 2D + 1)((a_n + b_n)x^n + \ldots + (a_1 + b_1)x + a_0 + b_0)$.

Addition is preserved .

$(D^2 - 2D + 1)(c(a_n x^n + \ldots + a_1 x + a_0))$

$= (D^2 - 2D + 1)(ca_n x^n + \ldots + ca_1 x + ca_0)$

$= n(n-1)ca_n x^{n-2} + \ldots + 6ca_3 x + 2ca_2 - 2(nca_n x^{n-1} + \ldots + 2ca_2 x + ca_1)$

$+ (ca_n x^n + \ldots + ca_1 x + ca_0) = c(n(n-1)a_n x^{n-2} + \ldots + 6a_3 x + 2a_2$

$- 2(na_n x^{n-1} + \ldots + 2a_2 x + a_1) + (a_n x^n + \ldots + a_1 x + a_0))$

$= c(D^2 - 2D + 1)(a_n x^n + \ldots + a_1 x + a_0).$

Scalar multiplication is preserved. Thus $(D^2 - 2D + 1)$ is linear.

$(D^2 - 2D + 1)(a_n x^n + \ldots + a_1 x + a_0) = a_n x^n + (a_{n-1} - 2na_n)x^{n-1}$

$+(a_{n-2} - 2(n-1)a_{n-1} + n(n-1)a_n)x^{n-2} + \ldots + (a_1 - 4a_2 + 6a_3)x + (a_0 - 2a_1 + 2a_2).$

Thus if

$(D^2 - 2D + 1)(a_n x^n + \ldots + a_1 x + a_0) = 12x - 4$, then $a_n = a_{n-1} = \ldots = a_2 = 0$, $a_1 = 12$,

and $a_0 = 20$. Thus the only polynomial mapped into $12x - 4$ is the polynomial $12x + 20$.

Chapter 5

Exercise Set 5.1

1. $(2,-3) = 2(1,0) + -3(0,1)$, so $\mathbf{u}_B = \begin{bmatrix} 2 \\ -3 \end{bmatrix}$.

2. $(8,-1) = a(3,-1) + b(2,1) = (3a+2b,-a+b)$, so $8 = 3a+2b$ and $-1 = -a+b$. This system of equations has the unique solution $a = 2$, $b = 1$, so $\mathbf{u}_B = \begin{bmatrix} 2 \\ 1 \end{bmatrix}$.

3. $(5,1) = 5(1,0) + 1(0,1)$, so $\mathbf{u}_B = \begin{bmatrix} 5 \\ 1 \end{bmatrix}$.

4. $(-7,-5) = a(1,-1) + b(3,1) = (a+3b,-a+b)$, so $-7 = a+3b$ and $-5 = -a+b$. This system of equations has the unique solution $a = 2$, $b = -3$, so $\mathbf{u}_B = \begin{bmatrix} 2 \\ -3 \end{bmatrix}$.

5. $(4,0,-2) = 4(1,0,0) + 0(0,1,0) + -2(0,0,1)$, so $\mathbf{u}_B = \begin{bmatrix} 4 \\ 0 \\ -2 \end{bmatrix}$.

6. $(6,-3,1) = a(1,-1,0) + b(2,1,-1) + c(2,0,0)$, so $6 = a+2b+2c$, $-3 = -a+b$, and $1 = -b$. This system of equations has the unique solution $a = 2$, $b = -1$, $c = 3$, so $\mathbf{u}_B = \begin{bmatrix} 2 \\ -1 \\ 3 \end{bmatrix}$.

7. $(3,-1,7) = a(2,0,-1) + b(0,1,3) + c(1,1,1)$, so $3 = 2a+c$, $-1 = b+c$, and $7 = -a+3b+c$. This system of equations has the unique solution $a = 16/3$, $b = 20/3$, and $c = -23/3$, so
$$\mathbf{u}_B = \begin{bmatrix} 16/3 \\ 20/3 \\ -23/3 \end{bmatrix}.$$

8. $(1,-6,-8) = a(1,2,3) + b(1,-1,0) + c(0,1,-2)$, so $1 = a+b$, $-6 = 2a-b+c$, and $-8 = 3a-2c$.

This system of equations has the unique solution a = −2, b = 3, c = 1, so $\mathbf{u}_B = \begin{bmatrix} -2 \\ 3 \\ 1 \end{bmatrix}$.

9. 7x − 3 = 7(x) + −3(1), so $\mathbf{u}_B = \begin{bmatrix} 7 \\ -3 \end{bmatrix}$.

10. 2x − 6 = a(3x − 5) + b(x − 1), so 2 = 3a+b and −6 = −5a−b. This system of equations has the unique solution a = 2, b = −4, so $\mathbf{u}_B = \begin{bmatrix} 2 \\ -4 \end{bmatrix}$.

11. $4x^2 - 5x + 2 = 4(x^2) + -5(x) + 2(1)$, so $\mathbf{u}_B = \begin{bmatrix} 4 \\ -5 \\ 2 \end{bmatrix}$.

12. $3x^2 - 6x - 2 = a(x^2) + b(x - 1) + c(2x)$, so 3 = a, −6 = b+2c, and −2 = −b. This system of equations has the unique solution a = 3, b = 2, c = −4, so $\mathbf{u}_B = \begin{bmatrix} 3 \\ 2 \\ -4 \end{bmatrix}$.

13. **u**=(5, 0, -10). **u**·(0,−1,0) = 0, **u**·(3/5,0,−4/5) = 11, **u**·(4/5,0,3/5) = −2, so $\mathbf{u}_B = \begin{bmatrix} 0 \\ 11 \\ -2 \end{bmatrix}$.

14. **u** = (1, 2, 3). **u**·(1/3,2/3,2/3) = 11/3, **u**·(2/3,−2/3,1/3) = 1/3, **u**·(2/3,1/3,−2/3) = −2/3,

so $\mathbf{u}_B = \begin{bmatrix} 11/3 \\ 1/3 \\ -2/3 \end{bmatrix}$.

15. **u** = (2, 1, 4). **u**·(1,0,0) = 2, $\mathbf{u}\cdot(0,1/\sqrt{2},1/\sqrt{2}) = 5/\sqrt{2}$, $\mathbf{u}\cdot(0,1/\sqrt{2},-1/\sqrt{2}) = -3/\sqrt{2}$,

so $\mathbf{u}_B = \begin{bmatrix} 2 \\ 5/\sqrt{2} \\ -3/\sqrt{2} \end{bmatrix}$.

16. $P = \begin{bmatrix} 2 & 1 \\ 3 & 2 \end{bmatrix}$, and $\mathbf{u}_{B'} = P\mathbf{u}_B = \begin{bmatrix} 4 \\ 7 \end{bmatrix}$, $\mathbf{v}_{B'} = P\mathbf{v}_B = \begin{bmatrix} 5 \\ 7 \end{bmatrix}$, and $\mathbf{w}_{B'} = P\mathbf{w}_B = \begin{bmatrix} 8 \\ 14 \end{bmatrix}$.

17. $P = \begin{bmatrix} 4 & 3 \\ 1 & -1 \end{bmatrix}$, and $\mathbf{u}_{B'} = P\mathbf{u}_B = \begin{bmatrix} 12 \\ 3 \end{bmatrix}$, $\mathbf{v}_{B'} = P\mathbf{v}_B = \begin{bmatrix} 11 \\ 1 \end{bmatrix}$, and $\mathbf{w}_{B'} = P\mathbf{w}_B = \begin{bmatrix} -5 \\ -3 \end{bmatrix}$.

18. $P = \begin{bmatrix} 1 & 2 \\ 1 & -3 \end{bmatrix}$, and $\mathbf{u}_{B'} = P\mathbf{u}_B = \begin{bmatrix} 2 \\ -3 \end{bmatrix}$, $\mathbf{v}_{B'} = P\mathbf{v}_B = \begin{bmatrix} -3 \\ 7 \end{bmatrix}$, and $\mathbf{w}_{B'} = P\mathbf{w}_B = \begin{bmatrix} 5 \\ 0 \end{bmatrix}$.

19. $P = \begin{bmatrix} 3 & 2 \\ 2 & 1 \end{bmatrix}$, and $\mathbf{u}_{B'} = P\mathbf{u}_B = \begin{bmatrix} 4 \\ 3 \end{bmatrix}$, $\mathbf{v}_{B'} = P\mathbf{v}_B = \begin{bmatrix} -2 \\ -1 \end{bmatrix}$, and $\mathbf{w}_{B'} = P\mathbf{w}_B = \begin{bmatrix} 21 \\ 13 \end{bmatrix}$.

20. $P = \begin{bmatrix} 2 & -3 \\ -3 & 4 \end{bmatrix}$, and $\mathbf{u}_{B'} = P\mathbf{u}_B = \begin{bmatrix} -1 \\ 1 \end{bmatrix}$, $\mathbf{v}_{B'} = P\mathbf{v}_B = \begin{bmatrix} 6 \\ -9 \end{bmatrix}$, and $\mathbf{w}_{B'} = P\mathbf{w}_B = \begin{bmatrix} 2 \\ -4 \end{bmatrix}$

21. The transition matrix from B' to B is $P = \begin{bmatrix} 5 & 3 \\ 3 & 2 \end{bmatrix}$, so the transition matrix from B to B' is

$P^{-1} = \begin{bmatrix} 2 & -3 \\ -3 & 5 \end{bmatrix}$, and $\mathbf{u}_{B'} = P^{-1}\,\mathbf{u}_B = \begin{bmatrix} -19 \\ 31 \end{bmatrix}$.

22. The transition matrix from B' to B is $P = \begin{bmatrix} 1 & -1 \\ 2 & -1 \end{bmatrix}$, so the transition matrix from B to B' is

$P^{-1} = \begin{bmatrix} -1 & 1 \\ -2 & 1 \end{bmatrix}$, and $\mathbf{u}_{B'} = P^{-1}\,\mathbf{u}_B = \begin{bmatrix} -5 \\ -13 \end{bmatrix}$.

23. The transition matrix from B to the standard basis is $R = \begin{bmatrix} 1 & 3 \\ 2 & 0 \end{bmatrix}$ and the transition matrix from B' to the standard basis is $Q = \begin{bmatrix} 2 & 3 \\ 1 & 2 \end{bmatrix}$. $P = Q^{-1} R = \begin{bmatrix} 2 & -3 \\ -1 & 2 \end{bmatrix}\begin{bmatrix} 1 & 3 \\ 2 & 0 \end{bmatrix}$ $= \begin{bmatrix} -4 & 6 \\ 3 & -3 \end{bmatrix}$, and $\mathbf{u}_{B'} = P\,\mathbf{u}_B = \begin{bmatrix} 24 \\ -15 \end{bmatrix}$.

24. The transition matrix from B to the standard basis is $R = \begin{bmatrix} 3 & 1 \\ -1 & 1 \end{bmatrix}$ and the transition matrix from B' to the standard basis is $Q = \begin{bmatrix} 2 & 1 \\ 5 & 2 \end{bmatrix}$. $P = Q^{-1} R = \begin{bmatrix} -2 & 1 \\ 5 & -2 \end{bmatrix}\begin{bmatrix} 3 & 1 \\ -1 & 1 \end{bmatrix}$ $= \begin{bmatrix} -7 & -1 \\ 17 & 3 \end{bmatrix}$, and $\mathbf{u}_{B'} = P\,\mathbf{u}_B = \begin{bmatrix} -47 \\ 113 \end{bmatrix}$.

25. $3x^2 = 3(x^2) + 0(x) + 0(1)$, $x - 1 = 0(x^2) + 1(x) + -1(1)$, and $4 = 0(x^2) + 0(x) + 4(1)$, so the transition matrix from B' to B is $P = \begin{bmatrix} 3 & 0 & 0 \\ 0 & 1 & 0 \\ 0 & -1 & 4 \end{bmatrix}$. The transition matrix from B to B' is $P^{-1} = \begin{bmatrix} 1/3 & 0 & 0 \\ 0 & 1 & 0 \\ 0 & 1/4 & 1/4 \end{bmatrix}$. The coordinate vectors relative to B' of $3x^2 + 4x + 8$, $6x^2 + 4$, $8x + 12$, and $3x^2 + 4x + 4$ are respectively $P^{-1}\begin{bmatrix} 3 \\ 4 \\ 8 \end{bmatrix} = \begin{bmatrix} 1 \\ 4 \\ 3 \end{bmatrix}$, $P^{-1}\begin{bmatrix} 6 \\ 0 \\ 4 \end{bmatrix} = \begin{bmatrix} 2 \\ 0 \\ 1 \end{bmatrix}$, $P^{-1}\begin{bmatrix} 0 \\ 8 \\ 12 \end{bmatrix} = \begin{bmatrix} 0 \\ 8 \\ 5 \end{bmatrix}$, and $P^{-1}\begin{bmatrix} 3 \\ 4 \\ 4 \end{bmatrix} = \begin{bmatrix} 1 \\ 4 \\ 2 \end{bmatrix}$.

26. $x + 2 = 1(x) + 2(1)$ and $3 = 0(x) + 3(1)$, so the transition matrix from B' to B is $P = \begin{bmatrix} 1 & 0 \\ 2 & 3 \end{bmatrix}$ and the transition matrix from B to B' is $P^{-1} = \frac{1}{3}\begin{bmatrix} 3 & 0 \\ -2 & 1 \end{bmatrix}$. The coordinate

vectors relative to B' of 3x + 3, 6x, 6x + 9, and 12x − 3 are respectively $P^{-1}\begin{bmatrix}3\\3\end{bmatrix} = \begin{bmatrix}3\\-1\end{bmatrix}$,

$P^{-1}\begin{bmatrix}6\\0\end{bmatrix} = \begin{bmatrix}6\\-4\end{bmatrix}$, $P^{-1}\begin{bmatrix}6\\9\end{bmatrix} = \begin{bmatrix}6\\-1\end{bmatrix}$, and $P^{-1}\begin{bmatrix}12\\-3\end{bmatrix} = \begin{bmatrix}12\\-9\end{bmatrix}$.

27. Let $T(\begin{bmatrix}a & 0\\0 & b\end{bmatrix})$= (a,b). $T(\begin{bmatrix}a & 0\\0 & b\end{bmatrix}+\begin{bmatrix}d & 0\\0 & e\end{bmatrix}) = T(\begin{bmatrix}a+d & 0\\0 & b+e\end{bmatrix})$= (a+d,b+e)

= (a,b) + (d,e) = $T(\begin{bmatrix}a & 0\\0 & b\end{bmatrix}) + T(\begin{bmatrix}d & 0\\0 & e\end{bmatrix})$ and $T(c\begin{bmatrix}a & 0\\0 & b\end{bmatrix}) = T(\begin{bmatrix}ca & 0\\0 & cb\end{bmatrix})$

= (ca,cb) = c(a,b) = c $T(\begin{bmatrix}a & 0\\0 & b\end{bmatrix})$ so T is linear. T is one-to-one because if

$T(\begin{bmatrix}x & 0\\0 & y\end{bmatrix})$= (a,b), then x = a and y = b, and T is onto because if (a,b) is an element of

$\mathbf{R}^2$ then $T(\begin{bmatrix}a & 0\\0 & b\end{bmatrix})$= (a,b). Thus T is an isomorphism.

28. Let $T(\begin{bmatrix}a & b\\b & c\end{bmatrix})$= (a,b,c). $T(\begin{bmatrix}a & b\\b & c\end{bmatrix}+\begin{bmatrix}f & g\\g & h\end{bmatrix}) = T(\begin{bmatrix}a+f & b+g\\b+g & c+h\end{bmatrix})$ = (a+f,b+g,c+h)

= (a,b,c) + (f,g,h) = $T(\begin{bmatrix}a & b\\b & c\end{bmatrix}) + T(\begin{bmatrix}f & g\\g & h\end{bmatrix})$ and $T(d\begin{bmatrix}a & b\\b & c\end{bmatrix}) = T(\begin{bmatrix}da & db\\db & dc\end{bmatrix})$

= (da,db,dc) = d(a,b,c) = d$T(\begin{bmatrix}a & b\\b & c\end{bmatrix})$ so T is linear. T is one-to-one because if

$T(\begin{bmatrix}x & y\\y & z\end{bmatrix})$= (a,b,c), then x = a, y = b, and z = c, and T is onto because if (a,b,c) is an

element of $\mathbf{R}^2$, then $T(\begin{bmatrix}a & b\\b & c\end{bmatrix})$= (a,b,c). Thus T is an isomorphism.

29. T is a linear transformation from $\mathbf{R}^n$ to $\mathbf{R}^n$. If T is an isomorphism, then T is onto $\mathbf{R}^n$. The range of T is $\mathbf{R}^n$, and since the columns of A are a basis for the range of T, this

means the rank of A is n. Thus A is nonsingular.

If A is nonsingular, the rank of A is n, so the columns of A are linearly independent and therefore are a basis for $\mathbf{R}^n$, so the range of A is $\mathbf{R}^n$; i.e., T is onto. Dim ker(T) = 0, since dim range(T) + dim ker(T) = dim domain(T) and n = dim range(T) = dim domain(T), so t is one-to-one. Thus the linear transformation T is an isomorphism.

30. $|A^{-1}|=1/|A|$ (Theorem 3.4(d)). Thus if A is nonsingular A^{-1} is also nonsingular. Therefore by Exercise 29, the linear transformation of $\mathbf{R}^n$ to $\mathbf{R}^n$ defined by A^{-1} is an isomorphism.

Exercise Set 5.2

1. T(0,1,−1) = T(0(1,0,0) + 1(0,1,0) − 1(0,0,1)) = T(0,1,0) − T(0,0,1) = (0,−2) − (−1,1)

= (1,−3). Alternatively, the matrix of T with respect to the standard basis is

$A = \begin{bmatrix} 2 & 0 & -1 \\ 1 & -2 & 1 \end{bmatrix}$. $A\begin{bmatrix} 0 \\ 1 \\ -1 \end{bmatrix} = \begin{bmatrix} 1 \\ -3 \end{bmatrix}$, so T(0,1,−1) = (1,−3).

2. T(3,−2) = T(3(1,0) − 2(0,1)) = 3T(1,0) − 2T(0,1) = 3(4) − 2(−3) = 18. Alternatively, the matrix of T with respect to the standard basis is [4 −3]. $[\,4\ \ -3\,]\begin{bmatrix} 3 \\ -2 \end{bmatrix}$ = 18, so T(3,−2) = 18.

3. T(2,1) = T(2(1,0) + (0,1)) = 2T(1,0) + T(0,1) = 2(2,5) + (1,−3) = (5,7). Alternatively, the matrix of T with respect to the standard basis is $A = \begin{bmatrix} 2 & 1 \\ 5 & -3 \end{bmatrix}$. $A\begin{bmatrix} 2 \\ 1 \end{bmatrix} = \begin{bmatrix} 5 \\ 7 \end{bmatrix}$, so T(2,1) = (5,7).

4. $T(3x^2 - 2x + 1) = T(3(x^2) - 2(x) + 1(1)) = 3T(x^2) - 2T(x) + T(1) = 3(3x+1) - 2(2) + 2x-5$

$= 9x + 3 - 4 + 2x - 5 = 11x - 6$. Alternatively, the matrix of T with respect to the standard basis is $A = \begin{bmatrix} 3 & 0 & 2 \\ 1 & 2 & -5 \end{bmatrix}$. $A\begin{bmatrix} 3 \\ -2 \\ 1 \end{bmatrix} = \begin{bmatrix} 11 \\ -6 \end{bmatrix}$, so $T(3x^2 - 2x + 1) = 11x - 6$.

5. $T(x^2 + 3x - 2) = T(x^2 + 3(x) - 2(1)) = T(x^2) + 3T(x) - 2T(1)$
$= x^2 + 3 + 3(2x^2 + 4x - 1) - 2(3x - 1) = x^2 + 3 + 6x^2 + 12x - 3 - 6x + 2 = 7x^2 + 6x + 2$.
Alternatively, the matrix of T with

respect to the standard basis is $A = \begin{bmatrix} 1 & 2 & 0 \\ 0 & 4 & 3 \\ 3 & -1 & -1 \end{bmatrix}$. $A\begin{bmatrix} 1 \\ 3 \\ -2 \end{bmatrix} = \begin{bmatrix} 7 \\ 6 \\ 2 \end{bmatrix}$, so

$T(x^2 + 3x - 2) = 7x^2 + 6x + 2$.

6. The coordinate vectors of $T(\mathbf{u}_1)$ and $T(\mathbf{u}_2)$ relative to the basis $\{\mathbf{v}_1, \mathbf{v}_2\}$ are $\begin{bmatrix} 2 \\ 3 \end{bmatrix}$ and $\begin{bmatrix} 4 \\ -1 \end{bmatrix}$, so $A = \begin{bmatrix} 2 & 4 \\ 3 & -1 \end{bmatrix}$. The coordinate vector of **u** relative to the basis $\{\mathbf{u}_1, \mathbf{u}_2\}$ is

$\begin{bmatrix} 2 \\ 5 \end{bmatrix}$, so the coordinate vector of $T(\mathbf{u})$ relative to $\{\mathbf{v}_1, \mathbf{v}_2\}$ is $A\begin{bmatrix} 2 \\ 5 \end{bmatrix} = \begin{bmatrix} 24 \\ 1 \end{bmatrix}$.
Thus $T(\mathbf{u}) = 24\mathbf{v}_1 + \mathbf{v}_2$.

7. The coordinate vectors of $T(\mathbf{u}_1)$ and $T(\mathbf{u}_2)$ relative to the basis $\{\mathbf{v}_1, \mathbf{v}_2, \mathbf{v}_3\}$ are $\begin{bmatrix} 2 \\ 1 \\ -3 \end{bmatrix}$

and $\begin{bmatrix} 1 \\ -2 \\ 3 \end{bmatrix}$, so $A = \begin{bmatrix} 2 & 1 \\ 1 & -2 \\ -3 & 3 \end{bmatrix}$. The coordinate vector of **u** relative to the basis $\{\mathbf{u}_1, \mathbf{u}_2\}$ is

$\begin{bmatrix} 4 \\ -7 \end{bmatrix}$, so the coordinate vector of $T(\mathbf{u})$ relative to $\{\mathbf{v}_1, \mathbf{v}_2, \mathbf{v}_3\}$ is $A\begin{bmatrix} 4 \\ -7 \end{bmatrix} = \begin{bmatrix} 1 \\ 18 \\ -33 \end{bmatrix}$. Thus

$T(\mathbf{u}) = \mathbf{v}_1 + 18\mathbf{v}_2 - 19\mathbf{v}_3$.

8. The coordinate vectors of $T(\mathbf{u}_1)$, $T(\mathbf{u}_2)$, and $T(\mathbf{u}_3)$ relative to the basis $\{\mathbf{v}_1, \mathbf{v}_2, \mathbf{v}_3\}$ are

$\begin{bmatrix}1\\1\\1\end{bmatrix}$, $\begin{bmatrix}3\\-2\\0\end{bmatrix}$, and $\begin{bmatrix}1\\2\\-1\end{bmatrix}$, so $A = \begin{bmatrix}1 & 3 & 1\\1 & -2 & 2\\1 & 0 & -1\end{bmatrix}$. The coordinate vector of **u** relative to the basis

$\{\mathbf{u}_1, \mathbf{u}_2, \mathbf{u}_3\}$ is $\begin{bmatrix}3\\2\\-5\end{bmatrix}$, so the coordinate vector of T(**u**) relative to $\{\mathbf{v}_1, \mathbf{v}_2, \mathbf{v}_3\}$ is $A\begin{bmatrix}3\\2\\-5\end{bmatrix}$

$= \begin{bmatrix}4\\-11\\8\end{bmatrix}$. Thus $T(\mathbf{u}) = 4\mathbf{v}_1 - 11\mathbf{v}_2 + 8\mathbf{v}_3$.

9. (a) T(1,0,0) = (1,0) = 1(1,0) + 0(0,1), T(0,1,0) = (0,0) = 0(1,0) + 0(0,1), and T(0,0,1)

= (0,1) = 0(1,0) + 1(0,1), so $A = \begin{bmatrix}1 & 0 & 0\\0 & 0 & 1\end{bmatrix}$. (1,2,3) = 1(1,0,0) + 2(0,1,0) + 3(0,0,1),

so the coordinate vector of T(1,2,3) relative to the standard basis of $\mathbf{R}^2$ is $A\begin{bmatrix}1\\2\\3\end{bmatrix}$

$= \begin{bmatrix}1\\3\end{bmatrix}$. Thus T(1,2,3) = 1(1,0) + 3(0,1) = (1,3).

(b) T(1,0,0) = (3,0) = 3(1,0) + 0(0,1), T(0,1,0) = (0,1) = 0(1,0) + 1(0,1), and T(0,0,1)

= (0,1) = 0(1,0) + 1(0,1), so $A = \begin{bmatrix}3 & 0 & 0\\0 & 1 & 1\end{bmatrix}$. (1,2,3) = 1(1,0,0) + 2(0,1,0) + 3(0,0,1),

so the coordinate vector of T(1,2,3) relative to the standard basis of $\mathbf{R}^2$ is $A\begin{bmatrix}1\\2\\3\end{bmatrix}$

$= \begin{bmatrix}3\\5\end{bmatrix}$. Thus T(1,2,3) = 3(1,0) + 5(0,1) = (3,5).

(c) T(1,0,0) = (1,2) = 1(1,0) + 2(0,1), T(0,1,0) = (1,−1) = 1(1,0) − 1(0,1),

and T(0,0,1)= (0,0) = 0(1,0) + 0(0,1), so $A = \begin{bmatrix}1 & 1 & 0\\2 & -1 & 0\end{bmatrix}$.

(1,2,3) = 1(1,0,0) + 2(0,1,0) + 3(0,0,1),

so the coordinate vector of T(1,2,3) relative to the standard basis of $\mathbf{R}^2$ is $A\begin{bmatrix}1\\2\\3\end{bmatrix}$

$= \begin{bmatrix}3\\0\end{bmatrix}$. Thus T(1,2,3) = 3(1,0) + 0(0,1) = (3,0).

10. (a) T(1,0,0) = (1,0,0) = 1(1,0,0) + 0(0,1,0) + 0(0,0,1), T(0,1,0) = (0,2,0)

= 0(1,0,0) + 2(0,1,0) + 0(0,0,1), and T(0,0,1) = (0,0,3) = 0(1,0,0) + 0(0,1,0) +

3(0,0,1), so $A = \begin{bmatrix}1 & 0 & 0\\0 & 2 & 0\\0 & 0 & 3\end{bmatrix}$.

(−1,5,2) = −1(1,0,0) + 5(0,1,0) + 2(0,0,1), so the coordinate vector of T(−1,5,2)

relative to the standard basis of $\mathbf{R}^3$ is $A\begin{bmatrix}-1\\5\\2\end{bmatrix} = \begin{bmatrix}-1\\10\\6\end{bmatrix}$.

Thus T(−1,5,2) = −1(1,0,0) + 10(0,1,0) + 6(0,0,1) = (−1,10,6).

(b) T(1,0,0) = (1,0,0), T(0,1,0) = (0,1,0), and T(0,0,1) = (0,0,1), so $A = I_3$. The coordinate vector of T(−1,5,2) relative to the standard basis of $\mathbf{R}^3$ is $A\begin{bmatrix}-1\\5\\2\end{bmatrix}$

$= \begin{bmatrix}-1\\5\\2\end{bmatrix}$, so T(−1,5,2) = (−1,5,2).

(c) T(1,0,0) = (1,0,0), T(0,1,0) = (0,0,0), and T(0,0,1) = (0,0,0), so $A = \begin{bmatrix}1 & 0 & 0\\0 & 0 & 0\\0 & 0 & 0\end{bmatrix}$.

The coordinate vector of T(−1,5,2) relative to the standard basis of $\mathbf{R}^3$ is $A\begin{bmatrix}-1\\5\\2\end{bmatrix}$

$= \begin{bmatrix} -1 \\ 0 \\ 0 \end{bmatrix}$, so T(−1,5,2) = (−1,0,0).

(d) T(1,0,0) = (1,0,1), T(0,1,0) = (1,3,2), and T(0,0,1) = (0,0,−4), so $A = \begin{bmatrix} 1 & 1 & 0 \\ 0 & 3 & 0 \\ 1 & 2 & -4 \end{bmatrix}$.

The coordinate vector of T(−1,5,2) relative to the standard basis of $\mathbf{R}^3$ is $A\begin{bmatrix} -1 \\ 5 \\ 2 \end{bmatrix}$

$= \begin{bmatrix} 4 \\ 15 \\ 1 \end{bmatrix}$, so T(−1,5,2) = (4,15,1).

11. $T(\mathbf{u}_1) = (2,1) = -2\mathbf{u}_1' + 1\mathbf{u}_2'$, $T(\mathbf{u}_2) = (2,3) = -2\mathbf{u}_1' + 3\mathbf{u}_2'$, and $T(\mathbf{u}_3) = (-1,2) = 1\mathbf{u}_1' + 2\mathbf{u}_2'$, so $A = \begin{bmatrix} -2 & -2 & 1 \\ 1 & 3 & 2 \end{bmatrix}$. $\mathbf{u} = (3,-4,0) = 2\mathbf{u}_1 + \mathbf{u}_2 - \mathbf{u}_3$, so the coordinate vector of $\mathbf{u}$ relative to the basis $\{\mathbf{u}_1, \mathbf{u}_2, \mathbf{u}_3\}$ is $\begin{bmatrix} 2 \\ 1 \\ -1 \end{bmatrix}$ and the coordinate vector of T(**u**) relative to the basis $\{\mathbf{u}_1', \mathbf{u}_2'\}$ is $A\begin{bmatrix} 2 \\ 1 \\ -1 \end{bmatrix} = \begin{bmatrix} -7 \\ 3 \end{bmatrix}$. Thus $T(\mathbf{u}) = -7\mathbf{u}_1' + 3\mathbf{u}_2' = -7(-1,0)+3(0,1) = (7,3)$.

12. $T(\mathbf{u}_1) = (1,4,6) = 1\mathbf{u}_1' + 2\mathbf{u}_2' - 6\mathbf{u}_3'$ and $T(\mathbf{u}_2) = (4,3,-2) = 4\mathbf{u}_1' + 3/2\mathbf{u}_2' + 2\mathbf{u}_3'$, so

$A = \begin{bmatrix} 1 & 4 \\ 2 & 3/2 \\ -6 & 2 \end{bmatrix}$. $\mathbf{u} = (9,1) = 1\mathbf{u}_1 + 2\mathbf{u}_2$, so the coordinate vector of $\mathbf{u}$ relative to the basis $\{\mathbf{u}_1, \mathbf{u}_2\}$ is $\begin{bmatrix} 1 \\ 2 \end{bmatrix}$ and the coordinate vector of T(**u**) relative to the basis $\{\mathbf{u}_1', \mathbf{u}_2', \mathbf{u}_3'\}$ is $A\begin{bmatrix} 1 \\ 2 \end{bmatrix} = \begin{bmatrix} 9 \\ 5 \\ -2 \end{bmatrix}$. Thus $T(\mathbf{u}) = 9\mathbf{u}_1' + 5\mathbf{u}_2' - 2\mathbf{u}_3' = 9(1,0,1)+5(0,2,0)-2(0,0,1) = (9,10,2)$.

13. $T(\mathbf{u}_1) = T(1,2) = (2,3) = 2\mathbf{u}_1 + 1\mathbf{u}_2$ and $T(\mathbf{u}_2) = T(0,-1) = (0,-1) = 0\mathbf{u}_1 + 1\mathbf{u}_2$, so

$A = \begin{bmatrix} 2 & 0 \\ 1 & 1 \end{bmatrix}$. $\mathbf{u} = (-1,3) = -1\mathbf{u}_1 - 5\mathbf{u}_2$, so the coordinate vector of $\mathbf{u}$ relative to the

basis $\{\mathbf{u}_1, \mathbf{u}_2\}$ is $\begin{bmatrix} -1 \\ -5 \end{bmatrix}$ and the coordinate vector of $T(\mathbf{u})$ relative to the basis $\{\mathbf{u}_1, \mathbf{u}_2\}$ is

$A\begin{bmatrix} -1 \\ -5 \end{bmatrix} = \begin{bmatrix} -2 \\ -6 \end{bmatrix}$. Thus $T(\mathbf{u}) = -2\mathbf{u}_1 - 6\mathbf{u}_2 = (-2,2)$.

14. $D(2x^2) = 4x = 0(2x^2) + 4(x) + 0\,(-1)$, $D(x) = 1 = 0(2x^2) + 0(x) - (-1)$, and $D(-1) = 0$

$= 0(2x^2) + 0(x) + 0\,(-1)$, so $A = \begin{bmatrix} 0 & 0 & 0 \\ 4 & 0 & 0 \\ 0 & -1 & 0 \end{bmatrix}$. $3x^2 - 2x + 4 = \frac{3}{2}(2x^2) - 2(x) - 4(-1)$, so

the coordinate vector of $3x^2 - 2x + 4$ relative to the given basis is $\begin{bmatrix} 3/2 \\ -2 \\ -4 \end{bmatrix}$, and the

coordinate vector of $T(3x^2 - 2x + 4)$ is $A\begin{bmatrix} 3/2 \\ -2 \\ -4 \end{bmatrix} = \begin{bmatrix} 0 \\ 6 \\ 2 \end{bmatrix}$. Thus $T(3x^2 - 2x + 4)$

$= 0(2x^2) + 6(x) + 2(-1) = 6x - 2$.

15. $(D^2 + 2D + 1)(\sin x) = 2\cos x = 0(\sin x) + 2(\cos x)$ and $(D^2 + 2D + 1)(\cos x) = -2\sin x$

$= -2(\sin x) + 0(\cos x)$, so $A = \begin{bmatrix} 0 & -2 \\ 2 & 0 \end{bmatrix}$. $3\sin x + \cos x$ has coordinate vector $\begin{bmatrix} 3 \\ 1 \end{bmatrix}$, so

the coordinate vector of $(D^2 + 2D + 1)(3\sin x + \cos x)$ is $A\begin{bmatrix} 3 \\ 1 \end{bmatrix} = \begin{bmatrix} -2 \\ 6 \end{bmatrix}$. Thus

$(D^2 + 2D + 1)(3\sin x + \cos x) = -2\sin x + 6\cos x$.

16. (a) $T(x^2) = 0 = 0(x^2) + 0(x) + 0(1)$, $T(x) = x^2 + x = 1(x^2) + 1(x) + 0(1)$,

and $T(1) = x^2 - x = 1(x^2) - 1(x) + 0(1)$, so $A = \begin{bmatrix} 0 & 1 & 1 \\ 0 & 1 & -1 \\ 0 & 0 & 0 \end{bmatrix}$.

(b) $T(x) = x = 0(x^2) + 1(x) + 0(1)$ and $T(1) = x^2 + 1 = 1(x^2) + 0(x) + 1(1)$, so $A = \begin{bmatrix} 0 & 1 \\ 1 & 0 \\ 0 & 1 \end{bmatrix}$.

(c) $T(x^2) = -1 = 0(x) - 1(1)$, $T(x) = 1 = 0(x) + 1(1)$, and $T(1) = 2x = 2(x) + 0(1)$, so

$A = \begin{bmatrix} 0 & 0 & 2 \\ -1 & 1 & 0 \end{bmatrix}$.

17. $T(x^2 + x) = x = 1(x) + 0(1)$, $T(x) = 0 = 0(x) + 0(1)$, and $T(1) = 1 = 0(x) + 1(1)$, so

$A = \begin{bmatrix} 1 & 0 & 0 \\ 0 & 0 & 1 \end{bmatrix}$. $3x^2 + 2x - 1 = 3(x^2 + x) - 1(x) - 1(1)$, so the coordinate vector for

$3x^2 + 2x - 1$ relative to the basis $\{x^2 + x, x, 1\}$ is $\begin{bmatrix} 3 \\ -1 \\ -1 \end{bmatrix}$, and the coordinate vector for

$T(3x^2 + 2x - 1)$ relative to the basis $\{x, 1\}$ is $A\begin{bmatrix} 3 \\ -1 \\ -1 \end{bmatrix} = \begin{bmatrix} 3 \\ -1 \end{bmatrix}$.

Thus $T(3x^2 + 2x - 1) = 3(x) - 1(1) = 3x - 1$.

18. $T(x+1) = x - 1 = 1(x+1) - 1(2)$ and $T(2) = 2x = 2(x+1) - 1(2)$, so $A = \begin{bmatrix} 1 & 2 \\ -1 & -1 \end{bmatrix}$.

$4x - 3 = 4(x+1) - 7/2(2)$, so the coordinate vector for $4x - 3$ relative to the given basis

is $\begin{bmatrix} 4 \\ -7/2 \end{bmatrix}$ and the coordinate vector for $T(4x - 3)$ is $A\begin{bmatrix} 4 \\ -7/2 \end{bmatrix} = \begin{bmatrix} -3 \\ -1/2 \end{bmatrix}$.

Thus $T(4x - 3) = -3(x+1) - 1/2(2) = -3x - 4$.

19. $T(x) = x = 1(x) + 0\,(1)$ and $T(1) = x - 1 = 1(x) - 1(1)$, so $A = \begin{bmatrix} 1 & 1 \\ 0 & -1 \end{bmatrix}$.

$x + 1 = 1(x) + 1(1)$ and $x - 1 = 1(x) - 1(1)$, so $P = \begin{bmatrix} 1 & 1 \\ 1 & -1 \end{bmatrix}$ and $P^{-1} = \frac{1}{2}\begin{bmatrix} 1 & 1 \\ 1 & -1 \end{bmatrix}$.

$P^{-1}AP = \frac{1}{2}\begin{bmatrix} 1 & 1 \\ 3 & -1 \end{bmatrix}$, which is the matrix of T with respect to the basis $\{x+1, x-1\}$.

20. $T(1,0) = (2,1) = 2(1,0) + 1(0,1)$ and $T(0,1) = (0,1) = 0(1,0) + 1(0,1)$, so $A = \begin{bmatrix} 2 & 0 \\ 1 & 1 \end{bmatrix}$.

$(1,1) = 1(1,0) + 1(0,1)$ and $(2,1) = 2(1,0) + 1(0,1)$, so $P = \begin{bmatrix} 1 & 2 \\ 1 & 1 \end{bmatrix}$ and $P^{-1} = \begin{bmatrix} -1 & 2 \\ 1 & -1 \end{bmatrix}$.

$P^{-1}AP = \begin{bmatrix} 2 & 2 \\ 0 & 1 \end{bmatrix}$, which is the matrix of T with respect to the basis $\{(1,1), (2,1)\}$.

21. $T(1,0) = (1,1) = 1(1,0) + 1(0,1)$ and $T(0,1) = (-1,1) = -1(1,0) + 1(0,1)$, so $A = \begin{bmatrix} 1 & -1 \\ 1 & 1 \end{bmatrix}$.

$(-2,1) = -2(1,0) + 1(0,1)$ and $(1,2) = 1(1,0) + 2(0,1)$, so $P = \begin{bmatrix} -2 & 1 \\ 1 & 2 \end{bmatrix}$ and
$P^{-1} = \frac{1}{5}\begin{bmatrix} -2 & 1 \\ 1 & 2 \end{bmatrix}$. $P^{-1}AP = \begin{bmatrix} 1 & 1 \\ -1 & 1 \end{bmatrix}$, which is the matrix of T with respect to the basis

$\{(-2,1), (1,2)\}$.

22. (a) Let $\{\mathbf{u}_1, \mathbf{u}_2, \ldots, \mathbf{u}_m\}$ be a basis for U, and let $\mathbf{v}_{m+1}, \ldots, \mathbf{v}_n$ be additional linearly independent vectors in V so that $\{\mathbf{u}_1, \mathbf{u}_2, \ldots, \mathbf{u}_m, \mathbf{v}_{m+1}, \ldots, \mathbf{v}_n\}$ is a basis for V. Let T be a linear transformation that maps each of $\mathbf{u}_1, \mathbf{u}_2, \ldots, \mathbf{u}_m$ into the zero vector of W. Then each element of U will be in the kernel of T. If $\dim(W) \geq n-m$, then $\mathbf{v}_{m+1}, \ldots, \mathbf{v}_n$ can be mapped by T into linearly independent vectors in W, and $U = \ker(T)$. If $\dim(W) < n-m$, then $T(\mathbf{v}_{m+1}), \ldots, T(\mathbf{v}_n)$ will be linearly dependent and there will be constants $a_{m+1}, \ldots, a_n$, not all zero, such that $T(a_{m+1}\mathbf{v}_{m+1} + \ldots + a_n\mathbf{v}_n) = a_{m+1}T(\mathbf{v}_{m+1}) + \ldots + a_nT(\mathbf{v}_n) = \mathbf{0}$. The vector $a_{m+1}\mathbf{v}_{m+1} + \ldots + a_n\mathbf{v}_n$ is not in U, for if it were, there would be constants $b_1, b_2, \ldots, b_m$ such that $b_1\mathbf{u}_1 + b_2\mathbf{u}_2 + \ldots + b_m\mathbf{u}_m = a_{m+1}\mathbf{v}_{m+1} + \ldots + a_n\mathbf{v}_n$; i.e., $b_1\mathbf{u}_1 + b_2\mathbf{u}_2 + \ldots + b_m\mathbf{u}_m - a_{m+1}\mathbf{v}_{m+1} - \ldots - a_n\mathbf{v}_n = \mathbf{0}$. This cannot be because the vectors $\mathbf{u}_1, \mathbf{u}_2, \ldots, \mathbf{u}_m, \mathbf{v}_{m+1}, \ldots, \mathbf{v}_n$ are linearly independent. Thus in this case U is contained in but not equal to ker(T). Thus: yes, if $\dim(W) \geq \dim(V) - \dim(U)$, no, if $\dim(W) < \dim(V) - \dim(U)$.

(b) The set $\{(1,3,-1), (1,0,0), (0,1,0)\}$ is a basis for $\mathbf{R}^3$. Let T be the linear transformation given by $T(1,3,-1) = (0,0)$, $T(1,0,0) = (1,0)$, and $T(0,1,0) = (0,1)$. Let $\mathbf{u} = a_1(1,3,-1) + a_2(1,0,0) + a_3(0,1,0)$ and suppose $T(\mathbf{u}) = (0,0)$. Thus $a_1\, T(1,3,-1) + a_2\, T(1,0,0) + a_3\, T(0,1,0) = a_1(0,0) + a_2(1,0) + a_3(0,1) = (0,0)$, so that $a_2(1,0) + a_3(0,1) = (0,0)$. Since $(1,0)$ and $(0,1)$ are linearly independent, a_2 and a_3 must be zero and thus $\mathbf{u}$ is a multiple of $(1,3,-1)$.

(c) $\{(r+s,2r,-s)\} = \{r(1,2,0)+s(1,0,-1)\}$. The set $\{(1,2,0), (1,0,-1)\}$ is a basis for this subspace of $\mathbf{R}^3$ and $\{(1,2,0), (1,0,-1), (0,1,0)\}$ is a basis for $\mathbf{R}^3$. Let T be the linear transformation given by $T(1,2,0) = (0,0)$, $T(1,0,-1) = (0,0)$, and $T(0,1,0) = (1,0)$. Let $\mathbf{u} = a_1(1,2,0) + a_2(1,0,-1) + a_3(0,1,0)$ and supose $T(\mathbf{u}) = (0,0)$. Thus $a_1\, T(1,2,0) + a_2\, T(1,0,-1) + a_3\, T(0,1,0) = a_1(0,0) + a_2(0,0) + a_3(1,0) = (0,0)$, so that $a_3(1,0) = (0,0)$, and therefore $a_3 = 0$. Thus $\mathbf{u} = a_1(1,2,0) + a_2(1,0,-1)$.

23. The set $\{(2,-1), (1,0)\}$ is a basis for $\mathbf{R}^2$. Let T be the linear transformation given by $T(2,-1) = (0,0)$ and $T(1,0) = (1,0)$. Let $\mathbf{u} = a_1(2,-1) + a_2(1,0)$ and suppose $T(\mathbf{u}) = (0,0)$. Thus $a_1\, T(2,-1) + a_2\, T(1,0) = a_1(0,0) + a_2(1,0) = (0,0)$, so that $a_2(1,0) = (0,0)$, and therefore $a_2 = 0$. Thus $\mathbf{u}$ is a multiple of $(2,-1)$.

24. The set $\{(4,1), (1,0)\}$ is a basis for $\mathbf{R}^2$. Let T be the linear transformation given by $T(4,1) = (0,0,0)$ and $T(1,0) = (1,0,0)$. Let $\mathbf{u} = a_1(4,1) + a_2(1,0)$ and suppose $T(\mathbf{u}) = (0,0,0)$. Thus $a_1\, T(4,1) + a_2\, T(1,0) = a_1(0,0,0) + a_2(1,0,0) = (0,0,0)$, so that $a_2(1,0,0) = (0,0,0)$, and therefore $a_2 = 0$. Thus $\mathbf{u}$ is a multiple of $(4,1)$.

25. $T(\mathbf{u} + \mathbf{v}) = \mathbf{u} + \mathbf{v} = T(\mathbf{u}) + T(\mathbf{v})$ and $T(c\mathbf{u}) = c\mathbf{u} = cT(\mathbf{u})$, so T is linear. Let $B = \{\mathbf{u}_1, \mathbf{u}_2, \ldots, \mathbf{u}_n\}$ be a basis for U. $T(\mathbf{u}_i) = \mathbf{u}_i$, so if A is the matrix of T with respect to B, its ith column will have a 1 in the ith position and zeros everywhere else. Thus $A = I_n$.

26. $T(\mathbf{u} + \mathbf{v}) = \mathbf{0} = \mathbf{0} + \mathbf{0} = T(\mathbf{u}) + T(\mathbf{v})$ and $T(c\mathbf{u}) = \mathbf{0} = c\mathbf{0} = cT(\mathbf{u})$, so T is linear. Let $B = \{\mathbf{u}_1, \mathbf{u}_2, \ldots, \mathbf{u}_n\}$. $T(\mathbf{u}_i) = \mathbf{0}$, so if A is the matrix of T with respect to B, its ith column will consist of all zeros. Thus $A = O_n$, the zero matrix.

27. If $\mathbf{u}$ is an element of U and $T(\mathbf{u}) = \mathbf{v}$ and $L(T(\mathbf{u})) = \mathbf{w}$, and if $\mathbf{u}_B$, $\mathbf{v}_{B'}$, and $\mathbf{w}_{B''}$ are the coordinate vectors of $\mathbf{u}$, $\mathbf{v}$, and $\mathbf{w}$ relative to the bases B, B', and B'', then $P\mathbf{u}_B = \mathbf{v}_{B'}$ and $Q\mathbf{v}_{B'} = \mathbf{w}_{B''}$. Thus $QP\mathbf{u}_B = Q\mathbf{v}_{B'} = \mathbf{w}_{B''}$ and QP is the matrix of $L \circ T$ with respect to B and B''.

28. If **u** is an element of U and $\mathbf{u}_B$ is the coordinate vector of **u** relative to basis B, then the coordinate vector of T(**u**) relative to the basis B' is given by $A\mathbf{u}_B$, where A is the matrix of T with respect to the bases B and B'. If A is also the matrix of L with respect to the same two bases, then the coordinate vector of L(**u**) relative to the basis B' will also be given by $A\mathbf{u}_B$. But this means that the coordinate vector of L(**u**) relative to the basis B' is the same as the coordinate vector of T(**u**) relative to the basis B'. Thus L(**u**) = T(**u**) for every vector **u** in U and therefore L = T.

Exercise Set 5.3

1. (a) $C^{-1}AC = \begin{bmatrix} 3 & -5 \\ -1 & 2 \end{bmatrix}\begin{bmatrix} 1 & 2 \\ -1 & 3 \end{bmatrix}\begin{bmatrix} 2 & 5 \\ 1 & 3 \end{bmatrix} = \begin{bmatrix} 7 & 13 \\ -2 & -3 \end{bmatrix}$.

(b) $C^{-1}AC = \begin{bmatrix} -1 & 2 \\ 2 & -3 \end{bmatrix}\begin{bmatrix} -8 & 18 \\ -6 & 13 \end{bmatrix}\begin{bmatrix} 3 & 2 \\ 2 & 1 \end{bmatrix} = \begin{bmatrix} 4 & 0 \\ 0 & 1 \end{bmatrix}$.

(c) $C^{-1}AC = \begin{bmatrix} 4 & -1 \\ -7 & 2 \end{bmatrix}\begin{bmatrix} 0 & 4 \\ 3 & 2 \end{bmatrix}\begin{bmatrix} 2 & 1 \\ 7 & 4 \end{bmatrix} = \begin{bmatrix} 92 & 53 \\ -156 & -90 \end{bmatrix}$.

2. (a) $C^{-1}AC = \begin{bmatrix} -1 & 2 & 2 \\ 0 & 1 & 1 \\ 2 & 0 & -1 \end{bmatrix}\begin{bmatrix} 2 & 0 & 0 \\ -2 & 2 & 1 \\ 2 & 0 & 1 \end{bmatrix}\begin{bmatrix} -1 & 2 & 0 \\ 2 & -3 & 1 \\ -2 & 4 & -1 \end{bmatrix} = \begin{bmatrix} 2 & 0 & 0 \\ 0 & 2 & 0 \\ 0 & 0 & 1 \end{bmatrix}$.

(b) $C^{-1}AC = \begin{bmatrix} -1 & -3 & 2 \\ 1 & 2 & -1 \\ 1 & 1 & 1 \end{bmatrix}\begin{bmatrix} 1 & 2 & 3 \\ 0 & -1 & 2 \\ 1 & 1 & 0 \end{bmatrix}\begin{bmatrix} 3 & 5 & -1 \\ -2 & -3 & 1 \\ -1 & -2 & 1 \end{bmatrix} = \begin{bmatrix} 6 & 14 & -7 \\ -5 & -11 & 6 \\ -3 & -6 & 5 \end{bmatrix}$.

3. (a) The eigenvalues and eigenvectors of this matrix were found in Exercise 1 in Section 3.4. They are $\lambda = 6$ with eigenvectors $r\begin{bmatrix} 4 \\ 1 \end{bmatrix}$ and $\lambda = 1$ with eigenvectors $s\begin{bmatrix} -1 \\ 1 \end{bmatrix}$. $\frac{1}{5}\begin{bmatrix} 1 & 1 \\ -1 & 4 \end{bmatrix}\begin{bmatrix} 5 & 4 \\ 1 & 2 \end{bmatrix}\begin{bmatrix} 4 & -1 \\ 1 & 1 \end{bmatrix} = \begin{bmatrix} 6 & 0 \\ 0 & 1 \end{bmatrix}$.

(b) $\begin{vmatrix} 2-\lambda & 1 \\ 2 & 3-\lambda \end{vmatrix} = (2-\lambda)(3-\lambda) - 2 = \lambda^2 - 5\lambda + 4 = (\lambda-4)(\lambda-1)$, so the eigenvalues are

$\lambda = 4$ and $\lambda = 1$. For $\lambda = 4$, the eigenvectors are the solutions of

$\begin{bmatrix} -2 & 1 \\ 2 & -1 \end{bmatrix}\begin{bmatrix} x_1 \\ x_2 \end{bmatrix} = \mathbf{0}$, so the eigenvectors are vectors of the form $r\begin{bmatrix} 1 \\ 2 \end{bmatrix}$. For $\lambda = 1$,

the eigenvectors are the solutions of $\begin{bmatrix} 1 & 1 \\ 2 & 2 \end{bmatrix}\begin{bmatrix} x_1 \\ x_2 \end{bmatrix} = \mathbf{0}$, so the eigenvectors are

vectors of the form $s\begin{bmatrix} -1 \\ 1 \end{bmatrix}$. $\frac{1}{3}\begin{bmatrix} 1 & 1 \\ -2 & 1 \end{bmatrix}\begin{bmatrix} 2 & 1 \\ 2 & 3 \end{bmatrix}\begin{bmatrix} 1 & -1 \\ 2 & 1 \end{bmatrix} = \begin{bmatrix} 4 & 0 \\ 0 & 1 \end{bmatrix}$.

(c) $\begin{vmatrix} 1-\lambda & 1 \\ 0 & 1-\lambda \end{vmatrix} = (1-\lambda)(1-\lambda)$, so the only eigenvalue is $\lambda = 1$. The eigenvectors are

the solutions of $\begin{bmatrix} 0 & 1 \\ 0 & 0 \end{bmatrix}\begin{bmatrix} x_1 \\ x_2 \end{bmatrix} = \mathbf{0}$, so the eigenvectors are vectors of the form $r\begin{bmatrix} 1 \\ 0 \end{bmatrix}$.

There are not two linearly independent eigenvectors, so matrix cannot be diagonalized.

(d) $\begin{vmatrix} 4-\lambda & -1 \\ 2 & 1-\lambda \end{vmatrix} = (4-\lambda)(1-\lambda) + 2 = \lambda^2 - 5\lambda + 6 = (\lambda-2)(\lambda-3)$, so the eigenvalues are

$\lambda = 2$ and $\lambda = 3$. For $\lambda = 2$, the eigenvectors are the solutions of

$\begin{bmatrix} 2 & -1 \\ 2 & -1 \end{bmatrix}\begin{bmatrix} x_1 \\ x_2 \end{bmatrix} = \mathbf{0}$, so the eigenvectors are vectors of the form $r\begin{bmatrix} 1 \\ 2 \end{bmatrix}$. For $\lambda = 3$,

the eigenvectors are the solutions of $\begin{bmatrix} 1 & -1 \\ 2 & -2 \end{bmatrix}\begin{bmatrix} x_1 \\ x_2 \end{bmatrix} = \mathbf{0}$, so the eigenvectors are

vectors of the form $s\begin{bmatrix} 1 \\ 1 \end{bmatrix}$. $\begin{bmatrix} -1 & 1 \\ 2 & -1 \end{bmatrix}\begin{bmatrix} 4 & -1 \\ 2 & 1 \end{bmatrix}\begin{bmatrix} 1 & 1 \\ 2 & 1 \end{bmatrix} = \begin{bmatrix} 2 & 0 \\ 0 & 3 \end{bmatrix}$.

(e) $\begin{vmatrix} 4-\lambda & 1 \\ -1 & 2-\lambda \end{vmatrix} = (4-\lambda)(2-\lambda) + 1 = \lambda^2 - 6\lambda + 9 = (\lambda-3)(\lambda-3)$, so the only eigenvalue is $\lambda = 3$. The eigenvectors are the solutions of $\begin{bmatrix} 1 & 1 \\ -1 & -1 \end{bmatrix}\begin{bmatrix} x_1 \\ x_2 \end{bmatrix} = \mathbf{0}$, so the eigenvectors are vectors of the form $r\begin{bmatrix} 1 \\ -1 \end{bmatrix}$. Since there are not two linearly independent eigenvectors, the matrix cannot be diagonalized.

4. (a) $\begin{vmatrix} -7-\lambda & 10 \\ -5 & 8-\lambda \end{vmatrix} = (-7-\lambda)(8-\lambda) + 50 = \lambda^2 - \lambda - 6 = (\lambda+2)(\lambda-3)$, so the eigenvalues are $\lambda = -2$ and $\lambda = 3$. For $\lambda = -2$, the eigenvectors are the solutions of $\begin{bmatrix} -5 & 10 \\ -5 & 10 \end{bmatrix}\begin{bmatrix} x_1 \\ x_2 \end{bmatrix} = \mathbf{0}$, so the eigenvectors are vectors of the form $r\begin{bmatrix} 2 \\ 1 \end{bmatrix}$. For $\lambda = 3$, the eigenvectors are the solutions of $\begin{bmatrix} -10 & 10 \\ -5 & 5 \end{bmatrix}\begin{bmatrix} x_1 \\ x_2 \end{bmatrix} = \mathbf{0}$, so the eigenvectors are vectors of the form $s\begin{bmatrix} 1 \\ 1 \end{bmatrix}$. $\begin{bmatrix} 1 & -1 \\ -1 & 2 \end{bmatrix}\begin{bmatrix} -7 & 10 \\ -5 & 8 \end{bmatrix}\begin{bmatrix} 2 & 1 \\ 1 & 1 \end{bmatrix} = \begin{bmatrix} -2 & 0 \\ 0 & 3 \end{bmatrix}$.

(b) $\begin{vmatrix} 7-\lambda & 4 \\ -8 & -5-\lambda \end{vmatrix} = (7-\lambda)(-5-\lambda) + 32 = \lambda^2 - 2\lambda - 3 = (\lambda+1)(\lambda-3)$, so the eigenvalues are $\lambda = -1$ and $\lambda = 3$. For $\lambda = -1$, the eigenvectors are the solutions of $\begin{bmatrix} 8 & 4 \\ -8 & -4 \end{bmatrix}\begin{bmatrix} x_1 \\ x_2 \end{bmatrix} = \mathbf{0}$, so the eigenvectors are vectors of the form $r\begin{bmatrix} 1 \\ -2 \end{bmatrix}$. For $\lambda = 3$, the eigenvectors are the solutions of $\begin{bmatrix} 4 & 4 \\ -8 & -8 \end{bmatrix}\begin{bmatrix} x_1 \\ x_2 \end{bmatrix} = \mathbf{0}$, so the eigenvectors are vectors of the form $s\begin{bmatrix} 1 \\ -1 \end{bmatrix}$. $\begin{bmatrix} -1 & -1 \\ 2 & 1 \end{bmatrix}\begin{bmatrix} 7 & 4 \\ -8 & -5 \end{bmatrix}\begin{bmatrix} 1 & 1 \\ -2 & -1 \end{bmatrix} = \begin{bmatrix} -1 & 0 \\ 0 & 3 \end{bmatrix}$.

(c) $\begin{vmatrix} 1-\lambda & -2 \\ 2 & -3-\lambda \end{vmatrix} = (1-\lambda)(-3-\lambda) + 4 = \lambda^2 + 2\lambda + 1 = (\lambda+1)(\lambda+1)$, so the only eigenvalue is $\lambda = -1$. The eigenvectors are the solutions of $\begin{bmatrix} 2 & -2 \\ 2 & -2 \end{bmatrix}\begin{bmatrix} x_1 \\ x_2 \end{bmatrix} = \mathbf{0}$, so the eigenvectors are vectors of the form $r\begin{bmatrix} 1 \\ -1 \end{bmatrix}$. Since there are not two linearly independent eigenvectors, the matrix cannot be diagonalized.

(d) $\begin{vmatrix} 7-\lambda & -4 \\ 1 & 3-\lambda \end{vmatrix} = (7-\lambda)(3-\lambda) + 4 = \lambda^2 - 10\lambda + 25 = (\lambda-5)(\lambda-5)$, so the only eigenvalue is $\lambda = 5$. The eigenvectors are the solutions of $\begin{bmatrix} 2 & -4 \\ 1 & -2 \end{bmatrix}\begin{bmatrix} x_1 \\ x_2 \end{bmatrix} = \mathbf{0}$, so the

eigenvectors are vectors of the form $r\begin{bmatrix} 2 \\ 1 \end{bmatrix}$. Since there are not two linearly independent eigenvectors, the matrix cannot be diagonalized.

(e) $\begin{vmatrix} a-\lambda & b \\ 0 & a-\lambda \end{vmatrix} = (a-\lambda)(a-\lambda)$, so the only eigenvalue is $\lambda = a$. The eigenvectors are the solutions of $\begin{bmatrix} 0 & b \\ 0 & 0 \end{bmatrix}\begin{bmatrix} x_1 \\ x_2 \end{bmatrix} = \mathbf{0}$, so the eigenvectors are vectors of the form $r\begin{bmatrix} 1 \\ 0 \end{bmatrix}$.

There are not two linearly independent eigenvectors, so the matrix cannot be diagonallized.

5. (a) The eigenvalues and eigenvectors of this matrix were found in Exercise 13 in Section 3.4. They are $\lambda = 1$, $\lambda = 2$, $\lambda = 8$, with corresponding eigenvectors

$$r\begin{bmatrix} 1 \\ -1 \\ 1 \end{bmatrix}, s\begin{bmatrix} 0 \\ 1 \\ 1 \end{bmatrix}, t\begin{bmatrix} 1 \\ 0 \\ 1 \end{bmatrix}. \quad \begin{bmatrix} -1 & -1 & 1 \\ -1 & 0 & 1 \\ 2 & 1 & -1 \end{bmatrix}\begin{bmatrix} 15 & 7 & -7 \\ -1 & 1 & 1 \\ 13 & 7 & -5 \end{bmatrix}\begin{bmatrix} 1 & 0 & 1 \\ -1 & 1 & 0 \\ 1 & 1 & 1 \end{bmatrix} = \begin{bmatrix} 1 & 0 & 0 \\ 0 & 2 & 0 \\ 0 & 0 & 8 \end{bmatrix}.$$

(b) The eigenvalues and eigenvectors of this matrix were found in Exercise 14 in Section 3.4. They are $\lambda = 5$, $\lambda = 3$, $\lambda = 1$, with corresponding eigenvectors

$$r\begin{bmatrix} 1 \\ 1 \\ 1 \end{bmatrix}, s\begin{bmatrix} 1 \\ 0 \\ -1 \end{bmatrix}, t\begin{bmatrix} 0 \\ 1 \\ 1 \end{bmatrix}. \quad \begin{bmatrix} 1 & -1 & 1 \\ 0 & 1 & -1 \\ -1 & 2 & -1 \end{bmatrix}\begin{bmatrix} 5 & -2 & 2 \\ 4 & -3 & 4 \\ 4 & -6 & 7 \end{bmatrix}\begin{bmatrix} 1 & 1 & 0 \\ 1 & 0 & 1 \\ 1 & -1 & 1 \end{bmatrix} = \begin{bmatrix} 5 & 0 & 0 \\ 0 & 3 & 0 \\ 0 & 0 & 1 \end{bmatrix}.$$

(c) $\begin{vmatrix} 1-\lambda & 0 & 0 \\ -2 & 1-\lambda & 2 \\ -2 & 0 & 3-\lambda \end{vmatrix} = (1-\lambda)^2(3-\lambda)$, so the eigenvalues are $\lambda = 1$ and $\lambda = 3$.

For $\lambda = 1$, the eigenvectors are $r\begin{bmatrix} 1 \\ 0 \\ 1 \end{bmatrix} + s\begin{bmatrix} 0 \\ 1 \\ 0 \end{bmatrix}$. For $\lambda = 3$, the eigenvectors are

$t\begin{bmatrix} 0 \\ 1 \\ 1 \end{bmatrix}$. $\begin{bmatrix} 1 & 0 & 0 \\ 1 & 1 & -1 \\ -1 & 0 & 1 \end{bmatrix}\begin{bmatrix} 1 & 0 & 0 \\ -2 & 1 & 2 \\ -2 & 0 & 3 \end{bmatrix}\begin{bmatrix} 1 & 0 & 0 \\ 0 & 1 & 1 \\ 1 & 0 & 1 \end{bmatrix} = \begin{bmatrix} 1 & 0 & 0 \\ 0 & 1 & 0 \\ 0 & 0 & 3 \end{bmatrix}$.

(d) The eigenvalues are $\lambda = 3$, $\lambda = 2$, $\lambda = -4$, with corresponding eigenvectors

$r\begin{bmatrix} 1 \\ 1 \\ 0 \end{bmatrix}, s\begin{bmatrix} 0 \\ 1 \\ 0 \end{bmatrix}, t\begin{bmatrix} 0 \\ 0 \\ 1 \end{bmatrix}$. $\begin{bmatrix} 1 & 0 & 0 \\ -1 & 1 & 0 \\ 0 & 0 & 1 \end{bmatrix}\begin{bmatrix} 3 & 0 & 0 \\ 1 & 2 & 0 \\ 0 & 0 & -4 \end{bmatrix}\begin{bmatrix} 1 & 0 & 0 \\ 1 & 1 & 0 \\ 0 & 0 & 1 \end{bmatrix} = \begin{bmatrix} 3 & 0 & 0 \\ 0 & 2 & 0 \\ 0 & 0 & -4 \end{bmatrix}$.

6. (a) $\begin{vmatrix} 1-\lambda & 2 \\ 2 & 1-\lambda \end{vmatrix} = (1-\lambda)(1-\lambda) - 4 = \lambda^2 - 2\lambda - 3 = (\lambda-3)(\lambda+1)$, so eigenvalues are $\lambda = 3$ and $\lambda = -1$, with corresponding eigenvectors $r\begin{bmatrix} 1 \\ 1 \end{bmatrix}$ and $s\begin{bmatrix} -1 \\ 1 \end{bmatrix}$. The set $\left\{\begin{bmatrix} \frac{1}{\sqrt{2}} \\ \frac{1}{\sqrt{2}} \end{bmatrix}, \begin{bmatrix} \frac{-1}{\sqrt{2}} \\ \frac{1}{\sqrt{2}} \end{bmatrix}\right\}$ is an orthonormal basis. $\begin{bmatrix} \frac{1}{\sqrt{2}} & \frac{1}{\sqrt{2}} \\ \frac{-1}{\sqrt{2}} & \frac{1}{\sqrt{2}} \end{bmatrix}\begin{bmatrix} 1 & 2 \\ 2 & 1 \end{bmatrix}\begin{bmatrix} \frac{1}{\sqrt{2}} & \frac{-1}{\sqrt{2}} \\ \frac{1}{\sqrt{2}} & \frac{1}{\sqrt{2}} \end{bmatrix} =$ $\begin{bmatrix} 3 & 0 \\ 0 & -1 \end{bmatrix}$.

(b) $\begin{vmatrix} 11-\lambda & 2 \\ 2 & 14-\lambda \end{vmatrix} = (11-\lambda)(14-\lambda) - 4 = \lambda^2 - 25\lambda + 150 = (\lambda-15)(\lambda-10)$, so the eigenvalues are $\lambda = 15$ and $\lambda = 10$, with corresponding eigenvectors

$r\begin{bmatrix} 1 \\ 2 \end{bmatrix}$ and $s\begin{bmatrix} -2 \\ 1 \end{bmatrix}$. The set $\left\{\begin{bmatrix} \frac{1}{\sqrt{5}} \\ \frac{2}{\sqrt{5}} \end{bmatrix}, \begin{bmatrix} \frac{-2}{\sqrt{5}} \\ \frac{1}{\sqrt{5}} \end{bmatrix}\right\}$ is an orthonormal basis.

$$\begin{bmatrix} \frac{1}{\sqrt{5}} & \frac{2}{\sqrt{5}} \\ \frac{-2}{\sqrt{5}} & \frac{1}{\sqrt{5}} \end{bmatrix} \begin{bmatrix} 11 & 2 \\ 2 & 14 \end{bmatrix} \begin{bmatrix} \frac{1}{\sqrt{5}} & \frac{-2}{\sqrt{5}} \\ \frac{2}{\sqrt{5}} & \frac{1}{\sqrt{5}} \end{bmatrix} = \begin{bmatrix} 15 & 0 \\ 0 & 10 \end{bmatrix}.$$

(c) $\begin{vmatrix} 3-\lambda & 1 \\ 1 & 3-\lambda \end{vmatrix} = (3-\lambda)(3-\lambda) - 1 = \lambda^2 - 6\lambda + 8 = (\lambda-4)(\lambda-2)$, so the

eigenvalues are $\lambda = 4$ and $\lambda = 2$, with corresponding eigenvectors

$r\begin{bmatrix} 1 \\ 1 \end{bmatrix}$ and $s\begin{bmatrix} -1 \\ 1 \end{bmatrix}$. The set $\left\{\begin{bmatrix} \frac{1}{\sqrt{2}} \\ \frac{1}{\sqrt{2}} \end{bmatrix}, \begin{bmatrix} \frac{-1}{\sqrt{2}} \\ \frac{1}{\sqrt{2}} \end{bmatrix}\right\}$ is an orthonormal basis.

$$\begin{bmatrix} \frac{1}{\sqrt{2}} & \frac{1}{\sqrt{2}} \\ \frac{-1}{\sqrt{2}} & \frac{1}{\sqrt{2}} \end{bmatrix} \begin{bmatrix} 3 & 1 \\ 1 & 3 \end{bmatrix} \begin{bmatrix} \frac{1}{\sqrt{2}} & \frac{-1}{\sqrt{2}} \\ \frac{1}{\sqrt{2}} & \frac{1}{\sqrt{2}} \end{bmatrix} = \begin{bmatrix} 4 & 0 \\ 0 & 2 \end{bmatrix}.$$

(d) $\begin{vmatrix} -1-\lambda & -8 \\ -8 & 11-\lambda \end{vmatrix} = (-1-\lambda)(11-\lambda) - 64 = \lambda^2 - 10\lambda - 75 = (\lambda-15)(\lambda+5)$, so the

eigenvalues are $\lambda = -5$ and $\lambda = 15$, with corresponding eigenvectors

$r\begin{bmatrix} 2 \\ 1 \end{bmatrix}$ and $s\begin{bmatrix} -1 \\ 2 \end{bmatrix}$. The set $\left\{\begin{bmatrix} \frac{2}{\sqrt{5}} \\ \frac{1}{\sqrt{5}} \end{bmatrix}, \begin{bmatrix} \frac{-1}{\sqrt{5}} \\ \frac{2}{\sqrt{5}} \end{bmatrix}\right\}$ is an orthonormal basis.

$$\begin{bmatrix} \frac{2}{\sqrt{5}} & \frac{1}{\sqrt{5}} \\ \frac{-1}{\sqrt{5}} & \frac{2}{\sqrt{5}} \end{bmatrix} \begin{bmatrix} -1 & -8 \\ -8 & 11 \end{bmatrix} \begin{bmatrix} \frac{2}{\sqrt{5}} & \frac{-1}{\sqrt{5}} \\ \frac{1}{\sqrt{5}} & \frac{2}{\sqrt{5}} \end{bmatrix} = \begin{bmatrix} -5 & 0 \\ 0 & 15 \end{bmatrix}.$$

7. (a) $\begin{vmatrix} 1-\lambda & 5 \\ 5 & 1-\lambda \end{vmatrix} = (1-\lambda)(1-\lambda) - 25 = \lambda^2 - 2\lambda - 24 = (\lambda-6)(\lambda+4)$, so the

eigenvalues are $\lambda = 6$ and $\lambda = -4$, with corresponding eigenvectors

$r\begin{bmatrix}1\\1\end{bmatrix}$ and $s\begin{bmatrix}-1\\1\end{bmatrix}$. The set $\left\{\begin{bmatrix}\frac{1}{\sqrt{2}}\\\frac{1}{\sqrt{2}}\end{bmatrix},\begin{bmatrix}\frac{-1}{\sqrt{2}}\\\frac{1}{\sqrt{2}}\end{bmatrix}\right\}$ is an orthonormal basis.

$$\begin{bmatrix}\frac{1}{\sqrt{2}} & \frac{1}{\sqrt{2}}\\\frac{-1}{\sqrt{2}} & \frac{1}{\sqrt{2}}\end{bmatrix}\begin{bmatrix}1 & 5\\5 & 1\end{bmatrix}\begin{bmatrix}\frac{1}{\sqrt{2}} & \frac{-1}{\sqrt{2}}\\\frac{1}{\sqrt{2}} & \frac{1}{\sqrt{2}}\end{bmatrix}=\begin{bmatrix}6 & 0\\0 & -4\end{bmatrix}.$$

(b) $\begin{vmatrix}9-\lambda & 2\\2 & 6-\lambda\end{vmatrix} = (9-\lambda)(6-\lambda) - 4 = \lambda^2 - 15\lambda + 50 = (\lambda-10)(\lambda-5)$, so the

eigenvalues are $\lambda = 10$ and $\lambda = 5$, with corresponding eigenvectors

$r\begin{bmatrix}2\\1\end{bmatrix}$ and $s\begin{bmatrix}-1\\2\end{bmatrix}$. The set $\left\{\begin{bmatrix}\frac{2}{\sqrt{5}}\\\frac{1}{\sqrt{5}}\end{bmatrix},\begin{bmatrix}\frac{-1}{\sqrt{5}}\\\frac{2}{\sqrt{5}}\end{bmatrix}\right\}$ is an orthonormal basis.

$$\begin{bmatrix}\frac{2}{\sqrt{5}} & \frac{1}{\sqrt{5}}\\\frac{-1}{\sqrt{5}} & \frac{2}{\sqrt{5}}\end{bmatrix}\begin{bmatrix}9 & 2\\2 & 6\end{bmatrix}\begin{bmatrix}\frac{2}{\sqrt{5}} & \frac{-1}{\sqrt{5}}\\\frac{1}{\sqrt{5}} & \frac{2}{\sqrt{5}}\end{bmatrix}=\begin{bmatrix}10 & 0\\0 & 5\end{bmatrix}.$$

(c) $\begin{vmatrix}1-\lambda & 3\\3 & 9-\lambda\end{vmatrix} = (1-\lambda)(9-\lambda) - 9 = \lambda^2 - 10\lambda = (\lambda-10)\lambda$, so the eigenvalues

are $\lambda = 10$ and $\lambda = 0$, with corresponding eigenvectors $r\begin{bmatrix}1\\3\end{bmatrix}$ and $s\begin{bmatrix}-3\\1\end{bmatrix}$.

The set $\left\{\begin{bmatrix}\frac{1}{\sqrt{10}}\\\frac{3}{\sqrt{10}}\end{bmatrix},\begin{bmatrix}\frac{-3}{\sqrt{10}}\\\frac{1}{\sqrt{10}}\end{bmatrix}\right\}$ is an orthonormal basis.

$$\begin{bmatrix} \frac{1}{\sqrt{10}} & \frac{3}{\sqrt{10}} \\ \frac{-3}{\sqrt{10}} & \frac{1}{\sqrt{10}} \end{bmatrix} \begin{bmatrix} 1 & 3 \\ 3 & 9 \end{bmatrix} \begin{bmatrix} \frac{1}{\sqrt{10}} & \frac{-3}{\sqrt{10}} \\ \frac{3}{\sqrt{10}} & \frac{1}{\sqrt{10}} \end{bmatrix} = \begin{bmatrix} 10 & 0 \\ 0 & 0 \end{bmatrix}.$$

(d) $\begin{vmatrix} 1.5-\lambda & -.5 \\ -.5 & 1.5-\lambda \end{vmatrix} = (1.5-\lambda)(1.5-\lambda) - .25 = \lambda^2 - 3\lambda + 2 = (\lambda-1)(\lambda-2)$, so the eigenvalues are $\lambda = 1$ and $\lambda = 2$, with corresponding eigenvectors

$r\begin{bmatrix} 1 \\ 1 \end{bmatrix}$ and $s\begin{bmatrix} -1 \\ 1 \end{bmatrix}$. The set $\left\{\begin{bmatrix} \frac{1}{\sqrt{2}} \\ \frac{1}{\sqrt{2}} \end{bmatrix}, \begin{bmatrix} \frac{-1}{\sqrt{2}} \\ \frac{1}{\sqrt{2}} \end{bmatrix}\right\}$ is an orthonormal basis.

$$\begin{bmatrix} \frac{1}{\sqrt{2}} & \frac{1}{\sqrt{2}} \\ \frac{-1}{\sqrt{2}} & \frac{1}{\sqrt{2}} \end{bmatrix} \begin{bmatrix} 1.5 & -.5 \\ -.5 & 1.5 \end{bmatrix} \begin{bmatrix} \frac{1}{\sqrt{2}} & \frac{-1}{\sqrt{2}} \\ \frac{1}{\sqrt{2}} & \frac{1}{\sqrt{2}} \end{bmatrix} = \begin{bmatrix} 1 & 0 \\ 0 & 2 \end{bmatrix}.$$

8. (a) $\begin{vmatrix} -\lambda & 2 & 0 \\ 2 & -\lambda & 0 \\ 0 & 0 & 1-\lambda \end{vmatrix} = \lambda^2(1-\lambda) - 4(1-\lambda)$, so the eigenvalues are $\lambda = 1$ and $\lambda = \pm 2$.

For $\lambda = 1$, the eigenvectors are $r\begin{bmatrix} 0 \\ 0 \\ 1 \end{bmatrix}$, for $\lambda = 2$, the eigenvectors are $s\begin{bmatrix} 1 \\ 1 \\ 0 \end{bmatrix}$, and for $\lambda = -2$, the eigenvectors are $t\begin{bmatrix} 1 \\ -1 \\ 0 \end{bmatrix}$.

$$\begin{bmatrix} 0 & 0 & 1 \\ 1/\sqrt{2} & 1/\sqrt{2} & 0 \\ 1/\sqrt{2} & -1/\sqrt{2} & 0 \end{bmatrix} \begin{bmatrix} 0 & 2 & 0 \\ 2 & 0 & 0 \\ 0 & 0 & 1 \end{bmatrix} \begin{bmatrix} 0 & 1/\sqrt{2} & 1/\sqrt{2} \\ 0 & 1/\sqrt{2} & -1/\sqrt{2} \\ 1 & 0 & 0 \end{bmatrix} = \begin{bmatrix} 1 & 0 & 0 \\ 0 & 2 & 0 \\ 0 & 0 & -2 \end{bmatrix}.$$

(b) $\begin{vmatrix} 9-\lambda & -3 & 3 \\ -3 & 6-\lambda & -6 \\ 3 & -6 & 6-\lambda \end{vmatrix} = (9-\lambda)(6-\lambda)^2 + 54\lambda - 324 = -\lambda(15-\lambda)(6-\lambda)$, so the

eigenvalues are $\lambda = 0$, $\lambda = 15$, and $\lambda = 6$. For $\lambda = 0$, the eigenvectors are $r\begin{bmatrix} 0 \\ 1 \\ 1 \end{bmatrix}$,

for $\lambda = 15$, the eigenvectors are $s\begin{bmatrix} 1 \\ -1 \\ 1 \end{bmatrix}$, and for $\lambda = 6$, the eigenvectors are $t\begin{bmatrix} 2 \\ 1 \\ -1 \end{bmatrix}$.

$$\begin{bmatrix} 0 & 1/\sqrt{2} & 1/\sqrt{2} \\ 1/\sqrt{3} & -1/\sqrt{3} & 1/\sqrt{3} \\ 2/\sqrt{6} & 1/\sqrt{6} & -1/\sqrt{6} \end{bmatrix} \begin{bmatrix} 9 & -3 & 3 \\ -3 & 6 & -6 \\ 3 & -6 & 6 \end{bmatrix} \begin{bmatrix} 0 & 1/\sqrt{3} & 2/\sqrt{6} \\ 1/\sqrt{2} & -1/\sqrt{3} & 1/\sqrt{6} \\ 1/\sqrt{2} & 1/\sqrt{3} & -1/\sqrt{6} \end{bmatrix}$$

$$= \begin{bmatrix} 0 & 0 & 0 \\ 0 & 15 & 0 \\ 0 & 0 & 6 \end{bmatrix}.$$

(c) $\begin{vmatrix} 1-\lambda & 2 & -2 \\ 2 & 4-\lambda & -4 \\ -2 & -4 & 4-\lambda \end{vmatrix} = (1-\lambda)(4-\lambda)^2 + 24\lambda - 16 = \lambda^2(9-\lambda)$, so the eigenvalues are

$\lambda = 0$ and $\lambda = 9$. For $\lambda = 0$, the eigenvectors are $r\begin{bmatrix} 0 \\ 1 \\ 1 \end{bmatrix} + s\begin{bmatrix} 2 \\ -1 \\ 0 \end{bmatrix}$, and for $\lambda = 9$, the

eigenvectors are $t\begin{bmatrix} 1 \\ 2 \\ -2 \end{bmatrix}$.

$$\begin{bmatrix} 0 & 1/\sqrt{2} & 1/\sqrt{2} \\ 2/\sqrt{5} & -1/\sqrt{5} & 0 \\ 1/3 & 2/3 & -2/3 \end{bmatrix} \begin{bmatrix} 1 & 2 & -2 \\ 2 & 4 & -4 \\ -2 & -4 & 4 \end{bmatrix} \begin{bmatrix} 0 & 2/\sqrt{5} & 1/3 \\ 1/\sqrt{2} & -1/\sqrt{5} & 2/3 \\ 1/\sqrt{2} & 0 & -2/3 \end{bmatrix} = \begin{bmatrix} 0 & 0 & 0 \\ 0 & 0 & 0 \\ 0 & 0 & 9 \end{bmatrix}.$$

9. (a) $\begin{bmatrix} 1 & 5 \\ 5 & 1 \end{bmatrix}^8 = \begin{bmatrix} \frac{1}{\sqrt{2}} & \frac{-1}{\sqrt{2}} \\ \frac{1}{\sqrt{2}} & \frac{1}{\sqrt{2}} \end{bmatrix} \begin{bmatrix} 6 & 0 \\ 0 & -4 \end{bmatrix}^8 \begin{bmatrix} \frac{1}{\sqrt{2}} & \frac{1}{\sqrt{2}} \\ \frac{-1}{\sqrt{2}} & \frac{1}{\sqrt{2}} \end{bmatrix}$

$$= \frac{1}{2}\begin{bmatrix} 1 & -1 \\ 1 & 1 \end{bmatrix}\begin{bmatrix} 1679616 & 0 \\ 0 & 65536 \end{bmatrix}\begin{bmatrix} 1 & 1 \\ -1 & 1 \end{bmatrix} = \begin{bmatrix} 872576 & 807040 \\ 807040 & 872576 \end{bmatrix}.$$

(b) $$\begin{bmatrix} 9 & 2 \\ 2 & 6 \end{bmatrix}^5 = \begin{bmatrix} \frac{2}{\sqrt{5}} & \frac{-1}{\sqrt{5}} \\ \frac{1}{\sqrt{5}} & \frac{2}{\sqrt{5}} \end{bmatrix}\begin{bmatrix} 10 & 0 \\ 0 & 5 \end{bmatrix}^5\begin{bmatrix} \frac{2}{\sqrt{5}} & \frac{1}{\sqrt{5}} \\ \frac{-1}{\sqrt{5}} & \frac{2}{\sqrt{5}} \end{bmatrix}$$

$$= \frac{1}{5}\begin{bmatrix} 2 & -1 \\ 1 & 2 \end{bmatrix}\begin{bmatrix} 100000 & 0 \\ 0 & 3125 \end{bmatrix}\begin{bmatrix} 2 & 1 \\ -1 & 2 \end{bmatrix} = \begin{bmatrix} 80625 & 38750 \\ 38750 & 22500 \end{bmatrix}.$$

(c) $$\begin{bmatrix} 1 & 3 \\ 3 & 9 \end{bmatrix}^6 = \begin{bmatrix} \frac{1}{\sqrt{10}} & \frac{-3}{\sqrt{10}} \\ \frac{3}{\sqrt{10}} & \frac{1}{\sqrt{10}} \end{bmatrix}\begin{bmatrix} 10 & 0 \\ 0 & 0 \end{bmatrix}^6\begin{bmatrix} \frac{1}{\sqrt{10}} & \frac{3}{\sqrt{10}} \\ \frac{-3}{\sqrt{10}} & \frac{1}{\sqrt{10}} \end{bmatrix}$$

$$= \frac{1}{10}\begin{bmatrix} 1 & -3 \\ 3 & 1 \end{bmatrix}\begin{bmatrix} 1000000 & 0 \\ 0 & 0 \end{bmatrix}\begin{bmatrix} 1 & 3 \\ -3 & 1 \end{bmatrix} = \begin{bmatrix} 100000 & 300000 \\ 300000 & 900000 \end{bmatrix}.$$

(d) $$\begin{bmatrix} 1.5 & -.5 \\ -.5 & 1.5 \end{bmatrix}^{16} = \begin{bmatrix} \frac{1}{\sqrt{2}} & \frac{-1}{\sqrt{2}} \\ \frac{1}{\sqrt{2}} & \frac{1}{\sqrt{2}} \end{bmatrix}\begin{bmatrix} 1 & 0 \\ 0 & 2 \end{bmatrix}^{16}\begin{bmatrix} \frac{1}{\sqrt{2}} & \frac{1}{\sqrt{2}} \\ \frac{-1}{\sqrt{2}} & \frac{1}{\sqrt{2}} \end{bmatrix}$$

$$= \frac{1}{2}\begin{bmatrix} 1 & -1 \\ 1 & 1 \end{bmatrix}\begin{bmatrix} 1 & 0 \\ 0 & 65536 \end{bmatrix}\begin{bmatrix} 1 & 1 \\ -1 & 1 \end{bmatrix} = \begin{bmatrix} 32768.5 & -32767.5 \\ -32767.5 & 32768.5 \end{bmatrix}.$$

10. (a) $$\begin{bmatrix} 0 & 2 & 0 \\ 2 & 0 & 0 \\ 0 & 0 & 1 \end{bmatrix}^6 = \begin{bmatrix} 0 & 1/\sqrt{2} & 1/\sqrt{2} \\ 0 & 1/\sqrt{2} & -1/\sqrt{2} \\ 1 & 0 & 0 \end{bmatrix}\begin{bmatrix} 1 & 0 & 0 \\ 0 & 2 & 0 \\ 0 & 0 & -2 \end{bmatrix}^6\begin{bmatrix} 0 & 0 & 1 \\ 1/\sqrt{2} & 1/\sqrt{2} & 0 \\ 1/\sqrt{2} & -1/\sqrt{2} & 0 \end{bmatrix}$$

$$= \frac{1}{2}\begin{bmatrix} 0 & 1 & 1 \\ 0 & 1 & -1 \\ \sqrt{2} & 0 & 0 \end{bmatrix}\begin{bmatrix} 1 & 0 & 0 \\ 0 & 64 & 0 \\ 0 & 0 & 64 \end{bmatrix}\begin{bmatrix} 0 & 0 & \sqrt{2} \\ 1 & 1 & 0 \\ 1 & -1 & 0 \end{bmatrix} = \begin{bmatrix} 64 & 0 & 0 \\ 0 & 64 & 0 \\ 0 & 0 & 1 \end{bmatrix}.$$

(b) $$\begin{bmatrix} 9 & -3 & 3 \\ -3 & 6 & -6 \\ 3 & -6 & 6 \end{bmatrix}^5 =$$

$$\begin{bmatrix} 0 & 1/\sqrt{3} & 2/\sqrt{6} \\ 1/\sqrt{2} & -1/\sqrt{3} & 1/\sqrt{6} \\ 1/\sqrt{2} & 1/\sqrt{3} & -1/\sqrt{6} \end{bmatrix}\begin{bmatrix} 0 & 0 & 0 \\ 0 & 15 & 0 \\ 0 & 0 & 6 \end{bmatrix}^5\begin{bmatrix} 0 & 1/\sqrt{2} & 1/\sqrt{2} \\ 1/\sqrt{3} & -1/\sqrt{3} & 1/\sqrt{3} \\ 2/\sqrt{6} & 1/\sqrt{6} & -1/\sqrt{6} \end{bmatrix}$$

$$= \frac{1}{6}\begin{bmatrix} 0 & \sqrt{2} & 2 \\ \sqrt{3} & -\sqrt{2} & 1 \\ \sqrt{3} & \sqrt{2} & -1 \end{bmatrix}\begin{bmatrix} 0 & 0 & 0 \\ 0 & 759375 & 0 \\ 0 & 0 & 7776 \end{bmatrix}\begin{bmatrix} 0 & \sqrt{3} & \sqrt{3} \\ \sqrt{2} & -\sqrt{2} & \sqrt{2} \\ 2 & 1 & -1 \end{bmatrix}$$

$$= \begin{bmatrix} 258309 & -250533 & 250533 \\ -250533 & 254421 & -254421 \\ 250533 & -254421 & 254421 \end{bmatrix}.$$

(c) $$\begin{bmatrix} 1 & 2 & -2 \\ 2 & 4 & -4 \\ -2 & -4 & 4 \end{bmatrix}^4 = \begin{bmatrix} 0 & 2/\sqrt{5} & 1/3 \\ 1/\sqrt{2} & -1/\sqrt{5} & 2/3 \\ 1/\sqrt{2} & 0 & -2/3 \end{bmatrix}\begin{bmatrix} 0 & 0 & 0 \\ 0 & 0 & 0 \\ 0 & 0 & 9 \end{bmatrix}^4\begin{bmatrix} 0 & 1/\sqrt{2} & 1/\sqrt{2} \\ 2/\sqrt{5} & -1/\sqrt{5} & 0 \\ 1/3 & 2/3 & -2/3 \end{bmatrix}$$

$$= \begin{bmatrix} 0 & 2/\sqrt{5} & 1/3 \\ 1/\sqrt{2} & -1/\sqrt{5} & 2/3 \\ 1/\sqrt{2} & 0 & -2/3 \end{bmatrix}\begin{bmatrix} 0 & 0 & 0 \\ 0 & 0 & 0 \\ 0 & 0 & 6561 \end{bmatrix}\begin{bmatrix} 0 & 1/\sqrt{2} & 1/\sqrt{2} \\ 2/\sqrt{5} & -1/\sqrt{5} & 0 \\ 1/3 & 2/3 & -2/3 \end{bmatrix}$$

$$= \begin{bmatrix} 729 & 1458 & -1458 \\ 1458 & 2916 & -2916 \\ -1458 & -2916 & 2916 \end{bmatrix}.$$

11. (a) If A and B are similar, then $B = C^{-1}AC$, so that $|B| = |C^{-1}AC| = |C^{-1}||A||C|$.

By Theorem 3.4 (d) $|C^{-1}| = 1/|C|$. Thus $|B| = |A|$.

(b) If A and B are similar, they represent the same linear operator T with respect to two different bases on the vector space. Thus rank(A) = dim range(T) and rank(B) = dim range(T), so rank(A) = rank(B).

(c) We know from Exercise 16, Section 2.3 that for nxn matrices E and F, tr(EF) = tr(FE). Thus if $B = C^{-1}AC$, then $tr(B) = tr(C^{-1}(AC)) = tr((AC)C^{-1}) = tr(A(CC^{-1})) = tr(A)$.

(d) If $B = C^{-1}AC$ then $B^t = (C^{-1}AC)^t = (AC)^t(C^{-1})^t = C^tA^t(C^{-1})^t = C^tA^t(C^t)^{-1}$. Thus $B^t = E^{-1}A^tE$, where $E = (C^{-1})^t$, so B^t and A^t are similar.

(e) If A and B are similar, then A is nonsingular if and only if $|A| \neq 0$ if and only if $|B| \neq 0$ (since $|A| = |B|$) if and only if B is nonsingular.

(f) If $B = C^{-1}AC$ and A is nonsingular, then B is also nonsingular and $B^{-1} = (C^{-1}AC)^{-1} = C^{-1}A^{-1}C$, so B^{-1} is similar to A^{-1}.

12. Let A be a symmetric matrix and λ_1 and λ_2 distinct eigenvalues of A with corresponding eigenvectors X and Y. $X \cdot Y = X^tY = Y^tX$, $AX = \lambda_1X$, and $AY = \lambda_2Y$. Thus $X^tAY = X^t\lambda_2Y = \lambda_2X^tY = \lambda_2X \cdot Y$ and $Y^tAX = Y^t\lambda_1X = \lambda_1Y^tX = \lambda_1X \cdot Y$. However, since Y^tAX is a real number, $Y^tAX = (Y^tAX)^t = X^tAY$, so $\lambda_1X \cdot Y = \lambda_2X \cdot Y$. Since $\lambda_1 \neq \lambda_2$ we conclude that $X \cdot Y = 0$.

13. (a) Let $A = C^{-1}BC$ and let D be the diagonal matrix $D = E^{-1}AE$. Then $D = E^{-1}(C^{-1}BC)E = (CE)^{-1}B(CE)$, so B is diagonalizable to the same diagonal matrix D.

(b) If $A = C^{-1}BC$ then $A + kI = C^{-1}BC + kC^{-1}IC = C^{-1}BC + C^{-1}kIC = C^{-1}(B+kI)C$, so $B + kI$ and $A + kI$ are similar for any scalar k.

14. Let $D = C^{-1}AC$. Then $D^2 = (C^{-1}AC)(C^{-1}AC) = C^{-1}A^2C$ and
$$D^n = \underbrace{(C^{-1}AC)(C^{-1}AC)\ldots(C^{-1}AC)(C^{-1}AC)}_{n \text{ terms}} = C^{-1}A^nC.$$

15. Consider the symmetric matrix $\begin{bmatrix}1 & 2\\ 2 & 1\end{bmatrix}$ of Exercise 6(a). It has eigenvalues $\lambda = 3$, $\lambda = -1$ with corresponding eigenvectors $r\begin{bmatrix}1\\1\end{bmatrix}$, $s\begin{bmatrix}-1\\1\end{bmatrix}$. Let $C = \begin{bmatrix}1/\sqrt{2} & -1/\sqrt{2}\\ 1/\sqrt{2} & 1/\sqrt{2}\end{bmatrix}$. Then $C^{-1}AC = \begin{bmatrix}3 & 0\\ 0 & -1\end{bmatrix}$.

However, if we let $C = \begin{bmatrix}-1/\sqrt{2} & 1/\sqrt{2}\\ 1/\sqrt{2} & 1/\sqrt{2}\end{bmatrix}$, reversing the order of the normalized eigenvectors, then $C^{-1}AC = \begin{bmatrix} -1 & 0 \\ 0 & 3 \end{bmatrix}$. Thus the diagonal matrix is not unique.

An eigenvalue λ_i will occupy the ith diagonal position in the diagonal matrix if the ith column of C is a normalized eigenvector corresponding to λ_i.

16. Let $D = C^{-1}AC$ be the diagonalization of A. If the eigenvalues of A are $\lambda_1, \lambda_2, \ldots, \lambda_n$, then $|D| = \lambda_1 \lambda_2 \ldots \lambda_n$. However $|A| = |D|$, so $|A| = \lambda_1 \lambda_2 \ldots \lambda_n$.

17. If $B = C^{-1}AC = C^tAC$, then $B^t = (C^tAC)^t = C^tA^t(C^t)^t = C^tA^tC = C^tAC = B$. Thus B is symmetric.

18. $B = E^tAE$ and $C = F^tBF$, so $C = F^t(E^tAE)F = (EF)^tA(EF)$. The product of orthogonal matrices is orthogonal, so C and A are orthogonally similar.

19. (a) $T(1,0) = (4,2) = 4(1,0) + 2(0,1)$ and $T(0,1) = (2,4) = 2(1,0) + 4(0,1)$, so the matrix representation of T relative to the standard basis is $A = \begin{bmatrix} 4 & 2 \\ 2 & 4 \end{bmatrix}$.

$\begin{vmatrix} 4-\lambda & 2 \\ 2 & 4-\lambda \end{vmatrix} = (4-\lambda)(4-\lambda) - 4 = \lambda^2 - 8\lambda + 12 = (\lambda-6)(\lambda-2)$, so the eigenvalues are $\lambda = 6$ and $\lambda = 2$ with corresponding eigenvectors $r\begin{bmatrix} 1 \\ 1 \end{bmatrix}$ and $s\begin{bmatrix} -1 \\ 1 \end{bmatrix}$. Orthonormal eigenvectors are $\begin{bmatrix} \frac{1}{\sqrt{2}} \\ \frac{1}{\sqrt{2}} \end{bmatrix}$ and $\begin{bmatrix} \frac{-1}{\sqrt{2}} \\ \frac{1}{\sqrt{2}} \end{bmatrix}$. Let B' be the basis $\left\{\left(\frac{1}{\sqrt{2}}, \frac{1}{\sqrt{2}}\right), \left(\frac{-1}{\sqrt{2}}, \frac{1}{\sqrt{2}}\right)\right\}$.

The matrix representation of T relative to B' is $A' = \begin{bmatrix} 6 & 0 \\ 0 & 2 \end{bmatrix}$. The transition matrix from B' to the standard basis is $P = \begin{bmatrix} \frac{1}{\sqrt{2}} & \frac{-1}{\sqrt{2}} \\ \frac{1}{\sqrt{2}} & \frac{1}{\sqrt{2}} \end{bmatrix}$ and the orthogonal transformation is $A' = P^tAP$.

The standard basis defines an xy coordinate system and the basis B' defines an x'y' coordinate system, rotated 45° counterclockwise from the xy system. The transformation T is a scaling in the x'y' system with factor 6 in the x' direction and factor 2 in the y' direction.

(b) T(1,0) = (5,3) = 5(1,0) + 3(0,1) and T(0,1) = (3,5) = 3(1,0) + 5(0,1), so the matrix representation of T relative to the standard basis is $A = \begin{bmatrix} 5 & 3 \\ 3 & 5 \end{bmatrix}$.

$\begin{vmatrix} 5-\lambda & 3 \\ 3 & 5-\lambda \end{vmatrix} = (5-\lambda)(5-\lambda) - 9 = \lambda^2 - 10\lambda + 16 = (\lambda-8)(\lambda-2)$, so the eigenvalues are $\lambda = 8$ and $\lambda = 2$ with corresponding eigenvectors $r\begin{bmatrix} 1 \\ 1 \end{bmatrix}$ and $s\begin{bmatrix} -1 \\ 1 \end{bmatrix}$. Orthonormal eigenvectors are $\begin{bmatrix} \frac{1}{\sqrt{2}} \\ \frac{1}{\sqrt{2}} \end{bmatrix}$ and $\begin{bmatrix} \frac{-1}{\sqrt{2}} \\ \frac{1}{\sqrt{2}} \end{bmatrix}$. Let B' be the basis $\left\{\left(\frac{1}{\sqrt{2}}, \frac{1}{\sqrt{2}}\right), \left(\frac{-1}{\sqrt{2}}, \frac{1}{\sqrt{2}}\right)\right\}$.

The matrix representation of T relative to B' is $A' = \begin{bmatrix} 8 & 0 \\ 0 & 2 \end{bmatrix}$. The transition matrix from B' to the standard basis is $P = \begin{bmatrix} \frac{1}{\sqrt{2}} & \frac{-1}{\sqrt{2}} \\ \frac{1}{\sqrt{2}} & \frac{1}{\sqrt{2}} \end{bmatrix}$ and the orthogonal transformation is $A' = P^tAP$.

The standard basis defines an xy coordinate system and the basis B' defines an x'y' coordinate system, rotated 45° counterclockwise from the xy system. The transformation T is a scaling in the x'y' system with factor 8 in the x' direction and factor 2 in the y' direction.

(c) T(1,0) = (9,2) = 9(1,0) + 2(0,1) and T(0,1) = (2,4) = 2(1,0) + 6(0,1), so the matrix representation of T relative to the standard basis is $A = \begin{bmatrix} 9 & 2 \\ 2 & 6 \end{bmatrix}$.

A has eigenvalues $\lambda = 10$ and $\lambda = 5$ with corresponding eigenvectors $r\begin{bmatrix} 2 \\ 1 \end{bmatrix}$ and $s\begin{bmatrix} -1 \\ 2 \end{bmatrix}$.

Orthonormal eigenvectors are $\begin{bmatrix} \frac{2}{\sqrt{5}} \\ \frac{1}{\sqrt{5}} \end{bmatrix}$ and $\begin{bmatrix} \frac{-1}{\sqrt{5}} \\ \frac{2}{\sqrt{5}} \end{bmatrix}$. Let B' be the basis

$\left\{\left(\frac{2}{\sqrt{5}}, \frac{1}{\sqrt{5}}\right), \left(\frac{-1}{\sqrt{5}}, \frac{2}{\sqrt{5}}\right)\right\}$. The matrix representation of T relative to B' is A' =

$\begin{bmatrix} 10 & 0 \\ 0 & 5 \end{bmatrix}$. The transition matrix from B' to the standard basis is P = $\begin{bmatrix} \frac{2}{\sqrt{5}} & \frac{-1}{\sqrt{5}} \\ \frac{1}{\sqrt{5}} & \frac{2}{\sqrt{5}} \end{bmatrix}$

and the orthogonal transformation is $A' = P^tAP$.

The standard basis defines an xy coordinate system and the basis B' defines an x'y' coordinate system, rotated counterclockwise through an angle θ from the xy system, where $\cos\theta = 2/\sqrt{5}$ and $\sin\theta = 1/\sqrt{5}$. The transformation T is a scaling in the x'y' system with factor 10 in the x' direction and factor 5 in the y' direction.

(d) T(1,0) = (14,2) = 14(1,0) + 2(0,1) and T(0,1) = (2,11) = 2(1,0) + 11(0,1), so the matrix representation of T relative to the standard basis is A = $\begin{bmatrix} 14 & 2 \\ 2 & 11 \end{bmatrix}$.

$\begin{vmatrix} 14-\lambda & 2 \\ 2 & 11-\lambda \end{vmatrix} = (14-\lambda)(11-\lambda) - 4 = \lambda^2 - 25\lambda + 150 = (\lambda-15)(\lambda-10)$, so the

eigenvalues are $\lambda = 15$ and $\lambda = 10$ with corresponding eigenvectors $r\begin{bmatrix} 2 \\ 1 \end{bmatrix}$ and $s\begin{bmatrix} -1 \\ 2 \end{bmatrix}$.

Orthonormal eigenvectors are $\begin{bmatrix} \frac{2}{\sqrt{5}} \\ \frac{1}{\sqrt{5}} \end{bmatrix}$ and $\begin{bmatrix} \frac{-1}{\sqrt{5}} \\ \frac{2}{\sqrt{5}} \end{bmatrix}$. Let B' be the basis

$\left\{\left(\frac{2}{\sqrt{5}}, \frac{1}{\sqrt{5}}\right), \left(\frac{-1}{\sqrt{5}}, \frac{2}{\sqrt{5}}\right)\right\}$. The matrix representation of T relative to B' is A' =

$\begin{bmatrix} 15 & 0 \\ 0 & 10 \end{bmatrix}$. The transition matrix from B' to the standard basis is P = $\begin{bmatrix} \frac{2}{\sqrt{5}} & \frac{-1}{\sqrt{5}} \\ \frac{1}{\sqrt{5}} & \frac{2}{\sqrt{5}} \end{bmatrix}$

and the orthogonal transformation is is $A' = P^tAP$.

The standard basis defines an xy coordinate system and the basis B' defines an x'y' coordinate system, rotated counterclockwise through an angle θ from the xy system, where $\cos\theta = 2/\sqrt{5}$ and $\sin\theta = 1/\sqrt{5}$. The transformation T is a scaling in the x'y' system with factor 15 in the x' direction and factor 10 in the y' direction.

20. (a) T(1,0) = (8,9) = 8(1,0) + 9(0,1) and T(0,1) = (−6,−7) = −6(1,0) − 7(0,1), so the matrix representation of T relative to the standard basis is $A = \begin{bmatrix} 8 & -6 \\ 9 & -7 \end{bmatrix}$.

$\begin{vmatrix} 8-\lambda & -6 \\ 9 & -7-\lambda \end{vmatrix} = (8-\lambda)(-7-\lambda) + 54 = \lambda^2 - \lambda - 2 = (\lambda-2)(\lambda+1)$, so the eigenvalues are $\lambda = 2$ and $\lambda = -1$ with corresponding eigenvectors $r\begin{bmatrix} 1 \\ 1 \end{bmatrix}$ and $s\begin{bmatrix} 2 \\ 3 \end{bmatrix}$. Let B' be the basis {(1,1), (2,3)}. The matrix representation of T relative to B' is

$A' = \begin{bmatrix} 2 & 0 \\ 0 & -1 \end{bmatrix}$. The transition matrix from B' to the standard basis is $P = \begin{bmatrix} 1 & 2 \\ 1 & 3 \end{bmatrix}$ and $P^{-1}AP = A'$.

The standard basis defines an xy coordinate system and the basis B' defines an x'y' coordinate system, which is not rectangular. The transformation T is a scaling in the x'y' system with factor 2 in the x' direction and factor 1 in the y' direction followed by a reflection about the x' axis.

(b) T(1,0) = (−2,−10) = −2(1,0) − 10(0,1) and T(0,1) = (2,7) = 2(1,0) + 7(0,1), so the matrix representation of T relative to the standard basis is $A = \begin{bmatrix} -2 & 2 \\ -10 & 7 \end{bmatrix}$.

$\begin{vmatrix} -2-\lambda & 2 \\ -10 & 7-\lambda \end{vmatrix} = (-2-\lambda)(7-\lambda) + 20 = \lambda^2 - 5\lambda + 6 = (\lambda-2)(\lambda-3)$, so the eigenvalues are $\lambda = 2$ and $\lambda = 3$ with corresponding eigenvectors $r\begin{bmatrix} 1 \\ 2 \end{bmatrix}$ and $s\begin{bmatrix} 2 \\ 5 \end{bmatrix}$. Let B' be the basis {(1,2), (2,5)}. The matrix representation of T relative to B' is

$A' = \begin{bmatrix} 2 & 0 \\ 0 & 3 \end{bmatrix}$. The transition matrix from B' to the standard basis is $P = \begin{bmatrix} 1 & 2 \\ 2 & 5 \end{bmatrix}$ and $P^{-1}AP = A'$.

The standard basis defines an xy coordinate system and the basis B' defines an x'y' coordinate system, which is not rectangular. The transformation T is a scaling in the x'y' system with factor 2 in the x' direction and factor 3 in the y' direction.

(c) T(1,0) = (3,2) = 3(1,0) + 2(0,1) and T(0,1) = (−4,−3) = −4(1,0) − 3(0,1), so the matrix

representation of T relative to the standard basis is $A = \begin{bmatrix} 3 & -4 \\ 2 & -3 \end{bmatrix}$.

$\begin{vmatrix} 3-\lambda & -4 \\ 2 & -3-\lambda \end{vmatrix} = (3-\lambda)(-3-\lambda) + 8 = \lambda^2 - 1 = (\lambda-1)(\lambda+1)$, so the eigenvalues are

$\lambda = 1$ and $\lambda = -1$ with corresponding eigenvectors $r\begin{bmatrix} 2 \\ 1 \end{bmatrix}$ and $s\begin{bmatrix} 1 \\ 1 \end{bmatrix}$. Let B' be the basis

{(2,1), (1,1)}. The matrix representation of T relative to B' is $A' = \begin{bmatrix} 1 & 0 \\ 0 & -1 \end{bmatrix}$. The

transition matrix from B' to the standard basis is $P = \begin{bmatrix} 2 & 1 \\ 1 & 1 \end{bmatrix}$ and $P^{-1}AP = A'$.

The standard basis defines an xy coordinate system and the basis B' defines an x'y' coordinate system, which is not rectangular. The transformation T is a reflection about the x' axis.

(d) T(1,0) = (7,−10) = 7(1,0) − 10(0,1) and T(0,1) = (5,−8) = 5(1,0) − 8(0,1), so the matrix

representation of T relative to the standard basis is $A = \begin{bmatrix} 7 & 5 \\ -10 & -8 \end{bmatrix}$.

$\begin{vmatrix} 7-\lambda & 5 \\ -10 & -8-\lambda \end{vmatrix} = (7-\lambda)(-8-\lambda) + 50 = \lambda^2 + \lambda - 6 = (\lambda-2)(\lambda+3)$, so the

eigenvalues are $\lambda = 2$ and $\lambda = -3$ with corresponding eigenvectors $r\begin{bmatrix} -1 \\ 1 \end{bmatrix}$ and $s\begin{bmatrix} 1 \\ -2 \end{bmatrix}$.

Let B' be the basis {(−1,1), (1,2)}. The matrix representation of T relative to B' is $A' = \begin{bmatrix} 2 & 0 \\ 0 & -3 \end{bmatrix}$. The transition matrix from B' to the standard basis is

$P = \begin{bmatrix} -1 & 1 \\ 1 & -2 \end{bmatrix}$ and $P^{-1}AP = A'$.

The standard basis defines an xy coordinate system and the basis B' defines an x'y' coordinate system, which is not rectangular. The transformation T is a scaling in the x'y' system with factor 2 in the x' direction and factor 3 in the y' direction followed by a reflection about the x' axis.

Exercise Set 5.4

1. (a) $x^2 + 4xy + 2y^2 = [x\ \ y]\begin{bmatrix} 1 & 2 \\ 2 & 2 \end{bmatrix}\begin{bmatrix} x \\ y \end{bmatrix}.$

(b) $3x^2 + 2xy - 4y^2 = [x\ \ y]\begin{bmatrix} 3 & 1 \\ 1 & -4 \end{bmatrix}\begin{bmatrix} x \\ y \end{bmatrix}.$

(c) $7x^2 - 6xy - y^2 = [x\ \ y]\begin{bmatrix} 7 & -3 \\ -3 & -1 \end{bmatrix}\begin{bmatrix} x \\ y \end{bmatrix}.$

(d) $2x^2 + 5xy + 3y^2 = [x\ \ y]\begin{bmatrix} 2 & 5/2 \\ 5/2 & 3 \end{bmatrix}\begin{bmatrix} x \\ y \end{bmatrix}.$

(e) $-3x^2 - 7xy + 4y^2 = [x\ \ y]\begin{bmatrix} -3 & -7/2 \\ -7/2 & 4 \end{bmatrix}\begin{bmatrix} x \\ y \end{bmatrix}.$

(f) $5x^2 + 3xy - 2y^2 = [x\ \ y]\begin{bmatrix} 5 & 3/2 \\ 3/2 & -2 \end{bmatrix}\begin{bmatrix} x \\ y \end{bmatrix}.$

2. (a) $11x^2 + 4xy + 14y^2 - 60 = [x\ \ y]\begin{bmatrix} 11 & 2 \\ 2 & 14 \end{bmatrix}\begin{bmatrix} x \\ y \end{bmatrix} - 60$. From Exercise 6(b) in Section 5.3, $C^t\begin{bmatrix} 11 & 2 \\ 2 & 14 \end{bmatrix}C = \begin{bmatrix} 15 & 0 \\ 0 & 10 \end{bmatrix}$, where $C = \begin{bmatrix} \frac{1}{\sqrt{5}} & \frac{-2}{\sqrt{5}} \\ \frac{2}{\sqrt{5}} & \frac{1}{\sqrt{5}} \end{bmatrix}$, so the given equation becomes $[x\ \ y]\,C\begin{bmatrix} 15 & 0 \\ 0 & 10 \end{bmatrix}C^t\begin{bmatrix} x \\ y \end{bmatrix} - 60 = 0$, or $[x'\ \ y']\begin{bmatrix} 15 & 0 \\ 0 & 10 \end{bmatrix}\begin{bmatrix} x' \\ y' \end{bmatrix} -60 = 0$, where $\begin{bmatrix} x' \\ y' \end{bmatrix} = C^t\begin{bmatrix} x \\ y \end{bmatrix}$. Thus $15x'^2 + 10y'^2 = 60$; i.e., $\frac{x'^2}{4} + \frac{y'^2}{6} = 1$. The graph is an ellipse with the lines $y = 2x$ and $x = -2y$ as axes.

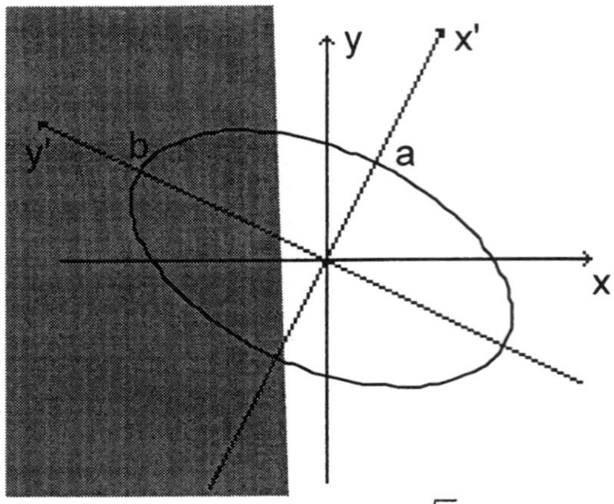

$a = 2,\ b = \sqrt{6}$

Figure for 2(a)

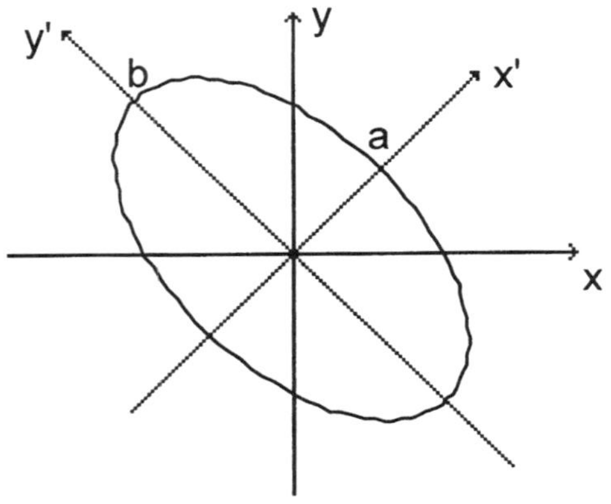

$a = \sqrt{3}\ ,\ b = \sqrt{6}$

Figure for 2(b)

(b) $3x^2 + 2xy + 3y^2 - 12 = [x\ \ y]\begin{bmatrix} 3 & 1 \\ 1 & 3 \end{bmatrix}\begin{bmatrix} x \\ y \end{bmatrix} - 12$. From Exercise 6(c) in Section 5.3, $C^t\begin{bmatrix} 3 & 1 \\ 1 & 3 \end{bmatrix}C = \begin{bmatrix} 4 & 0 \\ 0 & 2 \end{bmatrix}$, where $C = \begin{bmatrix} \frac{1}{\sqrt{2}} & \frac{-1}{\sqrt{2}} \\ \frac{1}{\sqrt{2}} & \frac{1}{\sqrt{2}} \end{bmatrix}$, so the given equation becomes $[x\ \ y]\,C\begin{bmatrix} 4 & 0 \\ 0 & 2 \end{bmatrix}C^t\begin{bmatrix} x \\ y \end{bmatrix} - 12 = 0$, or $[x'\ \ y']\begin{bmatrix} 4 & 0 \\ 0 & 2 \end{bmatrix}\begin{bmatrix} x' \\ y' \end{bmatrix} - 12 = 0$, where $\begin{bmatrix} x' \\ y' \end{bmatrix} = C^t\begin{bmatrix} x \\ y \end{bmatrix}$. Thus $4x'^2 + 2y'^2 = 12$; i.e., $\frac{x'^2}{3} + \frac{y'^2}{6} = 1$. The graph is an ellipse with the lines $y = x$ and $y = -x$ as axes.

(c) $x^2 - 6xy + y^2 - 8 = [x\ \ y]\begin{bmatrix} 1 & -3 \\ -3 & 1 \end{bmatrix}\begin{bmatrix} x \\ y \end{bmatrix} - 8$. The symmetric matrix has eigenvalues $\lambda = -2$ and $\lambda = 4$ with corresponding orthonormal eigenvectors $\begin{bmatrix} \frac{1}{\sqrt{2}} \\ \frac{1}{\sqrt{2}} \end{bmatrix}$ and $\begin{bmatrix} \frac{-1}{\sqrt{2}} \\ \frac{1}{\sqrt{2}} \end{bmatrix}$, so $C^t\begin{bmatrix} 1 & -3 \\ -3 & 1 \end{bmatrix}C = \begin{bmatrix} -2 & 0 \\ 0 & 4 \end{bmatrix}$, where $C = \begin{bmatrix} \frac{1}{\sqrt{2}} & \frac{-1}{\sqrt{2}} \\ \frac{1}{\sqrt{2}} & \frac{1}{\sqrt{2}} \end{bmatrix}$, so the given equation becomes $[x\ \ y]\, C \begin{bmatrix} -2 & 0 \\ 0 & 4 \end{bmatrix} C^t \begin{bmatrix} x \\ y \end{bmatrix} - 8 = 0$, or

$[x'\ \ y'] \begin{bmatrix} -2 & 0 \\ 0 & 4 \end{bmatrix} \begin{bmatrix} x' \\ y' \end{bmatrix} - 8 = 0$, where $\begin{bmatrix} x' \\ y' \end{bmatrix} = C^t \begin{bmatrix} x \\ y \end{bmatrix}$. Thus $-2x'^2 + 4y'^2 = 8$; i.e.,

$\frac{-x'^2}{4} + \frac{y'^2}{2} = 1$. The graph is a hyperbola with the lines y = x and y = −x as axes.

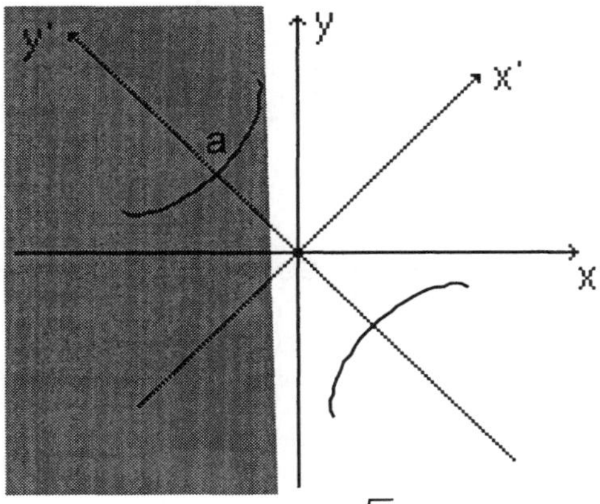

$a = \sqrt{2}$

Figure for 2(c)

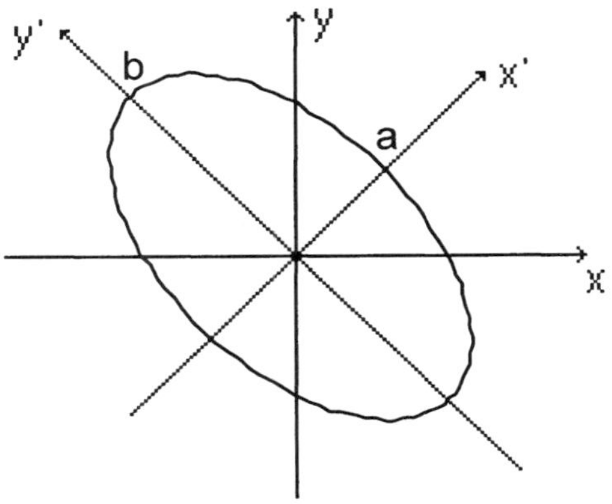

$a = \sqrt{5/6}$, $b = \sqrt{5/2}$

Figure for 2(d)

(d) $4x^2 + 4xy + 4y^2 - 5 = [x\ \ y] \begin{bmatrix} 4 & 2 \\ 2 & 4 \end{bmatrix} \begin{bmatrix} x \\ y \end{bmatrix} - 5$. The symmetric matrix has

eigenvalues $\lambda = 6$ and $\lambda = 2$ with corresponding orthonormal eigenvectors

$\begin{bmatrix} \frac{1}{\sqrt{2}} \\ \frac{1}{\sqrt{2}} \end{bmatrix}$ and $\begin{bmatrix} \frac{-1}{\sqrt{2}} \\ \frac{1}{\sqrt{2}} \end{bmatrix}$, so $C^t \begin{bmatrix} 4 & 2 \\ 2 & 4 \end{bmatrix} C = \begin{bmatrix} 6 & 0 \\ 0 & 2 \end{bmatrix}$, where $C = \begin{bmatrix} \frac{1}{\sqrt{2}} & \frac{-1}{\sqrt{2}} \\ \frac{1}{\sqrt{2}} & \frac{1}{\sqrt{2}} \end{bmatrix}$, so

the given equation becomes $[x\ \ y]\, C \begin{bmatrix} 6 & 0 \\ 0 & 2 \end{bmatrix} C^t \begin{bmatrix} x \\ y \end{bmatrix} - 5 = 0$, or

$[x'\ \ y'] \begin{bmatrix} 6 & 0 \\ 0 & 2 \end{bmatrix} \begin{bmatrix} x' \\ y' \end{bmatrix} - 5 = 0$, where $\begin{bmatrix} x' \\ y' \end{bmatrix} = C^t \begin{bmatrix} x \\ y \end{bmatrix}$. Thus $6x'^2 + 2y'^2 = 5$,

i.e. $\frac{6x'^2}{5} + \frac{2y'^2}{5} = 1$. The graph is an ellipse with the lines y = x and y = −x as axes.

3. (a) $a_n = a_{n-1} + 2a_{n-2}$, $a_1 = 1$, and $a_2 = 2$. Let $b_n = a_{n-1}$. Thus $\begin{bmatrix} a_n \\ b_n \end{bmatrix} = \begin{bmatrix} 1 & 2 \\ 1 & 0 \end{bmatrix}\begin{bmatrix} a_{n-1} \\ b_{n-1} \end{bmatrix}$

The matrix has eigenvalues $\lambda = -1$ and $\lambda = 2$ with corresponding eigenvectors

$\begin{bmatrix} 1 \\ -1 \end{bmatrix}$ and $\begin{bmatrix} 2 \\ 1 \end{bmatrix}$. Let $C = \begin{bmatrix} 1 & 2 \\ -1 & 1 \end{bmatrix}$. $C^{-1} = \frac{1}{3}\begin{bmatrix} 1 & -2 \\ 1 & 1 \end{bmatrix}$, and

$\begin{bmatrix} a_n \\ b_n \end{bmatrix} = C\begin{bmatrix} (-1)^{n-2} & 0 \\ 0 & 2^{n-2} \end{bmatrix}C^{-1}\begin{bmatrix} 2 \\ 1 \end{bmatrix} = \frac{1}{3}\begin{bmatrix} 2^n + 2^{n-1} \\ 2^{n-1} + 2^{n-2} \end{bmatrix}$, so $a_n = \frac{1}{3}(2^n + 2^{n-1}) = 2^{n-1}$. $a_{10} = 2^9 = 512$.

(b) $a_n = 2a_{n-1} + 3a_{n-2}$, $a_1 = 1$, and $a_2 = 3$. Let $b_n = a_{n-1}$. Thus $\begin{bmatrix} a_n \\ b_n \end{bmatrix} = \begin{bmatrix} 2 & 3 \\ 1 & 0 \end{bmatrix}\begin{bmatrix} a_{n-1} \\ b_{n-1} \end{bmatrix}$

The matrix has eigenvalues $\lambda = -1$ and $\lambda = 3$ with corresponding eigenvectors

$\begin{bmatrix} 1 \\ -1 \end{bmatrix}$ and $\begin{bmatrix} 3 \\ 1 \end{bmatrix}$. Let $C = \begin{bmatrix} 1 & 3 \\ -1 & 1 \end{bmatrix}$. $C^{-1} = \frac{1}{4}\begin{bmatrix} 1 & -3 \\ 1 & 1 \end{bmatrix}$, and

$\begin{bmatrix} a_n \\ b_n \end{bmatrix} = C\begin{bmatrix} (-1)^{n-2} & 0 \\ 0 & 3^{n-2} \end{bmatrix}C^{-1}\begin{bmatrix} 3 \\ 1 \end{bmatrix} = \frac{1}{4}\begin{bmatrix} 3^n + 3^{n-1} \\ 3^{n-1} + 3^{n-2} \end{bmatrix}$, so $a_n = \frac{1}{4}(3^n + 3^{n-1}) = 3^{n-1}$. $a_9 = 3^8 = 6561$.

(c) $a_n = 3a_{n-1} + 4a_{n-2}$, $a_1 = 1$, and $a_2 = -1$. Let $b_n = a_{n-1}$. Thus $\begin{bmatrix} a_n \\ b_n \end{bmatrix} = \begin{bmatrix} 3 & 4 \\ 1 & 0 \end{bmatrix}\begin{bmatrix} a_{n-1} \\ b_{n-1} \end{bmatrix}$

The matrix has eigenvalues $\lambda = -1$ and $\lambda = 4$ with corresponding eigenvectors

$\begin{bmatrix} 1 \\ -1 \end{bmatrix}$ and $\begin{bmatrix} 4 \\ 1 \end{bmatrix}$. Let $C = \begin{bmatrix} 1 & 4 \\ -1 & 1 \end{bmatrix}$. $C^{-1} = \frac{1}{5}\begin{bmatrix} 1 & -4 \\ 1 & 1 \end{bmatrix}$, and

$\begin{bmatrix} a_n \\ b_n \end{bmatrix} = C\begin{bmatrix} (-1)^{n-2} & 0 \\ 0 & 4^{n-2} \end{bmatrix} C^{-1}\begin{bmatrix} -1 \\ 1 \end{bmatrix} = \begin{bmatrix} (-1)^{n-1} \\ (-1)^{n-2} \end{bmatrix}$, so $a_n = (-1)^{n-1}$.

$a_{12} = (-1)^{11} = -1$.

4. $a_n = a_{n-1} + a_{n-2}$, $a_1 = 1$, and $a_2 = 1$. Let $b_n = a_{n-1}$. Thus $\begin{bmatrix} a_n \\ b_n \end{bmatrix} = \begin{bmatrix} 1 & 1 \\ 1 & 0 \end{bmatrix}\begin{bmatrix} a_{n-1} \\ b_{n-1} \end{bmatrix}$.

The matrix has characteristic polynomial $\lambda^2 - \lambda - 1$, so the eigenvalues are $\lambda = \frac{1 \pm \sqrt{5}}{2}$

with corresponding eigenvectors $\begin{bmatrix} \frac{1+\sqrt{5}}{2} \\ 1 \end{bmatrix}$ and $\begin{bmatrix} \frac{1-\sqrt{5}}{2} \\ 1 \end{bmatrix}$. Let $C = \begin{bmatrix} \frac{1+\sqrt{5}}{2} & \frac{1-\sqrt{5}}{2} \\ 1 & 1 \end{bmatrix}$.

$C^{-1} = \frac{1}{\sqrt{5}}\begin{bmatrix} 1 & \frac{-1+\sqrt{5}}{2} \\ -1 & \frac{1-\sqrt{5}}{2} \end{bmatrix}$, and $\begin{bmatrix} a_n \\ b_n \end{bmatrix} = C\begin{bmatrix} \left(\frac{1+\sqrt{5}}{2}\right)^{n-2} & 0 \\ 0 & \left(\frac{1-\sqrt{5}}{2}\right)^{n-2} \end{bmatrix} C^{-1}\begin{bmatrix} 1 \\ 1 \end{bmatrix}$

$= \frac{1}{\sqrt{5}}\begin{bmatrix} \left(\frac{1+\sqrt{5}}{2}\right)^{n} - \left(\frac{1-\sqrt{5}}{2}\right)^{n} \\ \left(\frac{1+\sqrt{5}}{2}\right)^{n-1} - \left(\frac{1-\sqrt{5}}{2}\right)^{n-1} \end{bmatrix}$, so $a_n = \frac{1}{\sqrt{5}}\left(\left(\frac{1+\sqrt{5}}{2}\right)^{n} - \left(\frac{1-\sqrt{5}}{2}\right)^{n}\right)$ and

$\frac{a_n}{a_{n-1}} = \frac{\left(\frac{1+\sqrt{5}}{2}\right)^{n} - \left(\frac{1-\sqrt{5}}{2}\right)^{n}}{\left(\frac{1+\sqrt{5}}{2}\right)^{n-1} - \left(\frac{1-\sqrt{5}}{2}\right)^{n-1}}$. We multiply the numerator by $\frac{\left(\frac{1+\sqrt{5}}{2}\right)}{\left(\frac{1+\sqrt{5}}{2}\right)^{n}}$

and the denominator by $\frac{1}{\left(\frac{1+\sqrt{5}}{2}\right)^{n-1}}$ to get $\frac{a_n}{a_{n-1}} = \frac{1+\sqrt{5}}{2}\ \frac{1 - \left(\frac{-3+\sqrt{5}}{2}\right)^{n}}{1 - \left(\frac{-3+\sqrt{5}}{2}\right)^{n-1}}$.

As n increases, $\left(\frac{-3+\sqrt{5}}{2}\right)^n$ approaches zero, so $\frac{a_n}{a_{n-1}}$ approaches $\frac{1+\sqrt{5}}{2}$.

5. $m\ddot{x}_1 = \frac{1}{a}(-2Tx_1 + Tx_2)$ and $m\ddot{x}_2 = \frac{1}{a}(Tx_1 - 2Tx_2)$, so $\begin{bmatrix} \ddot{x}_1 \\ \ddot{x}_2 \end{bmatrix} = \frac{T}{ma}\begin{bmatrix} -2 & 1 \\ 1 & -2 \end{bmatrix}\begin{bmatrix} x_1 \\ x_2 \end{bmatrix}$

The matrix has eigenvalues $\lambda = -3$ and $\lambda = -1$ with corresponding eigenvectors $\begin{bmatrix} 1 \\ -1 \end{bmatrix}$

and $\begin{bmatrix} 1 \\ 1 \end{bmatrix}$. $C^{-1}AC = \begin{bmatrix} -3 & 0 \\ 0 & -1 \end{bmatrix}$, where $C = \begin{bmatrix} 1 & 1 \\ -1 & 1 \end{bmatrix}$. Let $\begin{bmatrix} x_1' \\ x_2' \end{bmatrix} = C^{-1}\begin{bmatrix} x_1 \\ x_2 \end{bmatrix}$.

$\begin{bmatrix} \ddot{x}_1' \\ \ddot{x}_2' \end{bmatrix} = \frac{T}{ma}\begin{bmatrix} -3 & 0 \\ 0 & -1 \end{bmatrix}\begin{bmatrix} x_1' \\ x_2' \end{bmatrix}$, so that $\ddot{x}_1' = \frac{-3T}{ma}x_1'$ and $\ddot{x}_2' = \frac{-T}{ma}x_2'$.

Solutions of these equations are $x_1' = b_1\cos(\alpha_1 t + \gamma_1)$ and $x_2' = b_2\cos(\alpha_2 t + \gamma_2)$, where $\alpha_1 = (3T/ma)^{1/2}$ and $\alpha_2 = (T/ma)^{1/2}$. We now must solve for x_1 and x_2.

$\begin{bmatrix} x_1 \\ x_2 \end{bmatrix} = C\begin{bmatrix} x_1' \\ x_2' \end{bmatrix} = \begin{bmatrix} 1 & 1 \\ -1 & 1 \end{bmatrix}\begin{bmatrix} x_1' \\ x_2' \end{bmatrix} = x_1'\begin{bmatrix} 1 \\ -1 \end{bmatrix} + x_2'\begin{bmatrix} 1 \\ 1 \end{bmatrix}$

$= b_1\cos(\alpha_1 t + \gamma_1)\begin{bmatrix} 1 \\ -1 \end{bmatrix} + b_2\cos(\alpha_2 t + \gamma_2)\begin{bmatrix} 1 \\ 1 \end{bmatrix}$. Thus the normal modes are

mode 1: $\begin{bmatrix} x_1 \\ x_2 \end{bmatrix} = \cos(\alpha_1 t + \gamma_1)\begin{bmatrix} 1 \\ -1 \end{bmatrix}$, where $\alpha_1 = \left(\frac{3T}{ma}\right)^{1/2}$, and

mode 2: $\begin{bmatrix} x_1 \\ x_2 \end{bmatrix} = \cos(\alpha_2 t + \gamma_2)\begin{bmatrix} 1 \\ 1 \end{bmatrix}$, where $\alpha_2 = \left(\frac{T}{ma}\right)^{1/2}$.

6. $M\ddot{x}_1 = -5x_1 + 2x_2$ and $M\ddot{x}_2 = 2x_1 - 2x_2$, so $\begin{bmatrix} \ddot{x}_1 \\ \ddot{x}_2 \end{bmatrix} = \frac{1}{M}\begin{bmatrix} -5 & 2 \\ 2 & -2 \end{bmatrix}\begin{bmatrix} x_1 \\ x_2 \end{bmatrix}$.

The matrix has eigenvalues $\lambda = -1$ and $\lambda = -6$ with corresponding eigenvectors $\begin{bmatrix} 1 \\ 2 \end{bmatrix}$

and $\begin{bmatrix} 2 \\ -1 \end{bmatrix}$. $C^{-1}AC = \begin{bmatrix} -1 & 0 \\ 0 & -6 \end{bmatrix}$, where $C = \begin{bmatrix} 1 & 2 \\ 2 & -1 \end{bmatrix}$. Let $\begin{bmatrix} x_1' \\ x_2' \end{bmatrix} = C^{-1}\begin{bmatrix} x_1 \\ x_2 \end{bmatrix}$.

$\begin{bmatrix} \ddot{x}_1' \\ \ddot{x}_2' \end{bmatrix} = \frac{1}{M}\begin{bmatrix} -1 & 0 \\ 0 & -6 \end{bmatrix}\begin{bmatrix} x_1' \\ x_2' \end{bmatrix}$, so that $\ddot{x}_1' = \frac{-1}{M}\ x_1'$ and $\ddot{x}_2' = \frac{-6}{M}\ x_2'$.

Solutions of these equations are $x_1' = b_1 \cos(\alpha_1 t + \gamma_1)$ and $x_2' = b_2 \cos(\alpha_2 t + \gamma_2)$, where $\alpha_1 = (1/M)^{1/2}$ and $\alpha_2 = (6/M)^{1/2}$. We now must solve for x_1 and x_2.

$\begin{bmatrix} x_1 \\ x_2 \end{bmatrix} = C\begin{bmatrix} x_1' \\ x_2' \end{bmatrix} = \begin{bmatrix} 1 & 2 \\ 2 & -1 \end{bmatrix}\begin{bmatrix} x_1' \\ x_2' \end{bmatrix} = x_1'\begin{bmatrix} 1 \\ 2 \end{bmatrix} + x_2'\begin{bmatrix} 2 \\ -1 \end{bmatrix}$

$= b_1 \cos(\alpha_1 t + \gamma_1)\begin{bmatrix} 1 \\ 2 \end{bmatrix} + b_2 \cos(\alpha_2 t + \gamma_2)\begin{bmatrix} 2 \\ -1 \end{bmatrix}$.

7. $m\ddot{x}_1 = -3x_1 + 2x_2$ and $m\ddot{x}_2 = 2x_1 - 5x_2$, so $\begin{bmatrix} \ddot{x}_1 \\ \ddot{x}_2 \end{bmatrix} = \frac{1}{m}\begin{bmatrix} -3 & 2 \\ 2 & -5 \end{bmatrix}\begin{bmatrix} x_1 \\ x_2 \end{bmatrix}$.

The matrix has eigenvalues $\lambda = -4+\sqrt{5}$ and $\lambda = -4-\sqrt{5}$ with corresponding

eigenvectors $\begin{bmatrix} 1 \\ \frac{-1+\sqrt{5}}{2} \end{bmatrix}$ and $\begin{bmatrix} 1 \\ \frac{-1-\sqrt{5}}{2} \end{bmatrix}$. $C^{-1}AC = \begin{bmatrix} -4+\sqrt{5} & 0 \\ 0 & -4-\sqrt{5} \end{bmatrix}$, where

$C = \begin{bmatrix} 1 & 1 \\ \frac{-1+\sqrt{5}}{2} & \frac{-1-\sqrt{5}}{2} \end{bmatrix}$. Let $\begin{bmatrix} x_1' \\ x_2' \end{bmatrix} = C^{-1}\begin{bmatrix} x_1 \\ x_2 \end{bmatrix}$.

$\begin{bmatrix} \ddot{x}_1' \\ \ddot{x}_2' \end{bmatrix} = \frac{1}{m}\begin{bmatrix} -4+\sqrt{5} & 0 \\ 0 & -4-\sqrt{5} \end{bmatrix}\begin{bmatrix} x_1' \\ x_2' \end{bmatrix}$, so that $\ddot{x}_1' = \frac{-4+\sqrt{5}}{m}\ x_1'$ and $\ddot{x}_2' = \frac{-4-\sqrt{5}}{m}\ x_2'$.

Solutions of these equations are $x_1' = b_1 \cos(\alpha_1 t + \gamma_1)$ and $x_2' = b_2 \cos(\alpha_2 t + \gamma_2)$, where $\alpha_1 = \left(\frac{4-\sqrt{5}}{m}\right)^{1/2}$ and $\alpha_2 = \left(\frac{4+\sqrt{5}}{m}\right)^{1/2}$. We now must solve for x_1 and x_2.

$$\begin{bmatrix} x_1 \\ x_2 \end{bmatrix} = C\begin{bmatrix} x_1' \\ x_2' \end{bmatrix} = \begin{bmatrix} 1 & 1 \\ \frac{-1+\sqrt{5}}{2} & \frac{-1-\sqrt{5}}{2} \end{bmatrix}\begin{bmatrix} x_1' \\ x_2' \end{bmatrix} = x_1'\begin{bmatrix} 1 \\ \frac{-1+\sqrt{5}}{2} \end{bmatrix} + x_2'\begin{bmatrix} 1 \\ \frac{-1-\sqrt{5}}{2} \end{bmatrix}$$

$$= b_1 \cos(\alpha_1 t + \gamma_1)\begin{bmatrix} 1 \\ \frac{-1+\sqrt{5}}{2} \end{bmatrix} + b_2 \cos(\alpha_2 t + \gamma_2)\begin{bmatrix} 1 \\ \frac{-1-\sqrt{5}}{2} \end{bmatrix}.$$

8. In the Figure below:
$x' = OA' = OC + CA' = OC + DP = OB\sin\theta + BP\cos\theta = y\sin\theta + x\cos\theta$
$y' = OB' = PA' = BC - BD = OB\cos\theta - BP\sin\theta = y\cos\theta - x\sin\theta$

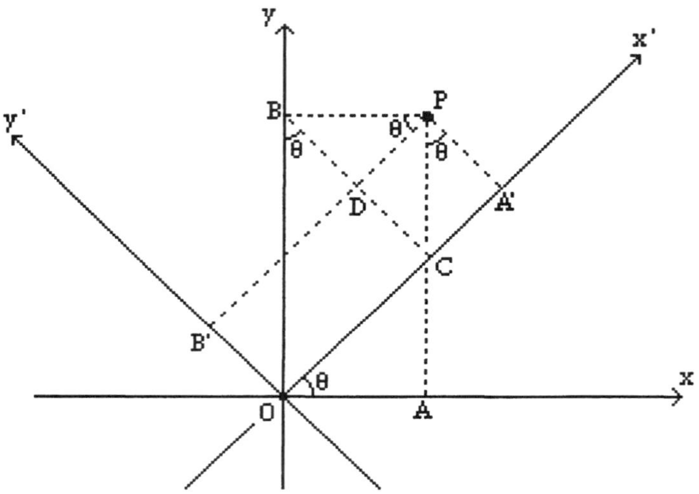

Figure for Exercise 8

Chapter 5 Review Exercises

1. $(-1,18) = a(1,3) + b(-1,4)$. Thus, $-1 = a - b$ and $18 = 3a + 4b$.

This system of equations has the unique solution $a = 2$, $b = 3$, so the coordinate vector is $\begin{bmatrix} a \\ b \end{bmatrix} = \begin{bmatrix} 2 \\ 3 \end{bmatrix}$.

2. $3x^2 + 2x - 13 = a(x^2 + 1) + b(x + 2) + c(x - 3)$.

Thus, 3 = a, 2 = b + c, and −13 = a + 2b − 3c. This system of equations has the unique solution a = 3, b = −2, c = 4, so the coordinate vector is $\begin{bmatrix} a \\ b \\ c \end{bmatrix} = \begin{bmatrix} 3 \\ -2 \\ 4 \end{bmatrix}$.

3. (0,5,−15)·(0,1,0) = 5, (0,5,−15)·(−3/5,0,4/5) = −12, (0,5,−15)·(4/5,0,3/5) = −9.

Thus the coordinate vector is $\begin{bmatrix} 5 \\ -12 \\ -9 \end{bmatrix}$.

4. $P = \begin{bmatrix} 1 & 5 \\ 3 & 2 \end{bmatrix}$, and $\mathbf{u}_{B'} = P\mathbf{u}_B = \begin{bmatrix} 8 \\ 11 \end{bmatrix}$,

$\mathbf{v}_{B'} = P\mathbf{v}_B = \begin{bmatrix} -5 \\ 11 \end{bmatrix}$, and $\mathbf{w}_{B'} = P\mathbf{w}_B = \begin{bmatrix} 9 \\ 14 \end{bmatrix}$.

5. The transition matrix from B to the standard basis is $R = \begin{bmatrix} -1 & 2 \\ 2 & 1 \end{bmatrix}$ and the transition matrix from B' to the standard basis is $Q = \begin{bmatrix} 4 & -3 \\ 3 & 2 \end{bmatrix}$.

$P = Q^{-1}R = \frac{1}{17}\begin{bmatrix} 2 & 3 \\ -3 & 4 \end{bmatrix}\begin{bmatrix} -1 & 2 \\ 2 & 1 \end{bmatrix} = \frac{1}{17}\begin{bmatrix} 4 & 7 \\ 11 & -2 \end{bmatrix}$.

$\mathbf{u}_{B'} = P\,\mathbf{u}_B = \frac{1}{17}\begin{bmatrix} 4 & 7 \\ 11 & -2 \end{bmatrix}\begin{bmatrix} 4 \\ 1 \end{bmatrix} = \frac{1}{17}\begin{bmatrix} 23 \\ 42 \end{bmatrix}$.

6. The matrix for the standard basis is $A = \begin{bmatrix} 3 & -1 \\ 2 & 4 \end{bmatrix}$. $A\begin{bmatrix} 2 \\ 7 \end{bmatrix} = \begin{bmatrix} -1 \\ 32 \end{bmatrix}$.
Thus, T(2,7) = (−1,32).

7. The matrix of T with respect to the given bases is $A = \begin{bmatrix} 1 & 3 \\ 5 & -1 \\ -2 & 2 \end{bmatrix}$.

$A\begin{bmatrix} 2 \\ -3 \end{bmatrix} = \begin{bmatrix} -7 \\ 13 \\ -10 \end{bmatrix}$. Thus, $T(\mathbf{u}) = -7\mathbf{v}_1 + 13\mathbf{v}_2 - 10\mathbf{v}_3$.

8. $T(1,0,0) = (2,0)$, $T(0,1,0) = (0,-3)$, and $T(0,0,1) = (0,0)$. Matrix of T with respect to the standard bases is $A = \begin{bmatrix} 2 & 0 & 0 \\ 0 & -3 & 0 \end{bmatrix}$. $A\begin{bmatrix} 1 \\ 2 \\ 3 \end{bmatrix} = \begin{bmatrix} 2 \\ -6 \end{bmatrix}$. Thus, $T(1,2,3) = (2,-6)$.

9. $T(x^2) = x^2$, $T(x) = -x^2$, and $T(1) = 2x$, so $A = \begin{bmatrix} 1 & -1 & 0 \\ 0 & 0 & 2 \\ 0 & 0 & 0 \end{bmatrix}$. $A\begin{bmatrix} 2 \\ -1 \\ 3 \end{bmatrix} = \begin{bmatrix} 3 \\ 6 \\ 0 \end{bmatrix}$.

Thus, $T(2x^2 - x + 3) = 3x^2 + 6x$.

10. $T(1,0) = (3,1)$ & $T(0,1) = (0,-1)$. Matrix of T for standard basis is $A = \begin{bmatrix} 3 & 0 \\ 1 & -1 \end{bmatrix}$.

The transition matrix from the basis {(1,2), (2,3)} to the standard basis is $P = \begin{bmatrix} 1 & 2 \\ 2 & 3 \end{bmatrix}$.

Thus $B = P^{-1}AP = \begin{bmatrix} -3 & 2 \\ 2 & -1 \end{bmatrix}\begin{bmatrix} 3 & 0 \\ 1 & -1 \end{bmatrix}\begin{bmatrix} 1 & 2 \\ 2 & 3 \end{bmatrix} = \begin{bmatrix} -11 & -20 \\ 7 & 13 \end{bmatrix}$ is the matrix of T with respect to the basis {(1,2), (2,3)}.

11. $C^{-1}AC = \begin{bmatrix} 1 & -1 \\ -1 & 2 \end{bmatrix}\begin{bmatrix} 4 & -2 \\ 1 & 1 \end{bmatrix}\begin{bmatrix} 2 & 1 \\ 1 & 1 \end{bmatrix} = \begin{bmatrix} 3 & 0 \\ 0 & 2 \end{bmatrix}$.

12. $\begin{vmatrix} 1-\lambda & 1 \\ -2 & 4-\lambda \end{vmatrix} = (3-\lambda)(2-\lambda)$, so the eigenvalues are $\lambda = 3$ and $\lambda = 2$. For $\lambda = 3$, the eigenvectors are $r\begin{bmatrix} 1 \\ 2 \end{bmatrix}$ and for $\lambda = 2$, the eigenvectors are $s\begin{bmatrix} 1 \\ 1 \end{bmatrix}$. Let $C = \begin{bmatrix} 1 & 1 \\ 2 & 1 \end{bmatrix}$.

$C^{-1}AC = \begin{bmatrix} -1 & 1 \\ 2 & -1 \end{bmatrix}\begin{bmatrix} 1 & 1 \\ -2 & 4 \end{bmatrix}\begin{bmatrix} 1 & 1 \\ 2 & 1 \end{bmatrix} = \begin{bmatrix} 3 & 0 \\ 0 & 2 \end{bmatrix}$.

13. $\begin{vmatrix} 7-\lambda & -2 & 1 \\ -2 & 10-\lambda & -2 \\ 1 & -2 & 7-\lambda \end{vmatrix} = (6-\lambda)(6-\lambda)(12-\lambda)$, so the eigenvalues are $\lambda = 6$ and $\lambda = 12$.

For $\lambda = 6$, the eigenvectors are vectors of the form $r\begin{bmatrix} 1 \\ 0 \\ -1 \end{bmatrix} + s\begin{bmatrix} 1 \\ 1 \\ 1 \end{bmatrix}$. For $\lambda = 12$, the eigenvectors are vectors of the form $t\begin{bmatrix}1\\-2\\1\end{bmatrix}$. Orthonormal eigenvectors are

$\begin{bmatrix}1/\sqrt{2}\\0\\-1/\sqrt{2}\end{bmatrix}$, $\begin{bmatrix}1/\sqrt{3}\\1/\sqrt{3}\\1/\sqrt{3}\end{bmatrix}$, $\begin{bmatrix}1/\sqrt{6}\\-2/\sqrt{6}\\1/\sqrt{6}\end{bmatrix}$. Let $C = \begin{bmatrix}1/\sqrt{2} & 1/\sqrt{3} & 1/\sqrt{6}\\0 & 1/\sqrt{3} & -2/\sqrt{6}\\-1/\sqrt{2} & 1/\sqrt{3} & 1/\sqrt{6}\end{bmatrix}$.

$$\begin{bmatrix}1/\sqrt{2} & 0 & -1/\sqrt{2}\\1/\sqrt{3} & 1/\sqrt{3} & 1/\sqrt{3}\\1/\sqrt{6} & -2/\sqrt{6} & 1/\sqrt{6}\end{bmatrix}\begin{bmatrix}7 & -2 & 1\\-2 & 10 & -2\\1 & -2 & 7\end{bmatrix}\begin{bmatrix}1/\sqrt{2} & 1/\sqrt{3} & 1/\sqrt{6}\\0 & 1/\sqrt{3} & -2/\sqrt{6}\\-1/\sqrt{2} & 1/\sqrt{3} & 1/\sqrt{6}\end{bmatrix} = \begin{bmatrix}6 & 0 & 0\\0 & 6 & 0\\0 & 0 & 12\end{bmatrix}.$$

14. $\begin{vmatrix}a-\lambda & b\\b & c-\lambda\end{vmatrix} = (a-\lambda)(c-\lambda) - b^2 = \lambda^2 - (a+c)\lambda + ac - b^2$. The characteristic equation $\lambda^2 - (a+c)\lambda + ac - b^2 = 0$ has roots $\lambda = \frac{1}{2}(a+c \pm \sqrt{D})$ where $D = (a-c)^2 + 4b^2$ (from the quadratic formula). D is nonnegative for all values of a, b, and c, so the roots are real.

15. If A is symmetric then, A can be diagonalized, $D = C^{-1}AC$, where the diagonal elements of D are the eigenvalues of A. If A has only one eigenvalue, λ, then $D = \lambda I$, so $\lambda I = C^{-1}AC$. Thus $A = CC^{-1}ACC^{-1} = C\lambda IC^{-1} = \lambda CIC^{-1} = \lambda CC^{-1} = \lambda I$.

16. $-x^2 - 16xy + 11y^2 - 30 = [x\ y]\begin{bmatrix}-1 & -8\\-8 & 11\end{bmatrix}\begin{bmatrix}x\\y\end{bmatrix} - 30$. The symmetric matrix has eigenvalues $\lambda = 15$ and $\lambda = -5$ with corresponding orthonormal eigenvectors

$\begin{bmatrix}\frac{1}{\sqrt{5}}\\\frac{-2}{\sqrt{5}}\end{bmatrix}$ and $\begin{bmatrix}\frac{2}{\sqrt{5}}\\\frac{1}{\sqrt{5}}\end{bmatrix}$, so $C^t\begin{bmatrix}-1 & -8\\-8 & 11\end{bmatrix}C = \begin{bmatrix}15 & 0\\0 & -5\end{bmatrix}$, where $C = \begin{bmatrix}\frac{1}{\sqrt{5}} & \frac{2}{\sqrt{5}}\\\frac{-2}{\sqrt{5}} & \frac{1}{\sqrt{5}}\end{bmatrix}$, so the given equation becomes $[x\ y]\,C\begin{bmatrix}15 & 0\\0 & -5\end{bmatrix}C^t\begin{bmatrix}x\\y\end{bmatrix} - 30 = 0$, or

$[x'\ y']\begin{bmatrix}15 & 0\\0 & -5\end{bmatrix}\begin{bmatrix}x'\\y'\end{bmatrix} - 30 = 0$, where $[x'\ y'] = [x\ y]\,C$. Thus $15x'^2 - 5y'^2 = 30$;

i.e., $\frac{x'^2}{2} - \frac{y'^2}{6} = 1$. The graph is a hyperbola with axes y = −2x and x = 2y.

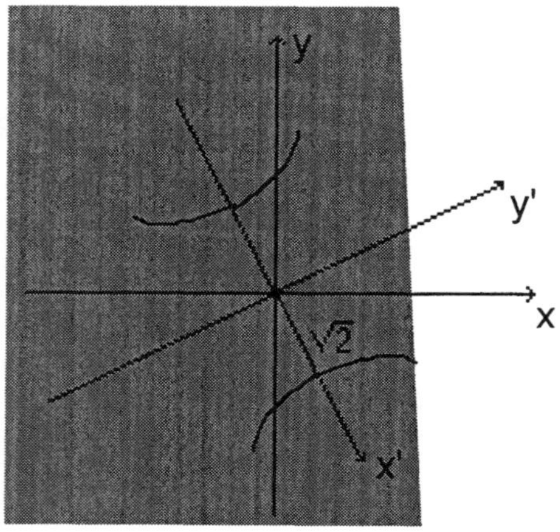

17. $a_n = 4a_{n-1} + 5a_{n-2}$, $a_1 = 3$, and $a_2 = 2$. Let $b_n = a_{n-1}$. Thus $\begin{bmatrix} a_n \\ b_n \end{bmatrix} = \begin{bmatrix} 4 & 5 \\ 1 & 0 \end{bmatrix}\begin{bmatrix} a_{n-1} \\ b_{n-1} \end{bmatrix}$.

The matrix has eigenvalues λ = −1 and λ = 5 with corresponding eigenvectors

$\begin{bmatrix} 1 \\ -1 \end{bmatrix}$ and $\begin{bmatrix} 5 \\ 1 \end{bmatrix}$. Let $C = \begin{bmatrix} 1 & 5 \\ -1 & 1 \end{bmatrix}$. $C^{-1} = \frac{1}{6}\begin{bmatrix} 1 & -5 \\ 1 & 1 \end{bmatrix}$, and

$\begin{bmatrix} a_n \\ b_n \end{bmatrix} = C\begin{bmatrix} (-1)^{n-2} & 0 \\ 0 & 5^{n-2} \end{bmatrix} C^{-1}\begin{bmatrix} 2 \\ 3 \end{bmatrix} = \frac{1}{6}\begin{bmatrix} 13(-1)^{n-1} + 5^n \\ 13(-1)^{n-2} + 5^{n-1} \end{bmatrix}$, so

$a_n = \frac{1}{6}(13(-1)^{n-1} + 5^n)$. $a_{12} = \frac{1}{6}(13(-1)^{11} + 5^{12}) = \frac{1}{6}(-13 + 244140625)$

$= 40{,}690{,}102$.

Chapter 6

Exercise Set 6.1

1. $\mathbf{u} = (x_1, x_2)$, $\mathbf{v} = (y_1, y_2)$, $\mathbf{w} = (z_1, z_2)$, and c is a scalar.

   $<\mathbf{u},\mathbf{v}> = 4x_1 y_1 + 9x_2 y_2 = 4y_1 x_1 + 9y_2 x_2 = <\mathbf{v},\mathbf{u}>$,

   $<\mathbf{u}+\mathbf{v},\mathbf{w}> = 4(x_1 + y_1)z_1 + 9(x_2 + y_2)z_2 = 4x_1 z_1 + 4y_1 z_1 + 9x_2 z_2 + 9y_2 z_2$

   $= 4x_1 z_1 + 9x_2 z_2 + 4y_1 z_1 + 9y_2 z_2 = <\mathbf{u},\mathbf{w}> + <\mathbf{v},\mathbf{w}>$,

   $<c\mathbf{u},\mathbf{v}> = 4cx_1 y_1 + 9cx_2 y_2 = c(4x_1 y_1 + 9x_2 y_2) = c<\mathbf{u},\mathbf{v}>$, and

   $<\mathbf{u},\mathbf{u}> = 4x_1 x_1 + 9x_2 x_2 = 4x_1^2 + 9x_2^2 \geq 0$ and equality holds if and only if $4x_1^2 = 0$ and $9x_2^2 = 0$, i.e., if and only if $x_1 = 0$ and $x_2 = 0$. Thus the given function is an inner product on $\mathbf{R}^2$.

2. $\mathbf{u} = (x_1, x_2, x_3)$, $\mathbf{v} = (y_1, y_2, y_3)$, $\mathbf{w} = (z_1, z_2, z_3)$, and c is a scalar.

   $<\mathbf{u},\mathbf{v}> = x_1 y_1 + 2x_2 y_2 + 4x_3 y_3 = y_1 x_1 + 2y_2 x_2 + 4y_3 x_3 = <\mathbf{v},\mathbf{u}>$,

   $<\mathbf{u}+\mathbf{v},\mathbf{w}> = (x_1 + y_1)z_1 + 2(x_2 + y_2)z_2 + 4(x_3 + y_3)z_3$

   $= x_1 z_1 + y_1 z_1 + 2x_2 z_2 + 2y_2 z_2 + 4x_3 z_3 + 4y_3 z_3$

   $= x_1 z_1 + 2x_2 z_2 + 4x_3 z_3 + y_1 z_1 + 2y_2 z_2 + 4y_3 z_3 = <\mathbf{u},\mathbf{w}> + <\mathbf{v},\mathbf{w}>$,

   $<c\mathbf{u},\mathbf{v}> = cx_1 y_1 + 2cx_2 y_2 + 4cx_3 y_3 = c(x_1 y_1 + 2x_2 y_2 + 4x_3 y_3) = c<\mathbf{u},\mathbf{v}>$, and

   $<\mathbf{u},\mathbf{u}> = x_1 x_1 + 2x_2 x_2 + 4x_3 x_3 = x_1^2 + 2x_2^2 + 4x_3^2 \geq 0$ and equality holds if and only if $4x_1^2 = 0$, $2x_2^2 = 0$, and $4x_3^2 = 0$, i.e., if and only if $x_1 = 0$, $x_2 = 0$, and $x_3 = 0$. Thus the given function is an inner product on $\mathbf{R}^3$.

3. $<\mathbf{u},\mathbf{u}> = 2x_1 x_1 - x_2 x_2 = 2x_1^2 - x_2^2 < 0$ for the vector $\mathbf{u} = (1,2)$, so condition 4 of the

definition is not satisfied.

4. $\mathbf{u} = \begin{bmatrix} a & b \\ c & d \end{bmatrix}$, $\mathbf{v} = \begin{bmatrix} e & f \\ g & h \end{bmatrix}$, and $\mathbf{w} = \begin{bmatrix} j & k \\ l & m \end{bmatrix}$.

$\langle \mathbf{u}+\mathbf{v},\mathbf{w} \rangle = (a+e)j + (b+f)k + (c+g)l + (d+h)m = aj + ej + bk + fk + cl + gl + dm + hm$

$= aj + bk + cl + dm + ej + fk + gl + hm = \langle \mathbf{u},\mathbf{w} \rangle + \langle \mathbf{v},\mathbf{w} \rangle$, so axiom 2 is satisfied.

$\langle \mathbf{u},\mathbf{u} \rangle = aa + bb + cc + dd = a^2 + b^2 + c^2 + d^2 \geq 0$ and equality holds if and only if

$a = b = c = d = 0$, so axiom 4 is satisfied.

5. $\mathbf{u} = \begin{bmatrix} a & b \\ c & d \end{bmatrix}$, $\mathbf{v} = \begin{bmatrix} e & f \\ g & h \end{bmatrix}$, $\mathbf{w} = \begin{bmatrix} j & k \\ l & m \end{bmatrix}$, and s is a scalar.

$\langle \mathbf{u},\mathbf{v} \rangle = ae + 2bf + 3cg + 4dh = ea + 2fb + 3gc + 4hd = \langle \mathbf{v},\mathbf{u} \rangle$,

$\langle \mathbf{u}+\mathbf{v},\mathbf{w} \rangle = (a+e)j + 2(b+f)k + 3(c+g)l + 4(d+h)m$

$= aj + ej + 2bk + 2fk + 3cl + 3gl + 4dm + 4hm$

$= aj + 2bk + 3cl + 4dm + ej + 2fk + 3gl + 4hm = \langle \mathbf{u},\mathbf{w} \rangle + \langle \mathbf{v},\mathbf{w} \rangle$,

$\langle s\mathbf{u},\mathbf{v} \rangle = sae + 2sbf + 3scg + 4sdh = s(ea + 2fb + 3gc + 4hd) = s\langle \mathbf{u},\mathbf{v} \rangle$, and

$\langle \mathbf{u},\mathbf{u} \rangle = aa + 2bb + 3cc + 4dd = a^2 + 2b^2 + 3c^2 + 4d^2 \geq 0$ and equality holds if and only if

$a = b = c = d = 0$. Thus the given function is an inner product on M_{22}.

(a) $\langle \mathbf{u},\mathbf{v} \rangle = 1x4 + 2(2x1) + 3(0x-3) + 4(-3x2) = 4 + 4 + 0 - 24 = -16$.

(b) $\langle \mathbf{u},\mathbf{v} \rangle = -2x5 + 2(4x-2) + 3(1x0) + 4(0x-3) = -10 - 16 + 0 + 0 = -26$.

6. $\langle cf,g \rangle = \int_0^1 cf(x)g(x)\,dx = c\int_0^1 f(x)g(x)\,dx = c\langle f,g \rangle$, so axiom 3 is satisfied.

$\langle f,f \rangle = \int_0^1 f(x)f(x)\,dx = \int_0^1 f(x)^2\,dx \geq 0$ and equality holds if and only if f(x) is the zero function, so axiom 4 is satisfied.

7. $\langle f,g\rangle = \int_a^b f(x)g(x)dx = \int_a^b g(x)f(x)dx = \langle g,f\rangle$,

$\langle f+g,h\rangle = \int_a^b (f(x)+g(x))h(x)dx = \int_a^b f(x)h(x)dx + \int_a^b g(x)h(x)dx = \langle f,h\rangle + \langle g,h\rangle$,

$\langle cf,g\rangle = \int_a^b cf(x)g(x)dx = c\int_a^b f(x)g(x)dx = c\langle f,g\rangle$, and

$\langle f,f\rangle = \int_a^b f(x)f(x)dx = \int_a^b f(x)^2 dx \geq 0$ and equality holds if and only if f(x) is the zero function. Thus the given function with a < b defines an inner product on P_n.

If a = b, the integral is zero for all functions f and g, violating axiom 4.

8. (a) $\langle f,g\rangle = \int_0^1 (2x+1)(3x-2)\ dx = \int_0^1 (6x^2-x-2)\ dx = \left[2x^3 - \frac{1}{2}x^2 - 2x\right]_0^1 = -\frac{1}{2}$.

(b) $\langle f,g\rangle = \int_0^1 (x^2+2)(3)\ dx = \int_0^1 (3x^2+6)\ dx = [x^3 + 6x]_0^1 = 7$.

(c) $\langle f,g\rangle = \int_0^1 (x^2+3x-2)(x+1)\ dx = \int_0^1 (x^3+4x^2+x-2)\ dx = \left[\frac{1}{4}x^4 + \frac{4}{3}x^3 + \frac{1}{2}x^2 - 2x\right]_0^1$

$= \frac{1}{12}$.

(d) $\langle f,g\rangle = \int_0^1 (x^3+2x-1)(3x^2-4x+2)\ dx = \int_0^1 (3x^5-4x^4+8x^3-11x^2+8x-2)\ dx$

$= \left[\frac{1}{2}x^6 - \frac{4}{5}x^5 + 2x^4 - \frac{11}{3}x^3 + 4x^2 - 2x\right]_0^1 = \frac{1}{30}$.

9. (a) $\|f\|^2 = \int_0^1 (4x-2)^2\ dx = \int_0^1 (16x^2-16x+4)\ dx = \left[\frac{16}{3}x^3 - 8x^2 + 4x\right]_0^1 = \frac{4}{3}$,

so $\|f\| = \frac{2}{\sqrt{3}}$.

(b) $\|f\|^2 = \int_0^1 (7x^3)^2\, dx = \int_0^1 (49x^6)\, dx = [\,7x^7\,]_0^1 = 7$, so $\|f\| = \sqrt{7}$.

(c) $\|f\|^2 = \int_0^1 (3x^2+2)^2\, dx = \int_0^1 (9x^4+12x^2+4)\, dx = \left[\frac{9}{5}x^5 + 4x^3 + 4x\right]_0^1 = \frac{49}{5}$, so $\|f\| = \frac{7}{\sqrt{5}}$.

(d) $\|g\|^2 = \int_0^1 (x^2+x+1)^2\, dx = \int_0^1 (x^4+2x^3+3x^2+2x+1)\, dx$

$= \left[\frac{1}{5}x^5 + \frac{1}{2}x^4 + x^3 + x^2 + x\right]_0^1 = \frac{37}{10}$, so $\|g\| = \sqrt{\frac{37}{10}}$.

10. $\langle f,g\rangle = \int_0^1 (x^2)(4x-3)\, dx = \int_0^1 (4x^3-3x^2)\, dx = [\,x^4 - x^3\,]_0^1 = 0$, so the functions are orthogonal.

11. $\langle f,g\rangle = \int_0^1 (1)\left(\frac{1}{2}-x\right) dx = \int_0^1 \left(\frac{1}{2}-x\right) dx = \left[\frac{1}{2}x - \frac{1}{2}x^2\right]_0^1 = 0$, so the functions are orthogonal.

12. $\langle f,g\rangle = \int_0^1 (5x^2)(9x)\, dx = \int_0^1 (45x^3)\, dx = \left[\frac{45}{4}x^4\right]_0^1 = \frac{45}{4}$,

$\|f\|^2 = \int_0^1 (5x^2)^2\, dx = \int_0^1 (25x^4)\, dx = [5x^5\,]_0^1 = 5$, so $\|f\| = \sqrt{5}$, and

$\|g\|^2 = \int_0^1 (9x)^2\, dx = \int_0^1 (81x^2)\, dx = [27x^3\,]_0^1 = 27$, so $\|g\| = 3\sqrt{3}$. Thus

$\cos\theta = \frac{45}{4\sqrt{5}\,3\sqrt{3}} = \frac{3x15}{4x3\sqrt{15}} = \frac{\sqrt{15}}{4}$.

13. $\langle f,g\rangle = \int_0^1 (6x+12)(ax+b)\ dx = \int_0^1 (6ax^2+12ax+6bx+12b)\ dx$

$= \left[2ax^3 + 6ax^2 + 3bx^2 + 12bx\right]_0^1 = 2a+6a+3b+12b$, so any choice of a and b that makes

$8a+15b = 0$ will do. Let $a = 15$ and $b = -8$. $g(x) = 15x - 8$ is orthogonal to $f(x)$.

14. $d(f,g) = \| f-g \| = \|2x+4\|$. $\|2x+4\|^2 = \int_0^1 (2x+4)^2\ dx = \int_0^1 (4x^2+16x+16)\ dx$

$= \left[\frac{4}{3}x^3 + 8x^2 + 16x\right]_0^1 = \frac{4}{3} + 8 + 16 = \frac{76}{3}$, so $d(f,g) = \sqrt{\frac{76}{3}}$. $\left(= 2\sqrt{\frac{19}{3}}\right)$

15. $d(f,g)^2 = \| f-g \|^2 = \|-2x+3\|^2 = \int_0^1 (-2x+3)^2\ dx = \int_0^1 (4x^2-12x+9)\ dx$

$= \left[\frac{4}{3}x^3 - 6x^2 + 9x\right]_0^1 = \frac{4}{3} - 6 + 9 = \frac{13}{3}$, and $d(f,h)^2 = \| f-h \|^2 = \|3x-4\|^2 = \int_0^1 (3x-4)^2\ dx$

$= \int_0^1 (9x^2-24x+16)\ dx = \left[3x^3 - 12x^2 + 16x\right]_0^1 = 7 > \frac{13}{3}$, so g is closer to f.

16. (a) $\left\langle \begin{bmatrix} 1 & 2 \\ 3 & 4 \end{bmatrix}, \begin{bmatrix} -2 & 0 \\ -3 & 5 \end{bmatrix} \right\rangle = 1\text{x}-2 + 2\text{x}0 + 3\text{x}-3 + 4\text{x}5 = -2 + 0 - 9 + 20 = 9.$

(b) $\left\langle \begin{bmatrix} 0 & -3 \\ 2 & 5 \end{bmatrix}, \begin{bmatrix} 3 & 6 \\ -2 & -7 \end{bmatrix} \right\rangle = 0\text{x}3 + -3\text{x}6 + 2\text{x}-2 + 5\text{x}-7 = 0 - 18 - 4 - 35 = -57.$

17. (a) $\left\|\begin{bmatrix} 1 & 2 \\ 3 & 4 \end{bmatrix}\right\|^2 = 1^2 + 2^2 + 3^2 + 4^2 = 1 + 4 + 9 + 16 = 30$, so $\left\|\begin{bmatrix} 1 & 2 \\ 3 & 4 \end{bmatrix}\right\| = \sqrt{30}$.

(b) $\left\|\begin{bmatrix} 0 & 1 \\ -1 & 3 \end{bmatrix}\right\|^2 = 0 + 1^2 + (-1)^2 + 3^2 = 0 + 1 + 1 + 9 = 11$, so $\left\|\begin{bmatrix} 0 & 1 \\ -1 & 3 \end{bmatrix}\right\| = \sqrt{11}$.

(c) $\left\|\begin{bmatrix} 5 & -2 \\ -1 & 6 \end{bmatrix}\right\|^2 = 5^2 + (-2)^2 + (-1)^2 + 6^2 = 25 + 4 + 1 + 36 = 66$, so $\left\|\begin{bmatrix} 5 & -2 \\ -1 & 6 \end{bmatrix}\right\| = \sqrt{66}$

(d) $\left\|\begin{bmatrix} 4 & -2 \\ -1 & -3 \end{bmatrix}\right\|^2 = 4^2 + (-2)^2 + (-1)^2 + (-3)^2 = 16 + 4 + 1 + 9 = 30$, so

$\left\|\begin{bmatrix} 4 & -2 \\ -1 & -3 \end{bmatrix}\right\| = \sqrt{30}$.

18. (a) $<\begin{bmatrix} 1 & 2 \\ -1 & 1 \end{bmatrix}, \begin{bmatrix} 2 & 4 \\ 3 & -7 \end{bmatrix}> = 1x2 + 2x4 + -1x3 + 1x-7 = 2 + 8 - 3 - 7 = 0$, so the matrices are orthogonal.

(b) $<\begin{bmatrix} 5 & 2 \\ -3 & 2 \end{bmatrix}, \begin{bmatrix} -1 & 6 \\ 1 & -2 \end{bmatrix}< = 5x-1 + 2x6 + -3x1 + 2x-2 = -5 + 12 - 3 - 4 = 0$, so the matrices are orthogonal.

19. $<\begin{bmatrix} 1 & 2 \\ 3 & 4 \end{bmatrix}, \begin{bmatrix} a & b \\ c & d \end{bmatrix}> = a + 2b + 3c + 4d$, so any choice of a,b,c, and d satisfying the condition $a + 2b + 3c + 4d = 0$ will do. One such choice is $a = -3$, $b = 2$, $c = 1$, $d = -1$.

20. (a) $d\left(\begin{bmatrix} 4 & 0 \\ -1 & 3 \end{bmatrix}, \begin{bmatrix} 1 & 1 \\ 1 & 1 \end{bmatrix}\right) = \left\|\begin{bmatrix} 3 & -1 \\ -2 & 2 \end{bmatrix}\right\|. \left\|\begin{bmatrix} 3 & -1 \\ -2 & 2 \end{bmatrix}\right\|^2 = 18$, so

$d\left(\begin{bmatrix} 4 & 0 \\ -1 & 3 \end{bmatrix}, \begin{bmatrix} 1 & 1 \\ 1 & 1 \end{bmatrix}\right) = \sqrt{18} = 3\sqrt{2}$.

(b) $d\left(\begin{bmatrix} 2 & -3 \\ -1 & 4 \end{bmatrix}, \begin{bmatrix} -3 & 2 \\ 1 & 0 \end{bmatrix}\right) = \left\|\begin{bmatrix} 5 & -5 \\ -2 & 4 \end{bmatrix}\right\|. \left\|\begin{bmatrix} 5 & -5 \\ -2 & 4 \end{bmatrix}\right\|^2 = 70$, so

$d\left(\begin{bmatrix} 2 & -3 \\ -1 & 4 \end{bmatrix}, \begin{bmatrix} -3 & 2 \\ 1 & 0 \end{bmatrix}\right) = \sqrt{70}$.

21. (a) $\langle \mathbf{u},\mathbf{v}\rangle = (2-i)(3+2i) + (3+2i)(2-i) = 2(6 + 2 - 3i + 4i) = 16 + 2i.$

$||\mathbf{u}||^2 = (2-i)(2+i) + (3+2i)(3-2i) = 4 + 1 + 9 + 4 = 18$, so $||\mathbf{u}|| = \sqrt{18} = 3\sqrt{2}$.

$||\mathbf{v}||^2 = (3-2i)(3+2i) + (2+i)(2-i) = 18$, so $||\mathbf{v}|| = \sqrt{18} = 3\sqrt{2}$.

$d(\mathbf{u},\mathbf{v})^2 = ||\mathbf{u}-\mathbf{v}||^2 = ||(-1+i, 1+i)||^2 = (-1+i)(-1-i) + (1+i)(1-i) = 4$, so $d(\mathbf{u},\mathbf{v}) = 2$.

u and **v** are not orthogonal since $\langle \mathbf{u},\mathbf{v}\rangle \neq 0$.

(b) $\langle \mathbf{u},\mathbf{v}\rangle = (4+3i)(2-i) + (1-i)(4+5i) = 8 + 3 - 4i + 6i + 4 + 5 + 5i - 4i = 20 + 3i.$

$||\mathbf{u}||^2 = (4+3i)(4-3i) + (1-i)(1+i) = 16 + 9 + 1 + 1 = 27$, so $||\mathbf{u}|| = \sqrt{27} = 3\sqrt{3}$.

$||\mathbf{v}||^2 = (2+i)(2-i) + (4-5i)(4+5i) = 4 + 1 + 16 + 25 = 46$, so $||\mathbf{v}|| = \sqrt{46}$.

$d(\mathbf{u},\mathbf{v})^2 = ||\mathbf{u}-\mathbf{v}||^2 = ||(2+2i,-3+4i)||^2 = (2+2i)(2-2i) + (-3+4i)(-3-4i) = 33$, so

$d(\mathbf{u},\mathbf{v}) = \sqrt{33}$. **u** and **v** are not orthogonal since $\langle \mathbf{u},\mathbf{v}\rangle \neq 0$.

(c) $\langle \mathbf{u},\mathbf{v}\rangle = (2+3i)(i) + (-1)(3-2i) = 2i - 3 - 3 + 2i = -6 + 4i.$

$||\mathbf{u}||^2 = (2+3i)(2-3i) + (-1)(-1) = 4 + 9 + 1 = 14$, so $||\mathbf{u}|| = \sqrt{14}$.

$||\mathbf{v}||^2 = (-i)(i) + (3+2i)(3-2i) = 1 + 9 + 4 = 14$, so $||\mathbf{v}|| = \sqrt{14}$.

$d(\mathbf{u},\mathbf{v})^2 = ||\mathbf{u}-\mathbf{v}||^2 = ||(2+4i,-4-2i)||^2 = (2+4i)(2-4i) + (-4-2i)(-4+2i) = 40$, so

$d(\mathbf{u},\mathbf{v}) = \sqrt{40} = 2\sqrt{10}$. **u** and **v** are not orthogonal since $\langle \mathbf{u},\mathbf{v}\rangle \neq 0$.

(d) $\langle \mathbf{u},\mathbf{v}\rangle = (2-3i)(1) + (-2+3i)(1) = 2 - 3i - 2 + 3i = 0.$

$||\mathbf{u}||^2 = (2-3i)(2+3i) + (-2+3i)(-2-3i) = 4 + 9 + 4 + 9 = 26$, so $||\mathbf{u}|| = \sqrt{26}$.

$||\mathbf{v}||^2 = (1)(1) + (1)(1) = 2$, so $||\mathbf{v}|| = \sqrt{2}$.

$d(\mathbf{u},\mathbf{v})^2 = ||\mathbf{u}-\mathbf{v}||^2 = ||(1-3i,-3+3i)||^2 = (1-3i)(1+3i) + (-3+3i)(-3-3i) = 28$, so

$d(\mathbf{u},\mathbf{v}) = \sqrt{28} = 2\sqrt{7}$. **u** and **v** are orthogonal since $\langle \mathbf{u},\mathbf{v}\rangle = 0$.

22. (a) $\langle \mathbf{u},\mathbf{v}\rangle = (1+4i)(1-i) + (1+i)(-4-i) = 1 + 4 - i + 4i - 4 + 1 - i - 4i = 2-2i$.

$||\mathbf{u}||^2 = (1+4i)(1-4i) + (1+i)(1-i) = 1 + 16 + 1 + 1 = 19$, so $||\mathbf{u}|| = \sqrt{19}$.

$||\mathbf{v}||^2 = (1+i)(1-i) + (-4+i)(-4-i) = 1 + 1 + 16 + 1 = 19$, so $||\mathbf{v}|| = \sqrt{19}$.

$d(\mathbf{u},\mathbf{v})^2 = ||\mathbf{u}-\mathbf{v}||^2 = ||(3i,5)||^2 = (3i)(-3i) + (5)(5\ ) = 34$, so

$d(\mathbf{u},\mathbf{v}) = \sqrt{34}$. **u** and **v** are not orthogonal since $\langle \mathbf{u},\mathbf{v}\rangle \neq 0$.

(b) $\langle \mathbf{u},\mathbf{v}\rangle = (2+7i)(3+4i) + (1+i)(2-5i) = 6 - 28 + 21i + 8i + 2 + 5 + 2i - 5i = -15+26i$.

$||\mathbf{u}||^2 = (2+7i)(2-7i) + (1+i)(1-i) = 4 + 49 + 1 + 1 = 55$, so $||\mathbf{u}|| = \sqrt{55}$.

$||\mathbf{v}||^2 = (3-4i)(3+4i) + (2+5i)(2-5i) = 9 + 16 + 4 + 25 = 54$, so $||\mathbf{v}|| = \sqrt{54} = 3\sqrt{6}$.

$d(\mathbf{u},\mathbf{v})^2 = ||\mathbf{u}-\mathbf{v}||^2 = ||(-1+11i,-1-4i)||^2 = (-1+11i)(-1-11i) + (-1-4i)(-1+4i) = 139$,

so $d(\mathbf{u},\mathbf{v}) = \sqrt{139}$. **u** and **v** are not orthogonal since $\langle \mathbf{u},\mathbf{v}\rangle \neq 0$.

(c) $\langle \mathbf{u},\mathbf{v}\rangle = (1-3i)(2+i) + (1+i)(-5i) = 2 + 3 - 6i + i + 5 - 5i = 10-10i$.

$||\mathbf{u}||^2 = (1-3i)(1+3i) + (1+i)(1-i) = 1 + 9 + 1 + 1 = 12$, so $||\mathbf{u}|| = \sqrt{12} = 2\sqrt{3}$.

$||\mathbf{v}||^2 = (2-i)(2+i) + (5i)(-5i) = 4 + 1 + 25 = 30$, so $||\mathbf{v}|| = \sqrt{30}$.

$d(\mathbf{u},\mathbf{v})^2 = ||\mathbf{u}-\mathbf{v}||^2 = ||(-1-2i,1-4i)||^2 = (-1-2i)(-1+2i) + (1-4i)(1+4i) = 22$,

so $d(\mathbf{u},\mathbf{v}) = \sqrt{22}$. **u** and **v** are not orthogonal since $\langle \mathbf{u},\mathbf{v}\rangle \neq 0$.

(d) $\langle \mathbf{u},\mathbf{v}\rangle = (3+i)(2-i) + (2+2i)\left(-2-\frac{3}{2}i\right) = 6 + 1 - 3i + 2i - 4 + 3 - 4i - 3i = 6-8i$.

$||\mathbf{u}||^2 = (3+i)(3-i) + (2+2i)(2-2i) = 9 + 1 + 4 + 4 = 18$, so $||\mathbf{u}|| = \sqrt{18} = 3\sqrt{2}$.

$||\mathbf{v}||^2 = (2+i)(2-i) + \left(-2+\frac{3}{2}i\right)\left(-2-\frac{3}{2}i\right) = 4 + 1 + 4 + \frac{9}{4} = \frac{45}{4}$, so $||\mathbf{v}|| = \frac{3}{2}\sqrt{5}$.

$$d(\mathbf{u},\mathbf{v})^2 = ||\mathbf{u}-\mathbf{v}||^2 = \left|\left|\left(1, 4+\frac{1}{2}i\right)\right|\right|^2 = (1)(1) + \left(4+\frac{1}{2}i\right)\left(4-\frac{1}{2}i\right) = \frac{69}{4}\text{, so}$$

$d(\mathbf{u},\mathbf{v}) = \frac{1}{2}\sqrt{69}$. **u** and **v** are not orthogonal since $\langle\mathbf{u},\mathbf{v}\rangle \neq 0$.

23. $\mathbf{u} = (x_1, \ldots, x_n)$, $\mathbf{v} = (y_1, \ldots, y_n)$, $\mathbf{w} = (z_1, \ldots, z_n)$, and c is a scalar.

$\langle\mathbf{u},\mathbf{v}\rangle = x_1\bar{y}_1 + \ldots + x_n\bar{y}_n = \bar{y}_1 x_1 + \ldots + \bar{y}_n x_n = \bar{y}_1\bar{\bar{x}}_1 + \ldots + \bar{y}_n\bar{\bar{x}}_n = \langle\bar{\mathbf{v}},\bar{\mathbf{u}}\rangle$
$= \overline{\langle\mathbf{v},\mathbf{u}\rangle}$,

$\langle\mathbf{u}+\mathbf{v},\mathbf{w}\rangle = (x_1 + y_1)\bar{z}_1 + \ldots + (x_n + y_n)\bar{z}_n = x_1\bar{z}_1 + y_1\bar{z}_1 + \ldots + x_n\bar{z}_n +$
$y_n\bar{z}_n$

$= x_1\bar{z}_1 + \ldots + x_n\bar{z}_n + y_1\bar{z}_1 + \ldots + y_n\bar{z}_n = \langle\mathbf{u},\mathbf{w}\rangle + \langle\mathbf{v},\mathbf{w}\rangle$,

$\langle c\mathbf{u},\mathbf{v}\rangle = cx_1\bar{y}_1 + \ldots + cx_n\bar{y}_n = c(x_1\bar{y}_1 + \ldots + x_n\bar{y}_n) = c\langle\mathbf{u},\mathbf{v}\rangle$, and

$\langle\mathbf{u},\mathbf{u}\rangle = x_1\bar{x}_1 + \ldots + x_n\bar{x}_n \geq 0$ since all $x_i\bar{x}_i \geq 0$, and equality holds if and only if all $x_i = 0$. Thus the given function is an inner product on $\mathbf{C}^n$.

24. $\langle\mathbf{u},k\mathbf{v}\rangle = x_1\bar{k}\bar{y}_1 + \ldots + x_n\bar{k}\bar{y}_n = \bar{k}x_1\bar{y}_1 + \ldots + \bar{k}x_n\bar{y}_n =$
$\bar{k}(x_1\bar{y}_1 + \ldots + x_n\bar{y}_n) = \bar{k}\langle\mathbf{u},\mathbf{v}\rangle$.

25. (a) Let **u** be any nonzero vector in the inner product space. Then using axiom 1, the fact that $0\mathbf{u} = \mathbf{0}$, axiom 3, and the fact that zero times any real number is zero we have $\langle\mathbf{v},\mathbf{0}\rangle = \langle\mathbf{0},\mathbf{v}\rangle = \langle 0\mathbf{u},\mathbf{v}\rangle = 0\langle\mathbf{u},\mathbf{v}\rangle = 0$.

(b) From axiom 1, axiom 2, and axiom 1 again $\langle\mathbf{u},\mathbf{v}+\mathbf{w}\rangle = \langle\mathbf{v}+\mathbf{w},\mathbf{u}\rangle = \langle\mathbf{v},\mathbf{u}\rangle + \langle\mathbf{w},\mathbf{u}\rangle = \langle\mathbf{u},\mathbf{v}\rangle + \langle\mathbf{u},\mathbf{w}\rangle$.

(c) From axiom 1, axiom 3, and axiom 1 again $\langle\mathbf{u},c\mathbf{v}\rangle = \langle c\mathbf{v},\mathbf{u}\rangle = c\langle\mathbf{v},\mathbf{u}\rangle = c\langle\mathbf{u},\mathbf{v}\rangle$.

26. (a) $\langle \mathbf{u},\mathbf{v}\rangle = \mathbf{u}A\mathbf{v}^t = (\mathbf{u}A)\cdot\mathbf{v} = \mathbf{v}\cdot(\mathbf{u}A) = \mathbf{v}(\mathbf{u}A)^t = \mathbf{v}A\mathbf{u}^t = \langle \mathbf{v},\mathbf{u}\rangle$,

$\langle \mathbf{u}+\mathbf{v},\mathbf{w}\rangle = (\mathbf{u}+\mathbf{v})A\mathbf{w}^t = \mathbf{u}A\mathbf{w}^t + \mathbf{v}A\mathbf{w}^t = \langle \mathbf{u},\mathbf{w}\rangle + \langle \mathbf{v},\mathbf{w}\rangle$,

$\langle c\mathbf{u},\mathbf{v}\rangle = (c\mathbf{u})A\mathbf{v}^t = c(\mathbf{u}A\mathbf{v}^t) = c\langle \mathbf{u},\mathbf{v}\rangle$, and

$\langle \mathbf{u},\mathbf{u}\rangle = \mathbf{u}A\mathbf{u}^t > 0$ if $\mathbf{u} \neq \mathbf{0}$ (given) and $\mathbf{u}A\mathbf{u}^t = 0$ if $\mathbf{u} = \mathbf{0}$.

(b) Let $\mathbf{u} = (x_1, \ldots, x_n)$ and $\mathbf{v} = (y_1, \ldots, y_n)$. $\mathbf{u}\cdot\mathbf{v} = x_1 y_1 + \ldots + x_n y_n = \mathbf{u}\mathbf{v}^t = \mathbf{u}I_n\mathbf{v}^t$.

(c) $\langle \mathbf{u},\mathbf{v}\rangle = \mathbf{u}A\mathbf{v}^t = [1\ 0]\begin{bmatrix} 2 & 0 \\ 0 & 3 \end{bmatrix}\begin{bmatrix} 0 \\ 1 \end{bmatrix} = [2\ 0]\begin{bmatrix} 0 \\ 1 \end{bmatrix} = 0.$

$\|\mathbf{u}\|^2 = \mathbf{u}A\mathbf{u}^t = [1\ 0]\begin{bmatrix} 2 & 0 \\ 0 & 3 \end{bmatrix}\begin{bmatrix} 1 \\ 0 \end{bmatrix} = [2\ 0]\begin{bmatrix} 1 \\ 0 \end{bmatrix} = 2$, so $\|\mathbf{u}\| = \sqrt{2}$.

$\|\mathbf{v}\|^2 = \mathbf{v}A\mathbf{v}^t = [0\ 1]\begin{bmatrix} 2 & 0 \\ 0 & 3 \end{bmatrix}\begin{bmatrix} 0 \\ 1 \end{bmatrix} = [0\ 3]\begin{bmatrix} 0 \\ 1 \end{bmatrix} = 3$, so $\|\mathbf{v}\| = \sqrt{3}$.

$d(\mathbf{u},\mathbf{v})^2 = \|\mathbf{u}-\mathbf{v}\|^2 = [1\ {-1}]\begin{bmatrix} 2 & 0 \\ 0 & 3 \end{bmatrix}\begin{bmatrix} 1 \\ -1 \end{bmatrix} = [2\ {-3}]\begin{bmatrix} 1 \\ -1 \end{bmatrix} = 5$, so $d(\mathbf{u},\mathbf{v}) = \sqrt{5}$.

$\langle \mathbf{u},\mathbf{v}\rangle = \mathbf{u}B\mathbf{v}^t = [1\ 0]\begin{bmatrix} 1 & 2 \\ 2 & 3 \end{bmatrix}\begin{bmatrix} 0 \\ 1 \end{bmatrix} = [1\ 2]\begin{bmatrix} 0 \\ 1 \end{bmatrix} = 2.$

$\|\mathbf{u}\|^2 = \mathbf{u}B\mathbf{u}^t = [1\ 0]\begin{bmatrix} 1 & 2 \\ 2 & 3 \end{bmatrix}\begin{bmatrix} 1 \\ 0 \end{bmatrix} = [1\ 2]\begin{bmatrix} 1 \\ 0 \end{bmatrix} = 1$, so $\|\mathbf{u}\| = 1$.

$\|\mathbf{v}\|^2 = \mathbf{v}B\mathbf{v}^t = [0\ 1]\begin{bmatrix} 1 & 2 \\ 2 & 3 \end{bmatrix}\begin{bmatrix} 0 \\ 1 \end{bmatrix} = [2\ 3]\begin{bmatrix} 0 \\ 1 \end{bmatrix} = 3$, so $\|\mathbf{v}\| = \sqrt{3}$.

$d(\mathbf{u},\mathbf{v})^2 = \|\mathbf{u}-\mathbf{v}\|^2 = [1\ {-1}]\begin{bmatrix} 1 & 2 \\ 2 & 3 \end{bmatrix}\begin{bmatrix} 1 \\ -1 \end{bmatrix} = [-1\ {-1}]\begin{bmatrix} 1 \\ -1 \end{bmatrix} = 0$, so $d(\mathbf{u},\mathbf{v}) = 0$.

$\langle \mathbf{u},\mathbf{v}\rangle = \mathbf{u}C\mathbf{v}^t = [1\ 0]\begin{bmatrix} 2 & 3 \\ 3 & 5 \end{bmatrix}\begin{bmatrix} 0 \\ 1 \end{bmatrix} = [2\ 3]\begin{bmatrix} 0 \\ 1 \end{bmatrix} = 3.$

$\|\mathbf{u}\|^2 = \mathbf{u}C\mathbf{u}^t = [1\ 0]\begin{bmatrix} 2 & 3 \\ 3 & 5 \end{bmatrix}\begin{bmatrix} 1 \\ 0 \end{bmatrix} = [2\ 3]\begin{bmatrix} 1 \\ 0 \end{bmatrix} = 2$, so $\|\mathbf{u}\| = \sqrt{2}$.

$$\|\mathbf{v}\|^2 = \mathbf{v}C\mathbf{v}^t = [0\ 1]\begin{bmatrix} 2 & 3 \\ 3 & 5 \end{bmatrix}\begin{bmatrix} 0 \\ 1 \end{bmatrix} = [3\ 5]\begin{bmatrix} 0 \\ 1 \end{bmatrix} = 5, \text{ so } \|\mathbf{v}\| = \sqrt{5}\ .$$

$$d(\mathbf{u},\mathbf{v})^2 = \|\mathbf{u}-\mathbf{v}\|^2 = [1\ -1]\begin{bmatrix} 2 & 3 \\ 3 & 5 \end{bmatrix}\begin{bmatrix} 1 \\ -1 \end{bmatrix} = [-1\ -2]\begin{bmatrix} 1 \\ -1 \end{bmatrix} = 1, \text{ so } d(\mathbf{u},\mathbf{v}) = 1.$$

(d) $u = (x_1, x_2)$, and $v = (y_1, y_2)$. Let $<u, v> = x_1y_1 + 4x_2y_2 = [x_1\ \ x_2]\begin{bmatrix} a & b \\ c & d \end{bmatrix}\begin{bmatrix} y_1 \\ y_2 \end{bmatrix}$

(d) $\mathbf{u} = (x_1, x_2)$ and $\mathbf{v} = (y_1, y_2)$. $<\mathbf{u},\mathbf{v}> = x_1 y_1 + 4x_2 y_2 = [x_1\ \ x_2]$

$= [\,ax_1 + cx_2 \quad bx_1 + d\,x_2]\begin{bmatrix} y_1 \\ y_2 \end{bmatrix} = (ax_1 + cx_2)y_1 + (bx_1 + d\,x_2)y_2 = ax_1y_1 + cx_2y_1 + bx_1y_2 +$

$d\,x_2y_2$. Equating coefficients, a=1, b= 0, c=0, d = 4. Thus $A = \begin{bmatrix} 1 & 0 \\ 0 & 4 \end{bmatrix}$.

27. Let $\mathbf{w}_1$ and $\mathbf{w}_2$ be elements of W and c a scalar. Then $<\mathbf{w}_1,\mathbf{v}>=0$ and $<\mathbf{w}_2,\mathbf{v}> = 0$. Thus $<\mathbf{w}_1,\mathbf{v}>+<\mathbf{w}_2,\mathbf{v}> = 0$ and $c<\mathbf{w}_1,\mathbf{v}>=0$. Properties of inner product give: $<\mathbf{w}_1+\mathbf{w}_2,\mathbf{v}> = 0$ and $<c\mathbf{w}_1,\mathbf{v}>=0$. Thus $\mathbf{w}_1+\mathbf{w}_2$ and $c\mathbf{w}_1$ are in W. W is closed under vector addition and under scalar multiplication. Thus it is a subspace of V.

28. (a) True: $<\mathbf{u},\mathbf{v}>=0$, Thus $c<\mathbf{u},\mathbf{v}>=0$, $<\mathbf{u},c\mathbf{v}>=0$.

(b) True: For arbitrary vectors $\mathbf{u}=(a,b)$, $\mathbf{v}=(c,d)$ let $<\mathbf{u},\mathbf{v}> = p(ac)+q(bd)$. This defines an inner product if p and q are positive. If $\mathbf{u}=(1,3)$, $\mathbf{v}=(-1,2)$, then $<(1,3),(-1,2)> = p(-1)+q(6)$. Let p=6 and q=1; then $<(1,3),(-1,2)> = 0$. $<\mathbf{u},\mathbf{v}> = 6ac+bd$ is an appropriate inner product.

(c) False: $<f, g> = \int_0^1 f(x)g(x)dx$. $<3x, 3x> = \int_0^1 9x^2dx = [3x^3]_0^1 = 3$. $\|3x\| = \sqrt{3}$.

(d) True: $<A,B>=0$, $<A,C>=0$. By props of inner product, $p<A,B>=0$, $q<A,C>=0$.

$<A,pB>=0$, $<A,qC>=0$. $<A,pB>+<A,qC>=0$. $<A,pB+qC>=0$.

(e) True: Let V be an inner product space with distance defined $d(\mathbf{u}, \mathbf{v}) = \|\mathbf{u} - \mathbf{v}\|$.

Then $\|\mathbf{u} - \mathbf{v}\| = \sqrt{<\mathbf{u}-\mathbf{v},\mathbf{u}-\mathbf{v}>} = \sqrt{<(-1)(\mathbf{v}-\mathbf{u}),(-1)(\mathbf{v}-\mathbf{u})>} = \sqrt{(-1)^2<(\mathbf{v}-\mathbf{u}),(\mathbf{v}-\mathbf{u})>}$

$= \sqrt{(-1)^2<(\mathbf{v}-\mathbf{u}),(\mathbf{v}-\mathbf{u})>} = d(\mathbf{v}, \mathbf{u})$.

Exercise Set 6.2

1. $d((x_1, x_2), (0,0))^2 = ||(x_1, x_2)||^2 = x_1^2 + 4x_2^2$, so the equation of the circle with radius 1 and center at the origin is $x_1^2 + 4x_2^2 = 1$.

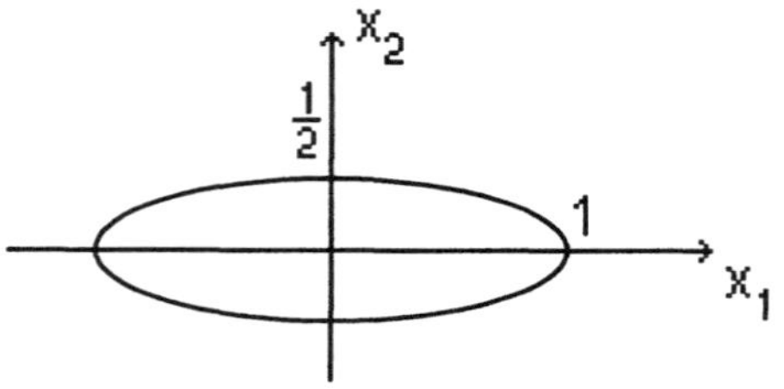

2. (a) $||(1,0)|| = \sqrt{4x1x1+9x0x0} = 2$, $||(0,1)|| = \sqrt{4x0x0+9x1x1} = 3$,

$||(1,1)|| = \sqrt{4x1x1+9x1x1} = \sqrt{13}$, $||(2,3)|| = \sqrt{4x2x2+9x3x3} = \sqrt{97}$.

(b) $<(2,1),(-9,8)> = 4x2x-9 + 9x1x8 = 0$, so the vectors are orthogonal in this space.

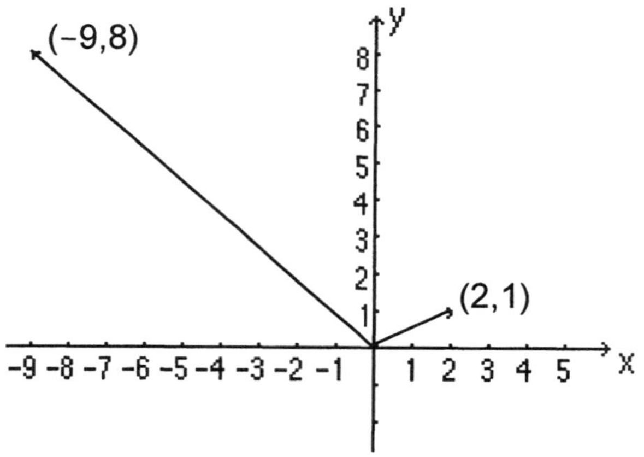

(c) $d((1,0),(0,1)) = ||(1,-1)|| = \sqrt{4x1x1+9x-1x-1} = \sqrt{13}$.

(d) $d((x_1, x_2), (0,0))^2 = ||(x_1, x_2)||^2 = 4x_1^2 + 9x_2^2$, so the equation of the circle with radius 1 and center at the origin is $4x_1^2 + 9x_2^2 = 1$.

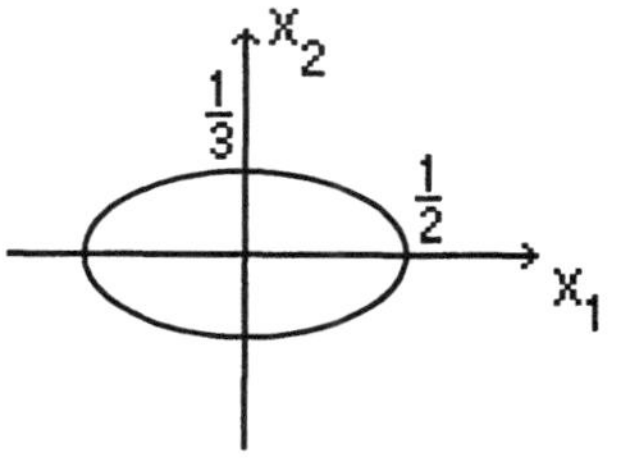

3. (a) $||(1,0)|| = \sqrt{1x1+16x0x0} = 1$, $||(0,1)|| = \sqrt{0x0+16x1x1} = 4$, and

$||(1,1)|| = \sqrt{1x1+16x1x1} = \sqrt{17}$.

(b) $<(1,1),(-16,1)> = 1x-16+ 16x1x1 = 0$, so the vectors are orthogonal.

(c) $d((5,0),(0,4)) = ||(5,-4)|| = \sqrt{5x5+16x\text{-}4x\text{-}4} = \sqrt{281}$.
In Euclidean space the distance is $\sqrt{5^2+4^2} = \sqrt{41}$.

(d) $d((x_1 ,x_2), (0,0))^2 = ||(x_1 ,x_2)||^2 = x_1{}^2 + 16x_2{}^2$, so the equation of the circle with radius 1 and center at the origin is $x_1{}^2 + 16x_2{}^2 = 1$.

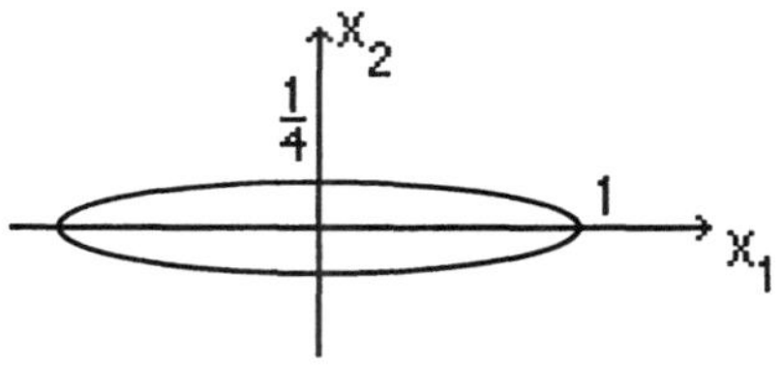

4. $<(x_1 , x_2),(y_1 , y_2)> = \frac{1}{4} x_1 y_1 + \frac{1}{25} x_2 y_2$.

5. If $X = (1,0,0,1)$ then $<X,X> = |-1^2 -0^2 -0^2 +1^2 | = 0$.

6. $d(R,Q)^2 = ||(4,0,0,-5)||^2 = |-16 - 0 - 0 + 25 | = 9$, so $d(R,Q) = 3$.

7. $d(M,P)^2 = ||(1,0,0,1)||^2 = |-1 - 0 - 0 + 1| = 0$, so $d(M,P) = 0$.

8. $<(2,0,0,1),(1,0,0,2)> = -2 - 0 - 0 + 2 = 0$, so the vectors are orthogonal.

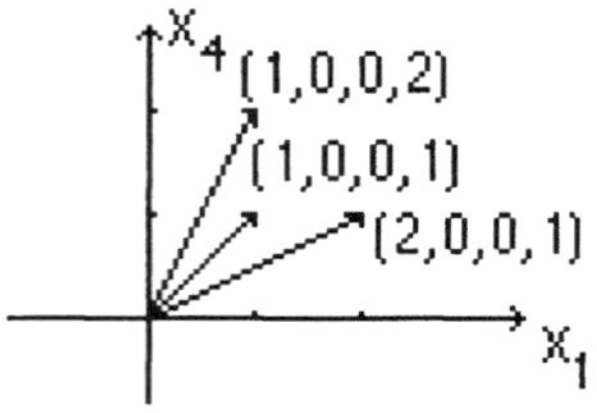

$<(a,0,0,b),(b,0,0,a)> = -ab - 0 - 0 + ba = 0$, so the vectors are orthogonal.

9. $d((x_1, 0, 0, x_4), (0,0,0,0))^2 = ||(x_1, 0, 0, x_4)||^2 = |-x_1^{\ 2} + x_4^{\ 2}|$, so the equations of the circles with radii a = 1,2,3 and center at the origin are $|-x_1^{\ 2} + x_4^{\ 2}| = a^2$.

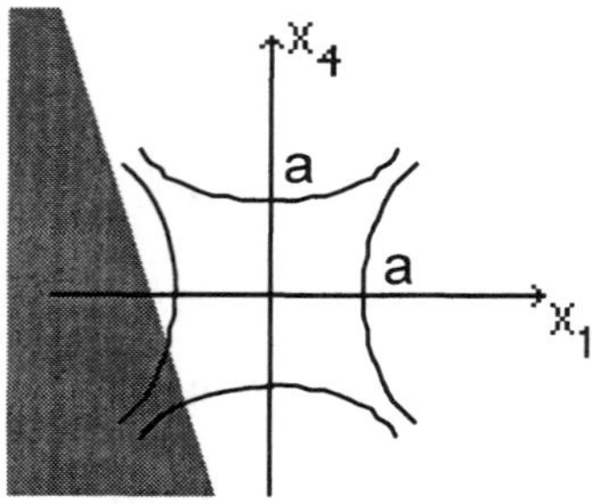

We use the following illustration for the remaining exercises in this section.

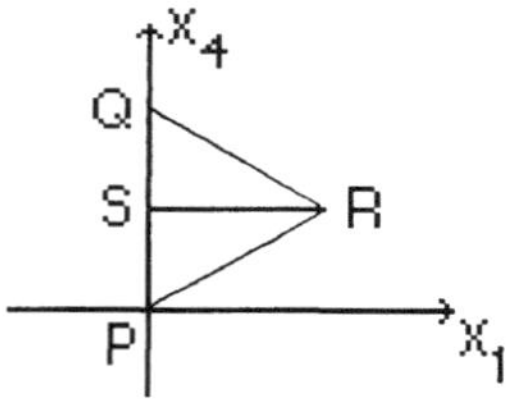

10. PQ = (0,0,0,20), so PS = (0,0,0,10) and PR = (8,0,0,10).

$||PR|| = \sqrt{-64-0-0+100} = 6$, so the duration of the voyage for a person on the space ship is 2x6 = 12 years.

Duration of voyage relative to Earth = $\dfrac{\text{distance in light years}}{\text{speed}}$, so speed = $\dfrac{16}{20}$ = 0.8 speed of light.

11. PQ = (0,0,0,120), so PS = (0,0,0,60) and PR = (45,0,0,60).

$||PR|| = \sqrt{-(45)^2-0-0+(60)^2} = 15\sqrt{7}$, so the duration of the voyage for a person on the spaceship is $30\sqrt{7}$ years.

12. Let PS = (0,0,0,t). Then PR = (410,0,0,t) and $||PR||^2 = |-(410)^2 - 0 - 0 + t^2| = 400$. We assume the speed of the traveler is less than the speed of light, so t > 410.

$-(410)^2 + t^2 = 400$, so $t^2 = 168{,}500$ and t = 410.49 years. Therefore the duration of the

voyage from Earth's point of view is 820.98 years. More than 8 centuries will have passed on Earth.

13. Duration of voyage relative to Earth = $\dfrac{\text{distance in light years}}{\text{speed}} = \dfrac{1030}{.99999} = 1030.0103$ years. PR = (515,0,0,515.00515), so the duration of the voyage relative to the traveler is $2||PR|| = 2\sqrt{-(515)^2+(515.00515)^2} = 2\sqrt{5.3045} = 4.606$ years.

Exercise Set 6.3

1. The set $\left\{\dfrac{1}{\sqrt{2}}, \dfrac{\sqrt{3}}{\sqrt{2}}x\right\}$ is an orthonormal basis for P_1 [−1,1] (see the text).

$$<x^2, \frac{1}{\sqrt{2}}> = \frac{1}{\sqrt{2}}\int_{-1}^{1} x^2\,dx = \left[\frac{1}{3\sqrt{2}}x^3\right]_{-1}^{1} = \frac{2}{3\sqrt{2}} = \frac{\sqrt{2}}{3} \quad \text{and}$$

$$<x^2, \frac{\sqrt{3}}{\sqrt{2}}x> = \frac{\sqrt{3}}{\sqrt{2}}\int_{-1}^{1} x^3\,dx = \left[\frac{\sqrt{3}}{\sqrt{2}}\frac{x^4}{4}\right]_{-1}^{1} = 0, \text{ so } \operatorname{proj}_{P_1[0,1]} x^2 = \frac{\sqrt{2}}{3}\frac{1}{\sqrt{2}} = \frac{1}{3}$$

is the least squares linear approximation to $f(x) = x^2$ over the interval [−1,1].

2. The set $\left\{1, x-\frac{1}{2}\right\}$ is an orthogonal basis for P_1 [0,1], $||1|| = 1$, and

$$||x - \tfrac{1}{2}||^2 = \int_0^1 \left(x^2 - x + \tfrac{1}{4}\right) dx = \left[\tfrac{1}{3}x^3 - \tfrac{1}{2}x^2 + \tfrac{1}{4}x\right]_0^1 = \tfrac{1}{3} - \tfrac{1}{2} + \tfrac{1}{4} = \tfrac{1}{12}$$, so the set

$\left\{1, 2\sqrt{3}\left(x-\frac{1}{2}\right)\right\}$ is an orthonormal basis for P_1 [0,1].

$<e^x, 1> = \int_0^1 e^x\, dx = [e^x]_0^1 = e - 1$, and $<e^x, 2\sqrt{3}\,x - \sqrt{3}> = \int_0^1 (2\sqrt{3}\,xe^x - \sqrt{3}\,e^x)\, dx$

$= [2\sqrt{3}\,(xe^x - e^x) - \sqrt{3}\, e^x]_0^1 = -\sqrt{3}\, e + 3\sqrt{3}$, so

$\text{proj}_{P_1[0,1]}\, e^x = e - 1 + (-\sqrt{3}\, e + 3\sqrt{3})2\sqrt{3}\left(x-\frac{1}{2}\right) = 4e - 10 + (18 - 6e)x$ is the least squares linear approximation to e^x over the interval [0,1].

3. The set $\left\{1, 2\sqrt{3}\left(x-\frac{1}{2}\right)\right\}$ is an orthonormal basis for P_1 [0,1] (see Exercise 2).

$<\sqrt{x}, 1> = \int_0^1 \sqrt{x}\, dx = \left[\frac{2}{3}x^{\frac{3}{2}}\right]_0^1 = \frac{2}{3}$, and $<\sqrt{x}, 2\sqrt{3}\,x - \sqrt{3}> = \int_0^1 (2\sqrt{3}\,x^{\frac{3}{2}} - \sqrt{3}\,x^{\frac{1}{2}})\, dx$

$$= \left[\frac{4\sqrt{3}}{5}x^{\frac{5}{2}} - \frac{2\sqrt{3}}{3}x^{\frac{3}{2}}\right]_0^1 = \frac{4\sqrt{3}}{5} - \frac{2\sqrt{3}}{3} = \frac{2\sqrt{3}}{15}$$, so that

$\text{proj}_{P_1[0,1]}\sqrt{x} = \frac{2}{3} + \frac{2\sqrt{3}}{15}\, 2\sqrt{3}\left(x-\frac{1}{2}\right) = \frac{4}{5}x + \frac{4}{15}$ is the least squares linear approximation to $\sqrt{x}$ over the interval [0,1].

4. The set $\left\{1, x-\frac{\pi}{2}\right\}$ is an orthogonal basis for P_1 $[0,\pi]$, $||1|| = \pi$, and

$$||x - \tfrac{\pi}{2}|| = \int_0^\pi \left(x^2 - \pi x + \tfrac{\pi^2}{4}\right) dx = \left[\tfrac{1}{3}x^3 - \tfrac{1}{2}\pi x^2 + \tfrac{\pi^2}{4}x\right]_0^\pi = \left(\tfrac{1}{3} - \tfrac{1}{2} + \tfrac{1}{4}\right)\pi^3 = \tfrac{\pi^3}{12}$$, so

the set $\left\{\pi^{-\frac{1}{2}}, 2\sqrt{3}\,\pi^{-\frac{3}{2}}\left(x-\frac{\pi}{2}\right)\right\}$ is an orthonormal basis for $P_1\ [0,\pi]$.

$<\cos x, \pi^{-\frac{1}{2}}> = \pi^{-\frac{1}{2}} \int_0^{\pi} \cos x \ dx = \pi^{-\frac{1}{2}} \ [\sin x]_0^{\pi} = 0$, and

$<\cos x,\ 2\sqrt{3}\ \pi^{-\frac{3}{2}}\left(x-\frac{\pi}{2}\right)> = 2\sqrt{3}\ \pi^{-\frac{3}{2}} \int_0^{\pi} \left(x\cos x - \frac{\pi}{2}\cos x\right)\ dx$

$= 2\sqrt{3}\ \pi^{-\frac{3}{2}} \left[x\sin x + \cos x - \frac{\pi}{2}\sin x\right]_0^{\pi} = 2\sqrt{3}\ \pi^{-\frac{3}{2}}\ [-1-1] = -4\sqrt{3}\ \pi^{-\frac{3}{2}}$, so that

$\text{proj}_{P_1[0,\pi]} \cos x = -4\sqrt{3}\ \pi^{-\frac{3}{2}}\ 2\sqrt{3}\ \pi^{-\frac{3}{2}} \left(x-\frac{\pi}{2}\right) = -24\pi^{-3}\ x + 12\pi^{-2}$ is the least

squares linear approximation to cos x over the interval $[0,\pi]$.

5. (a) The set $\left\{\frac{1}{\sqrt{2}}, \frac{\sqrt{3}}{\sqrt{2}}x, \frac{3\sqrt{5}}{2\sqrt{2}}\left(x^2-\frac{1}{3}\right)\right\}$ is an orthonormal basis for $P_2\ [-1,1]$.

$< e^x, \frac{1}{\sqrt{2}} > = \frac{1}{\sqrt{2}} \int_{-1}^{1} e^x \ dx = \left[\frac{1}{\sqrt{2}} e^x\right]_{-1}^{1} = \frac{1}{\sqrt{2}}\ (e - e^{-1})$, $< e^x, \frac{\sqrt{3}}{\sqrt{2}}\ x >$

$= \frac{\sqrt{3}}{\sqrt{2}} \int_{-1}^{1} xe^x \ dx = \frac{\sqrt{3}}{\sqrt{2}}\ [xe^x - e^x]_{-1}^{1} = \frac{\sqrt{3}}{\sqrt{2}}(2e^{-1})$, and $\left\langle e^x, \frac{3\sqrt{5}}{2\sqrt{2}}\left(x^2-\frac{1}{3}\right)\right\rangle$

$= \frac{3\sqrt{5}}{2\sqrt{2}} \int_{-1}^{1} \left(x^2e^x - \frac{1}{3}e^x\right)\ dx = \frac{3\sqrt{5}}{2\sqrt{2}} \left[x^2e^x - 2xe^x + 2e^x - \frac{1}{3}e^x\right]_{-1}^{1}$

$= \frac{3\sqrt{5}}{2\sqrt{2}} \left(\frac{2}{3}e - \frac{14}{3}e^{-1}\right)$, so that $\text{proj}_{P_2[-1,1]}\ e^x$

$= \frac{1}{\sqrt{2}}\ (e - e^{-1})\frac{1}{\sqrt{2}} + \frac{\sqrt{3}}{\sqrt{2}}(2e^{-1})\frac{\sqrt{3}}{\sqrt{2}}\ x + \frac{3\sqrt{5}}{2\sqrt{2}}\left(\frac{2}{3}e - \frac{14}{3}e^{-1}\right)\frac{3\sqrt{5}}{2\sqrt{2}}\left(x^2-\frac{1}{3}\right)$

$= -\frac{3}{4}\ e + \frac{33}{4}\ e^{-1} + 3e^{-1}\ x + \frac{15}{4}(e - 7e^{-1})\ x^2$ is the least squares quadratic approximation to $f(x) = e^x$ over the interval [−1,1].

(b) The set $\left\{1,\ 2\sqrt{3}\left(x-\frac{1}{2}\right),\ 6\sqrt{5}\left(x^2-x+\frac{1}{6}\right)\right\}$ is an orthonormal basis for P_2 [0,1]. From Exercise 2, $<e^x, 1> = e-1$ and $<e^x, 2\sqrt{3}\ x-\sqrt{3}> = -\sqrt{3}\ e+3\sqrt{3}$.

$< e^x, 6\sqrt{5}\left(x^2-x+\frac{1}{6}\right)> = \int_0^1 (6\sqrt{5}\, x^2 e^x - 6\sqrt{5}\, x e^x + \sqrt{5}\, e^x)\, dx$

$= [\ 6\sqrt{5}\ (x^2e^x - 2xe^x + 2e^x) - 6\sqrt{5}\ (xe^x - e^x) + \sqrt{5}\ e^x\ ]_0^1 = 7\sqrt{5}\ e - 19\sqrt{5}$,so that

$\text{proj}_{P_2[0,1]}\ e^x = e - 1 + (-\sqrt{3}\ e + 3\sqrt{3}\ )2\sqrt{3}\left(x-\frac{1}{2}\right) + (7\sqrt{5}\ e - 19\sqrt{5}\ )6\sqrt{5}\left(x^2-x+\frac{1}{6}\right)$

$= e - 1 + (-6e + 18)\left(x-\frac{1}{2}\right) + (210e - 570)\left(x^2-x+\frac{1}{6}\right)$

$= 39e - 105 + (-216e + 588)x + (210e - 570)x^2$ is the least squares quadratic

approximation to $f(x) = e^x$ over the interval [0,1].

6. The set $\left\{1,\ 2\sqrt{3}\left(x-\frac{1}{2}\right),\ 6\sqrt{5}\left(x^2-x+\frac{1}{6}\right)\right\}$ is an orthonormal basis for P_2 [0,1].

$<\sqrt{x}, 1> = \frac{2}{3}$, $<\sqrt{x}, 2\sqrt{3}\ x - \sqrt{3}> = \frac{2\sqrt{3}}{15}$, and $<\sqrt{x},\ 6\sqrt{5}\left(x^2-x+\frac{1}{6}\right)>$

$= 6\sqrt{5}\int_0^1 (x^{\frac{5}{2}} - x^{\frac{3}{2}} + \frac{1}{6}x^{\frac{1}{2}})dx = 6\sqrt{5}\left[\frac{2}{7}x^{\frac{7}{2}} - \frac{2}{5}\ x^{\frac{5}{2}} + \frac{1}{9}\ x^{\frac{3}{2}}\right]_0^1 = 6\sqrt{5}(\frac{2}{7} - \frac{2}{5} + \frac{1}{9})$

$= -\frac{2\sqrt{5}}{105}$, so that $\text{proj}_{P_2[0,1]}\sqrt{x} = \frac{2}{3} + \frac{2\sqrt{3}}{15}\ 2\sqrt{3}\left(x-\frac{1}{2}\right) - \frac{2\sqrt{5}}{105}\ 6\sqrt{5}\left(x^2-x+\frac{1}{6}\right)$

$= -\frac{4}{7}x^2 + \frac{48}{35}x + \frac{6}{35}$ is the least squares quadratic approximation to $f(x) = \sqrt{x}$ over the

interval [0,1].

7. $f(x) = x^2$ is in P_2 [0,1], so the least squares quadratic approximation to $f(x) = x^2$ is x^2 .

8. The set $\left\{\pi^{-\frac{1}{2}}, 2\sqrt{3}\,\pi^{-\frac{3}{2}}\left(x-\frac{\pi}{2}\right), 6\sqrt{5}\,\pi^{-\frac{5}{2}}\left(x^2 - \pi x + \frac{\pi^2}{6}\right)\right\}$ is an orthonormal

basis for P_2 $[0,\pi]$. $\langle \sin x, \pi^{-\frac{1}{2}} \rangle = \pi^{-\frac{1}{2}} \int_0^{\pi} \sin x \; dx = \pi^{-\frac{1}{2}} [-\cos x]_0^{\pi} = 2\pi^{-\frac{1}{2}}$,

$\langle \sin x,\ 2\sqrt{3}\ \pi^{-\frac{3}{2}}\left(x - \frac{\pi}{2}\right)\rangle = 2\sqrt{3}\ \pi^{-\frac{3}{2}} \int_0^{\pi} (x\sin x - \frac{\pi}{2}\sin x)dx$

$= 2\sqrt{3}\ \pi^{-\frac{3}{2}} \left[-x\cos x + \sin x + \frac{\pi}{2}\cos x\right]_0^{\pi}$ = 0, and

$\left\langle \sin x,\ 6\sqrt{5}\ \pi^{-\frac{5}{2}}\left(x^2 - \pi x + \frac{\pi^2}{6}\right) \right\rangle$

$= 6\sqrt{5}\ \pi^{-\frac{5}{2}} \int_0^{\pi} (x^2\sin x - \pi x\sin x + \frac{\pi^2}{6}\sin x)dx$

$= 6\sqrt{5}\ \pi^{-\frac{5}{2}} \left[-x^2\cos x + 2x\sin x + 2\cos x + \pi x\cos x - \pi\sin x - \frac{\pi^2}{6}\cos x\right]_0^{\pi}$

$= 6\sqrt{5}\ \pi^{-\frac{5}{2}} \left(-4 + \frac{\pi^2}{3}\right)$, so

$\text{proj}_{P_2[0,\pi]} \sin x = 2\pi^{-\frac{1}{2}}\ \pi^{-\frac{1}{2}} + 6\sqrt{5}\ \pi^{-\frac{5}{2}}\left(-4 + \frac{\pi^2}{3}\right) 6\sqrt{5}\ \pi^{-\frac{5}{2}}\left(x^2 - \pi x + \frac{\pi^2}{6}\right)$

$= 180\pi^{-5}\left(-4 + \frac{\pi^2}{3}\right) x^2 - 180\pi^{-4}\left(-4 + \frac{\pi^2}{3}\right) x - 120\pi^{-3} + 12\pi^{-1}$ is the least squares

quadratic approximation to cos x over the interval $[0,\pi]$.

9. The vectors that form an orthonormal basis over $[-\pi, \pi]$ also form an orthonormal basis over $[0,2\pi]$, so the Fourier approximations can be found in the same way, only with integration over $[0,2\pi]$.

$$a_0 = \frac{1}{2\pi}\int_0^{2\pi} x\,dx = \left[\frac{1}{4\pi}x^2\right]_0^{2\pi} = \pi,$$

$$a_k = \frac{1}{\pi}\int_0^{2\pi} x\cos kx\,dx = \frac{1}{\pi}\left[\frac{x}{k}\sin kx + \frac{1}{k^2}\cos kx\right]_0^{2\pi} = 0, \text{ and}$$

$$b_k = \frac{1}{\pi}\int_0^{2\pi} x\sin kx\,dx = \frac{1}{\pi}\left[\frac{x}{k}\cos kx + \frac{1}{k^2}\sin kx\right]_0^{2\pi} = \frac{1}{\pi}\left(-\frac{2\pi}{k}\cos 2k\pi\right) = \frac{-2}{k},$$

so the fourth-order Fourier approximation to f(x) over $[0,2\pi]$ is

$$g(x) = \pi + \sum_{k=1}^{4}\frac{-2}{k}\sin kx = \pi - 2\left(\sin x + \frac{1}{2}\sin 2x + \frac{1}{3}\sin 3x + \frac{1}{4}\sin 4x\right).$$

10. The fourth-order Fourier approximation to $f(x) = 1 + x$ over the interval $[-\pi,\pi]$ is $1 + g(x)$, where g(x) is the fourth-order Fourier approximation to $h(x) = x$, which was found in the text. $1 + g(x) = 1 + 2\left(\sin x - \frac{1}{2}\sin 2x + \frac{1}{3}\sin 3x - \frac{1}{4}\sin 4x\right)$.

11. $$a_0 = \frac{1}{2\pi}\int_0^{2\pi} x^2\,dx = \left[\frac{1}{6\pi}x^3\right]_0^{2\pi} = \frac{4}{3}\pi^2,$$

$$a_k = \frac{1}{\pi}\int_0^{2\pi} x^2\cos kx\,dx = \frac{1}{\pi}\left[\frac{x^2}{k}\sin kx - \frac{2}{k}\left(\frac{-x}{k}\cos kx + \frac{1}{k^2}\sin kx\right)\right]_0^{2\pi} = \frac{4}{k^2}, \text{ and}$$

$$b_k = \frac{1}{\pi}\int_0^{2\pi} x^2\sin kx\,dx = \frac{1}{\pi}\left[\frac{-x^2}{k}\cos kx - \frac{2}{k}\left(\frac{x}{k}\sin kx + \frac{1}{k^2}\cos kx\right)\right]_0^{2\pi} = -\frac{4\pi}{k}, \text{ so}$$

$$g(x) = \frac{4}{3}\pi^2 + \sum_{k=1}^{4}\left(\frac{4}{k^2}\cos kx - \frac{4\pi}{k}\sin kx\right)$$

$$= \frac{4}{3}\pi^2 + 4\left(\cos x + \frac{1}{4}\cos 2x + \frac{1}{9}\cos 3x + \frac{1}{16}\cos 4x\right)$$

$$- 4\pi\left(\sin x + \frac{1}{2}\sin 2x + \frac{1}{3}\sin 3x + \frac{1}{4}\sin 4x\right).$$

12. $$<1,\sin nx> = \int_{-\pi}^{\pi} \sin nx \; dx = \left[-\frac{1}{n}\cos nx\right]_{-\pi}^{\pi} = 0,$$

$$<1 \cos nx> = \int_{-\pi}^{\pi} \cos nx \; dx = \left[\frac{1}{n}\sin nx\right]_{-\pi}^{\pi} = 0,$$

$$<\sin mx, \sin nx> = \int_{-\pi}^{\pi} \sin mx \sin nx \; dx = \left[\frac{\sin (m-n)x}{2(m-n)} - \frac{\sin (m+n)x}{2(m+n)}\right]_{-\pi}^{\pi} = 0,$$

$$<\cos mx, \cos nx> = \int_{-\pi}^{\pi} \cos mx \cos nx \; dx = \left[\frac{\sin (m-n)x}{2(m-n)} + \frac{\sin (m+n)x}{2(m+n)}\right]_{-\pi}^{\pi} = 0, \text{ and}$$

$$<\sin mx, \cos nx> = \int_{-\pi}^{\pi} \sin mx \cos nx \; dx = \left[\frac{\cos (m-n)x}{2(m-n)} - \frac{\cos (m+n)x}{2(m+n)}\right]_{-\pi}^{\pi} = 0, \text{ so}$$

the vectors are mutually orthogonal in $C[-\pi,\pi]$.

13. (1,0,0,0,0,1,1), (0,1,0,0,1,0,1), (0,0,1,0,1,1,0), (0,0,0,1,1,1,1),
(0,1,1,1,1,0,0), (1,0,1,1,0,1,0), (1,1,0,1,0,0,1), (1,1,1,0,0,0,0),
(1,1,0,0,1,1,0), (1,0,1,0,1,0,1), (1,0,0,1,1,0,0), (1,1,1,1,1,1,1),
(0,0,1,1,0,0,1), (0,1,0,1,0,1,0), (0,1,1,0,0,1,1), (0,0,0,0,0,0,0)

14. (a) center = (0,0,1,0,1,1,0)
(1,0,1,0,1,1,0), (0,1,1,0,1,1,0), (0,0,0,0,1,1,0), (0,0,1,1,1,1,0), (0,0,1,0,0,1,0), (0,0,1,0,1,0,0), (0,0,1,0,1,1,1)

(b) center = (1,1,0,1,0,0,1)
(0,1,0,1,0,0,1), (1,0,0,1,0,0,1), (1,1,1,1,0,0,1), (1,1,0,0,0,0,1), (1,1,0,1,1,0,1), (1,1,0,1,0,1,1), (1,1,0,1,0,0,0)

(c) center = (1,1,1,1,1,1,1)
(0,1,1,1,1,1,1), (1,0,1,1,1,1,1), (1,1,0,1,1,1,1), (1,1,1,0,1,1,1), (1,1,1,1,0,1,1), (1,1,1,1,1,0,1), (1,1,1,1,1,1,0)

(d) center = (0,0,0,0,0,0,0)
(1,0,0,0,0,0,0), (0,1,0,0,0,0,0), (0,0,1,0,0,0,0), (0,0,0,1,0,0,0), (0,0,0,0,1,0,0), (0,0,0,0,0,1,0), (0,0,0,0,0,0,1)

15. (a) (0,0,0,0,0,0,0) (b) (0,0,1,1,0,0,1) (c) (0,0,1,0,1,1,0)
(d) (1,0,1,1,0,1,0)

16. (a) Since each vector in V_{23} has 23 components and there are 2 possible values (zero and 1) for each component, there are 2^{23} vectors.

(b) Each vector in $C_{23,12}$ is a linear combination of 12 basis vectors and each coefficient is either zero or 1. Thus there are 2^{12} = 4096 vectors.

(c) There are 23 ways to change exactly one component of the center vector, 23x22/2 = 253 ways to change exactly two components, and 23x22x21/6 = 1771 ways to change exactly three components. 1 + 23 + 253 + 1771 = 2048 vectors.

Exercise Set 6.4

1. $$\text{Pinv}\begin{bmatrix} 1 & 3 \\ 0 & 1 \\ -1 & 2 \end{bmatrix} = \left(\begin{bmatrix} 1 & 0 & -1 \\ 3 & 1 & 2 \end{bmatrix}\begin{bmatrix} 1 & 3 \\ 0 & 1 \\ -1 & 2 \end{bmatrix}\right)^{-1}\begin{bmatrix} 1 & 0 & -1 \\ 3 & 1 & 2 \end{bmatrix} = \begin{bmatrix} 2 & 1 \\ 1 & 14 \end{bmatrix}^{-1}\begin{bmatrix} 1 & 0 & -1 \\ 3 & 1 & 2 \end{bmatrix}$$

$$= \frac{1}{27}\begin{bmatrix} 14 & -1 \\ -1 & 2 \end{bmatrix}\begin{bmatrix} 1 & 0 & -1 \\ 3 & 1 & 2 \end{bmatrix} = \frac{1}{27}\begin{bmatrix} 11 & -1 & -16 \\ 5 & 2 & 5 \end{bmatrix}.$$

2. $$\text{Pinv}\begin{bmatrix} 1 & 1 \\ 2 & 3 \\ 0 & 1 \end{bmatrix} = \left(\begin{bmatrix} 1 & 2 & 0 \\ 1 & 3 & 1 \end{bmatrix}\begin{bmatrix} 1 & 1 \\ 2 & 3 \\ 0 & 1 \end{bmatrix}\right)^{-1}\begin{bmatrix} 1 & 2 & 0 \\ 1 & 3 & 1 \end{bmatrix} = \frac{1}{6}\begin{bmatrix} 4 & 1 & -7 \\ -2 & 1 & 5 \end{bmatrix}.$$

3. The pseudoinverse does not exist because $\left(\begin{bmatrix} 1 & 2 & 3 \\ 2 & 4 & 6 \end{bmatrix}\begin{bmatrix} 1 & 2 \\ 2 & 4 \\ 3 & 6 \end{bmatrix}\right)^{-1}$ does not exist.

4. $$\text{Pinv}\begin{bmatrix} 1 & 2 \\ 3 & 4 \\ 5 & 6 \end{bmatrix} = \left(\begin{bmatrix} 1 & 3 & 5 \\ 2 & 4 & 6 \end{bmatrix}\begin{bmatrix} 1 & 2 \\ 3 & 4 \\ 5 & 6 \end{bmatrix}\right)^{-1}\begin{bmatrix} 1 & 3 & 5 \\ 2 & 4 & 6 \end{bmatrix} = \frac{1}{24}\begin{bmatrix} -32 & -8 & 16 \\ 26 & 8 & -10 \end{bmatrix}$$

$$= \frac{1}{12}\begin{bmatrix} -16 & -4 & 8 \\ 13 & 4 & -5 \end{bmatrix}.$$

5. Pinv $\begin{bmatrix} 1 & 3 \\ 2 & 0 \\ 1 & -1 \end{bmatrix} = \left(\begin{bmatrix} 1 & 2 & 1 \\ 3 & 0 & -1 \end{bmatrix} \begin{bmatrix} 1 & 3 \\ 2 & 0 \\ 1 & -1 \end{bmatrix} \right)^{-1} \begin{bmatrix} 1 & 2 & 1 \\ 3 & 0 & -1 \end{bmatrix} = \frac{1}{56} \begin{bmatrix} 4 & 20 & 12 \\ 16 & -4 & -8 \end{bmatrix}$

$= \frac{1}{14} \begin{bmatrix} 1 & 5 & 3 \\ 4 & -1 & -2 \end{bmatrix}$.

6. There is no pseudoinverse. This is not the coefficient matrix of an overdetermined system of equations.

7. There is no pseudoinverse. This is not the coefficient matrix of an overdetermined system of equations.

8. Pinv $\begin{bmatrix} 1 & 2 \\ -3 & 5 \end{bmatrix} = \left(\begin{bmatrix} 1 & -3 \\ 2 & 5 \end{bmatrix} \begin{bmatrix} 1 & 2 \\ -3 & 5 \end{bmatrix} \right)^{-1} \begin{bmatrix} 1 & -3 \\ 2 & 5 \end{bmatrix} = \frac{1}{11} \begin{bmatrix} 5 & -2 \\ 3 & 1 \end{bmatrix}$.

9. The equation of the line will be y = a + bx. The data points give 0 = a + b, 4 = a + 2b, and 7 = a + 3b. Thus the system of equations is given by $A \begin{bmatrix} a \\ b \end{bmatrix} = \begin{bmatrix} 0 \\ 4 \\ 7 \end{bmatrix}$, where $A = \begin{bmatrix} 1 & 1 \\ 1 & 2 \\ 1 & 3 \end{bmatrix}$.

Pinv $A = \left(\begin{bmatrix} 1 & 1 & 1 \\ 1 & 2 & 3 \end{bmatrix} \begin{bmatrix} 1 & 1 \\ 1 & 2 \\ 1 & 3 \end{bmatrix} \right)^{-1} \begin{bmatrix} 1 & 1 & 1 \\ 1 & 2 & 3 \end{bmatrix} = \frac{1}{6} \begin{bmatrix} 8 & 2 & -4 \\ -3 & 0 & 3 \end{bmatrix}$.

Pinv $A \begin{bmatrix} 0 \\ 4 \\ 7 \end{bmatrix} = \frac{1}{6} \begin{bmatrix} -20 \\ 21 \end{bmatrix}$, so $a = \frac{-10}{3}$, $b = \frac{7}{2}$, and the least squares line is $y = \frac{-10}{3} + \frac{7}{2}x$.

10. The system of equations is given by $A \begin{bmatrix} a \\ b \end{bmatrix} = \begin{bmatrix} 1 \\ 3 \\ 7 \end{bmatrix}$, with A given in Exercise 9.

Pinv $A \begin{bmatrix} 1 \\ 3 \\ 7 \end{bmatrix} = \frac{1}{6} \begin{bmatrix} 8 & 2 & -4 \\ -3 & 0 & 3 \end{bmatrix} \begin{bmatrix} 1 \\ 3 \\ 7 \end{bmatrix} = \frac{1}{6} \begin{bmatrix} -14 \\ 18 \end{bmatrix}$, so

$a = -7/3$, $b = 3$, and the least squares line is $y = -7/3 + 3x$.

11. The system of equations is given by $A\begin{bmatrix} a \\ b \end{bmatrix} = \begin{bmatrix} 1 \\ 5 \\ 9 \end{bmatrix}$, with A given in Exercise 9.

$$\text{Pinv } A\begin{bmatrix} 1 \\ 5 \\ 9 \end{bmatrix} = \frac{1}{6}\begin{bmatrix} 8 & 2 & -4 \\ -3 & 0 & 3 \end{bmatrix}\begin{bmatrix} 1 \\ 5 \\ 9 \end{bmatrix} = \frac{1}{6}\begin{bmatrix} -18 \\ 24 \end{bmatrix},$$ so $a = -3$, $b = 4$, and the least squares line is $y = -3 + 4x$.

12. The system of equations is given by $A\begin{bmatrix} a \\ b \end{bmatrix} = \begin{bmatrix} 1 \\ 6 \\ 9 \end{bmatrix}$, with A given in Exercise 9.

$$\text{Pinv } A\begin{bmatrix} 1 \\ 6 \\ 9 \end{bmatrix} = \frac{1}{6}\begin{bmatrix} 8 & 2 & -4 \\ -3 & 0 & 3 \end{bmatrix}\begin{bmatrix} 1 \\ 6 \\ 9 \end{bmatrix} = \frac{1}{6}\begin{bmatrix} -16 \\ 24 \end{bmatrix},$$ so

$a = -8/3$, $b = 4$, and the least squares line is $y = -8/3 + 4x$.

13. The system of equations is given by $A\begin{bmatrix} a \\ b \end{bmatrix} = \begin{bmatrix} 7 \\ 2 \\ 0 \end{bmatrix}$, with A given in Exercise 9.

$$\text{Pinv } A\begin{bmatrix} 7 \\ 2 \\ 0 \end{bmatrix} = \frac{1}{6}\begin{bmatrix} 8 & 2 & -4 \\ -3 & 0 & 3 \end{bmatrix}\begin{bmatrix} 7 \\ 2 \\ 0 \end{bmatrix} = \frac{1}{6}\begin{bmatrix} 60 \\ -21 \end{bmatrix},$$ so

$a = 10$, $b = -7/2$, and the least squares line is $y = 10 - 7x/2$.

14. The system of equations is given by $A\begin{bmatrix} a \\ b \end{bmatrix} = \begin{bmatrix} 12 \\ 5 \\ 1 \end{bmatrix}$, with A given in Exercise 9.

$$\text{Pinv } A\begin{bmatrix} 12 \\ 5 \\ 1 \end{bmatrix} = \frac{1}{6}\begin{bmatrix} 8 & 2 & -4 \\ -3 & 0 & 3 \end{bmatrix}\begin{bmatrix} 12 \\ 5 \\ 1 \end{bmatrix} = \frac{1}{6}\begin{bmatrix} 102 \\ -33 \end{bmatrix},$$ so

$a = 17$, $b = -11/2$, and the least squares line is $y = 17 - 11x/2$.

15. The equation of the line will be y = a + bx. The data points give 0 = a + b, 3 = a + 2b, 5 = a + 3b, and 6 = a + 4b. Thus the system of equations is given by $A\begin{bmatrix} a \\ b \end{bmatrix} = \begin{bmatrix} 0 \\ 3 \\ 5 \\ 6 \end{bmatrix}$,

where $A = \begin{bmatrix} 1 & 1 \\ 1 & 2 \\ 1 & 3 \\ 1 & 4 \end{bmatrix}$. Pinv $A = \frac{1}{10}\begin{bmatrix} 10 & 5 & 0 & -5 \\ -3 & -1 & 1 & 3 \end{bmatrix}$. Pinv $A\begin{bmatrix} 0 \\ 3 \\ 5 \\ 6 \end{bmatrix} = \frac{1}{10}\begin{bmatrix} -15 \\ 20 \end{bmatrix}$, so

a = −3/2, b = 2, and the least squares line is y = −3/2 + 2x.

16. The system of equations is given by $A\begin{bmatrix} a \\ b \end{bmatrix} = \begin{bmatrix} 1 \\ 3 \\ 5 \\ 9 \end{bmatrix}$, with A given in Exercise 15.

Pinv $A\begin{bmatrix} 1 \\ 3 \\ 5 \\ 9 \end{bmatrix} = \frac{1}{10}\begin{bmatrix} 10 & 5 & 0 & -5 \\ -3 & -1 & 1 & 3 \end{bmatrix}\begin{bmatrix} 1 \\ 3 \\ 5 \\ 9 \end{bmatrix} = \frac{1}{10}\begin{bmatrix} -20 \\ 26 \end{bmatrix}$, so

a = −2, b = 13/5, and the least squares line is y = −2 + 13x/5.

17. The system of equations is given by $A\begin{bmatrix} a \\ b \end{bmatrix} = \begin{bmatrix} 9 \\ 7 \\ 3 \\ 2 \end{bmatrix}$, with A given in Exercise 15.

Pinv $A\begin{bmatrix} 9 \\ 7 \\ 3 \\ 2 \end{bmatrix} = \frac{1}{10}\begin{bmatrix} 10 & 5 & 0 & -5 \\ -3 & -1 & 1 & 3 \end{bmatrix}\begin{bmatrix} 9 \\ 7 \\ 3 \\ 2 \end{bmatrix} = \frac{1}{2}\begin{bmatrix} 23 \\ -5 \end{bmatrix}$, so

a = 23/2, b = −5/2, and the least squares line is y = 23/2 − 5x/2.

18. The system of equations is given by $A\begin{bmatrix} a \\ b \end{bmatrix} = \begin{bmatrix} 21 \\ 17 \\ 12 \\ 7 \end{bmatrix}$, with A given in Exercise 15.

$$\text{Pinv } A\begin{bmatrix}21\\17\\12\\7\end{bmatrix} = \frac{1}{10}\begin{bmatrix}10 & 5 & 0 & -5\\ -3 & -1 & 1 & 3\end{bmatrix}\begin{bmatrix}21\\17\\12\\7\end{bmatrix} = \frac{1}{10}\begin{bmatrix}260\\-47\end{bmatrix}, \text{ so}$$

a = 26, b = −47/10, and the least squares line is y = 26 − 47x/10.

19. The equation of the line will be y = a + bx. The data points give 0 = a + b, 1 = a + 2b, 4 = a + 3b, 7 = a + 4b, and 9 = a + 5b. Thus the system of equations is given by

$$A\begin{bmatrix}a\\b\end{bmatrix} = \begin{bmatrix}0\\1\\4\\7\\9\end{bmatrix}, \text{ where } A = \begin{bmatrix}1 & 1\\1 & 2\\1 & 3\\1 & 4\\1 & 5\end{bmatrix}.$$

$$\text{Pinv } A = \left(\begin{bmatrix}1 & 1 & 1 & 1 & 1\\1 & 2 & 3 & 4 & 5\end{bmatrix}\begin{bmatrix}1 & 1\\1 & 2\\1 & 3\\1 & 4\\1 & 5\end{bmatrix}\right)^{-1}\begin{bmatrix}1 & 1 & 1 & 1 & 1\\1 & 2 & 3 & 4 & 5\end{bmatrix}$$

$$= \frac{1}{10}\begin{bmatrix}8 & 5 & 2 & -1 & -4\\ -2 & -1 & 0 & 1 & 2\end{bmatrix}. \quad \text{Pinv } A\begin{bmatrix}0\\1\\4\\7\\9\end{bmatrix} = \frac{1}{10}\begin{bmatrix}-30\\24\end{bmatrix}, \text{ so}$$

a = −3, b = 12/5, and the least squares line is y = −3 + 12x/5.

20. The system of equations is given by $A\begin{bmatrix}a\\b\end{bmatrix} = \begin{bmatrix}1\\3\\4\\8\\11\end{bmatrix}$, with A given in Exercise 19.

$$\text{Pinv } A\begin{bmatrix}1\\3\\4\\8\\11\end{bmatrix} = \frac{1}{10}\begin{bmatrix}8 & 5 & 2 & -1 & -4\\ -2 & -1 & 0 & 1 & 2\end{bmatrix}\begin{bmatrix}1\\3\\4\\8\\11\end{bmatrix} = \frac{1}{10}\begin{bmatrix}-21\\25\end{bmatrix}, \text{ so}$$

a = −21/10, b = 5/2, and the least squares line is y = −21/10 + 5x/2.

21. The equation of the parabola will be $y = a + bx + cx^2$. The data points give 5 = a + b + c, 2 = a + 2b + 4c, 3 = a + 3b + 9c, and 8 = a + 4b + 16c. Thus the system of equations is

$$\text{given by } A\begin{bmatrix}a\\b\\c\end{bmatrix} = \begin{bmatrix}5\\2\\3\\8\end{bmatrix}, \text{ where } A = \begin{bmatrix}1 & 1 & 1\\1 & 2 & 4\\1 & 3 & 9\\1 & 4 & 16\end{bmatrix}.$$

$$\text{Pinv } A = \left(\begin{bmatrix}1 & 1 & 1 & 1\\1 & 2 & 3 & 4\\1 & 4 & 9 & 16\end{bmatrix}\begin{bmatrix}1 & 1 & 1\\1 & 2 & 4\\1 & 3 & 9\\1 & 4 & 16\end{bmatrix}\right)^{-1}\begin{bmatrix}1 & 1 & 1 & 1\\1 & 2 & 3 & 4\\1 & 4 & 9 & 16\end{bmatrix}$$

$$= \begin{bmatrix}4 & 10 & 30\\10 & 30 & 100\\30 & 100 & 354\end{bmatrix}^{-1}\begin{bmatrix}1 & 1 & 1 & 1\\1 & 2 & 3 & 4\\1 & 4 & 9 & 16\end{bmatrix} = \frac{1}{80}\begin{bmatrix}620 & -540 & 100\\-540 & 516 & -100\\100 & -100 & 20\end{bmatrix}\begin{bmatrix}1 & 1 & 1 & 1\\1 & 2 & 3 & 4\\1 & 4 & 9 & 16\end{bmatrix}$$

$$= \frac{1}{20}\begin{bmatrix}45 & -15 & -25 & 15\\-31 & 23 & 27 & -19\\5 & -5 & -5 & 5\end{bmatrix}. \quad \text{Pinv } A\begin{bmatrix}5\\2\\3\\8\end{bmatrix} = \begin{bmatrix}12\\-9\\2\end{bmatrix}, \text{ so}$$

a = 12, b = −9, c = 2, and the least squares parabola is $y = 12 - 9x + 2x^2$.

22. The system of equations is given by $A\begin{bmatrix}a\\b\\c\end{bmatrix} = \begin{bmatrix}4\\0\\3\\5\end{bmatrix}$ with A given in Exercise 21.

$$\text{Pinv A}\begin{bmatrix}4\\0\\3\\5\end{bmatrix} = \frac{1}{20}\begin{bmatrix}45 & -15 & -25 & 15\\-31 & 23 & 27 & -19\\5 & -5 & -5 & 5\end{bmatrix}\begin{bmatrix}4\\0\\3\\5\end{bmatrix} = \frac{1}{20}\begin{bmatrix}180\\-138\\30\end{bmatrix}, \text{ so}$$

a = 9, b = −69/10, c = 3/2, and the least squares parabola is $y = 9 - 69x/10 + 3x^2/2$.

23. The system of equations is given by $A\begin{bmatrix}a\\b\\c\end{bmatrix} = \begin{bmatrix}2\\5\\7\\1\end{bmatrix}$ with A given in Exercise 21.

$$\text{Pinv A}\begin{bmatrix}2\\5\\7\\1\end{bmatrix} = \frac{1}{20}\begin{bmatrix}45 & -15 & -25 & 15\\-31 & 23 & 27 & -19\\5 & -5 & -5 & 5\end{bmatrix}\begin{bmatrix}2\\5\\7\\1\end{bmatrix} = \frac{1}{20}\begin{bmatrix}-145\\223\\-45\end{bmatrix}, \text{ so } a = \frac{-29}{4},\ b = \frac{223}{20},$$

$c = \frac{-9}{4}$, and the least squares parabola is $y = \frac{-29}{4} + \frac{223}{20}x - \frac{9}{4}x^2$.

24. The system of equations is given by $A\begin{bmatrix}a\\b\\c\end{bmatrix} = \begin{bmatrix}23\\4\\7\\17\end{bmatrix}$ with A given in Exercise 21.

$$\text{Pinv A}\begin{bmatrix}23\\4\\7\\17\end{bmatrix} = \frac{1}{20}\begin{bmatrix}45 & -15 & -25 & 15\\-31 & 23 & 27 & -19\\5 & -5 & -5 & 5\end{bmatrix}\begin{bmatrix}23\\4\\7\\17\end{bmatrix} = \frac{1}{20}\begin{bmatrix}1055\\-755\\145\end{bmatrix}, \text{ so } a = \frac{211}{4},$$

$b = \frac{-151}{4}$, $c = \frac{29}{4}$, and the least squares parabola is $y = \frac{211}{4} - \frac{151}{4}x + \frac{29}{4}x^2$.

25. (a) Let line be L = a + bF (see example 5). Data points give 5.9 = a + 2b, 8 = a + 4b, 10.3 = a + 6b, and 12.2 = a + 8b. Thus the system of equations is given

by $A\begin{bmatrix} a \\ b \end{bmatrix} = \begin{bmatrix} 5.9 \\ 8 \\ 10.3 \\ 12.2 \end{bmatrix}$, where $A = \begin{bmatrix} 1 & 2 \\ 1 & 4 \\ 1 & 6 \\ 1 & 8 \end{bmatrix}$. $\text{pinv}(A) = \begin{bmatrix} 1 & .5 & 0 & -.5 \\ -.15 & -.05 & .05 & .15 \end{bmatrix}$,

as in example 5. $\text{pinv}(A)\begin{bmatrix} 5.9 \\ 8 \\ 10.3 \\ 12.2 \end{bmatrix} = \begin{bmatrix} 3.8 \\ 1.06 \end{bmatrix}$, so a = 3.8, b = 1.06, and the least

squares equation is L = 3.8 + 1.06F. If the force is 15 ounces, then the predicted

length of the spring is L = 3.8 + 1.06x15 = 19.7 inches.

(b) Let line be L = a + bF. Data points give 7.4 = a + 2b, 10.3 = a + 4b, 13.7 = a + 6b, and 16.7 = a + 8b. Thus the system of equations is

given by $A\begin{bmatrix} a \\ b \end{bmatrix} = \begin{bmatrix} 7.4 \\ 10.3 \\ 13.7 \\ 16.7 \end{bmatrix}$, where $A = \begin{bmatrix} 1 & 2 \\ 1 & 4 \\ 1 & 6 \\ 1 & 8 \end{bmatrix}$.

$\text{pinv}(A) = \begin{bmatrix} 1 & .5 & 0 & -.5 \\ -.15 & -.05 & .05 & .15 \end{bmatrix}$, as in example 5.

$\text{pinv}(A)\begin{bmatrix} 7.4 \\ 10.3 \\ 13.7 \\ 16.7 \end{bmatrix} = \begin{bmatrix} 4.2 \\ 1.565 \end{bmatrix}$, so a = 4.2, b = 1.565, and the least squares equation is L = 4.2 + 1.565F. If the force is 15 ounces, then the predicted length of the spring is L = 4.2 + 1.565x15 = 27.675 inches.

26. Let line be G = a + bS. The data points give 18.25 = a + 30b, 20 = a + 40b, 16.32 = a + 50b, 15.77 = a + 60b, and 13.61 = a + 70b. Thus the system of

equations is given by $A\begin{bmatrix} a \\ b \end{bmatrix} = \begin{bmatrix} 18.25 \\ 20.00 \\ 16.32 \\ 15.77 \\ 13.61 \end{bmatrix}$, where $A = \begin{bmatrix} 1 & 30 \\ 1 & 40 \\ 1 & 50 \\ 1 & 60 \\ 1 & 70 \end{bmatrix}$. $\text{pinv } A = (A^t A)^{-1} A^t$

$$= \frac{1}{5000}\begin{bmatrix} 13500 & -250 \\ -250 & 5 \end{bmatrix}\begin{bmatrix} 1 & 1 & 1 & 1 & 1 \\ 30 & 40 & 50 & 60 & 70 \end{bmatrix} = \frac{1}{50}\begin{bmatrix} 60 & 35 & 10 & -15 & -40 \\ 1 & .5 & 0 & .5 & 1 \end{bmatrix}.$$

$$\text{pinv } A\begin{bmatrix} 18.25 \\ 20.00 \\ 16.32 \\ 15.77 \\ 13.61 \end{bmatrix} = \frac{1}{50}\begin{bmatrix} 1177.25 \\ -6.755 \end{bmatrix},$$ so a = 23.545, b = −.1351, and the least squares

line is G = 23.545 − .1351S. The predicted mileage per gallon of this car at 55 mph is

G = 23.545 − .1351x55 = 16.1145 miles per gallon.

27. Let line be N = a + bX. The data points give 101 = a + 20b, 115 = a + 25b, 92 = a + 30b, 64 = a + 35b, 60 = a + 40b, 50 = a + 45b, and 49 = a + 50b.

Thus the system of equations is given by $A\begin{bmatrix} a \\ b \end{bmatrix} = \begin{bmatrix} 101 \\ 115 \\ 92 \\ 64 \\ 60 \\ 50 \\ 49 \end{bmatrix}$, where $A = \begin{bmatrix} 1 & 20 \\ 1 & 25 \\ 1 & 30 \\ 1 & 35 \\ 1 & 40 \\ 1 & 45 \\ 1 & 50 \end{bmatrix}$.

$$\text{pinv } A = (A^t A)^{-1} A^t = \frac{1}{4900}\begin{bmatrix} 9275 & -245 \\ -245 & 7 \end{bmatrix}\begin{bmatrix} 1 & 1 & 1 & 1 & 1 & 1 & 1 \\ 20 & 25 & 30 & 35 & 40 & 45 & 50 \end{bmatrix}$$

$$= \frac{1}{4900}\begin{bmatrix} 4375 & 3150 & 1925 & 700 & -525 & -1750 & -2975 \\ -105 & -70 & -35 & 0 & 35 & 70 & 105 \end{bmatrix}.$$

$$\text{pinv } A\begin{bmatrix} 101 \\ 115 \\ 92 \\ 64 \\ 60 \\ 50 \\ 49 \end{bmatrix} = \frac{1}{4900}\begin{bmatrix} 761250 \\ -11130 \end{bmatrix},$$ so $a = \frac{76125}{490}$, $b = \frac{-1113}{490}$, and the least

squares equation is $N = \frac{76125}{490} - \frac{1113}{490}X$. The predicted number of driver fatalities at age 22 is $\dfrac{76125 - 1113x22}{490} = 105$.

28. Let line be U = a + bD. The data points give 520 = a + 1000b, 540 = a + 1500b, 582 = a + 2000b, 600 = a + 2500b, 610 = a + 3000b, and 615 = a + 3500b.

Thus the system of equations is given by $A\begin{bmatrix} a \\ b \end{bmatrix} = \begin{bmatrix} 520 \\ 540 \\ 582 \\ 600 \\ 610 \\ 615 \end{bmatrix}$, where $A = \begin{bmatrix} 1 & 1000 \\ 1 & 1500 \\ 1 & 2000 \\ 1 & 2500 \\ 1 & 3000 \\ 1 & 3500 \end{bmatrix}$.

pinv $A = (A^t A)^{-1} A^t$

$$= \frac{1}{26250000}\begin{bmatrix} 34750000 & -13500 \\ -13500 & 6 \end{bmatrix}\begin{bmatrix} 1 & 1 & 1 & 1 & 1 & 1 \\ 1000 & 1500 & 2000 & 2500 & 3000 & 3500 \end{bmatrix}$$

$$= \frac{1}{262500}\begin{bmatrix} 212500 & 145000 & 77500 & 10000 & -57500 & -125000 \\ -75 & -45 & -15 & 15 & 45 & 75 \end{bmatrix}.$$

$$\text{pinv } A\begin{bmatrix} 520 \\ 540 \\ 582 \\ 600 \\ 610 \\ 615 \end{bmatrix} = \frac{1}{262500}\begin{bmatrix} 127955000 \\ 10545 \end{bmatrix}, \text{ so } a = \frac{127955000}{262500}, \; b = \frac{10545}{262500},$$

and the least squares line is $U = \dfrac{127955000 + 10545D}{262500}$. Predicted sales when

\$5000 is spent on advertising is $U = \dfrac{127955000 + 10545x5000}{262500} = 688$ units.

29. Let equation be G = a + bS. The system of equations is given by

$$A\begin{bmatrix} a \\ b \end{bmatrix} = C, \text{ where } C = \begin{bmatrix} 3.2 \\ 2.8 \\ 3.9 \\ 2.4 \\ 4.0 \\ 3.5 \\ 2.7 \\ 3.6 \\ 3.5 \\ 2.4 \end{bmatrix} \text{ and } A = \begin{bmatrix} 1 & 490 \\ 1 & 450 \\ 1 & 640 \\ 1 & 510 \\ 1 & 680 \\ 1 & 610 \\ 1 & 480 \\ 1 & 450 \\ 1 & 600 \\ 1 & 650 \end{bmatrix}. \quad \text{pinv}(A)C = (A^t A)^{-1} A^t C$$

$$= \frac{1}{70840}\begin{bmatrix} 316220 & -556 \\ -556 & 1 \end{bmatrix} A^t C = \frac{1}{70840}\begin{bmatrix} 122716 \\ 187 \end{bmatrix}, \text{ so the least squares line is}$$

given by $G = \dfrac{122716 + 187S}{70840}$. $G = 1.732298137 + 0.0026397516s$.

Prediction of the graduating GPA of a student entering with an SAT of $S = 670$ is $G = 3.5$.

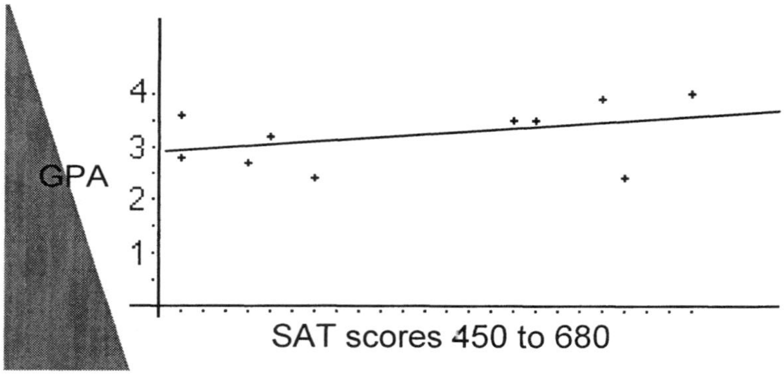

30. Let line be $\% = a + bY$. The system of equations is

$$\text{given by } A\begin{bmatrix} a \\ b \end{bmatrix} = C, \text{ where } C = \begin{bmatrix} .18 \\ .19 \\ .21 \\ .25 \\ .35 \\ .54 \\ .83 \end{bmatrix} \text{ and } A = \begin{bmatrix} 1 & 20 \\ 1 & 25 \\ 1 & 30 \\ 1 & 35 \\ 1 & 40 \\ 1 & 45 \\ 1 & 50 \end{bmatrix}. \quad \text{pinv } A = (A^t A)^{-1} A^t$$

$$= \frac{1}{4900}\begin{bmatrix} 4375 & 3150 & 1925 & 700 & -525 & -1750 & -2975 \\ -105 & -70 & -35 & 0 & 35 & 70 & 105 \end{bmatrix}, \text{ as in Exercise 27.}$$

$(\text{pinv } A)\ C = \dfrac{1}{4900}\begin{bmatrix} -1632.75 \\ 97.65 \end{bmatrix}$, so the least squares line is given by

$\% = \dfrac{-1632.75 + 97.65Y}{4900}$. The predicted percentage of deaths at 23 years of

age is $\dfrac{-1632.75 + 97.65\text{x}23}{4900} = 0.125$.

It is clear from the drawing below that the least squares line does not fit the data well. The parabola fits better. The equation of the least squares parabola will be

$\% = a + bY + cY^2$. The system of equations is given by $A\begin{bmatrix} a \\ b \\ c \end{bmatrix} = C$, where

$A = \begin{bmatrix} 1 & 20 & 400 \\ 1 & 25 & 625 \\ 1 & 30 & 900 \\ 1 & 35 & 1225 \\ 1 & 40 & 1600 \\ 1 & 45 & 2025 \\ 1 & 50 & 2500 \end{bmatrix}$ and C is given above. $(\text{pinv } A)C = \begin{bmatrix} .9364 \\ -.0591 \\ .0011 \end{bmatrix}$, so

the least squares parabola is $\% = .9364 - .0591Y + .0011Y^2$. The predicted percentage of deaths at 23 years of age is $.9364 - .0591\text{x}23 + .0011\text{x}529 = .159$.

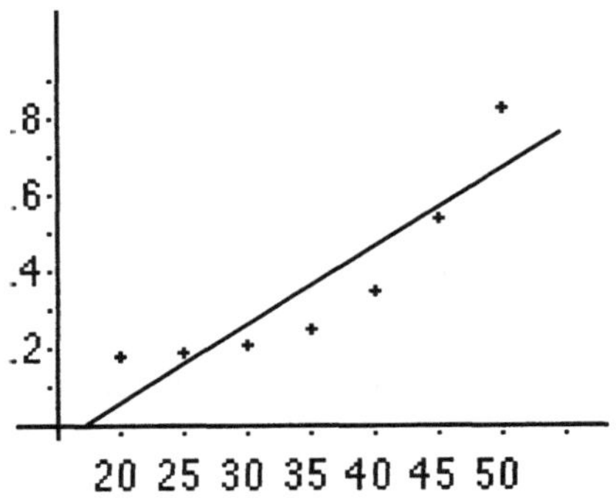

least squares line

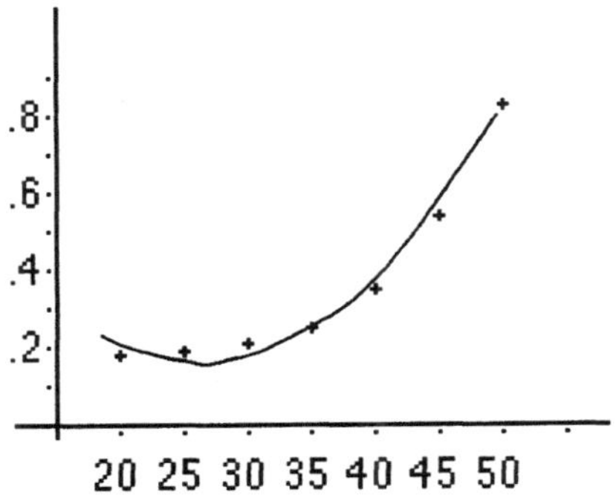

least squares parabola

31. Let cost-volume formula be $C = a + bV$. The system of equations is given by

$$A\begin{bmatrix} a \\ b \end{bmatrix} = D, \text{ where } D = \begin{bmatrix} 6300 \\ 6500 \\ 6670 \\ 6450 \\ 6100 \\ 5950 \end{bmatrix} \text{ and } A = \begin{bmatrix} 1 & 120 \\ 1 & 140 \\ 1 & 155 \\ 1 & 135 \\ 1 & 110 \\ 1 & 105 \end{bmatrix}. \quad (\text{pinv } A)D = (A^t A)^{-1} A^t D$$

$$= \frac{1}{11025}\begin{bmatrix} 99375 & -765 \\ -765 & 6 \end{bmatrix} A^t D = \frac{1}{441}\begin{bmatrix} 2020440 \\ 6042 \end{bmatrix},$$ so the least squares line is

given by $C = \dfrac{2020440 + 6042V}{441}$. For the next month, the predicted cost is

$$C = \frac{2020440 + 6042\text{x}165}{441} = \$6842.$$

32. Use example 6 as the model for handling both sets of data.

$$A = \begin{bmatrix} 1 & 0 & 1 \\ 1 & 10 & 100 \\ 1 & 20 & 400 \\ 1 & 30 & 900 \\ 1 & 31 & 961 \end{bmatrix}, \mathbf{y}_1 = \begin{bmatrix} 45.550 \\ 41.651 \\ 40.543 \\ 46.857 \\ 47.051 \end{bmatrix} \text{ and } \mathbf{y}_2 = \begin{bmatrix} 5.500 \\ 5.000 \\ 5.198 \\ 6.018 \\ 5.851 \end{bmatrix}$$

Get $[(A^tA)^{-1} A^t][\mathbf{y}_1 \ \ \mathbf{y}_2] = \begin{bmatrix} 2850.428571 & 12175 \\ 128.4285714 & 634 \end{bmatrix}$

The least squares parabolas are thus
Public: y = 2850.4286 + 128.4286x, Private: y = 12175 + 634x.
For 2005 (x=36), Public enrollment=51.5903 million, Private=6.6326 million
For 2010 (x=41), Public enrollment=56.9953 million, Private=7.3871 million.

33. (a) Let W be the plane given by the equation $x - y - z = 0$. For all points in this plane $x = y + z$, so $W = \{(y + z, y, z)\} = \{y(1,1,0) + z(1,0,1)\}$. Thus a basis for W is the set $\{(1,1,0), (1,0,1)\}$ and $A = \begin{bmatrix} 1 & 1 \\ 1 & 0 \\ 0 & 1 \end{bmatrix}$. $(A^t A)^{-1} = \frac{1}{3}\begin{bmatrix} 2 & -1 \\ -1 & 2 \end{bmatrix}$, and the projection

matrix $P = A(A^t A)^{-1} A^t = \frac{1}{3}\begin{bmatrix} 2 & 1 & 1 \\ 1 & 2 & -1 \\ 1 & -1 & 2 \end{bmatrix}$. $P\begin{bmatrix} 1 \\ 2 \\ 0 \end{bmatrix} = \frac{1}{3}\begin{bmatrix} 4 \\ 5 \\ -1 \end{bmatrix}$, so the projection of the vector (1,2,0) onto the plane $x - y - z = 0$ is the vector (4/3, 5/3, −1/3).

(b) Let W be the plane given by the equation $x - y + z = 0$. For all points in this plane $x = y - z$, so W = {(y − z, y, z)} = {y(1,1,0) + z(−1,0,1)}. Thus a basis for W is the set {(1,1,0), (−1,0,1)} and $A = \begin{bmatrix} 1 & -1 \\ 1 & 0 \\ 0 & 1 \end{bmatrix}$. $(A^t A)^{-1} = \frac{1}{3}\begin{bmatrix} 2 & 1 \\ 1 & 2 \end{bmatrix}$, and the projection

matrix $P = A(A^t A)^{-1} A^t = \frac{1}{3}\begin{bmatrix} 2 & 1 & -1 \\ 1 & 2 & 1 \\ -1 & 1 & 2 \end{bmatrix}$. $P\begin{bmatrix} 1 \\ 1 \\ 1 \end{bmatrix} = \frac{1}{3}\begin{bmatrix} 2 \\ 4 \\ 2 \end{bmatrix}$, so the projection of

the vector (1,1,1) onto the plane $x - y + z = 0$ is the vector (2/3, 4/3, 2/3).

(c) Let W be the plane given by the equation $x - 2y + z = 0$. For all points in this plane $x = 2y - z$, so W = {(2y − z, y, z)} = {y(2,1,0) + z(−1,0,1)}. Thus a basis for W is the set {(2,1,0), (−1,0,1)} and $A = \begin{bmatrix} 2 & -1 \\ 1 & 0 \\ 0 & 1 \end{bmatrix}$. $(A^t A)^{-1} = \frac{1}{6}\begin{bmatrix} 2 & 2 \\ 2 & 5 \end{bmatrix}$, and the projection

matrix $P = A(A^t A)^{-1} A^t = \frac{1}{6}\begin{bmatrix} 5 & 2 & -1 \\ 2 & 2 & 2 \\ -1 & 2 & 5 \end{bmatrix}$. $P\begin{bmatrix} 0 \\ 3 \\ 0 \end{bmatrix} = \frac{1}{6}\begin{bmatrix} 6 \\ 6 \\ 6 \end{bmatrix}$, so the projection of

the vector (0,3,0) onto the plane $x - 2y + z = 0$ is the vector (1,1,1).

34. (a) $A = \begin{bmatrix} 1 & 1 \\ -1 & 1 \\ 1 & -1 \end{bmatrix}$ and $(A^t A)^{-1} = \frac{1}{8}\begin{bmatrix} 3 & 1 \\ 1 & 3 \end{bmatrix}$. Thus the projection matrix

$P = A(A^t A)^{-1} A^t = \frac{1}{8}\begin{bmatrix} 8 & 0 & 0 \\ 0 & 4 & -4 \\ 0 & -4 & 4 \end{bmatrix}$. $P\begin{bmatrix} -2 \\ 1 \\ 3 \end{bmatrix} = \frac{1}{8}\begin{bmatrix} -16 \\ -8 \\ 8 \end{bmatrix}$, so the projection of the vector (−2,1,3) onto W is the vector (−2,−1,1).

(b) $A = \begin{bmatrix} 1 & 2 \\ 0 & 1 \\ 1 & 0 \end{bmatrix}$ and $(A^t A)^{-1} = \frac{1}{6}\begin{bmatrix} 5 & -2 \\ -2 & 2 \end{bmatrix}$. Thus the projection matrix

$P = A(A^t A)^{-1} A^t = \frac{1}{6}\begin{bmatrix} 5 & 2 & 1 \\ 2 & 2 & -2 \\ 1 & -2 & 5 \end{bmatrix}$. $P\begin{bmatrix} 6 \\ 0 \\ 12 \end{bmatrix} = \frac{1}{6}\begin{bmatrix} 42 \\ -12 \\ 66 \end{bmatrix}$, so the projection of

the vector (6,0,12) onto W is the vector (7,−2,11).

35. $A = \begin{bmatrix} 1 & 2 \\ 1 & 1 \\ 1 & 0 \end{bmatrix}$ and $(A^t A)^{-1} = \frac{1}{6}\begin{bmatrix} 5 & -3 \\ -3 & 3 \end{bmatrix}$. Thus the projection matrix

$P = A(A^t A)^{-1} A^t = \frac{1}{6}\begin{bmatrix} 5 & 2 & -1 \\ 2 & 2 & 2 \\ -1 & 2 & 5 \end{bmatrix}$. $P\begin{bmatrix} 3 \\ 0 \\ 1 \end{bmatrix} = \frac{1}{6}\begin{bmatrix} 14 \\ 8 \\ 2 \end{bmatrix}$, so the projection of the vector

(3,0,1) onto W is the vector (7/3,4/3,1/3). $P\begin{bmatrix} 6 \\ 3 \\ 0 \end{bmatrix} = \frac{1}{6}\begin{bmatrix} 36 \\ 18 \\ 0 \end{bmatrix} = \begin{bmatrix} 6 \\ 3 \\ 0 \end{bmatrix}$, so the projection

of the vector (6,3,0) onto W is the vector (6,3,0). This is because (6,3,0) is a multiple of a

basis vector and is therefore in W.

36. $A\mathbf{x} = \mathbf{y}$, so $A^t A\mathbf{x} = A^t\mathbf{y}$, and if $(A^t A)^{-1}$ exists then $\mathbf{x} = (A^t A)^{-1} A^t A\mathbf{x} = (A^t A)^{-1} A^t \mathbf{y}$.

$\mathbf{x} = (A^t A)^{-1} A^t \mathbf{y}$ solves the system $A^t A\mathbf{x} = A^t\mathbf{y}$ but does not solve the original system unless A^t (and therefore A) is an invertible matrix. The first step of multiplying by A^t is not reversible unless A^t is invertible. Least squares solution will be a solution if A is invertible.

37. If A is invertible then A^{-1} and $(A^t)^{-1}$ exist and $(A^t A)^{-1} = A^{-1}(A^t)^{-1}$ so that

$\text{pinv}(A) = (A^t A)^{-1} A^t = A^{-1}(A^t)^{-1} A^t = A^{-1}$.

38. $\text{pinv}(AB) = ((AB)^t(AB))^{-1}(AB)^t$.

If A is mxn and B is nxk then pinv(A) is nxm and pinv(B) is kxn, AB is mxk, and pinv (AB) is kxm. In general pinv(A)pinv(B) does not exist and pinv(B)pinv(A) is kxm, so (a) is not possible, but (b) is. In fact (b) obviously holds if A and B are invertible. (b) also holds for some fairly simple examples where A and B are not invertible, such as $A = \begin{bmatrix} 1 & 0 & 0 \\ 0 & 1 & 0 \\ 0 & 0 & 1 \\ 0 & 0 & 0 \end{bmatrix}$ and $B = \begin{bmatrix} 1 & 0 \\ 0 & 1 \\ 0 & 0 \end{bmatrix}$. However it is easy to find examples where (b) does not hold. If $A = \begin{bmatrix} 1 & 0 & 1 \\ 1 & 0 & 0 \\ 1 & 1 & 0 \\ 0 & 1 & 0 \end{bmatrix}$ and $B = \begin{bmatrix} 1 & 1 \\ 0 & 1 \\ 1 & 0 \end{bmatrix}$ then

$$\text{pinv}(B)\text{pinv}(A) = \frac{1}{3}\begin{bmatrix} 1 & -1 & 2 \\ 1 & 2 & -1 \end{bmatrix}\frac{1}{3}\begin{bmatrix} 0 & 2 & 1 & -1 \\ 0 & -1 & 1 & 2 \\ 3 & -2 & -1 & 1 \end{bmatrix} = \frac{1}{9}\begin{bmatrix} 6 & -1 & -2 & -1 \\ -3 & 2 & 4 & 2 \end{bmatrix} \text{ and}$$

$$\text{pinv}(AB) = \frac{1}{17}\begin{bmatrix} 9 & 2 & -3 & -5 \\ -4 & 1 & 7 & 6 \end{bmatrix}.$$

39. (a) $\text{Pinv}(cA) = ((cA)^t(cA))^{-1}(cA)^t = (c^2 A^t A)^{-1} cA^t = \frac{1}{c^2} (A^tA)^{-1} cA^t = \frac{1}{c} (A^tA)^{-1} A^t$.

(b) Pinv(A) exists if and only if $(A^t A)^{-1}$ exists, and $(A^t A)^{-1}$ exists if and only if $|A^t A| \neq 0$. Since A (and therefore A^t) is square, $|A^t A| = |A^t||A|$, so $|A^t A| \neq 0$ if and only if $|A| = |A^t| \neq 0$. Thus pinv(A) exists if and only if A is invertible, and therefore pinv(A) = A^{-1} (see Exercise 40) . Thus pinv(pinv(A)) = pinv(A^{-1}) = $(A^{-1})^{-1}$ = A.

(c) As in (b), pinv(A) = A^{-1} and pinv(A^t) = $(A^t)^{-1}$ = $(A^{-1})^t$ = $(\text{pinv}(A))^t$.

40. If A is mxn, then A^t is nxm and $A^t A$ is nxn. Pinv(A) exists if and only if $(A^t A)^{-1}$exists, and $(A^t A)^{-1}$exists if and only if $A^t A$ has rank n. We must show that $A^t A$ has rank n if and only

if A has rank n. A is the matrix of a linear transformation T from $\mathbf{R}^n$ to $\mathbf{R}^m$, A^t is the matrix of a linear transformation S from $\mathbf{R}^m$ to $\mathbf{R}^n$, and $A^t A$ is the matrix of the linear transformation ST from $\mathbf{R}^n$ to $\mathbf{R}^n$. Thus dim range(T) + dim ker(T) = dim domain(T) = dim($\mathbf{R}^n$) = n and dim range(ST) + dim ker(ST) = dim domain(ST) = dim($\mathbf{R}^n$) = n.

Dim range(T) = rank(A) and dim range(ST) = rank($A^t A$), so rank(A) + dim ker(T) = n = rank($A^t A$) + dim ker(ST). We show that ker(T) = ker(ST).

If **x** is in ker(T) then A**x** = **0** and $A^t A\mathbf{x} = \mathbf{0}$ so **x** is in ker(ST). If **x** is in ker(ST) then $A^t A\mathbf{x} = \mathbf{0}$ and $\mathbf{x}^t A^t A\mathbf{x} = 0$. But $\mathbf{x}^t A^t A\mathbf{x} = (A\mathbf{x})^t A\mathbf{x} = ||A\mathbf{x}||^2$ and if $||A\mathbf{x}||^2 = 0$ then $||A\mathbf{x}|| = 0$ and $A\mathbf{x} = \mathbf{0}$. Thus **x** is in ker(T). So dim ker(T) = 0 if and only if dim ker(ST) = 0 and rank(A) = n if and only if rank($A^t A$) = n.

If m < n, then rank(A) ≤ m < n, so pinv(A) does not exist.

41. $P = A(A^t A)^{-1} A^t$, where the columns of A are a basis for the vector space.

(a) $P^t = (A(A^t A)^{-1} A^t)^t = (A^t)^t ((A^tA)^{-1})^t A^t = A((A^t A)^t)^{-1} A^t = A(A^t A)^{-1} A^t = P$.

(b) $P^2 = (A(A^t A)^{-1} A^t)(A(A^t A)^{-1} A^t) = A(A^t A)^{-1}(A^tA)(A^tA)^{-1} A^t = A(A^t A)^{-1} A^t = P$.

<u>Chapter 6 Review Exercises</u>

1. $\mathbf{u} = (x_1, x_2)$, $\mathbf{v} = (y_1, y_2)$ $\mathbf{w} = (z_1, z_2)$, and c is a scalar.

$<\mathbf{u},\mathbf{v}> = 2x_1 y_1 + 3x_2 y_2 = 2y_1 x_1 + 3y_2 x_2 = <\mathbf{v},\mathbf{u}>$,

$<\mathbf{u}+\mathbf{v},\mathbf{w}> = 2(x_1 + y_1)z_1 + 3(x_2 + y_2)z_2 = 2x_1 z_1 + 2y_1 z_1 + 3x_2 z_2 + 3y_2 z_2$

$= 2x_1 z_1 + 3x_2 z_2 + 2y_1 z_1 + 3y_2 z_2 = <\mathbf{u},\mathbf{w}> + <\mathbf{v},\mathbf{w}>$,

$<c\mathbf{u},\mathbf{v}> = 2cx_1 y_1 + 3cx_2 y_2 = c(2x_1 y_1 + 3x_2 y_2) = c<\mathbf{u},\mathbf{v}>$, and

$<\mathbf{u},\mathbf{u}> = 2x_1x_1 + 3x_2x_2 = 2x_1^2 + 3x_2^2 \geq 0$ and equality holds if and only if $2x_1^2 = 0$ and $3x_2^2 = 0$, i.e. if and only if $x_1 = x_2 = 0$. Thus the given function is an inner product.

2. $<f,g> = \int_0^1 (3x-1)(5x+3)\ dx = \int_0^1 (15x^2+4x-3)\ dx = [\,5x^3 + 2x^2 - 3x]_0^1 = 4.$

$\|f\|^2 = \int_0^1 (3x-1)^2\ dx = \int_0^1 (9x^2-6x+1)\ dx = [\,3x^3 - 3x^2 + x\,]_0^1 = 1$, so $\|f\| = 1$.

$\|g\|^2 = \int_0^1 (5x+3)^2\ dx = \int_0^1 (25x^2+30x+9)\ dx = \left[\frac{25}{3}x^3 + 15x^2 + 9x\right]_0^1 = \frac{97}{3}$,

so $\|g\| = \frac{\sqrt{97}}{\sqrt{3}}$.

$d(f,g)^2 = \|f-g\|^2 = \|-2x-4\|^2 = \int_0^1 (-2x-4)^2\ dx = 4\int_0^1 (x^2+4x+4)\ dx$

$= 4\left[\frac{x^3}{3} + 2x^2 + 4x\right]_0^1 = 4(\frac{19}{3})$, so $d(f,g) = 2\frac{\sqrt{19}}{\sqrt{3}}$.

3. $\|(1,-2)\| = \sqrt{2(1)+3(4)} = \sqrt{14}$, $\|(3,2)\| = \sqrt{2(9)+3(4)} = \sqrt{30}$, and

$d((1,-2),(3,2)) = \|(-2,-4)\| = \sqrt{2(4)+3(16)} = \sqrt{56} = 2\sqrt{14}$.

4. $d((x_1, x_2), (0,0))^2 = \|(x_1, x_2)\|^2 = 2x_1^2 + 3x_2^2$, so the equation of the circle with radius 1 and center at the origin is $2x_1^2 + 3x_2^2 = 1$. In the sketch below, $a = 1/\sqrt{2}$ and $b = 1/\sqrt{3}$.

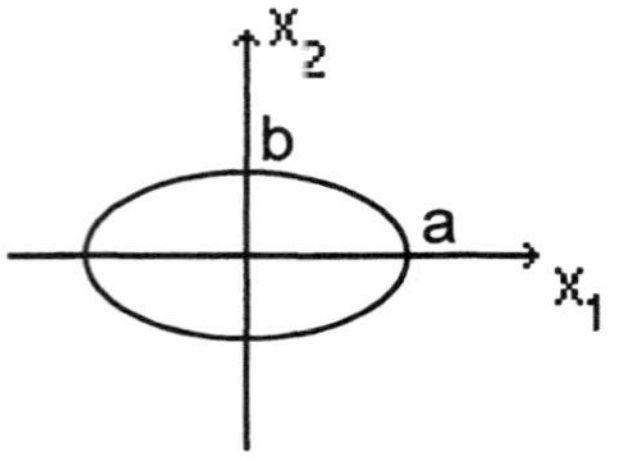

5. The set $\left\{\frac{1}{\sqrt{2}}, \frac{\sqrt{3}}{\sqrt{2}}x\right\}$ is an orthonormal basis for $P_1\ [-1,1]$ (see the text).

$$<x^2 + 2x - 1, \frac{1}{\sqrt{2}}> = \frac{1}{\sqrt{2}} \int_{-1}^{1} (x^2 + 2x - 1)\ dx = \frac{1}{\sqrt{2}} \left[\frac{x^3}{3} + x^2 - x\right]_{-1}^{1} = -\frac{4}{3\sqrt{2}} \text{ and}$$

$$<x^2 + 2x - 1, \frac{\sqrt{3}}{\sqrt{2}} x> = \frac{\sqrt{3}}{\sqrt{2}} \int_{-1}^{1} (x^3 + 2x^2 - x)\ dx = \frac{\sqrt{3}}{\sqrt{2}} \left[\frac{x^4}{4} + 2\frac{x^3}{3} - \frac{x^2}{2}\right]_{-1}^{1}$$

$$= \frac{\sqrt{3}}{\sqrt{2}} \frac{4}{3}, \text{ so that } \operatorname{proj}_{P_1[-1,1]} x^2 + 2x - 1 = -\frac{4}{3\sqrt{2}} \frac{1}{\sqrt{2}} + \frac{\sqrt{3}}{\sqrt{2}} \frac{4}{3} \frac{\sqrt{3}}{\sqrt{2}} x = 2x - \frac{2}{3}$$

is the least squares linear approximation to $f(x) = x^2 + 2x - 1$ over the interval $[-1,1]$.

6. $$a_0 = \frac{1}{2\pi}\int_0^{2\pi}(2x-1)dx = \frac{1}{2\pi}\ [x^2 - x]_0^{2\pi} = 2\pi - 1,$$

$$a_k = \frac{1}{\pi}\int_0^{2\pi}(2x-1)\cos kx\ dx = \frac{1}{\pi}[\frac{2x}{k}\sin kx + \frac{2}{k^2}\cos kx - \frac{1}{k}\sin kx)]_0^{2\pi} = 0, \text{ and}$$

$$b_k = \frac{1}{\pi}\int_0^{2\pi}(2x-1)\sin kx\ dx = \frac{1}{\pi}[\frac{2x}{k}\cos kx + \frac{2}{k^2}\sin kx + \frac{1}{k}\cos kx)]_0^{2\pi} = \frac{-4}{k}, \text{ so the}$$

fourth-order Fourier approximation to $f(x) = 2x-1$ over $[0,2\pi]$ is

$$g(x) = 2\pi - 1 + \sum_{k=1}^{4} \frac{-4}{k} \sin kx = 2\pi - 1 - 4\left(\sin x + \frac{1}{2} \sin 2x + \frac{1}{3} \sin 3x + \frac{1}{4} \sin 4x \right).$$

7. (1,1,1,0,0,1,1), (0,0,1,0,0,1,1), (0,1,0,0,0,1,1), (0,1,1,1,0,1,1), (0,1,1,0,1,1,1), (0,1,1,0,0,0,1), (0,1,1,0,0,1,0)

8. Pinv $\begin{bmatrix} 1 & 2 \\ 3 & 1 \\ 4 & 2 \end{bmatrix} = \left(\begin{bmatrix} 1 & 3 & 4 \\ 2 & 1 & 2 \end{bmatrix}\begin{bmatrix} 1 & 2 \\ 3 & 1 \\ 4 & 2 \end{bmatrix}\right)^{-1}\begin{bmatrix} 1 & 3 & 4 \\ 2 & 1 & 2 \end{bmatrix} = \frac{1}{65}\begin{bmatrix} -17 & 14 & 10 \\ 39 & -13 & 0 \end{bmatrix}$.

9. The equation of the parabola will be $y = a + bx + cx^2$. The data points give $6 = a + b + c$, $2 = a + 2b + 4c$, $5 = a + 3b + 9c$, and $9 = a + 4b + 16c$. Thus the system of equations is given by $A\begin{bmatrix} a \\ b \\ c \end{bmatrix} = \begin{bmatrix} 6 \\ 2 \\ 5 \\ 9 \end{bmatrix}$, where $A = \begin{bmatrix} 1 & 1 & 1 \\ 1 & 2 & 4 \\ 1 & 3 & 9 \\ 1 & 4 & 16 \end{bmatrix}$.

$$\text{Pinv } A = \left(\begin{bmatrix} 1 & 1 & 1 & 1 \\ 1 & 2 & 3 & 4 \\ 1 & 4 & 9 & 16 \end{bmatrix}\begin{bmatrix} 1 & 1 & 1 \\ 1 & 2 & 4 \\ 1 & 3 & 9 \\ 1 & 4 & 16 \end{bmatrix}\right)^{-1}\begin{bmatrix} 1 & 1 & 1 & 1 \\ 1 & 2 & 3 & 4 \\ 1 & 4 & 9 & 16 \end{bmatrix}$$

$$= \begin{bmatrix} 4 & 10 & 30 \\ 10 & 30 & 100 \\ 30 & 100 & 354 \end{bmatrix}^{-1}\begin{bmatrix} 1 & 1 & 1 & 1 \\ 1 & 2 & 3 & 4 \\ 1 & 4 & 9 & 16 \end{bmatrix} = \frac{1}{80}\begin{bmatrix} 620 & -540 & 100 \\ -540 & 516 & -100 \\ 100 & -100 & 20 \end{bmatrix}\begin{bmatrix} 1 & 1 & 1 & 1 \\ 1 & 2 & 3 & 4 \\ 1 & 4 & 9 & 16 \end{bmatrix}$$

$= \frac{1}{20}\begin{bmatrix} 45 & -15 & -25 & 15 \\ -31 & 23 & 27 & -19 \\ 5 & -5 & -5 & 5 \end{bmatrix}$. Pinv $A\begin{bmatrix} 6 \\ 2 \\ 5 \\ 9 \end{bmatrix} = \begin{bmatrix} 12.5 \\ -8.8 \\ 2 \end{bmatrix}$, so $a = 12.5$, $b = -8.8$, $c = 2$,

and the least squares parabola is $y = 12.5 - 8.8x + 2x^2$.

10. $B = \text{Pinv } A = (A^t A)^{-1} A^t$. If A is nxm, then B is mxn.

(a) $ABA = A(A^t A)^{-1} A^t A = AI_m = A$. (b) $BAB = (A^t A)^{-1} A^t AB = I_m B = B$.

(c) AB is the projection matrix for the subspace spanned by the columns of A. This matrix was shown to be symmetric in Exercise 44(a) of Section 6.4.

(d) $BA = (A^t A)^{-1} A^t A = I_m$ and is therefore symmetric.

Chapter 7

Exercise Set 7.1

1. (a) Yes. (b) Yes. (c) Yes.

   (d) No. The leading 1 of row three is not to the right of the leading 1 of row two.

2. (a) Yes. (b) No. The leading 1 in row 3 is not to the right of the leading 1 in row 2.

   (c) No. The leading nonzero number in row 2 is not a 1.

   (d) No. There is a row of zeros not at the bottom of the matrix.

3. $$\begin{bmatrix} 1 & 1 & 1 & 6 \\ 1 & -1 & 1 & 2 \\ 1 & 2 & 3 & 14 \end{bmatrix} \underset{\substack{R2+(-1)R1 \\ R3+(-1)R1}}{\approx} \begin{bmatrix} 1 & 1 & 1 & 6 \\ 0 & -2 & 0 & -4 \\ 0 & 1 & 2 & 8 \end{bmatrix} \underset{(-1/2)R2}{\approx} \begin{bmatrix} 1 & 1 & 1 & 6 \\ 0 & 1 & 0 & 2 \\ 0 & 1 & 2 & 8 \end{bmatrix}$$

$$\underset{R3+(-1)R2}{\approx} \begin{bmatrix} 1 & 1 & 1 & 6 \\ 0 & 1 & 0 & 2 \\ 0 & 0 & 2 & 6 \end{bmatrix} \underset{(1/2)R3}{\approx} \begin{bmatrix} 1 & 1 & 1 & 6 \\ 0 & 1 & 0 & 2 \\ 0 & 0 & 1 & 3 \end{bmatrix}$$

   (a) $x_1 + x_2 + x_3 = 6$, $x_2 = 2$, and $x_3 = 3$. Back substituting $x_2 = 2$ and $x_3 = 3$ into the first equation, $x_1 = 1$.

   (b) $\begin{bmatrix} 1 & 1 & 1 & 6 \\ 0 & 1 & 0 & 2 \\ 0 & 0 & 1 & 3 \end{bmatrix} \underset{R1+(-1)R3}{\approx} \begin{bmatrix} 1 & 1 & 0 & 3 \\ 0 & 1 & 0 & 2 \\ 0 & 0 & 1 & 3 \end{bmatrix} \underset{R1+(-1)R2}{} \begin{bmatrix} 1 & 0 & 0 & 1 \\ 0 & 1 & 0 & 2 \\ 0 & 0 & 1 & 3 \end{bmatrix}$, so again $x_1 = 1, x_2 = 2, x_3 = 3$.

4. $$\begin{bmatrix} 1 & -1 & -1 & 2 \\ 1 & -1 & 1 & 2 \\ 3 & -2 & 1 & 5 \end{bmatrix} \underset{\substack{R2+(-1)R1 \\ R3+(-3)R1}}{\approx} \begin{bmatrix} 1 & -1 & -1 & 2 \\ 0 & 0 & 2 & 0 \\ 0 & 1 & 4 & -1 \end{bmatrix} \underset{R2 \Leftrightarrow R3}{\approx} \begin{bmatrix} 1 & -1 & -1 & 2 \\ 0 & 1 & 4 & -1 \\ 0 & 0 & 2 & 0 \end{bmatrix}$$

$$\underset{(1/2)R3}{\approx}\begin{bmatrix}1 & -1 & -1 & 2\\ 0 & 1 & 4 & -1\\ 0 & 0 & 1 & 0\end{bmatrix}.$$

(a) $x_1 - x_2 - x_3 = 2$, $x_2 + 4x_3 = -1$, and $x_3 = 0$. Back substituting $x_3 = 0$ into the second equation, $x_2 = -1$, and substituting for x_2 and x_3 in the first equation, $x_1 = 1$.

(b) $$\begin{bmatrix}1 & -1 & -1 & 2\\ 0 & 1 & 4 & -1\\ 0 & 0 & 1 & 0\end{bmatrix}\underset{\substack{R1+R3\\ R2+(-4)R3}}{\approx}\begin{bmatrix}1 & -1 & 0 & 2\\ 0 & 1 & 0 & -1\\ 0 & 0 & 1 & 0\end{bmatrix}\underset{R1+R2}{\approx}\begin{bmatrix}1 & 0 & 0 & 1\\ 0 & 1 & 0 & -1\\ 0 & 0 & 1 & 0\end{bmatrix},$$ so again

$x_1 = 1$, $x_2 = -1$, $x_3 = 0$.

5. $$\begin{bmatrix}1 & -1 & 2 & 3\\ 2 & -2 & 5 & 4\\ 1 & 2 & -1 & -3\\ 0 & 2 & 2 & 1\end{bmatrix}\underset{\substack{R2+(-2)R1\\ R3+(-1)R1}}{\approx}\begin{bmatrix}1 & -1 & 2 & 3\\ 0 & 0 & 1 & -2\\ 0 & 3 & -3 & -6\\ 0 & 2 & 2 & 1\end{bmatrix}\underset{R2\Leftrightarrow R3}{\approx}\begin{bmatrix}1 & -1 & 2 & 3\\ 0 & 3 & -3 & -6\\ 0 & 0 & 1 & -2\\ 0 & 2 & 2 & 1\end{bmatrix}$$

$$\underset{(1/3)R2}{\approx}\begin{bmatrix}1 & -1 & 2 & 3\\ 0 & 1 & -1 & -2\\ 0 & 0 & 1 & -2\\ 0 & 2 & 2 & 1\end{bmatrix}\underset{R4+(-2)R2}{\approx}\begin{bmatrix}1 & -1 & 2 & 3\\ 0 & 1 & -1 & -2\\ 0 & 0 & 1 & -2\\ 0 & 0 & 4 & 5\end{bmatrix}.$$ There is no reason to

continue - the last two rows give inconsistent equations, so there is no solution.

6. $$\begin{bmatrix}1 & -1 & 1 & 2 & -2 & 1\\ 2 & -1 & -1 & 3 & -1 & 3\\ -1 & -1 & 5 & 0 & -4 & -3\end{bmatrix}\underset{\substack{R2+(-2)R1\\ R3+R1}}{\approx}\begin{bmatrix}1 & -1 & 1 & 2 & -2 & 1\\ 0 & 1 & -3 & -1 & 3 & 1\\ 0 & -2 & 6 & 2 & -6 & -2\end{bmatrix}$$

$$\underset{R3+2R2}{\approx}\begin{bmatrix}1 & -1 & 1 & 2 & -2 & 1\\ 0 & 1 & -3 & -1 & 3 & 1\\ 0 & 0 & 0 & 0 & 0 & 0\end{bmatrix}.$$

(a) $x_1 - x_2 + x_3 + 2x_4 - 2x_5 = 1$ and $x_2 - 3x_3 - x_4 + 3x_5 = 1$.
Back substituting $x_3 = r$, $x_4 = s$, $x_5 = t$ into the second equation, $x_2 = 1 + 3r + s - 3t$, and substituting for x_2, x_3, x_4, and x_5 in the first equation, $x_1 = 2 + 2r - s - t$.

(b) $\begin{bmatrix} 1 & -1 & 1 & 2 & -2 & 1 \\ 0 & 1 & -3 & -1 & 3 & 1 \\ 0 & 0 & 0 & 0 & 0 & 0 \end{bmatrix} \underset{R1+R2}{\approx} \begin{bmatrix} 1 & 0 & -2 & 1 & 1 & 2 \\ 0 & 1 & -3 & -1 & 3 & 1 \\ 0 & 0 & 0 & 0 & 0 & 0 \end{bmatrix}$, and again the general

solution is $x_1 = 2 + 2r - s - t$, $x_2 = 1 + 3r + s - 3t$, $x_3 = r$, $x_4 = s$, $x_5 = t$.

7 and 8. The augmented matrix is 4x(n+1), where n is the number of variables.

Gauss-Jordan elimination:

$$\begin{bmatrix} * & * & * & * & * & \cdots & * \\ * & * & * & * & * & \cdots & * \\ * & * & * & * & * & \cdots & * \\ * & * & * & * & * & \cdots & * \end{bmatrix} \underset{n\ mults}{\approx} \begin{bmatrix} 1 & * & * & * & * & \cdots & * \\ * & * & * & * & * & \cdots & * \\ * & * & * & * & * & \cdots & * \\ * & * & * & * & * & \cdots & * \end{bmatrix} \underset{\substack{3n\ mults \\ 3n\ adds}}{\approx} \begin{bmatrix} 1 & * & * & * & * & \cdots & * \\ 0 & * & * & * & * & \cdots & * \\ 0 & * & * & * & * & \cdots & * \\ 0 & * & * & * & * & \cdots & * \end{bmatrix}$$

$$\underset{n-1\ mults}{\approx} \begin{bmatrix} 1 & * & * & * & * & \cdots & * \\ 0 & 1 & * & * & * & \cdots & * \\ 0 & * & * & * & * & \cdots & * \\ 0 & * & * & * & * & \cdots & * \end{bmatrix} \underset{\substack{3(n-1)\ mults \\ 3(n-1)\ adds}}{\approx} \begin{bmatrix} 1 & 0 & * & * & * & \cdots & * \\ 0 & 1 & * & * & * & \cdots & * \\ 0 & 0 & * & * & * & \cdots & * \\ 0 & 0 & * & * & * & \cdots & * \end{bmatrix}$$

$$\underset{n-2\ mults}{\approx} \begin{bmatrix} 1 & 0 & * & * & * & \cdots & * \\ 0 & 1 & * & * & * & \cdots & * \\ 0 & 0 & 1 & * & * & \cdots & * \\ 0 & 0 & * & * & * & \cdots & * \end{bmatrix} \underset{\substack{3(n-2)\ mults \\ 3(n-2)\ adds}}{\approx} \begin{bmatrix} 1 & 0 & 0 & * & * & \cdots & * \\ 0 & 1 & 0 & * & * & \cdots & * \\ 0 & 0 & 1 & * & * & \cdots & * \\ 0 & 0 & 0 & * & * & \cdots & * \end{bmatrix}$$

$$\underset{n-3\ mults}{\approx} \begin{bmatrix} 1 & 0 & 0 & * & * & \cdots & * \\ 0 & 1 & 0 & * & * & \cdots & * \\ 0 & 0 & 1 & * & * & \cdots & * \\ 0 & 0 & 0 & 1 & * & \cdots & * \end{bmatrix} \underset{\substack{3(n-3)\ mults \\ 3(n-3)\ adds}}{\approx} \begin{bmatrix} 1 & 0 & 0 & 0 & * & \cdots & * \\ 0 & 1 & 0 & 0 & * & \cdots & * \\ 0 & 0 & 1 & 0 & * & \cdots & * \\ 0 & 0 & 0 & 1 & * & \cdots & * \end{bmatrix}.$$

number of mults: n + (n−1) + (n−2) + (n−3) + 3[n + (n−1) + (n−2) + (n−3)]
number of adds: 3[n + (n−1) + (n−2) + (n−3)]
total number of operations: 7[n + (n−1) + (n−2) + (n−3)] = 7(4n−6)

Gaussian elimination:

$$\begin{bmatrix} * & * & * & * & * & \cdots & * \\ * & * & * & * & * & \cdots & * \\ * & * & * & * & * & \cdots & * \\ * & * & * & * & * & \cdots & * \end{bmatrix} \underset{n\ mults}{\approx} \begin{bmatrix} 1 & * & * & * & * & \cdots & * \\ * & * & * & * & * & \cdots & * \\ * & * & * & * & * & \cdots & * \\ * & * & * & * & * & \cdots & * \end{bmatrix} \underset{\substack{3n\ mults \\ 3n\ adds}}{\approx} \begin{bmatrix} 1 & * & * & * & * & \cdots & * \\ 0 & * & * & * & * & \cdots & * \\ 0 & * & * & * & * & \cdots & * \\ 0 & * & * & * & * & \cdots & * \end{bmatrix}$$

$$\underset{n-1\ mults}{\approx} \begin{bmatrix} 1 & * & * & * & * & \cdots & * \\ 0 & 1 & * & * & * & \cdots & * \\ 0 & * & * & * & * & \cdots & * \\ 0 & * & * & * & * & \cdots & * \end{bmatrix} \underset{\substack{2(n-1)\ mults \\ 2(n-1)\ adds}}{\approx} \begin{bmatrix} 1 & * & * & * & * & \cdots & * \\ 0 & 1 & * & * & * & \cdots & * \\ 0 & 0 & * & * & * & \cdots & * \\ 0 & 0 & * & * & * & \cdots & * \end{bmatrix}$$

$$\underset{n-2\ mults}{\approx} \begin{bmatrix} 1 & * & * & * & * & \cdots & * \\ 0 & 1 & * & * & * & \cdots & * \\ 0 & 0 & 1 & * & * & \cdots & * \\ 0 & 0 & * & * & * & \cdots & * \end{bmatrix} \underset{\substack{n-2\ mults \\ n-2\ adds}}{\approx} \begin{bmatrix} 1 & * & * & * & * & \cdots & * \\ 0 & 1 & * & * & * & \cdots & * \\ 0 & 0 & 1 & * & * & \cdots & * \\ 0 & 0 & 0 & * & * & \cdots & * \end{bmatrix}$$

$$\underset{n-3\ mults}{\approx} \begin{bmatrix} 1 & * & * & * & * & \cdots & * \\ 0 & 1 & * & * & * & \cdots & * \\ 0 & 0 & 1 & * & * & \cdots & * \\ 0 & 0 & 0 & 1 & * & \cdots & * \end{bmatrix}.$$

number of mults: n + (n−1) + (n−2) + (n−3) + 3n + 2(n−1) + (n−2)
number of adds: 3n + 2(n−1) + (n−2)
total number of operations: 16n − 14

Back substitution:

$$\begin{bmatrix} 1 & * & * & * & * & \cdots & * \\ 0 & 1 & * & * & * & \cdots & * \\ 0 & 0 & 1 & * & * & \cdots & * \\ 0 & 0 & 0 & * & * & \cdots & * \end{bmatrix} \underset{\substack{3(n-3)\ mults \\ 3(n-3)\ adds}}{\approx} \begin{bmatrix} 1 & * & * & 0 & * & \cdots & * \\ 0 & 1 & * & 0 & * & \cdots & * \\ 0 & 0 & 1 & 0 & * & \cdots & * \\ 0 & 0 & 0 & 1 & * & \cdots & * \end{bmatrix}$$

$$\underset{\substack{2(n-3)\ mults \\ 2(n-3)\ adds}}{\approx} \begin{bmatrix} 1 & * & 0 & 0 & * & \cdots & * \\ 0 & 1 & 0 & 0 & * & \cdots & * \\ 0 & 0 & 1 & 0 & * & \cdots & * \\ 0 & 0 & 0 & 1 & * & \cdots & * \end{bmatrix} \underset{\substack{n-3\ mults \\ n-3\ adds}}{\approx} \begin{bmatrix} 1 & 0 & 0 & 0 & * & \cdots & * \\ 0 & 1 & 0 & 0 & * & \cdots & * \\ 0 & 0 & 1 & 0 & * & \cdots & * \\ 0 & 0 & 0 & 1 & * & \cdots & * \end{bmatrix}$$

number of mults: 3(n−3)+ 2(n−3) + (n−3), number of adds: 3(n−3)+ 2(n−3) + (n−3),
total number of operations: 12(n−3)

In both exercises, eight operations are saved using Gaussian elimination. One addition and one multiplication are saved in getting zeros in each of the positions above the diagonal in column 3 and two additions and two multiplications are saved in getting the zero above the diagonal in column 2. The basic reason for the savings is that in back substituting, one is working from right to left. This provides the same savings in operations above the diagonal that one has below the diagonal working from left to right.

Let n = 5 (Exercise 7).
For Gauss-Jordan elimination the number of operations is 7(4n−6) = 98.
For Gaussian elimination the total number of operations is 16n − 14 + 12(n−3) = 90.

Let n = 6 (Exercise 8).
For Gauss-Jordan elimination the number of operations is 7(4n−6) = 126.
For Gaussian elimination the total number of operations is 16n − 14 + 12(n−3) = 118.

9. Yes. Consider a 2x3 matrix with no zeros in the first column. Find the echelon form leaving the rows in the original order, then do it switching the order of the rows first.

10.
$$\begin{bmatrix} * & * & \cdots & * \\ * & * & \cdots & * \\ \vdots & \vdots & & \vdots \\ * & * & \cdots & * \end{bmatrix} \underset{n\ mults}{\approx} \begin{bmatrix} 1 & * & \cdots & * \\ * & * & \cdots & * \\ \vdots & \vdots & & \vdots \\ * & * & \cdots & * \end{bmatrix} \underset{\substack{(n-1)\ mults \\ (n-1)\ adds}}{\approx} \begin{bmatrix} 1 & * & \cdots & * \\ 0 & * & \cdots & * \\ \vdots & \vdots & & \vdots \\ 0 & * & \cdots & * \end{bmatrix}$$

$$\underset{n-1\ mults}{\approx} \begin{bmatrix} 1 & * & \cdots & * \\ 0 & 1 & \cdots & * \\ \vdots & \vdots & & \vdots \\ 0 & * & \cdots & * \end{bmatrix} \underset{\substack{(n-2)(n-1)\ mults \\ (n-2)(n-1)\ adds}}{\approx} \begin{bmatrix} 1 & * & \cdots & * \\ 0 & 1 & \cdots & * \\ \vdots & \vdots & & \vdots \\ 0 & 0 & \cdots & * \end{bmatrix} \underset{n-2\ mults}{\approx} \cdots$$

Back substitution

$$\underset{1\ mult}{\approx} \begin{bmatrix} 1 & * & \cdots & * & * \\ 0 & 1 & \cdots & * & * \\ \vdots & \vdots & & \vdots & \vdots \\ 0 & 0 & \cdots & 1 & * \end{bmatrix} \underset{\substack{n-1\ mults \\ n-1\ adds}}{\approx} \begin{bmatrix} 1 & * & \cdots & 0 & * \\ 0 & 1 & \cdots & 0 & * \\ \vdots & \vdots & & \vdots & \vdots \\ 0 & 0 & \cdots & 1 & * \end{bmatrix} \underset{\substack{n-2\ mults \\ n-2\ adds}}{\approx} \cdots \underset{\substack{1\ mult \\ 1\ add}}{\approx} \begin{bmatrix} 1 & 0 & \cdots & 0 & * \\ 0 & 1 & \cdots & 0 & * \\ \vdots & \vdots & & \vdots & \vdots \\ 0 & 0 & \cdots & 1 & * \end{bmatrix}$$

So the number of multiplications in Gaussian elimination is

$$n + (n-1)n + (n-2) + (n-2)(n-1) + \cdots + 2 + (1)2 + 1 = n^2 + (n-1)^2 + \cdots + 1$$

$= n(n+1)(2n+1)/6 = n^3/3 + n^2/2 + n/6$. The number of additions is

$(n-1)n + (n-2)(n-1) + \cdots + 2(3) + 1(2) = n^2 - n + (n-1)^2 - (n-1) + \cdots + 2^2 - 2 + (1-1)$

$= n^2 + (n-1)^2 + \cdots + 2^2 + 1 - [n + (n-1) + \cdots + 2 + 1] = n(n+1)(2n+1)/6 - n(n+1)/2$

$= n^3/3 + n^2/2 + n/6 - n^2/2 - n/2 = n^3/3 - n/3$.

Back substitution requires equal numbers of multiplications and additions:

$(n-1) + (n-2) + 2 + 1 = (n-1)n/2 = n^2/2 - n/2$. The total number of multiplications is

therefore $n^3/3 + n^2/2 + n/6 + n^2/2 - n/2 = n^3/3 + n^2 - n/3$ and the total number of

additions is $n^3/3 - n/3 + n^2/2 - n/2 = n^3/3 + n^2/2 - 5n/6$.

11.

	Gauss-Jordan		Gauss	
	mult	add	mult	add
n	$n^3/2 + n^2/2$	$n^3/2 - n/2$	$n^3/3 + n^2 - n/3$	$n^3/3 + n^2/2 - 5n/6$
2	6	3	6	3
3	18	12	17	7
4	40	30	36	26
5	75	60	65	50
6	126	105	106	85
7	196	168	161	133
8	288	252	232	196
9	405	360	321	276
10	550	495	430	375

Exercise Set 7.2

1. From the first equation $x_1 = 1$, and from the second equation $2 - x_2 = -2$, so $x_2 = 4$. From the third equation $3 + 4 - x_3 = 8$, so $x_3 = -1$.

2. From the first equation $x_1 = 2$, and from the second equation $-2 + x_2 = 1$, so $x_2 = 3$. From the third equation $2 + 3 + x_3 = 5$, so $x_3 = 0$.

3. From the first equation $x_1 = -2$, and from the second equation $-6 + x_2 = -5$, so $x_2 = 1$. From the third equation $-2 + 4 + 2x_3 = 4$, so $x_3 = 1$.

4. From the third equation $x_3 = 2$, and from the second equation $2x_2 + 2 = 3$, so $x_2 = 1/2$. From the first equation $x_1 + 2 + 1/2 = 3$, so $x_1 = 1/2$.

5. From the third equation $x_3 = 3$, and from the second equation $x_2 - 9 = -7$, so $x_2 = 2$. From the first equation $2x_1 - 2 + 3 = 3$, so $x_1 = 1$.

6. From the third equation $x_3 = 7$, and from the second equation $2x_2 + 28 = 26$, so $x_2 = -1$. From the first equation $3x_1 - 2 - 7 = -6$, so $x_1 = 1$.

7. The "rule" is Ri − aRj causes $a_{ij} = a$, where a_{ij} is the (i,j)th position in L.

(a) $L = \begin{bmatrix} 1 & 0 & 0 \\ 1 & 1 & 0 \\ -1 & -1 & 1 \end{bmatrix}$.

(b) $L = \begin{bmatrix} 1 & 0 & 0 \\ -5 & 1 & 0 \\ 2 & -7 & 1 \end{bmatrix}$.

(c) $L = \begin{bmatrix} 1 & 0 & 0 \\ -1/2 & 1 & 0 \\ -1/5 & 1 & 1 \end{bmatrix}$.

(d) $L = \begin{bmatrix} 1 & 0 & 0 & 0 \\ -4 & 1 & 0 & 0 \\ 2 & 2 & 1 & 0 \\ 8 & -2 & 6 & 1 \end{bmatrix}$.

8. (a) R2 − 2R1, R3 + 3R1, R3 − 5R2

(b) R2 + R1, R3 − 4R1, R3 − 2R2

(c) R3 − $\frac{1}{7}$ R1, R3 + $\frac{3}{4}$ R2

(d) R2 + 2R1, R3 − 3R1, R4 − 6R1, R3 − 5R2, R4 + 3R3

9. The matrix equation A**x** = **b** is $\begin{bmatrix} 1 & 2 & -1 \\ -2 & -1 & 3 \\ 1 & -1 & -4 \end{bmatrix}\begin{bmatrix} x_1 \\ x_2 \\ x_3 \end{bmatrix} = \begin{bmatrix} 2 \\ 3 \\ -7 \end{bmatrix}$.

$$\begin{bmatrix} 1 & 2 & -1 \\ -2 & -1 & 3 \\ 1 & -1 & -4 \end{bmatrix} \underset{\substack{R2+2R1 \\ R3+(-1)R1}}{\approx} \begin{bmatrix} 1 & 2 & -1 \\ 0 & 3 & 1 \\ 0 & -3 & -3 \end{bmatrix} \underset{R3+R2}{\approx} \begin{bmatrix} 1 & 2 & -1 \\ 0 & 3 & 1 \\ 0 & 0 & -2 \end{bmatrix} = U, \text{ and}$$

L = $\begin{bmatrix} 1 & 0 & 0 \\ -2 & 1 & 0 \\ 1 & -1 & 1 \end{bmatrix}$. L**y** = $\begin{bmatrix} 2 \\ 3 \\ -7 \end{bmatrix}$ gives **y** = $\begin{bmatrix} 2 \\ 7 \\ -2 \end{bmatrix}$, and U**x** = **y** gives **x** = $\begin{bmatrix} -1 \\ 2 \\ 1 \end{bmatrix}$.

10. The matrix equation A**x** = **b** is $\begin{bmatrix} 2 & 1 & 0 \\ 6 & 4 & -1 \\ 2 & 0 & 4 \end{bmatrix}\begin{bmatrix} x_1 \\ x_2 \\ x_3 \end{bmatrix} = \begin{bmatrix} 9 \\ 25 \\ 20 \end{bmatrix}$.

$$\begin{bmatrix} 2 & 1 & 0 \\ 6 & 4 & -1 \\ 2 & 0 & 4 \end{bmatrix} \underset{\substack{R2+(-3)R1 \\ R3+(-1)R1}}{\approx} \begin{bmatrix} 2 & 1 & 0 \\ 0 & 1 & -1 \\ 0 & -1 & 4 \end{bmatrix} \underset{R3+R2}{\approx} \begin{bmatrix} 2 & 1 & 0 \\ 0 & 1 & -1 \\ 0 & 0 & 3 \end{bmatrix} = U, \text{ and}$$

L = $\begin{bmatrix} 1 & 0 & 0 \\ 3 & 1 & 0 \\ 1 & -1 & 1 \end{bmatrix}$. L**y** = $\begin{bmatrix} 9 \\ 25 \\ 20 \end{bmatrix}$ gives **y** = $\begin{bmatrix} 9 \\ -2 \\ 9 \end{bmatrix}$, and U**x** = **y** gives **x** = $\begin{bmatrix} 4 \\ 1 \\ 3 \end{bmatrix}$.

11. The matrix equation A**x** = **b** is $\begin{bmatrix} 3 & -1 & 1 \\ -3 & 2 & 1 \\ 9 & 5 & -3 \end{bmatrix}\begin{bmatrix} x_1 \\ x_2 \\ x_3 \end{bmatrix} = \begin{bmatrix} 10 \\ -8 \\ 24 \end{bmatrix}$.

$$\begin{bmatrix} 3 & -1 & 1 \\ -3 & 2 & 1 \\ 9 & 5 & -3 \end{bmatrix} \underset{\substack{R2+R1 \\ R3+(-3)R1}}{\approx} \begin{bmatrix} 3 & -1 & 1 \\ 0 & 1 & 2 \\ 0 & 8 & -6 \end{bmatrix} \underset{R3+(-8)R2}{\approx} \begin{bmatrix} 3 & -1 & 1 \\ 0 & 1 & 2 \\ 0 & 0 & -22 \end{bmatrix} = U, \text{ and}$$

L = $\begin{bmatrix} 1 & 0 & 0 \\ -1 & 1 & 0 \\ 3 & 8 & 1 \end{bmatrix}$. L**y** = $\begin{bmatrix} 10 \\ -8 \\ 24 \end{bmatrix}$ gives **y** = $\begin{bmatrix} 10 \\ 2 \\ -22 \end{bmatrix}$, and U**x** = **y** gives **x** = $\begin{bmatrix} 3 \\ 0 \\ 1 \end{bmatrix}$.

12. The matrix equation A**x** = **b** is $\begin{bmatrix} 1 & 2 & 3 \\ 2 & 8 & 7 \\ -1 & 14 & -1 \end{bmatrix}\begin{bmatrix} x_1 \\ x_2 \\ x_3 \end{bmatrix} = \begin{bmatrix} -5 \\ -9 \\ 15 \end{bmatrix}$.

$$\begin{bmatrix} 1 & 2 & 3 \\ 2 & 8 & 7 \\ -1 & 14 & -1 \end{bmatrix} \underset{\substack{R2+(-2)R1 \\ R3+R1}}{\approx} \begin{bmatrix} 1 & 2 & 3 \\ 0 & 4 & 1 \\ 0 & 16 & 2 \end{bmatrix} \underset{R3+(-4)R2}{\approx} \begin{bmatrix} 1 & 2 & 3 \\ 0 & 4 & 1 \\ 0 & 0 & -2 \end{bmatrix}$$ = U, and

L = $\begin{bmatrix} 1 & 0 & 0 \\ 2 & 1 & 0 \\ -1 & 4 & 1 \end{bmatrix}$. L**y** = $\begin{bmatrix} -5 \\ -9 \\ 15 \end{bmatrix}$ gives **y** = $\begin{bmatrix} -5 \\ 1 \\ 6 \end{bmatrix}$, and U**x** = **y** gives **x** = $\begin{bmatrix} 2 \\ 1 \\ -3 \end{bmatrix}$.

13. The matrix equation A**x** = **b** is $\begin{bmatrix} 4 & 1 & 2 \\ -12 & -4 & -3 \\ 0 & -5 & 16 \end{bmatrix}\begin{bmatrix} x_1 \\ x_2 \\ x_3 \end{bmatrix} = \begin{bmatrix} 18 \\ -56 \\ -10 \end{bmatrix}$.

$$\begin{bmatrix} 4 & 1 & 2 \\ -12 & -4 & -3 \\ 0 & -5 & 16 \end{bmatrix} \underset{R2+3R1}{\approx} \begin{bmatrix} 4 & 1 & 2 \\ 0 & -1 & 3 \\ 0 & -5 & 16 \end{bmatrix} \underset{R3+(-5)R2}{\approx} \begin{bmatrix} 4 & 1 & 2 \\ 0 & -1 & 3 \\ 0 & 0 & 1 \end{bmatrix}$$ = U, and

L = $\begin{bmatrix} 1 & 0 & 0 \\ -3 & 1 & 0 \\ 0 & 5 & 1 \end{bmatrix}$. L**y** = $\begin{bmatrix} 18 \\ -56 \\ -10 \end{bmatrix}$ gives **y** = $\begin{bmatrix} 18 \\ -2 \\ 0 \end{bmatrix}$, and U**x** = **y** gives **x** = $\begin{bmatrix} 4 \\ 2 \\ 0 \end{bmatrix}$.

14. The matrix equation A**x** = **b** is $\begin{bmatrix} -2 & 0 & 3 \\ -4 & 3 & 5 \\ 8 & 9 & -11 \end{bmatrix}\begin{bmatrix} x_1 \\ x_2 \\ x_3 \end{bmatrix} = \begin{bmatrix} 3 \\ 11 \\ 7 \end{bmatrix}$.

$$\begin{bmatrix} -2 & 0 & 3 \\ -4 & 3 & 5 \\ 8 & 9 & -11 \end{bmatrix} \underset{\substack{R2+(-2)R1 \\ R3+4R1}}{\approx} \begin{bmatrix} -2 & 0 & 3 \\ 0 & 3 & -1 \\ 0 & 9 & 1 \end{bmatrix} \underset{R3+(-3)R2}{\approx} \begin{bmatrix} -2 & 0 & 3 \\ 0 & 3 & -1 \\ 0 & 0 & 4 \end{bmatrix}$$ = U, and

$L = \begin{bmatrix} 1 & 0 & 0 \\ 2 & 1 & 0 \\ -4 & 3 & 1 \end{bmatrix}$. $L\mathbf{y} = \begin{bmatrix} 3 \\ 11 \\ 7 \end{bmatrix}$ gives $\mathbf{y} = \begin{bmatrix} 3 \\ 5 \\ 4 \end{bmatrix}$, and $U\mathbf{x} = \mathbf{y}$ gives $\mathbf{x} = \begin{bmatrix} 0 \\ 2 \\ 1 \end{bmatrix}$.

15. The matrix equation $A\mathbf{x} = \mathbf{b}$ is $\begin{bmatrix} -2 & 0 & 3 \\ -4 & 3 & 5 \\ 8 & 9 & -11 \end{bmatrix}\begin{bmatrix} x_1 \\ x_2 \\ x_3 \end{bmatrix} = \begin{bmatrix} 14 \\ 30 \\ -34 \end{bmatrix}$. Since the matrix A is the same as in Exercise 14, so are L and U. $L\mathbf{y} = \begin{bmatrix} 14 \\ 30 \\ -34 \end{bmatrix}$ gives $\mathbf{y} = \begin{bmatrix} 14 \\ 2 \\ 16 \end{bmatrix}$, and

$U\mathbf{x} = \mathbf{y}$ gives $\mathbf{x} = \begin{bmatrix} -1 \\ 2 \\ 4 \end{bmatrix}$.

16. The matrix equation $A\mathbf{x} = \mathbf{b}$ is $\begin{bmatrix} 2 & -3 & 1 \\ 4 & -5 & 6 \\ -10 & 19 & 9 \end{bmatrix}\begin{bmatrix} x_1 \\ x_2 \\ x_3 \end{bmatrix} = \begin{bmatrix} -5 \\ 2 \\ 55 \end{bmatrix}$.

$$\begin{bmatrix} 2 & -3 & 1 \\ 4 & -5 & 6 \\ -10 & 19 & 9 \end{bmatrix} \underset{\substack{R2+(-2)R1 \\ R3+5R1}}{\approx} \begin{bmatrix} 2 & -3 & 1 \\ 0 & 1 & 4 \\ 0 & 4 & 14 \end{bmatrix} \underset{R3+(-4)R2}{\approx} \begin{bmatrix} 2 & -3 & 1 \\ 0 & 1 & 4 \\ 0 & 0 & -2 \end{bmatrix} = U, \text{ and}$$

$L = \begin{bmatrix} 1 & 0 & 0 \\ 2 & 1 & 0 \\ -5 & 4 & 1 \end{bmatrix}$. $L\mathbf{y} = \begin{bmatrix} -5 \\ 2 \\ 55 \end{bmatrix}$ gives $\mathbf{y} = \begin{bmatrix} -5 \\ 12 \\ -18 \end{bmatrix}$, and $U\mathbf{x} = \mathbf{y}$ gives $\mathbf{x} = \begin{bmatrix} -43 \\ -24 \\ 9 \end{bmatrix}$.

17. The matrix equation $A\mathbf{x} = \mathbf{b}$ is $\begin{bmatrix} 4 & 1 & 3 \\ 12 & 1 & 10 \\ -8 & -16 & 6 \end{bmatrix}\begin{bmatrix} x_1 \\ x_2 \\ x_3 \end{bmatrix} = \begin{bmatrix} 11 \\ 28 \\ -62 \end{bmatrix}$.

$$\begin{bmatrix} 4 & 1 & 3 \\ 12 & 1 & 10 \\ -8 & -16 & 6 \end{bmatrix} \underset{\substack{R2+(-3)R1 \\ R3+2R1}}{\approx} \begin{bmatrix} 4 & 1 & 3 \\ 0 & -2 & 1 \\ 0 & -14 & 12 \end{bmatrix} \underset{R3+(-7)R2}{\approx} \begin{bmatrix} 4 & 1 & 3 \\ 0 & -2 & 1 \\ 0 & 0 & 5 \end{bmatrix} = U, \text{ and}$$

$L = \begin{bmatrix} 1 & 0 & 0 \\ 3 & 1 & 0 \\ -2 & 7 & 1 \end{bmatrix}$. $L\mathbf{y} = \begin{bmatrix} 11 \\ 28 \\ -62 \end{bmatrix}$ gives $\mathbf{y} = \begin{bmatrix} 11 \\ -5 \\ -5 \end{bmatrix}$, and $U\mathbf{x} = \mathbf{y}$ gives $\mathbf{x} = \begin{bmatrix} 3 \\ 2 \\ -1 \end{bmatrix}$.

18. The matrix equation $A\mathbf{x} = \mathbf{b}$ is $\begin{bmatrix} 1 & 2 & -1 \\ 2 & 5 & 1 \\ -1 & -1 & 4 \end{bmatrix}\begin{bmatrix} x_1 \\ x_2 \\ x_3 \end{bmatrix} = \begin{bmatrix} 2 \\ 3 \\ -3 \end{bmatrix}$.

$$\begin{bmatrix} 1 & 2 & -1 \\ 2 & 5 & 1 \\ -1 & -1 & 4 \end{bmatrix} \underset{\substack{R2+(-2)R1 \\ R3+R1}}{\approx} \begin{bmatrix} 1 & 2 & -1 \\ 0 & 1 & 3 \\ 0 & 1 & 3 \end{bmatrix} \underset{R3+(-1)R2}{\approx} \begin{bmatrix} 1 & 2 & -1 \\ 0 & 1 & 3 \\ 0 & 0 & 0 \end{bmatrix} = U, \text{ and}$$

$L = \begin{bmatrix} 1 & 0 & 0 \\ 2 & 1 & 0 \\ -1 & 1 & 1 \end{bmatrix}$. $L\mathbf{y} = \begin{bmatrix} 2 \\ 3 \\ -3 \end{bmatrix}$ gives $\mathbf{y} = \begin{bmatrix} 2 \\ -1 \\ 0 \end{bmatrix}$, and $U\mathbf{x} = \mathbf{y}$ gives $\mathbf{x} = \begin{bmatrix} 4+7r \\ -1-3r \\ r \end{bmatrix}$.

19. The matrix equation $A\mathbf{x} = \mathbf{b}$ is $\begin{bmatrix} 4 & 1 & -2 \\ -4 & 2 & 3 \\ 8 & -7 & -7 \end{bmatrix}\begin{bmatrix} x_1 \\ x_2 \\ x_3 \end{bmatrix} = \begin{bmatrix} 3 \\ 1 \\ -2 \end{bmatrix}$.

$$\begin{bmatrix} 4 & 1 & -2 \\ -4 & 2 & 3 \\ 8 & -7 & -7 \end{bmatrix} \underset{\substack{R2+R1 \\ R3+(-2)R1}}{\approx} \begin{bmatrix} 4 & 1 & -2 \\ 0 & 3 & 1 \\ 0 & -9 & -3 \end{bmatrix} \underset{R3+3R2}{\approx} \begin{bmatrix} 4 & 1 & -2 \\ 0 & 3 & 1 \\ 0 & 0 & 0 \end{bmatrix} = U, \text{ and}$$

$L = \begin{bmatrix} 1 & 0 & 0 \\ -1 & 1 & 0 \\ 2 & -3 & 1 \end{bmatrix}$. $L\mathbf{y} = \begin{bmatrix} 3 \\ 1 \\ -2 \end{bmatrix}$ gives $\mathbf{y} = \begin{bmatrix} 3 \\ 4 \\ 4 \end{bmatrix}$, and $U\mathbf{x} = \mathbf{y}$ is a system with no solution.

20. The matrix equation $A\mathbf{x} = \mathbf{b}$ is $\begin{bmatrix} 1 & 1 & -1 & 2 \\ 1 & 3 & 2 & 2 \\ -1 & -3 & -4 & 6 \\ 0 & 4 & 7 & -2 \end{bmatrix}\begin{bmatrix} x_1 \\ x_2 \\ x_3 \\ x_4 \end{bmatrix} = \begin{bmatrix} 7 \\ 6 \\ 12 \\ -7 \end{bmatrix}$.

$$\begin{bmatrix} 1 & 1 & -1 & 2 \\ 1 & 3 & 2 & 2 \\ -1 & -3 & -4 & 6 \\ 0 & 4 & 7 & -2 \end{bmatrix} \underset{\substack{R2+(-1)R1 \\ R3+R1}}{\approx} \begin{bmatrix} 1 & 1 & -1 & 2 \\ 0 & 2 & 3 & 0 \\ 0 & -2 & -5 & 8 \\ 0 & 4 & 7 & -2 \end{bmatrix} \underset{\substack{R3+R2 \\ R4+(-2)R2}}{\approx} \begin{bmatrix} 1 & 1 & -1 & 2 \\ 0 & 2 & 3 & 0 \\ 0 & 0 & -2 & 8 \\ 0 & 0 & 1 & -2 \end{bmatrix}$$

$$\underset{R4+(1/2)R3}{\approx} \begin{bmatrix} 1 & 1 & -1 & 2 \\ 0 & 2 & 3 & 0 \\ 0 & 0 & -2 & 8 \\ 0 & 0 & 0 & 2 \end{bmatrix} = \text{U, and L} = \begin{bmatrix} 1 & 0 & 0 & 0 \\ 1 & 1 & 0 & 0 \\ -1 & -1 & 1 & 0 \\ 0 & 2 & -1/2 & 1 \end{bmatrix}.$$

$$\mathrm{L}\mathbf{y} = \begin{bmatrix} 7 \\ 6 \\ 12 \\ -7 \end{bmatrix} \text{ gives } \mathbf{y} = \begin{bmatrix} 7 \\ -1 \\ 18 \\ 4 \end{bmatrix}, \text{ and } \mathrm{U}\mathbf{x} = \mathbf{y} \text{ gives } \mathbf{x} = \begin{bmatrix} 1 \\ 1 \\ -1 \\ 2 \end{bmatrix}.$$

21. The matrix equation A**x** = **b** is $\begin{bmatrix} 2 & 0 & 1 & -1 \\ 6 & 3 & 2 & -1 \\ 4 & 3 & -2 & 3 \\ -2 & -6 & 2 & -14 \end{bmatrix} \begin{bmatrix} x_1 \\ x_2 \\ x_3 \\ x_4 \end{bmatrix} = \begin{bmatrix} 6 \\ 15 \\ 3 \\ 12 \end{bmatrix}$.

$$\begin{bmatrix} 2 & 0 & 1 & -1 \\ 6 & 3 & 2 & -1 \\ 4 & 3 & -2 & 3 \\ -2 & -6 & 2 & -14 \end{bmatrix} 2 \underset{\substack{R2+(-3)R2 \\ R3+(-2)R1 \\ R4+R1}}{\approx} \begin{bmatrix} 2 & 0 & 1 & -1 \\ 0 & 3 & -1 & 2 \\ 0 & 3 & -4 & 5 \\ 0 & -6 & 3 & -15 \end{bmatrix} \underset{\substack{R3+(-1)R2 \\ R4+2R}}{\approx} \begin{bmatrix} 2 & 0 & 1 & -1 \\ 0 & 3 & -1 & 2 \\ 0 & 0 & -3 & 3 \\ 0 & 0 & 1 & -11 \end{bmatrix}$$

$$\underset{R4+(1/3)R3}{\approx} \begin{bmatrix} 2 & 0 & 1 & -1 \\ 0 & 3 & -1 & 2 \\ 0 & 0 & -3 & 3 \\ 0 & 0 & 0 & -10 \end{bmatrix} = \text{U, and L} = \begin{bmatrix} 1 & 0 & 0 & 0 \\ 3 & 1 & 0 & 0 \\ 2 & 1 & 1 & 0 \\ -1 & -2 & -1/3 & 1 \end{bmatrix}$$

$$\mathrm{L}\mathbf{y} = \begin{bmatrix} 6 \\ 15 \\ 3 \\ 12 \end{bmatrix} \text{ gives } \mathbf{y} = \begin{bmatrix} 6 \\ -3 \\ -6 \\ 10 \end{bmatrix}, \text{ and } \mathrm{U}\mathbf{x} = \mathbf{y} \text{ gives } \mathbf{x} = \begin{bmatrix} 2 \\ 0 \\ 1 \\ -1 \end{bmatrix}.$$

22. Let $B = A^{-1}$. The (i,j)th element of the product $AB = I_n$ is $a_{i1} b_{1j} + a_{i2} b_{2j} + \ldots + a_{in} b_{nj}$.
For i = 1, $a_{11} b_{1j} + a_{12} b_{2j} + \ldots + a_{1n} b_{nj} = a_{11} b_{1j}$, because $a_{12} = a_{13} = \ldots = a_{1n} = 0$.
This means that $a_{11} b_{1j} = 0$ for j > 1. But $a_{11} \neq 0$ (because $|A| \neq 0$), so $b_{1j} = 0$ for j >1.
For i = 2, $a_{21} b_{1j} + a_{22} b_{2j} + \ldots + a_{2n-1} b_{nj} = a_{22} b_{2j}$, for j > 1, because
$b_{1j} = a_{23} = a_{24} = \ldots = a_{2n} = 0$. This means that $a_{22} b_{2j} = 0$ for j > 2.
But $a_{22} \neq 0$, so $b_{2j} = 0$ for j > 2.

In the same manner each row of B is shown to have zero in all positions for which the column number is greater than the row number. Thus B is lower triangular.

23. Let A and B be two nxn lower triangular matrices. The (i,j)th element of the product AB is $a_{i1} b_{1j} + a_{i2} b_{2j} + \ldots + a_{in} b_{nj}$. If $1 \leq i < j \leq n$, the elements $a_{ii+1}, a_{ii+2}, \ldots, a_{in}$ of A and
$b_{1j}, b_{2j}, \ldots, b_{j-1j}$ of B are all zero. Thus the first j − 1 terms of $a_{i1} b_{1j} + a_{i2} b_{2j} + \ldots + a_{in} b_{nj}$ are zero because $b_{1j} = b_{2j} = \ldots = b_{j-1j} = 0$ and the remainder of the terms are zero because j > i and therefore $a_{ij} = a_{ij+1} = \ldots = a_{in} = 0$. Thus if $1 \leq i < j \leq n$, the (i,j)th element of the product is zero, so the product is a lower triangular matrix.

24. $$\begin{bmatrix} 6 & -2 \\ 12 & 8 \end{bmatrix} = \begin{bmatrix} 1 & 0 \\ 2 & 1 \end{bmatrix}\begin{bmatrix} 6 & -2 \\ 0 & 12 \end{bmatrix} = \begin{bmatrix} 2 & 0 \\ 4 & 2 \end{bmatrix}\begin{bmatrix} 3 & -1 \\ 0 & 6 \end{bmatrix}.$$

25. If $\begin{bmatrix} 1 & 0 & 0 \\ 0 & 0 & 1 \\ 0 & 1 & 0 \end{bmatrix} = LU = \begin{bmatrix} a & 0 & 0 \\ b & c & 0 \\ d & e & f \end{bmatrix}\begin{bmatrix} p & q & r \\ 0 & s & t \\ 0 & 0 & u \end{bmatrix}$, then ap = 1, so $a \neq 0$ and $p \neq 0$.

aq = 0 and ar = 0, so q = r = 0. Also bp = 0 and dp = 0, so b = d = 0. Thus

$$\begin{bmatrix} 1 & 0 & 0 \\ 0 & 0 & 1 \\ 0 & 1 & 0 \end{bmatrix} = \begin{bmatrix} a & 0 & 0 \\ 0 & c & 0 \\ 0 & e & f \end{bmatrix}\begin{bmatrix} p & 0 & 0 \\ 0 & s & t \\ 0 & 0 & u \end{bmatrix} = \begin{bmatrix} 1 & 0 & 0 \\ 0 & cs & ct \\ 0 & es & et+fu \end{bmatrix}.$$ cs = 0, so one of c and s must be

zero, but ct = es = 1, so neither c nor s can be zero. Since no number can be both zero and nonzero, we must conclude that the matrices L and U do not exist.

26. The first pair of row operations below require three multiplications and two additions each. The last row operation requires two multiplications and one addition. Thus a total of thirteen arithmetic operations are required to obtain U.

$$\begin{bmatrix} a & b & c \\ d & e & f \\ g & h & i \end{bmatrix} \begin{matrix} \approx \\ R2+(-d/a)R1 \\ R3+(-g/a)R1 \end{matrix} \begin{bmatrix} a & b & c \\ 0 & j & k \\ 0 & m & n \end{bmatrix} \begin{matrix} \approx \\ R3+(-m/j)R2 \end{matrix} \begin{bmatrix} a & b & c \\ 0 & j & k \\ 0 & 0 & s \end{bmatrix}$$ = U. No arithmetic operations

are required for L since it can be written down with no additional calculation.

$$L = \begin{bmatrix} 1 & 0 & 0 \\ -p & 1 & 0 \\ -q & -r & 0 \end{bmatrix}, \text{ where } p = \frac{d}{a}, q = \frac{g}{a}, \text{ and } r = \frac{m}{j}.$$

To solve L**y** = **b** requires six operations as follows: $y_1 = b_1$ requires no operations, $y_2 = b_2 + pb_1$ requires one multiplication and one addition, and $y_3 = b_3 + qy_1 + ry_2$ requires two multiplications and two additions.

To solve U**x** = **y** requires nine operations because there is an additional multiplication at each stage: $x_3 = y_3/s$ requires one multiplication, $x_2 = (y_2 - kx_3)/j$ requires two multiplications and one addition, and $x_1 = (y_1 - bx_2 - cx_3)/a$ requires three multiplications and two additions.

Exercise Set 7.3

1. (a) max{1+3, 2+4} = max{4,6} = 6

   (b) max{5+1, 2+3} = max{6,5} = 6

   (c) max{1+3+6, 2+5+1, 0+4+2} = max{10,8,6} = 10

   (d) max{1+6+5, 24+247+7, 3+56+219} = max{12,278,278} = 278

2. (a) $A = \begin{bmatrix} 1 & 2 \\ 3 & 4 \end{bmatrix}$ and $A^{-1} = \begin{bmatrix} -2 & 1 \\ 3/2 & -1/2 \end{bmatrix}$, so ||A|| = max{4,6} = 6 and

   $||A^{-1}||$ = max{7/2, 3/2} = 7/2. Thus c(A) = 21.

   (b) $A = \begin{bmatrix} 2 & -2 \\ 3 & 1 \end{bmatrix}$ and $A^{-1} = \begin{bmatrix} 1/8 & 1/4 \\ -3/8 & 1/4 \end{bmatrix}$, so ||A|| = max{5,3} = 5 and

   $||A^{-1}||$ = max{1/2, 1/2} = 1/2. Thus c(A) = 5/2.

(c) $A = \begin{bmatrix} 5 & 2 \\ 8 & 3 \end{bmatrix}$ and $A^{-1} = \begin{bmatrix} -3 & 2 \\ 8 & -5 \end{bmatrix}$, so $||A|| = \max\{13,5\} = 13$ and

$||A^{-1}|| = \max\{11,7\} = 11$. Thus $c(A) = 143$.

(d) $A = \begin{bmatrix} 24 & 21 \\ 6 & 5 \end{bmatrix}$ and $A^{-1} = \begin{bmatrix} -5/6 & 7/2 \\ 1 & -4 \end{bmatrix}$, so $||A|| = \max\{30,26\} = 30$ and

$||A^{-1}|| = \max\{11/6, 15/2\} = 15/2$. Thus $c(A) = 225$.

(e) $A = \begin{bmatrix} 5.2 & 3.7 \\ 3.8 & 2.6 \end{bmatrix}$ and $A^{-1} = \begin{bmatrix} -2.6/.54 & 3.7/.54 \\ 3.8/.54 & -5.2/.54 \end{bmatrix}$, so $||A|| = \max\{9,6.3\} = 9$ and

$||A^{-1}|| = \max\{6.4/.54, 8.9/.54\} = 8.9/.54$. Thus $c(A) = 148\frac{1}{3}$.

3. (a) $A = \begin{bmatrix} 9 & 2 \\ 4 & 7 \end{bmatrix}$ and $A^{-1} = \begin{bmatrix} 7/55 & -2/55 \\ -4/55 & 9/55 \end{bmatrix}$, so $||A|| = \max\{13,9\} = 13$ and

$||A^{-1}|| = \max\{1/5, 1/5\} = 1/5$. Thus $c(A) = 13/5 = 2.6 = .26\text{x}10^1$.

The solution of a system of equations AX = B can have one fewer significant digit of accuracy than the elements of A.

(b) $A = \begin{bmatrix} 51 & 3 \\ -1 & 27 \end{bmatrix}$ and $A^{-1} = \begin{bmatrix} 27/1380 & -3/1380 \\ 1/1380 & 51/1380 \end{bmatrix}$, so $||A|| = \max\{52,30\} = 52$ and

$||A^{-1}|| = \max\{28/1380, 54/1380\} = 54/1380$. Thus $c(A) = 2.035 = .2035\text{x}10^1$.

The solution of a system of equations AX = B can have one fewer significant digit of accuracy than the elements of A.

(c) $A = \begin{bmatrix} 300 & 1001 \\ 75 & 250 \end{bmatrix}$ and $A^{-1} = \begin{bmatrix} -10/3 & 1001/75 \\ 1 & -4 \end{bmatrix}$, so $||A|| = \max\{375,1251\}$

$= 1251$ and $||A^{-1}|| = \max\{13/3, 1301/75\} = 1301/75$. Thus $c(A) = 21700.7$
$= .217\text{x}10^5$.

The solution of a system of equations AX = B can have five fewer significant digits of accuracy than the elements of A.

(d) $A = \begin{bmatrix} 3 & 4 \\ 1 & 2 \end{bmatrix}$ and $A^{-1} = \begin{bmatrix} 1 & -2 \\ -1/2 & 3/2 \end{bmatrix}$, so $||A|| = \max\{4,6\} = 6$ and

$||A^{-1}|| = \max\{3/2, 7/2\} = 7/2$. Thus $c(A) = 21 = .21\text{x}10^2$.

The solution of a system of equations AX = B can have two fewer significant digits of accuracy than the elements of A.

(e) $A = \begin{bmatrix} 5 & 3 \\ 4 & 2 \end{bmatrix}$ and $A^{-1} = \begin{bmatrix} -1 & 3/2 \\ 2 & -5/2 \end{bmatrix}$, so $||A|| = \max\{9,5\} = 9$ and

$||A^{-1}|| = \max\{3,4\} = 4$. Thus $c(A) = 36 = .36\text{x}10^2$.

The solution of a system of equations AX = B can have two fewer significant digits of accuracy than the elements of A.

4. (a) $A = \begin{bmatrix} 1 & 1 & -1 \\ 1 & 0 & 2 \\ 1 & -2 & 0 \end{bmatrix}$ and $A^{-1} = \begin{bmatrix} 1/2 & 1/4 & 1/4 \\ 1/4 & 1/8 & -3/8 \\ -1/4 & 3/8 & -1/8 \end{bmatrix}$, so $||A|| = \max\{3,3,3\} = 3$ and

$||A^{-1}|| = \max\{1, 3/4, 3/4\} = 1$. Thus $c(A) = 3 = .3\square 10^1$.

The solution of a system of equations AX = B can have one fewer significant digit of accuracy than the elements of A.

(b) Let $P = \begin{bmatrix} .8 & -.2 & -.4 \\ -.6 & .4 & 0 \\ 0 & 0 & .8 \end{bmatrix}$ (= (I – A) from Example 2 in Section 2.5) , and

$P^{-1} = \begin{bmatrix} 2 & 1 & 1 \\ 3 & 4 & 1.5 \\ 0 & 0 & 1.25 \end{bmatrix}$, so $||P|| = \max\{1.4, .6, 1.2\} = 1.4$ and

$||P^{-1}|| = \max\{5,5,3.75\} = 5$. Thus $c(P) = 1.4\square 5 = 7 = .7\square 10^1$.

The solution of a system of equations PX = D can have one fewer significant digit of accuracy than the elements of P.

5. $A = \begin{bmatrix} 1/3 & 1/4 & 1/5 \\ 1/4 & 1/5 & 1/6 \\ 1/5 & 1/6 & 1/7 \end{bmatrix}$ and $A^{-1} = \begin{bmatrix} 300 & -900 & 630 \\ -900 & 2880 & -2100 \\ 630 & -2100 & 1575 \end{bmatrix}$, so $||A|| = 47/60$ and

$||A^{-1}|| = 5880$. Thus $c(A) = 4606$.

6. $A = \begin{bmatrix} 1 & k \\ 1 & 1 \end{bmatrix}$, $k \neq 1$, and $A^{-1} = \begin{bmatrix} 1/(1-k) & -k/(1-k) \\ -1/(1-k) & 1/(1-k) \end{bmatrix}$, so $||A|| = \max\{2, 1+|k|\}$ and

$||A^{-1}|| = \max\{2/|1-k|, (1+|k|)/|1-k|\}$. Thus if $k > 1$, $c(A) = (1+k)^2 /k-1$. Solutions to $c(A) = 100$ are $k = 1.0417$ and $k = 96.9583$. $c(A) > 100$ if k is in the interval $(1, 1.0417)$ or $k > 96.9583$. If $-1 \le k < 1$, $c(A) = 4/(1-k)$, which equals 100 if $k = .96$. $c(A) > 100$ if $.96 < k < 1$. If $k < -1$, $c(A) = 1- k$, which is greater than 100 if $k < -99$.

7. (a) Since $||A||$ is the sum of nonnegative numbers, $||A||$ is nonnegative.

(b) For any j, $0 \le j \le n$, $|a_{1j}| + |a_{2j}| + \ldots + |a_{nj}| = 0$ if and only if $a_{ij} = 0$ for $i = 1,2,\ldots, n$.

$||A|| = 0$ if and only if $|a_{1j}| + |a_{2j}| + \ldots + |a_{nj}| = 0$ for $j = 1,2,\ldots, n$. Thus $||A|| = 0$ if and only if $a_{ij} = 0$ for $i = 1,2,\ldots, n$ and $j = 1,2,\ldots, n$, so $||A|| = 0$ if and only if A is the zero matrix.

(c) $||cA|| = \max\{|ca_{1j}| + |ca_{2j}| + \ldots + |ca_{nj}|\} = \max\{ |c|(|a_{1j}| + |a_{2j}| + \ldots + |a_{nj}|)\}$

$= |c| \max\{|a_{1j}| + |a_{2j}| + \ldots + |a_{nj}|\} = |c|\ ||A||$.

(d) $|a_{1j} +b_{1j}|+|a_{2j} +b_{2j}|+\ldots+|a_{nj} +b_{nj}| \le |a_{1j}|+|b_{1j}|+|a_{2j}|+|b_{2j}|+\ldots+|a_{nj}|+|b_{nj}|$

$= |a_{1j}| + |a_{2j}| + \ldots + |a_{nj}|+|b_{1j}| + |b_{2j}| + \ldots + |b_{nj}|$, for $j = 1,2,\ldots, n$, so

$||A+B|| = \max\{|a_{1j} +b_{1j}|+|a_{2j} +b_{2j}|+\ldots+|a_{nj} +b_{nj}|\}$

$\le \max\{|a_{1j}| + |a_{2j}| + \ldots + |a_{nj}|+|b_{1j}| + |b_{2j}| + \ldots + |b_{nj}|\}$

$\le \max\{|a_{1j}| + |a_{2j}| + \ldots + |a_{nj}|\} + \max\{|b_{1j}| + |b_{2j}| + \ldots + |b_{nj}|\} = ||A|| + ||B||$.

The sum of the absolute values of the elements of a column of AB is

$|a_{11}b_{1j}+a_{12}b_{2j}+...+a_{1n}b_{nj}|+|a_{21}b_{1j}+a_{22}b_{2j}+...+a_{2n}b_{nj}|+...+|a_{n1}b_{1j}+a_{n2}b_{2j}+...+a_{nn}b_{nj}|$

$\le |a_{11}b_{1j}|+|a_{12}b_{2j}|+...+|a_{1n}b_{nj}|+|a_{21}b_{1j}|+|a_{22}b_{2j}|+...+|a_{2n}b_{nj}|+...+|a_{n1}b_{1j}|+|a_{n2}b_{2j}|$

$+...+|a_{nn}b_{nj}| = (|a_{11}|+|a_{21}|+...+|a_{n1}|)|b_{1j}|+(|a_{12}|+|a_{22}|+...+|a_{n2}|)|b_{2j}|+...$

$+(|a_{1n}|+|a_{2n}|+...+|a_{nn}|)|b_{nj}| \le (\max\{|a_{1j}|+|a_{2j}|+...+|a_{nj}|\})(|b_{1j}|+|b_{2j}|+...+|b_{nj}|)$

$= ||A||(|b_{1j}|+|b_{2j}|+...+|b_{nj}|)$, so $||AB|| \le \max\{||A||(|b_{1j}|+|b_{2j}|+...+|b_{nj}|)\}$

$= ||A|| \max\{|b_{1j}|+|b_{2j}|+...+|b_{nj}|\} = ||A||\,||B||$.

8. (a) $I^{-1} = I$, so both have 1-norm of 1 and $C(I) = 1$.

(b) $1 = ||AA^{-1}|| \le ||A||\,||A^{-1}|| = c(A)$.

9. $c(kA) = ||kA||\,||(kA)^{-1}|| = ||kA||\,||\frac{1}{k}A^{-1}|| = k||A||\frac{1}{k}||A^{-1}|| = ||A||\,||A^{-1}|| = c(A)$. Thus we see that multiplying the system by a constant does not alter the expected accuracy.

10. If A is a diagonal matrix with diagonal elements a_{ii}, then A^{-1} is a diagonal matrix with diagonal elements $1/a_{ii}$. $||A|| = \max\{a_{ii}\}$ and $||A^{-1}|| = \max\{1/a_{ii}\} = 1/\min\{a_{ii}\}$, so $c(A) = (\max\{a_{ii}\})(1/\min\{a_{ii}\})$.

11. $\begin{bmatrix} 0 & -1 & 2 \\ 1 & 2 & -1 \\ 1 & 2 & 2 \end{bmatrix}\begin{bmatrix} x_1 \\ x_2 \\ x_3 \end{bmatrix} = \begin{bmatrix} 4 \\ 1 \\ 4 \end{bmatrix}$. $\begin{bmatrix} 0 & -1 & 2 & 4 \\ 1 & 2 & -1 & 1 \\ 1 & 2 & 2 & 4 \end{bmatrix} \underset{R1 \Leftrightarrow R2}{\approx} \begin{bmatrix} 1 & 2 & -1 & 1 \\ 0 & -1 & 2 & 4 \\ 1 & 2 & 2 & 4 \end{bmatrix}$

$$\underset{R3+(-1)R1}{\approx}\begin{bmatrix}1 & 2 & -1 & 1\\ 0 & -1 & 2 & 4\\ 0 & 0 & 3 & 3\end{bmatrix}\underset{\substack{(-1)R2\\(1/3)R3}}{\approx}\begin{bmatrix}1 & 2 & -1 & 1\\ 0 & 1 & -2 & -4\\ 0 & 0 & 1 & 1\end{bmatrix}\underset{R1+(2)R2}{\approx}\begin{bmatrix}1 & 0 & 3 & 9\\ 0 & 1 & -2 & -4\\ 0 & 0 & 1 & 1\end{bmatrix}$$

$$\underset{\substack{R1+(-3)R3\\R2+2R3}}{\approx}\begin{bmatrix}1 & 0 & 0 & 6\\ 0 & 1 & 0 & -2\\ 0 & 0 & 1 & 1\end{bmatrix}$$, so the exact solution is $\begin{bmatrix}x_1\\x_2\\x_3\end{bmatrix}=\begin{bmatrix}6\\-2\\1\end{bmatrix}$.

12. $\begin{bmatrix}-1 & 1 & 2\\ 2 & 4 & -1\\ 1 & 2 & 2\end{bmatrix}\begin{bmatrix}x_1\\x_2\\x_3\end{bmatrix}=\begin{bmatrix}8\\10\\2\end{bmatrix}$. $\begin{bmatrix}-1 & 1 & 2 & 8\\ 2 & 4 & -1 & 10\\ 1 & 2 & 2 & 2\end{bmatrix}\underset{R1\Leftrightarrow R2}{\approx}\begin{bmatrix}2 & 4 & -1 & 10\\ -1 & 1 & 2 & 8\\ 1 & 2 & 2 & 2\end{bmatrix}$

$$\underset{(1/2)R1}{\approx}\begin{bmatrix}1 & 2 & -1/2 & 5\\ -1 & 1 & 2 & 8\\ 1 & 2 & 2 & 2\end{bmatrix}\underset{\substack{R2+R1\\R3+(-1)R1}}{\approx}\begin{bmatrix}1 & 2 & -1/2 & 5\\ 0 & 3 & 3/2 & 13\\ 0 & 0 & 5/2 & -3\end{bmatrix}$$

$$\underset{\substack{(1/3)R2\\(2/5)R3}}{\approx}\begin{bmatrix}1 & 2 & -1/2 & 5\\ 0 & 1 & 1/2 & 4.33\\ 0 & 0 & 1 & -6/5\end{bmatrix}\underset{R1+(-2)R2}{\approx}\begin{bmatrix}1 & 0 & -3/2 & -3.66\\ 0 & 1 & 1/2 & 4.33\\ 0 & 0 & 1 & -1.2\end{bmatrix}$$

(Here is the first round-off error.) (The round-off error gets multiplied here.)

$$\underset{\substack{R1+(1.5)R3\\R2+(-.5)R3}}{\approx}\begin{bmatrix}1 & 0 & 0 & -5.46\\ 0 & 1 & 0 & 4.93\\ 0 & 0 & 1 & -1.2\end{bmatrix}$$, so $\begin{bmatrix}x_1\\x_2\\x_3\end{bmatrix}=\begin{bmatrix}-5.46\\4.93\\-1.2\end{bmatrix}$. Substituting into the original equations $5.46 + 4.93 - 2.4 = 7.99$, $-10.92 + 19.72 + 1.2 = 10$, and $-5.46 + 9.86 - 2.4 = 2$. The exact solution is $x_1 = -82/15$, $x_2 = 74/15$, $x_3 = -6/5$.

13. Here one should substitute $.001x_2 = y_2$, $x_1 = y_1$, $x_3 = y_3$, which gives the matrix equation

$\begin{bmatrix}-1 & 2 & 0\\ 1 & 0 & 2\\ 0 & 1 & 1\end{bmatrix}\begin{bmatrix}y_1\\y_2\\y_3\end{bmatrix}=\begin{bmatrix}0\\-2\\1\end{bmatrix}$. The method from here on is straightforward and will result in inconsistent equations. It is easy to see that the equations are inconsistent. The first equation gives $y_1 = 2y_2$ and the third equation gives $y_3 = 1 - y_2$. Substituting in the second equation $2y_2 + 2(1 - y_2) = 2$, not -2.

14. First multiply the second equation by 1/.0001. Then the matrix equation is

$$\begin{bmatrix} 2 & 0 & 1 \\ 1 & 2 & 4 \\ 1 & -2 & -3 \end{bmatrix}\begin{bmatrix} x_1 \\ x_2 \\ x_3 \end{bmatrix} = \begin{bmatrix} 1 \\ 4 \\ -3 \end{bmatrix}. \quad \begin{bmatrix} 2 & 0 & 1 & 1 \\ 1 & 2 & 4 & 4 \\ 1 & -2 & -3 & -3 \end{bmatrix} \underset{(1/2)R1}{\approx} \begin{bmatrix} 1 & 0 & 1/2 & 1/2 \\ 1 & 2 & 4 & 4 \\ 1 & -2 & -3 & -3 \end{bmatrix}$$

$$\underset{\substack{R2+(-1)R1 \\ R3+(-1)R1}}{\approx} \begin{bmatrix} 1 & 0 & 1/2 & 1/2 \\ 0 & 2 & 7/2 & 7/2 \\ 0 & -2 & -7/2 & -7/2 \end{bmatrix} \underset{\substack{(1/2)R2 \\ R3+2R2}}{\approx} \begin{bmatrix} 1 & 0 & 1/2 & 1/2 \\ 0 & 1 & 7/4 & 7/4 \\ 0 & 0 & 0 & 0 \end{bmatrix},$$ so there are many exact solutions: $x_1 = 1/2 - r/2$, $x_2 = 7/4 - 7r/4$, and $x_3 = r$.

15. First we multiply equation 2 by 1/.01, then let $.1x_3 = y_3$, $x_2 = y_2$, $x_1 = y_1$. The matrix equation becomes $\begin{bmatrix} 0 & 1 & -1 \\ 2 & 1 & 0 \\ 1 & -4 & 1 \end{bmatrix}\begin{bmatrix} y_1 \\ y_2 \\ y_3 \end{bmatrix} = \begin{bmatrix} 2 \\ 1 \\ 2 \end{bmatrix}$.

$$\begin{bmatrix} 0 & 1 & -1 & 2 \\ 2 & 1 & 0 & 1 \\ 1 & -4 & 1 & 2 \end{bmatrix} \underset{R1 \Leftrightarrow R2}{\approx} \begin{bmatrix} 2 & 1 & 0 & 1 \\ 0 & 1 & -1 & 2 \\ 1 & -4 & 1 & 2 \end{bmatrix} \underset{(1/2)R1}{\approx} \begin{bmatrix} 1 & 1/2 & 0 & 1/2 \\ 0 & 1 & -1 & 2 \\ 1 & -4 & 1 & 2 \end{bmatrix}$$

$$\underset{R3+(-1)R1}{\approx} \begin{bmatrix} 1 & 1/2 & 0 & 1/2 \\ 0 & 1 & -1 & 2 \\ 0 & -9/2 & 1 & 3/2 \end{bmatrix} \underset{R2 \Leftrightarrow R3}{\approx} \begin{bmatrix} 1 & 1/2 & 0 & 1/2 \\ 0 & -9/2 & 1 & 3/2 \\ 0 & 1 & -1 & 2 \end{bmatrix}$$

$$\underset{(-2/9)R2}{\approx} \begin{bmatrix} 1 & 1/2 & 0 & 1/2 \\ 0 & 1 & -.2222 & -.3333 \\ 0 & 1 & -1 & 2 \end{bmatrix} \underset{\substack{R1+(-1/2)R2 \\ R3+(-1)R2}}{\approx} \begin{bmatrix} 1 & 0 & .1111 & .6667 \\ 0 & 1 & -.2222 & -.3333 \\ 0 & 0 & -.7778 & 2.3333 \end{bmatrix}$$

(Four decimal places for **y** will result in three decimal places for **x**.)

$$\underset{(-1/.7778)R3}{\approx}\begin{bmatrix}1 & 0 & .1111 & .6667\\ 0 & 1 & -.2222 & -.3333\\ 0 & 0 & 1 & -2.9999\end{bmatrix}\underset{\substack{R1+(-.1111)R3\\ R2+(.2222)R3}}{\approx}\begin{bmatrix}1 & 0 & 0 & 1.0000\\ 0 & 1 & 0 & -.9999\\ 0 & 0 & 1 & -2.9999\end{bmatrix}$$

$\begin{bmatrix}y_1\\ y_2\\ y_3\end{bmatrix} = \begin{bmatrix}1.0000\\ -.9999\\ -2.9999\end{bmatrix}$. Therefore $x_1 = 1.000$, $x_2 = -1.000$, and $x_3 = -29.999$.

The exact solution is $x_1 = 1$, $x_2 = -1$, and $x_3 = -30$.

16. First multiply equations 2 and 3 by 10, then let $.01x_2 = y_2$, $x_1 = y_1$, $x_3 = y_3$. The matrix equation becomes $\begin{bmatrix}-1 & 0 & 2\\ 1 & 1 & 0\\ 1 & 2 & 0\end{bmatrix}\begin{bmatrix}y_1\\ y_2\\ y_3\end{bmatrix} = \begin{bmatrix}1\\ 1\\ 2\end{bmatrix}$.

$$\begin{bmatrix}-1 & 0 & 2 & 1\\ 1 & 1 & 0 & 1\\ 1 & 2 & 0 & 2\end{bmatrix}\underset{(-1)R1}{\approx}\begin{bmatrix}1 & 0 & -2 & -1\\ 1 & 1 & 0 & 1\\ 1 & 2 & 0 & 2\end{bmatrix}\underset{\substack{R2+(-1)R1\\ R3+(-1)R2}}{\approx}\begin{bmatrix}1 & 0 & -2 & -1\\ 0 & 1 & 2 & 2\\ 0 & 2 & 2 & 3\end{bmatrix}$$

$$\underset{R2\Leftrightarrow R3}{\approx}\begin{bmatrix}1 & 0 & -2 & -1\\ 0 & 2 & 2 & 3\\ 0 & 1 & 2 & 2\end{bmatrix}\underset{(1/2)R2}{\approx}\begin{bmatrix}1 & 0 & -2 & -1\\ 0 & 1 & 1 & 3/2\\ 0 & 1 & 2 & 2\end{bmatrix}\underset{R3+(-1)R2}{\approx}\begin{bmatrix}1 & 0 & -2 & -1\\ 0 & 1 & 1 & 3/2\\ 0 & 0 & 1 & 1/2\end{bmatrix}$$

$\underset{\substack{R1+2R3\\ R2+(-1)R3}}{\approx}\begin{bmatrix}1 & 0 & 0 & 0\\ 0 & 1 & 0 & 1\\ 0 & 0 & 1 & 1/2\end{bmatrix}$, so $\begin{bmatrix}y_1\\ y_2\\ y_3\end{bmatrix} = \begin{bmatrix}0\\ 1\\ 1/2\end{bmatrix}$. Thus $x_1 = 0$, $x_2 = 100$, and $x_3 = 1/2$.

17. $\begin{bmatrix}1 & -2 & -1\\ 1 & -.001 & -1\\ 1 & 3 & -.002\end{bmatrix}\begin{bmatrix}x_1\\ x_2\\ x_3\end{bmatrix} = \begin{bmatrix}1\\ 2\\ -1\end{bmatrix}$.

$$\begin{bmatrix} 1 & -2 & -1 & 1 \\ 1 & -.001 & -1 & 2 \\ 1 & 3 & -.002 & -1 \end{bmatrix} \underset{\substack{R2+(-1)R1 \\ R3+(-1)R1}}{\approx} \begin{bmatrix} 1 & -2 & -1 & 1 \\ 0 & 1.999 & 0 & 1 \\ 0 & 5 & .998 & -2 \end{bmatrix}$$

$$\underset{R2 \Leftrightarrow R3}{\approx} \begin{bmatrix} 1 & -2 & -1 & 1 \\ 0 & 5 & .998 & -2 \\ 0 & 1.999 & 0 & 1 \end{bmatrix} \underset{(1/5)R2}{\approx} \begin{bmatrix} 1 & -2 & -1 & 1 \\ 0 & 1 & .200 & -.4 \\ 0 & 1.999 & 0 & 1 \end{bmatrix}$$

(The first round-off error is here.)

$$\underset{\substack{R1+2R2 \\ R3+(-1.999)R2}}{\approx} \begin{bmatrix} 1 & 0 & -.600 & .2 \\ 0 & 1 & .200 & -.4 \\ 0 & 0 & -.400 & 1.800 \end{bmatrix} \underset{(-1/.400)R3}{\approx} \begin{bmatrix} 1 & 0 & -.600 & .2 \\ 0 & 1 & .200 & -.4 \\ 0 & 0 & 1 & -4.500 \end{bmatrix}$$

$$\underset{\substack{R1+.600R3 \\ R2+(-.200)R3}}{\approx} \begin{bmatrix} 1 & 0 & 0 & -2.500 \\ 0 & 1 & 0 & .500 \\ 0 & 0 & 1 & -4.500 \end{bmatrix}$$, so $\begin{bmatrix} x_1 \\ x_2 \\ x_3 \end{bmatrix} = \begin{bmatrix} -2.500 \\ .500 \\ -4.500 \end{bmatrix}$. Substituting in the original

equations $-2.500 - 1.000 + 4.500 = 1$, $-2.500 - .001 + 4.500 = 2 - .001 = 1.999$, and $-2.500 + 1.500 + .009 = -1 + .009 = -.991$.

Exercise Set 7.4

For Exercises 1–6 we give iterations found using a computer program. We stop when for all variables two successive iterations have given the same value to two decimal places.

1.

x	y	z
1	2	3
2.25	0	1.9375
2.484375	0.425	1.98515625
2.390039063	0.4059375	2.003974609
2.399509277	0.3984101562	1.99972522
2.400328766	0.4001099121	1.999945287

Thus to two decimal places x = 2.40, y = 0.40, and z = 2.00. In fact this is the exact solution.

2.

x	y	z
1	2	3

2	0.75	1.75
1.6875	0.59375	1.78125
1.6484375	0.62109375	1.79453125
1.655273438	0.6209960937	1.793144531
1.655249023	0.6206616211	1.79308252

Thus to two decimal places x = 1.66, y = 0.62, and z = 1.79. These values are the two-decimal-place approximations to the exact solution x = 144/87, y = 54/87, z = 156/87.

3.

x	y	z
0	0	0
4	5.5	2.875
4.525	5.2375	2.678125
4.511875	5.2440625	2.683046875
4.512203125	5.243898438	2.682923828

Thus to two decimal places x = 4.51, y = 5.24, and z = 2.68. These values are the two-decimal-place approximations to the exact solution x = 185/41, y = 215/41, z = 110/41.

4.

x	y	z
5	6	7
4.166666667	1.270833333	3.29375
5.125347222	2.317230903	3.206653646
4.762031973	2.293590585	3.276952664
4.781628582	2.278465407	3.271520824
4.785765002	2.280340419	3.270881042
4.785033367	2.280408911	3.271034218

Thus to two decimal places x = 4.79, y = 2.28, and z = 3.27. These values are the two-decimal-place approximations to the exact solution x = 512/107, y = 244/107, z = 350/107.

5.

x	y	z
20	30	−40
30	−22.5	11.3
−1.02	5.835	−0.571
9.3954	−2.34045	3.14717
6.273042	0.1502715	2.0245541
7.22023266	−0.603977805	2.364842093

6.933267602	−0.3754232776	2.261738176
7.020220074	−0.4446754931	2.292979113
6.993873256	−0.4236918496	2.283513021
7.001856422	−0.4300499555	2.286381275
6.999437499	−0.4281234305	2.285512186

Thus to two decimal places x = 7.00, y = −0.43, and z = 2.29. These values are the two-decimal-place approximations to the exact solution x = 7, y = −3/7, z = 16/7.

6.

x	y	z	w
0	0	0	0
3.333333333	3.333333333	6.666666667	2.333333333
2.388888888	1.784722222	5.989583333	2.319097222
2.246006944	1.971853299	5.939326534	2.332596571
2.283321699	1.979753154	5.939729234	2.32525443
2.285794915	1.979343327	5.942657121	2.325638153
2.285174675	1.978688885	5.942534914	2.325725539

Thus to two decimal places x = 2.29, y = 1.98, z = 5.94, and w = 2.33. These values are the two-decimal-place approximations to the exact solution x = 7310/3199, y = 6330/3199, z = 19010/3199, w = 7440/3199.

7.

x	y	z
0	0	0
1.6	−2.4	0.1
3.54	−0.36	−2.86
1.316	−5.544	1.956
6.4264	4.3824	−8.1176
−3.52944	−15.64704	11.85296
16.488224	24.341184	−28.158816
−23.5047104	−55.6635264	51.8364736
56.49811584	104.3345894	−
108.1654106		
−103.5007537	−215.6661642	211.8338358
216.4996985	424.3335343	−428.1664657

The method doesn't work because the system doesn't satisfy the conditions needed for the method to work. The solution is x = 19/6, y = −7/3, z = −3/2.

Exercise Set 7.5

In Exercises 1–4 we choose $\mathbf{x} = \begin{bmatrix} 1 \\ 2 \\ 1 \end{bmatrix}$. Results are recorded to five decimal places. The process is stopped when successive results agree to three decimal places.

1.

$\frac{A\mathbf{x}\cdot\mathbf{x}}{\mathbf{x}\cdot\mathbf{x}}$	Scaled $A\mathbf{x}$
2	$\begin{bmatrix} 1 \\ 0.5 \\ -0.5 \end{bmatrix}$
8	$\begin{bmatrix} 1 \\ 0.125 \\ -0.875 \end{bmatrix}$
8.31579	$\begin{bmatrix} 1 \\ 0.03125 \\ -0.96875 \end{bmatrix}$
8.09063	$\begin{bmatrix} 1 \\ 0.00781 \\ -0.99219 \end{bmatrix}$
8.02325	$\begin{bmatrix} 1 \\ 0.00195 \\ -0.99805 \end{bmatrix}$
8.00585	$\begin{bmatrix} 1 \\ 0.00049 \\ -0.99951 \end{bmatrix}$
8.00146	$\begin{bmatrix} 1 \\ 0.00012 \\ -0.99988 \end{bmatrix}$
8.00037	$\begin{bmatrix} 1 \\ 0.00003 \\ -0.99997 \end{bmatrix}$

Hence $\lambda = 8$ and the dominant eigenvector is $r\begin{bmatrix} 1 \\ 0 \\ -1 \end{bmatrix}$.

2.

$\frac{A\mathbf{x}\cdot\mathbf{x}}{\mathbf{x}\cdot\mathbf{x}}$	Scaled $A\mathbf{x}$
5	$\begin{bmatrix} 1 \\ 0.15385 \\ 1 \end{bmatrix}$
5.56140	$\begin{bmatrix} 1 \\ 0.02740 \\ 1 \end{bmatrix}$
5.10805	$\begin{bmatrix} 1 \\ 0.00536 \\ 1 \end{bmatrix}$
5.02139	$\begin{bmatrix} 1 \\ 0.00107 \\ 1 \end{bmatrix}$
5.00427	$\begin{bmatrix} 1 \\ 0.00021 \\ 1 \end{bmatrix}$
5.00085	$\begin{bmatrix} 1 \\ 0.00004 \\ 1 \end{bmatrix}$

Hence $\lambda = 5$ and the dominant eigenvector is $r\begin{bmatrix} 1 \\ 0 \\ 1 \end{bmatrix}$.

3.

$\dfrac{A\mathbf{x}\cdot\mathbf{x}}{\mathbf{x}\cdot\mathbf{x}}$	Scaled A**x**
6	$\begin{bmatrix} 1 \\ -0.1 \\ 1 \end{bmatrix}$
5.26866	$\begin{bmatrix} 1 \\ 0.01887 \\ 1 \end{bmatrix}$
6.13081	$\begin{bmatrix} 1 \\ -0.00308 \\ 1 \end{bmatrix}$
5.97843	$\begin{bmatrix} 1 \\ 0.00051 \\ 1 \end{bmatrix}$
6.00360	$\begin{bmatrix} 1 \\ -0.00008 \\ 1 \end{bmatrix}$
5.99940	$\begin{bmatrix} 1 \\ 0.00001 \\ 1 \end{bmatrix}$

Hence $\lambda = 6$ and the dominant eigenvector is $r\begin{bmatrix} 1 \\ 0 \\ 1 \end{bmatrix}$.

4.

$\dfrac{A\mathbf{x}\cdot\mathbf{x}}{\mathbf{x}\cdot\mathbf{x}}$	Scaled A**x**
9	$\begin{bmatrix} 1 \\ 0.08 \\ 1 \end{bmatrix}$
9.61244	$\begin{bmatrix} 1 \\ 0.00829 \\ 1 \end{bmatrix}$
9.06611	$\begin{bmatrix} 1 \\ 0.00092 \\ 1 \end{bmatrix}$
9.00732	$\begin{bmatrix} 1 \\ 0.00010 \\ 1 \end{bmatrix}$
9.00081	$\begin{bmatrix} 1 \\ 0.00001 \\ 1 \end{bmatrix}$
9.00009	$\begin{bmatrix} 1 \\ 0.00000 \\ 1 \end{bmatrix}$

Hence $\lambda = 9$ and the dominant eigenvector is $r\begin{bmatrix} 1 \\ 0 \\ 1 \end{bmatrix}$.

In Exercises 5–8 we choose $\mathbf{x} = \begin{bmatrix} 1 \\ 2 \\ 1 \\ 2 \end{bmatrix}$. Results are recorded to five decimal places. The process is stopped when successive results agree to three decimal places.

5.

$\frac{A\mathbf{x}\cdot\mathbf{x}}{\mathbf{x}\cdot\mathbf{x}}$	Scaled $A\mathbf{x}$
4.8	$\begin{bmatrix} 0.28571 \\ 0.14286 \\ 0.85714 \\ 1 \end{bmatrix}$
5.46667	$\begin{bmatrix} 0.10526 \\ 0.05263 \\ 0.94737 \\ 1 \end{bmatrix}$
5.86087	$\begin{bmatrix} 0.03636 \\ 0.01818 \\ 0.98182 \\ 1 \end{bmatrix}$
5.95964	$\begin{bmatrix} 0.01227 \\ 0.00613 \\ 0.99387 \\ 1 \end{bmatrix}$
5.98728	$\begin{bmatrix} 0.00411 \\ 0.00205 \\ 0.99795 \\ 1 \end{bmatrix}$
5.99584	$\begin{bmatrix} 0.00137 \\ 0.00069 \\ 0.99931 \\ 1 \end{bmatrix}$
5.99862	$\begin{bmatrix} 0.00046 \\ 0.00023 \\ 0.99977 \\ 1 \end{bmatrix}$
5.99954	$\begin{bmatrix} 0.00015 \\ 0.00008 \\ 0.99992 \\ 1 \end{bmatrix}$

6.

$\frac{A\mathbf{x}\cdot\mathbf{x}}{\mathbf{x}\cdot\mathbf{x}}$	Scaled $A\mathbf{x}$
6.4	$\begin{bmatrix} 0 \\ 0.09091 \\ 0.72727 \\ 1 \end{bmatrix}$
9.56989	$\begin{bmatrix} -0.04255 \\ 0.10638 \\ 0.97872 \\ 1 \end{bmatrix}$
10.42444	$\begin{bmatrix} -0.02449 \\ 0.02857 \\ 0.98776 \\ 1 \end{bmatrix}$
10.13543	$\begin{bmatrix} -0.01134 \\ 0.01377 \\ 0.99919 \\ 1 \end{bmatrix}$
10.07516	$\begin{bmatrix} -0.00482 \\ 0.00498 \\ 0.99952 \\ 1 \end{bmatrix}$
10.02848	$\begin{bmatrix} -0.00199 \\ 0.00208 \\ 0.99997 \\ 1 \end{bmatrix}$
10.01222	$\begin{bmatrix} -0.00081 \\ 0.00081 \\ 0.99998 \\ 1 \end{bmatrix}$
10.00483	$\begin{bmatrix} -0.00032 \\ 0.00033 \\ 1.00000 \\ 1 \end{bmatrix}$

Hence $\lambda = 6$ and the dominant eigenvector is $r\begin{bmatrix}0\\0\\1\\1\end{bmatrix}$.

Hence $\lambda = 10$ and the dominant eigenvector is $r\begin{bmatrix}0\\0\\1\\1\end{bmatrix}$.

7.

$\frac{A\mathbf{x}\cdot\mathbf{x}}{\mathbf{x}\cdot\mathbf{x}}$	Scaled $A\mathbf{x}$
14.2	$\begin{bmatrix}1\\0\\0.03030\\0.20202\end{bmatrix}$
82.22351	$\begin{bmatrix}1\\0.00976\\-0.00940\\0.04785\end{bmatrix}$
84.30649	$\begin{bmatrix}1\\0.01118\\-0.01117\\0.04011\end{bmatrix}$
84.31129	$\begin{bmatrix}1\\0.01125\\-0.01125\\0.03976\end{bmatrix}$
84.31129	$\begin{bmatrix}1\\0.01126\\-0.01126\\0.03975\end{bmatrix}$

Hence $\lambda = 84.311$ and the dominant

8.

$\frac{A\mathbf{x}\cdot\mathbf{x}}{\mathbf{x}\cdot\mathbf{x}}$	Scaled $A\mathbf{x}$
6.8	$\begin{bmatrix}1\\0.04762\\0.14286\\1\end{bmatrix}$
9.33184	$\begin{bmatrix}1\\0.02513\\0.00503\\0.98995\end{bmatrix}$
10.08787	$\begin{bmatrix}1\\0.00350\\0.00150\\0.99700\end{bmatrix}$
10.00593	$\begin{bmatrix}1\\0.00085\\0.00005\\0.99930\end{bmatrix}$
10.00330	$\begin{bmatrix}1\\0.00015\\0.00001\\0.99985\end{bmatrix}$
10.00054	$\begin{bmatrix}1\\0.00003\\0.00000\\0.99997\end{bmatrix}$

eigenvector is $r\begin{bmatrix} 1 \\ 0.011 \\ -0.011 \\ 0.040 \end{bmatrix}$.

Hence $\lambda = 10$ and the dominant eigenvector is $r\begin{bmatrix} 1 \\ 0 \\ 0 \\ 1 \end{bmatrix}$.

In exercises 9–11 we choose $\mathbf{x} = \begin{bmatrix} 1 \\ 2 \\ 1 \end{bmatrix}$. Results are recorded to five decimal places. The process is stopped when successive results agree to three decimal places.

9.

$\dfrac{A\mathbf{x}\cdot\mathbf{x}}{\mathbf{x}\cdot\mathbf{x}}$	9.16667	9.99083	9.99991	10.00000	10.00000
Scaled $A\mathbf{x}$	$\begin{bmatrix} 0.9375 \\ 1 \\ 0.5 \end{bmatrix}$	$\begin{bmatrix} 0.99359 \\ 1 \\ 0.5 \end{bmatrix}$	$\begin{bmatrix} 0.99936 \\ 1 \\ 0.5 \end{bmatrix}$	$\begin{bmatrix} 0.99994 \\ 1 \\ 0.5 \end{bmatrix}$	$\begin{bmatrix} 0.99999 \\ 1 \\ 0.5 \end{bmatrix}$

Thus the dominant eigenvalue is $\lambda = 10$ and the dominant eigenvector is $r\begin{bmatrix} 1 \\ 1 \\ 0.5 \end{bmatrix}$.

The unit eigenvector is $\mathbf{y} = \begin{bmatrix} 2/3 \\ 2/3 \\ 1/3 \end{bmatrix}$, and $B = A - \lambda\mathbf{y}\mathbf{y}^t = \frac{1}{9}\begin{bmatrix} 5 & -4 & -2 \\ -4 & 5 & -2 \\ -2 & -2 & 8 \end{bmatrix}$.

Let $C = 9B$.

$\dfrac{C\mathbf{x}\cdot\mathbf{x}}{\mathbf{x}\cdot\mathbf{x}}$	0.83333	9	9
$C\mathbf{x}$	$\begin{bmatrix} -1 \\ 0.8 \\ 0.4 \end{bmatrix}$	$\begin{bmatrix} -1 \\ 0.8 \\ 0.4 \end{bmatrix}$	$\begin{bmatrix} -1 \\ 0.8 \\ 0.4 \end{bmatrix}$

Thus the dominant eigenvalue of C is $\lambda = 9$ with eigenvector $\begin{bmatrix} -1 \\ 0.8 \\ 0.4 \end{bmatrix}$, so the dominant eigenvalue of B (and therefore an eigenvalue of A) is 1 with eigenvector $\begin{bmatrix} -1 \\ 0.8 \\ 0.4 \end{bmatrix}$.

The eigenspace of $\lambda = 1$ has dimension 2 with basis $\begin{bmatrix} -1 \\ 1 \\ 0 \end{bmatrix}$ and $\begin{bmatrix} 0 \\ -1 \\ 2 \end{bmatrix}$, so there are no more eigenvalues.

10. Using $\mathbf{x} = \begin{bmatrix} 1 \\ 2 \\ 1 \end{bmatrix}$, it takes far too many iterations to find dominant eigenvalue $\lambda = 6$ and dominant eigenvector r$\begin{bmatrix} 1 \\ 0 \\ 1 \end{bmatrix}$. The unit eigenvector is $\mathbf{y} = \begin{bmatrix} 1/\sqrt{2} \\ 0 \\ 1/\sqrt{2} \end{bmatrix}$, and

$B = A - \lambda \mathbf{y}\mathbf{y}^t = \begin{bmatrix} -1 & 0 & 1 \\ 0 & 4 & 0 \\ 1 & 0 & -1 \end{bmatrix}$.

$\frac{B\mathbf{x}\cdot\mathbf{x}}{\mathbf{x}\cdot\mathbf{x}}$	2.66667	4	4
B**x**	$\begin{bmatrix} 0 \\ 1 \\ 0 \end{bmatrix}$	$\begin{bmatrix} 0 \\ 1 \\ 0 \end{bmatrix}$	$\begin{bmatrix} 0 \\ 1 \\ 0 \end{bmatrix}$

Thus $\lambda = 4$ with unit eigenvector $\mathbf{z} = \begin{bmatrix} 0 \\ 1 \\ 0 \end{bmatrix}$. $C = B - \lambda \mathbf{z}\mathbf{z}^t = \begin{bmatrix} -1 & 0 & 1 \\ 0 & 0 & 0 \\ 1 & 0 & -1 \end{bmatrix}$. It is clear that $\begin{bmatrix} 1 \\ 2 \\ 1 \end{bmatrix}$ is not a good choice for **x**, since $C\begin{bmatrix} 1 \\ 2 \\ 1 \end{bmatrix} = \mathbf{0}$, so we choose $\mathbf{x} = \begin{bmatrix} 2 \\ 1 \\ 1 \end{bmatrix}$.

Three iterations produce $\lambda = 2$ with eigenvector $r\begin{bmatrix}1\\0\\-1\end{bmatrix}$.

11.

$\frac{A\mathbf{x}\cdot\mathbf{x}}{\mathbf{x}\cdot\mathbf{x}}$	5.33333	11.47368	11.98459	11.99957	11.99999	12.00000
Scaled A**x**	$\begin{bmatrix}1\\0.33333\\1\end{bmatrix}$	$\begin{bmatrix}1\\0.05556\\1\end{bmatrix}$	$\begin{bmatrix}1\\0.00926\\1\end{bmatrix}$	$\begin{bmatrix}1\\0.00154\\1\end{bmatrix}$	$\begin{bmatrix}1\\0.00026\\1\end{bmatrix}$	$\begin{bmatrix}1\\0.00004\\1\end{bmatrix}$

Thus $\lambda = 12$ with dominant eigenvector $r\begin{bmatrix}1\\0\\1\end{bmatrix}$. The unit eigenvector is $\mathbf{y} = \begin{bmatrix}1/\sqrt{2}\\0\\1/\sqrt{2}\end{bmatrix}$, and

$B = A - \lambda\mathbf{y}\mathbf{y}^t = \begin{bmatrix}1 & 0 & -1\\0 & 2 & 0\\-1 & 0 & 1\end{bmatrix}$. Three iterations of the process confirm that $\lambda = 2$ with dominant eigenvector $\begin{bmatrix}0\\1\\0\end{bmatrix}$. The eigenspace of $\lambda = 2$ has dimension 2 with basis vectors $\begin{bmatrix}0\\1\\0\end{bmatrix}$ and $\begin{bmatrix}1\\0\\-1\end{bmatrix}$, so there are no additional eigenvalues of A.

In Exercises 12 and 13 we choose $\mathbf{x} = \begin{bmatrix}1\\2\\1\\2\end{bmatrix}$. We record to five decimal places successive values of $\frac{A\mathbf{x}\cdot\mathbf{x}}{\mathbf{x}\cdot\mathbf{x}}$.

12. $\frac{A\mathbf{x}\cdot\mathbf{x}}{\mathbf{x}\cdot\mathbf{x}}$: 3.8, 9.36066, 9.92232, 9.97852, 9.99251, 9.99731, 9.99903, 9.99965.

$\lambda = 10$ and the dominant eigenvector is $r\begin{bmatrix}1\\0\\0\\1\end{bmatrix}$. The unit eigenvector is $\mathbf{y} = \begin{bmatrix}1/\sqrt{2}\\0\\0\\1/\sqrt{2}\end{bmatrix}$

and $B = A - \lambda\mathbf{y}\mathbf{y}^t = \begin{bmatrix}-1 & 0 & 0 & 1\\0 & 2 & -4 & 0\\0 & -4 & 2 & 0\\1 & 0 & 0 & -1\end{bmatrix}$.

$\dfrac{B\mathbf{x}\cdot\mathbf{x}}{\mathbf{x}\cdot\mathbf{x}}$: −0.7, 1.78947, 5.12088, 5.89175, 5.98783, 5.99865, 5.99985, 5.99998.

$\lambda = 6$ and the dominant eigenvector is $r\begin{bmatrix}0\\1\\-1\\0\end{bmatrix}$. The unit eigenvector is $\mathbf{z} = \begin{bmatrix}0\\1/\sqrt{2}\\-1/\sqrt{2}\\0\end{bmatrix}$

and $C = B - \lambda\mathbf{z}\mathbf{z}^t = \begin{bmatrix}-1 & 0 & 0 & 1\\0 & -1 & -1 & 0\\0 & -1 & -1 & 0\\1 & 0 & 0 & -1\end{bmatrix}$. $\dfrac{C\mathbf{x}\cdot\mathbf{x}}{\mathbf{x}\cdot\mathbf{x}}$: −1, −2, −2.

$\lambda = -2$ with dominant eigenvector $r\begin{bmatrix}-1/3\\1\\1\\1/3\end{bmatrix}$. The eigenspace of $\lambda = -2$ is two-dimensional with basis vectors $\begin{bmatrix}-1\\0\\0\\1\end{bmatrix}$ and $\begin{bmatrix}0\\1\\1\\0\end{bmatrix}$, so there are no more eigenvalues.

13. $\dfrac{A\mathbf{x}\cdot\mathbf{x}}{\mathbf{x}\cdot\mathbf{x}}$: 8.7, 9.84072, 9.96468, 9.99130, 9.99783, 9.99946, 9.99986.

$\lambda = 10$ and the dominant eigenvector is $r\begin{bmatrix} 1 \\ 1 \\ 1/2 \\ 1/2 \end{bmatrix}$. The unit eigenvector is $\mathbf{y} = \begin{bmatrix} 2/\sqrt{10} \\ 2/\sqrt{10} \\ 1/\sqrt{10} \\ 1/\sqrt{10} \end{bmatrix}$

and $B = A - \lambda \mathbf{y}\mathbf{y}^t = \begin{bmatrix} 1 & 0 & -1 & -1 \\ 0 & 1 & -1 & -1 \\ -1 & -1 & 3 & 1 \\ -1 & -1 & 1 & 3 \end{bmatrix}$.

$\dfrac{B\mathbf{x}\cdot\mathbf{x}}{\mathbf{x}\cdot\mathbf{x}}$: .6, 4.68, 4.95447, 4.99305, 4.99890, 4.99983, 4.99997.

$\lambda = 5$ and the dominant eigenvector is $r\begin{bmatrix} -1/2 \\ -1/2 \\ 1 \\ 1 \end{bmatrix}$. The unit eigenvector is $\mathbf{z} = \begin{bmatrix} -1/\sqrt{10} \\ -1/\sqrt{10} \\ 2/\sqrt{10} \\ 2/\sqrt{10} \end{bmatrix}$

and $C = B - \lambda \mathbf{z}\mathbf{z}^t = \begin{bmatrix} .5 & -.5 & 0 & 0 \\ -.5 & .5 & 0 & 0 \\ 0 & 0 & 1 & -1 \\ 0 & 0 & -1 & 1 \end{bmatrix}$.

$\dfrac{C\mathbf{x}\cdot\mathbf{x}}{\mathbf{x}\cdot\mathbf{x}}$: .15, 1.8, 1.94118, 1.98462, 1.99611, 1.99902, 1.99976.

$\lambda = 2$ with dominant eigenvector $r\begin{bmatrix} 0 \\ 0 \\ -1 \\ 1 \end{bmatrix}$. The unit eigenvector is $\mathbf{u} = \begin{bmatrix} 0 \\ 0 \\ -1/\sqrt{2} \\ 1/\sqrt{2} \end{bmatrix}$

and $D = C - \lambda \mathbf{u}\mathbf{u}^t = \begin{bmatrix} .5 & -.5 & 0 & 0 \\ -.5 & .5 & 0 & 0 \\ 0 & 0 & 0 & 0 \\ 0 & 0 & 0 & 0 \end{bmatrix}$. $\dfrac{D\mathbf{x}\cdot\mathbf{x}}{\mathbf{x}\cdot\mathbf{x}}$: 0.05, 1, 1.

$\lambda = 1$ with dominant eigenvector $r\begin{bmatrix} -1 \\ 1 \\ 0 \\ 0 \end{bmatrix}$.

14. (a) $B = A + I = \begin{bmatrix} 1 & 1 & 0 & 0 \\ 1 & 1 & 1 & 0 \\ 0 & 1 & 1 & 1 \\ 0 & 0 & 1 & 1 \end{bmatrix}$ and $\mathbf{x} = \begin{bmatrix} 1 \\ 2 \\ 2 \\ 1 \end{bmatrix}$.

$\frac{B\mathbf{x}\cdot\mathbf{x}}{\mathbf{x}\cdot\mathbf{x}}$	2.6	2.6176	2.6180
Scaled B**x**	$\begin{bmatrix} 0.6 \\ 1 \\ 1 \\ 0.6 \end{bmatrix}$	$\begin{bmatrix} 0.615 \\ 1 \\ 1 \\ 0.615 \end{bmatrix}$	$\begin{bmatrix} 0.618 \\ 1 \\ 1 \\ 0.618 \end{bmatrix}$

The sum of the entries is 3.236, so the Gould accessibility indices are .618/3.236 = .191, 1/3.236 = .309, 1/3.236 = .309, and .618/3.236 = .191. The given initial vector yields three decimal place accuracy after only three iterations.

(b) $B = A + I = \begin{bmatrix} 1 & 0 & 1 & 0 & 0 & 0 \\ 0 & 1 & 1 & 0 & 0 & 0 \\ 1 & 1 & 1 & 1 & 0 & 0 \\ 0 & 0 & 1 & 1 & 1 & 1 \\ 0 & 0 & 0 & 1 & 1 & 0 \\ 0 & 0 & 0 & 1 & 0 & 1 \end{bmatrix}$ and $\mathbf{x} = \begin{bmatrix} 1 \\ 1 \\ 3 \\ 3 \\ 1 \\ 1 \end{bmatrix}$.

$\frac{B\mathbf{x}\cdot\mathbf{x}}{\mathbf{x}\cdot\mathbf{x}}$	2.90909	3	3
Scaled B**x**	$\begin{bmatrix} 0.5 \\ 0.5 \\ 1 \\ 1 \\ 0.5 \\ 0.5 \end{bmatrix}$	$\begin{bmatrix} 0.5 \\ 0.5 \\ 1 \\ 1 \\ 0.5 \\ 0.5 \end{bmatrix}$	$\begin{bmatrix} 0.5 \\ 0.5 \\ 1 \\ 1 \\ 0.5 \\ 0.5 \end{bmatrix}$

The sum of the entries is 4, so the Gould accessibility indices are 1/8,1/8,1/4,1/4,1/8, and 1/8. The given initial vector yields an eigenvector for the largest positive eigenvalue immediately.

(c) $B = A + I = \begin{bmatrix} 1 & 1 & 1 & 1 & 1 \\ 1 & 1 & 0 & 0 & 0 \\ 1 & 0 & 1 & 0 & 0 \\ 1 & 0 & 0 & 1 & 0 \\ 1 & 0 & 0 & 0 & 1 \end{bmatrix}$ and $\mathbf{x} = \begin{bmatrix} 2 \\ 1 \\ 1 \\ 1 \\ 1 \end{bmatrix}$.

$\frac{B\mathbf{x}\cdot\mathbf{x}}{\mathbf{x}\cdot\mathbf{x}}$	3	3	3
Scaled B**x**	$\begin{bmatrix} 1 \\ 0.5 \\ 0.5 \\ 0.5 \\ 0.5 \end{bmatrix}$	$\begin{bmatrix} 1 \\ 0.5 \\ 0.5 \\ 0.5 \\ 0.5 \end{bmatrix}$	$\begin{bmatrix} 1 \\ 0.5 \\ 0.5 \\ 0.5 \\ 0.5 \end{bmatrix}$

The sum of the entries is 3, so the Gould accessibility indices are 1/3, 1/6, 1/6, 1/6, and 1/6. In this case the initial vector is an eigenvector for the largest positive eigenvalue.

15. $A = \begin{bmatrix} 0 & 1 & 0 & 0 & 0 \\ 1 & 0 & 0 & 0 & 1 \\ 0 & 0 & 0 & 1 & 1 \\ 0 & 0 & 1 & 0 & 1 \\ 0 & 1 & 1 & 1 & 0 \end{bmatrix}$, $B = \begin{bmatrix} 1 & 1 & 0 & 0 & 0 \\ 1 & 1 & 0 & 0 & 1 \\ 0 & 0 & 1 & 1 & 1 \\ 0 & 0 & 1 & 1 & 1 \\ 0 & 1 & 1 & 1 & 1 \end{bmatrix}$.

Power method gives $\lambda = 3.2143$, **x**=(.26, .57, .82, .82, 1.00). Normalized vector is (.07, .16, .24, .24, .29), giving connectivities. Note that vertices 3 and 4 have same connect.
Add: 1-3, $\lambda = 3.4812$; 1-4, $\lambda = 3.4812$, 1-5, $\lambda = 3.5616$, 2-3, $\lambda = 3.6412$, 2-4, $\lambda = 3.6412$.
Best to put a road 2-3 or 2-4. Both lead to same connectivity increase.
For 2-3, **x**=(0.33, 0.88, 1.00, 0.76, 1.00). For 2-4, **x**=(0.33, 0.88, 1.00, 0.76, 1.00).
Normalized vectors are, 2-3, (0.8, 0.22, 0.25, 0.19, 0.25). 2-4, (0.8, 0.22, 0.19, 0.25, 0.25).
Note, while vertex 5 is still among the most connected, it's relative importance has decreased in both cases. Whether we put 2-3 or 2-4 depends on whether the most influential politician favors 3 or 4!

16. North Island: A =
Wa Ne H A T G Na We

B=
Wa Ne H A T G Na We

$$\begin{array}{c} \text{Wa} \\ \text{Ne} \\ \text{H} \\ \text{A} \\ \text{T} \\ \text{G} \\ \text{Na} \\ \text{We} \end{array} \begin{array}{c} \begin{bmatrix} 0 & 1 & 0 & 0 & 0 & 0 & 1 & 1 \\ 1 & 0 & 1 & 0 & 0 & 0 & 0 & 0 \\ 0 & 1 & 0 & 1 & 1 & 0 & 0 & 0 \\ 0 & 0 & 1 & 0 & 0 & 0 & 0 & 0 \end{bmatrix} \\ \begin{bmatrix} 0 & 0 & 1 & 0 & 0 & 0 & 0 & 0 \\ 0 & 0 & 0 & 0 & 0 & 0 & 1 & 0 \\ 1 & 0 & 0 & 0 & 0 & 1 & 0 & 1 \\ 1 & 0 & 0 & 0 & 0 & 0 & 1 & 0 \end{bmatrix} \end{array}, \quad \begin{array}{c} \text{Wa} \\ \text{Ne} \\ \text{H} \\ \text{A} \\ \text{T} \\ \text{G} \\ \text{Na} \\ \text{We} \end{array} \begin{array}{c} \begin{bmatrix} 1 & 1 & 0 & 0 & 0 & 0 & 1 & 1 \\ 1 & 1 & 1 & 0 & 0 & 0 & 0 & 0 \\ 0 & 1 & 1 & 1 & 1 & 0 & 0 & 0 \\ 0 & 0 & 1 & 1 & 0 & 0 & 0 & 0 \end{bmatrix} \\ \begin{bmatrix} 0 & 0 & 1 & 0 & 1 & 0 & 0 & 0 \\ 0 & 0 & 0 & 0 & 0 & 1 & 1 & 0 \\ 1 & 0 & 0 & 0 & 0 & 1 & 1 & 1 \\ 1 & 0 & 0 & 0 & 0 & 0 & 1 & 1 \end{bmatrix} \end{array},$$

Power method gives $\lambda = 3.3577$, $\mathbf{x}=(1,0.59,0.39,0.17,0.17,0.40,0.94,0.82)$. Normalized vector is (0.22,0.13,0.09,0.04,0.04,0.09,0.21,0.18), giving connectivities. Wanganui has the highest connectivity, 0.22.

South Island: A = ... B =

$$\begin{array}{c} \\ \text{Ri} \\ \text{Cr} \\ \text{D} \\ \text{T} \\ \text{Ch} \\ \text{G} \\ \text{Re} \\ \text{N} \end{array} \begin{array}{c} \begin{array}{cccccccc} \text{Ri} & \text{Cr} & \text{D} & \text{T} & \text{Ch} & \text{G} & \text{Re} & \text{N} \end{array} \\ \begin{bmatrix} 0 & 0 & 1 & 0 & 0 & 0 & 0 & 0 \\ 0 & 0 & 1 & 0 & 0 & 0 & 0 & 0 \\ 1 & 1 & 0 & 1 & 0 & 0 & 0 & 0 \\ 0 & 0 & 1 & 0 & 1 & 0 & 0 & 0 \end{bmatrix} \\ \begin{bmatrix} 0 & 0 & 0 & 1 & 0 & 1 & 0 & 1 \\ 0 & 0 & 0 & 0 & 1 & 0 & 1 & 0 \\ 0 & 0 & 0 & 0 & 0 & 1 & 0 & 0 \\ 0 & 0 & 0 & 0 & 1 & 0 & 0 & 0 \end{bmatrix} \end{array}, \quad \begin{array}{c} \\ \text{Ri} \\ \text{Cr} \\ \text{D} \\ \text{T} \\ \text{Ch} \\ \text{G} \\ \text{Re} \\ \text{N} \end{array} \begin{array}{c} \begin{array}{cccccccc} \text{Ri} & \text{Cr} & \text{D} & \text{T} & \text{Ch} & \text{G} & \text{Re} & \text{N} \end{array} \\ \begin{bmatrix} 1 & 0 & 1 & 0 & 0 & 0 & 0 & 0 \\ 0 & 1 & 1 & 0 & 0 & 0 & 0 & 0 \\ 1 & 1 & 1 & 1 & 0 & 0 & 0 & 0 \\ 0 & 0 & 1 & 1 & 1 & 0 & 0 & 0 \end{bmatrix} \\ \begin{bmatrix} 0 & 0 & 0 & 1 & 1 & 1 & 0 & 1 \\ 0 & 0 & 0 & 0 & 1 & 1 & 1 & 0 \\ 0 & 0 & 0 & 0 & 0 & 1 & 1 & 0 \\ 0 & 0 & 0 & 0 & 1 & 0 & 0 & 1 \end{bmatrix} \end{array},$$

Power method gives $\lambda = 3.0420$, $\mathbf{x}=(0.41,0.41,0.84,0.90,1,0.65,0.32,0.50)$. Normalized vector is (0.08,0.08,0.17,0.18,0.20,0.13,0.06,0.10)), giving connectivities. Christchurch has the highest connectivity, 0.20. The North Island (with an index of 3.36) is more connected than the South Island (3.04).

Chapter 7 Review Exercises

1. From the first equation $x_1 = 2$, and from the second equation $6 - x_2 = 7$, so $x_2 = -1$.

   From the third equation $2 - 3 + 2x_3 = 1$, so $x_3 = 1$.

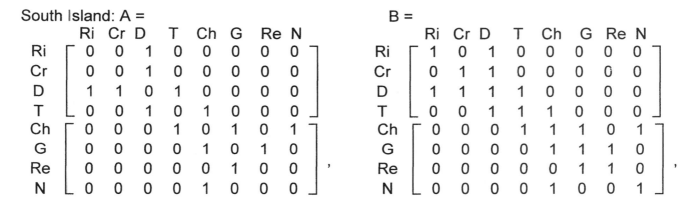

2. The "rule" is Ri − aRj causes the (i,j)th position in L to be a. $L = \begin{bmatrix} 1 & 0 & 0 \\ 3/2 & 1 & 0 \\ -2/7 & 4 & 1 \end{bmatrix}$.

3. The matrix equation A**x** = **b** is $\begin{bmatrix} 6 & 1 & -1 \\ -6 & 1 & 1 \\ 12 & 12 & 1 \end{bmatrix}\begin{bmatrix} x_1 \\ x_2 \\ x_3 \end{bmatrix} = \begin{bmatrix} 5 \\ 1 \\ 52 \end{bmatrix}$.

$$\begin{bmatrix} 6 & 1 & -1 \\ -6 & 1 & 1 \\ 12 & 12 & 1 \end{bmatrix} \underset{\substack{R2+R1 \\ R3+(-2)R1}}{\approx} \begin{bmatrix} 6 & 1 & -1 \\ 0 & 2 & 0 \\ 0 & 10 & 3 \end{bmatrix} \underset{R3+(-5)R2}{\approx} \begin{bmatrix} 6 & 1 & -1 \\ 0 & 2 & 0 \\ 0 & 0 & 3 \end{bmatrix} = U, \text{ and}$$

$L = \begin{bmatrix} 1 & 0 & 0 \\ -1 & 1 & 0 \\ 2 & 5 & 1 \end{bmatrix}$. $L\mathbf{y} = \begin{bmatrix} 5 \\ 1 \\ 52 \end{bmatrix}$ gives $\mathbf{y} = \begin{bmatrix} 5 \\ 6 \\ 12 \end{bmatrix}$, and $U\mathbf{x} = \mathbf{y}$ gives $\mathbf{x} = \begin{bmatrix} 1 \\ 3 \\ 4 \end{bmatrix}$.

4. The arrays below show the numbers of multiplications and additions needed to compute each entry in the matrix LU = A.

$$\begin{matrix} 1 & 1 & 1 & \dots & 1 & 1 \\ 1 & 2 & 2 & \dots & 2 & 2 \\ 1 & 2 & 3 & \dots & 3 & 3 \\ \vdots & \vdots & \vdots & & \vdots & \vdots \\ 1 & 2 & 3 & \dots & n-1 & n-1 \\ 1 & 2 & 3 & \dots & n-1 & n \end{matrix} \qquad \begin{matrix} 0 & 0 & 0 & \dots & 0 & 0 \\ 0 & 1 & 1 & \dots & 1 & 1 \\ 0 & 1 & 2 & \dots & 2 & 2 \\ \vdots & \vdots & \vdots & & \vdots & \vdots \\ 0 & 1 & 2 & \dots & n-2 & n-2 \\ 0 & 1 & 2 & \dots & n-2 & n-1 \end{matrix}$$

multiplications additions

In the multiplication array, for any, is the sum of the numbers in the ith column down to a_{ii} and the numbers in the ith row to and including a_{ii} is i^2 :

$$a_{i1} + a_{1i} + a_{i2} + a_{2i} + \dots + a_{i\,i-1} + a_{i-1\,i} + a_{ii} = 2x1 + 2x2 + \dots + 2(i-1) + i$$

$= 2(1+2+\dots+(i-1)) + i = 2\frac{(i-1)i}{2} + i = i^2$. Thus the number of multiplications needed to calculate LU is the sum of the squares of the numbers from 1 to n:

$1^2 + 2^2 + \dots + (n-1)^2 + n^2 = \frac{n(n+1)(2n+1)}{6} = \frac{2n^3+3n^2+n}{6}$. Likewise, the number of additions is the sum of the squares of the numbers from 1 to n−1:

$1^2 + 2^2 + \ldots + (n-2)^2 + (n-1)^2 = \dfrac{(n-1)n(2n-1)}{6} = \dfrac{2n^3-3n^2+n}{6}$. Thus the total

number of operations is $\dfrac{2n^3+3n^2+n}{6} + \dfrac{2n^3-3n^2+n}{6} = \dfrac{2n^3+n}{3}$.

If one assumes the elements on the diagonal of L are all 1, the number of multiplications for each term on or above the diagonal is decreased by 1. There are

$n + (n-1) + \ldots + 2 + 1 = \dfrac{n(n+1)}{2}$ such terms, so the number of multiplications in this

case is $\dfrac{2n^3+3n^2+n}{6} - \dfrac{n(n+1)}{2} = \dfrac{n^3-n}{3}$ and the total number of operations is

$$\frac{2n^3-3n^2+n}{6} + \frac{n^3-n}{3} = \frac{4n^3-3n^2-n}{6}.$$

5. $A = \begin{bmatrix} 250 & 401 \\ 125 & 201 \end{bmatrix}$ and $A^{-1} = \begin{bmatrix} 201/125 & -401/125 \\ -1 & 2 \end{bmatrix}$, so $||A|| = \max\{375,602\} = 602$ and

$||A^{-1}|| = \max\{326/125, 651/125\} = 651/125$. Thus $c(A) = 3135.216 = .3135216\text{x}10^4$.

The solution of a system of equations AX = B can have four fewer significant digits of accuracy than the elements of A or B.

6. $A = \begin{bmatrix} 1 & 1 & -1 \\ 1 & 0 & 2 \\ 0 & 3 & 2 \end{bmatrix}$ and $A^{-1} = \begin{bmatrix} 6/11 & 5/11 & -2/11 \\ 2/11 & -2/11 & 3/11 \\ -3/11 & 3/11 & 1/11 \end{bmatrix}$, so $||A|| = \max\{2,4,5\} = 5$ and

$||A^{-1}|| = \max\{1,10/11,6/11\} = 1$. Thus $c(A) = 5 = .5\text{x}10^1$. The solution of a system of equations AX = B can have one fewer significant digit of accuracy than the elements of A.

$$A = \begin{bmatrix} 1 & -1 & -1 & 0 & 0 \\ 0 & 0 & 1 & -1 & -1 \\ 1 & 1 & 0 & 0 & 0 \\ 1 & 0 & 0 & 2 & 0 \\ 0 & 0 & 0 & 2 & 2 \end{bmatrix} \text{ and } A^{-1} = \begin{bmatrix} 1/2 & 1/2 & 1/2 & 0 & 1/4 \\ -1/2 & -1/2 & 1/2 & 0 & -1/4 \\ 0 & 1 & 0 & 0 & 1/2 \\ -1/4 & -1/4 & -1/4 & 1/2 & -1/8 \\ 1/4 & 1/4 & 1/4 & -1/2 & 5/8 \end{bmatrix},$$

so $||A|| = \max\{3,2,2,5,3\} = 5$ and $||A^{-1}|| = \max\{3/2,5/2,3/2,1,7/4\} = 5/2$. Thus $c(A) = 25/2 = 12.5 = .125 \times 10^2$. The solution of a system of equations $AX = B$ can have two fewer significant digits of accuracy than the elements of A.

7. $c(A) = ||A||\,||A^{-1}|| = ||A^{-1}||\,||A|| = ||A^{-1}||\,||(A^{-1})^{-1}|| = c(A^{-1})$.

8. $c(A)$ is defined only for nonsingular matrices, so it is not a mapping of M_{nn}.

9. (a) Multiply the first equation by 10, then let $100x_1 = y_1$, $x_2 = y_2$, $.01x_3 = y_3$. The matrix equation becomes $\begin{bmatrix} 0 & -4 & 2 \\ 1 & 1 & -1 \\ 2 & 2 & -3 \end{bmatrix}\begin{bmatrix} y_1 \\ y_2 \\ y_3 \end{bmatrix} = \begin{bmatrix} 2 \\ 1 \\ 1 \end{bmatrix}$.

$$\begin{bmatrix} 0 & -4 & 2 & 2 \\ 1 & 1 & -1 & 1 \\ 2 & 2 & -3 & 1 \end{bmatrix} \underset{R1 \Leftrightarrow R3}{\approx} \begin{bmatrix} 2 & 2 & -3 & 1 \\ 1 & 1 & -1 & 1 \\ 0 & -4 & 2 & 2 \end{bmatrix} \underset{(1/2)R1}{\approx} \begin{bmatrix} 1 & 1 & -3/2 & 1/2 \\ 1 & 1 & -1 & 1 \\ 0 & -4 & 2 & 2 \end{bmatrix}$$

$$\underset{R2+(-1)R1}{\approx} \begin{bmatrix} 1 & 1 & -3/2 & 1/2 \\ 0 & 0 & 1/2 & 1/2 \\ 0 & -4 & 2 & 2 \end{bmatrix} \underset{R2 \Leftrightarrow R3}{\approx} \begin{bmatrix} 1 & 1 & -3/2 & 1/2 \\ 0 & -4 & 2 & 2 \\ 0 & 0 & 1/2 & 1/2 \end{bmatrix}$$

$$\underset{(-1/4)R2}{\approx} \begin{bmatrix} 1 & 1 & -3/2 & 1/2 \\ 0 & 1 & -1/2 & -1/2 \\ 0 & 0 & 1/2 & 1/2 \end{bmatrix} \underset{R1+(-1)R2}{\approx} \begin{bmatrix} 1 & 0 & -1 & 1 \\ 0 & 1 & -1/2 & -1/2 \\ 0 & 0 & 1/2 & 1/2 \end{bmatrix}$$

$$\underset{2R3}{\approx} \begin{bmatrix} 1 & 0 & -1 & 1 \\ 0 & 1 & -1/2 & -1/2 \\ 0 & 0 & 1 & 1 \end{bmatrix} \underset{\substack{R1+R3 \\ R2+(1/2)R3}}{\approx} \begin{bmatrix} 1 & 0 & 0 & 2 \\ 0 & 1 & 0 & 0 \\ 0 & 0 & 1 & 1 \end{bmatrix}, \text{ so } \begin{bmatrix} y_1 \\ y_2 \\ y_3 \end{bmatrix} = \begin{bmatrix} 2 \\ 0 \\ 1 \end{bmatrix}.$$

Thus $x_1 = 1/50$, $x_2 = -0$, and $x_3 = 100$.

(b) Multiply the second equation by 10 and the third equation by 100, then let

$.001x_3 = y_3$, $x_1 = y_1$, $x_2 = y_2$. The matrix equation becomes

$$\begin{bmatrix} 0 & -1 & 1 \\ 1 & 0 & 2 \\ 1 & 1 & 3 \end{bmatrix}\begin{bmatrix} y_1 \\ y_2 \\ y_3 \end{bmatrix} = \begin{bmatrix} 6 \\ -2 \\ 2 \end{bmatrix}. \quad \begin{bmatrix} 0 & -1 & 1 & 6 \\ 1 & 0 & 2 & -2 \\ 1 & 1 & 3 & 2 \end{bmatrix} \underset{R1 \Leftrightarrow R2}{\approx} \begin{bmatrix} 1 & 0 & 2 & -2 \\ 0 & -1 & 1 & 6 \\ 1 & 1 & 3 & 2 \end{bmatrix}$$

$$\underset{R3+(-1)R2}{\approx} \begin{bmatrix} 1 & 0 & 2 & -2 \\ 0 & -1 & 1 & 6 \\ 0 & 1 & 1 & 4 \end{bmatrix} \underset{(-1)R2}{\approx} \begin{bmatrix} 1 & 0 & 2 & -2 \\ 0 & 1 & -1 & -6 \\ 0 & 1 & 1 & 4 \end{bmatrix} \underset{R3+(-1)R2}{\approx} \begin{bmatrix} 1 & 0 & 2 & -2 \\ 0 & 1 & -1 & -6 \\ 0 & 0 & 2 & 10 \end{bmatrix}$$

$$\underset{(1/2)R3}{\approx} \begin{bmatrix} 1 & 0 & 2 & -2 \\ 0 & 1 & -1 & -6 \\ 0 & 0 & 1 & 5 \end{bmatrix} \underset{\substack{R1+(-2)R3 \\ R2+R3}}{\approx} \begin{bmatrix} 1 & 0 & 0 & -1/2 \\ 0 & 1 & 0 & -1 \\ 0 & 0 & 1 & 5 \end{bmatrix}, \text{ so } \begin{bmatrix} y_1 \\ y_2 \\ y_3 \end{bmatrix} = \begin{bmatrix} -12 \\ -1 \\ 5 \end{bmatrix}.$$

Thus $x_1 = -12$, $x_2 = -1$, and $x_3 = 5000$.

10. These iterations were found using a computer program. We stop when for all variables two successive iterations have given the same value to two decimal places.

x	y	z
1	2	-3
3.666666667	-1.047619048	3.845238095
5.31547619	1.414965986	2.817389456
4.733737245	1.140670554	3.03139805
4.815121249	1.20913595	2.9939357
4.797466625	1.198124836	3.001102135
4.800496217	1.200330567	2.999793304

Thus to two decimal places x = 4.80, y = 1.20, and z = 3.00. Substitution of these values into the given equations will show that, in fact, this is the exact solution.

11. We choose $\mathbf{x} = \begin{bmatrix} 1 \\ 2 \\ 1 \end{bmatrix}$. Results are recorded to 5 decimal places. The process is stopped when successive results agree to three significant digits.

$\frac{A\mathbf{x}\cdot\mathbf{x}}{\mathbf{x}\cdot\mathbf{x}}$	Scaled $A\mathbf{x}$

$$9.83333 \qquad \begin{bmatrix} 0.61111 \\ 0.83333 \\ 1 \end{bmatrix}$$

$$11.71045 \qquad \begin{bmatrix} 0.64 \\ 1 \\ 0.955 \end{bmatrix}$$

$$11.65458 \qquad \begin{bmatrix} 0.64954 \\ 0.98520 \\ 1 \end{bmatrix}$$

$$11.72116 \qquad \begin{bmatrix} 0.65327 \\ 1 \\ 0.99933 \end{bmatrix}$$

$$11.71197 \qquad \begin{bmatrix} 0.65228 \\ 0.99646 \\ 1 \end{bmatrix}$$

$$11.71629 \qquad \begin{bmatrix} 0.65256 \\ 0.99752 \\ 1 \end{bmatrix}$$

$$11.71567 \qquad \begin{bmatrix} 0.65246 \\ 0.99723 \\ 1 \end{bmatrix}$$

$$11.71594 \qquad \begin{bmatrix} 0.65248 \\ 0.99731 \\ 1 \end{bmatrix}$$

Hence $\lambda = 11.716$ and the dominant eigenvector is $r\begin{bmatrix} 0.652 \\ 0.997 \\ 1 \end{bmatrix}$.

12. We choose $\mathbf{x} = \begin{bmatrix} 1 \\ 2 \\ 1 \end{bmatrix}$. Values of $\dfrac{A\mathbf{x}\cdot\mathbf{x}}{\mathbf{x}\cdot\mathbf{x}}$ are 4, 5.5, 5.74194, 5.87645, 5.94314, 5.97432, 5.98851, 5.99487, 5.99772, 5.99899. Thus $\lambda = 6$. The dominant eigenvector is $r\begin{bmatrix} 1 \\ 2 \\ -1 \end{bmatrix}$ and the unit eigenvector is $\mathbf{y} = \begin{bmatrix} 1/\sqrt{6} \\ 2/\sqrt{6} \\ -1/\sqrt{6} \end{bmatrix}$.

$B = A - \lambda\mathbf{y}\mathbf{y}^t = \begin{bmatrix} 2 & 0 & 2 \\ 0 & 0 & 0 \\ 2 & 0 & 2 \end{bmatrix}$. Three iterations of the process confirm that $\lambda = 4$ with dominant eigenvector s$\begin{bmatrix} 1 \\ 0 \\ 1 \end{bmatrix}$. The unit eigenvector is $\mathbf{z} = \begin{bmatrix} 1/\sqrt{2} \\ 0 \\ 1/\sqrt{2} \end{bmatrix}$, and

$C = A - \lambda \mathbf{z}\mathbf{z}^t$ is the zero matrix. Thus the remaining eigenvalue is zero. The

corresponding eigenvector is t$\begin{bmatrix} -1 \\ 1 \\ 1 \end{bmatrix}$.

Chapter 8

Exercise Set 8.1

For Exercises 1 - 14 in this section, we refer to the drawing at the right. The points A, B, C, E, and F will be identified, and the objective function will be evaluated at the three nonzero vertices A, B, and C of the feasible region.

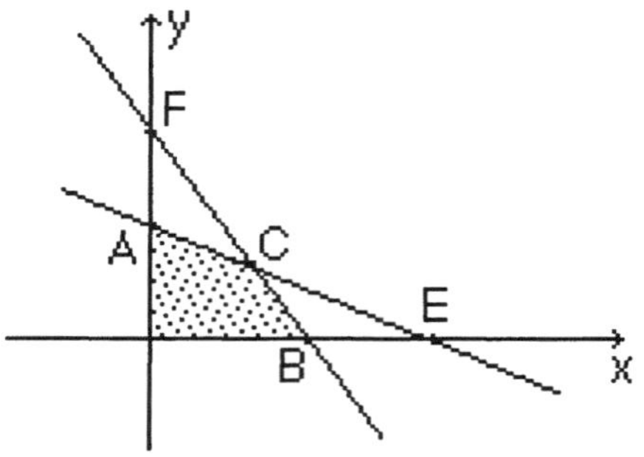

1. $4x + y = 36$ $\quad F = (0,36)$ $\quad B = (9,0)$

   $4x + 3y = 60$ $\quad A = (0,20)$ $\quad E = (15,0)$ $\quad C = (6,12)$

   At A, $f = 2(0) + 20 = 20$; at B, $f = 2(9) + 0 = 18$; and at C, $f = 2(6) + 12 = 24$, so the maximum value of f is 24 and it occurs at the point C.

2. $x + 2y = 4$ $\quad F = (0,2)$ $\quad B = (4,0)$

   $x + 6y = 8$ $\quad A = (0,4/3)$ $\quad E = (8,0)$ $\quad C = (2,1)$

   At A, $f = 0 - 4(4/3) = -16/3$; at B, $f = 4 - 4(0) = 4$; and at C, $f = 2 - 4(1) = -2$, so the maximum value of f is 4 and it occurs at the point B.

3. $x + 3y = 15$ $\quad A = (0,5)$ $\quad E = (15,0)$

   $2x + y = 10$ $\quad F = (0,10)$ $\quad B = (5,0)$ $\quad C = (3,4)$

   At A, $f = 4(0) + 2(5) = 10$; at B, $f = 4(5) + 2(0) = 20$; and at C, $f = 4(3) + 2(4) = 20$, so the maximum value of f is 20 and it occurs at every point on the line segment joining B and C.

4. $4x + y = 16$ $F = (0,16)$ $B = (4,0)$

$x + y = 7$ $A = (0,7)$ $E = (7,0)$ $C = (3,4)$

At A, $f = 2(0) + 7 = 7$; at B, $f = 2(4) + 0 = 8$; and at C, $f = 2(3) + 4 = 10$, so the maximum value of f is 10 and it occurs at the point C.

5. $2x + y = 4$ $A = (0,4)$ $E = (2,0)$

$6x + y = 8$ $F = (0,8)$ $B = (4/3,0)$ $C = (1,2)$

At A, $f = 4(0) + 4 = 4$; at B, $f = 4(4/3) + 0 = 16/3$; and at C, $f = 4(1) + 2 = 6$, so the maximum value of f is 6 and it occurs at the point C.

6. $x + y = 150$ $A = (0,150)$ $E = (150,0)$

$4x + y = 450$ $F = (0,450)$ $B = (112.5,0)$ $C = (100,50)$

At A, $-f = 3(0) + 150 = 150$; at B, $-f = 3(112.5) + 0 = 337.5$; and at C, $-f = 3(100) + 50 = 350$, so the maximum value of $-f$ is 350 and it occurs at the point C. Thus the minimum value of f is -350 and it occurs at the point C.

7. $2x + y = 440$ $A = (0,440)$ $E = (220,0)$

$4x + y = 680$ $F = (0,680)$ $B = (170,0)$ $C = (120,200)$

At A, $-f = 2(0) - 440 = -440$; at B, $-f = 2(170) - 0 = 340$; and at C, $-f = 2(120) - 200 = 40$, so the maximum value of $-f$ is 340 and it occurs at the point B. Thus the minimum value of f is -340 and it occurs at the point B.

8. $x + 2y = 4$ $F = (0,2)$ $B = (4,0)$

$x + 4y = 6$ $A = (0,3/2)$ $E = (6,0)$ $C = (2,1)$

At A, $-f = 0 - 2(3/2) = -3$; at B, $-f = 4 - 2(0) = 4$; and at C,

$-f = 2 - 2(1) = 0$, so the maximum value of $-f$ is 4 and it occurs at the point B.

Thus the minimum value of f is -4 and it occurs at the point B.

9.

	hours	cost($)	profit($)	number to manufacture
C1	1	30	10	x
C2	4	20	8	y
totals	≤1600	≤18000		

We are to maximize profit, $f = 10x + 8y$, subject to the constraints

$x + 4y \le 1600$ (hours),

$30x + 20y \le 18000$ (divide by 10: $3x + 2y \le 1800$) (cost),

$x \ge 0$, and $y \ge 0$.

$x + 4y = 1600$	A = (0,400)	E = (1600,0)	
$3x + 2y = 1800$	F = (0,900)	B = (600,0)	C = (400,300)

At A, $f = 10(0) + 8(400) = 3200$; at B, $f = 10(600) + 8(0) = 6000$; and at C, $f = 10(400) + 8(300) = 6400$, so the maximum value of f is 6400 and it occurs at the point C. To ensure the maximum profit of \$6400, the company should manufacture 400 of model C1 and 300 of model C2.

10.

	machine I time(min)	machine II time(min)	profit($)	number
X	3	3	15	x
Y	1	2	7	y
totals	≤3000	≤4500		

We are to maximize profit, $f = 15x + 7y$, subject to the constraints

$3x + y \le 3000$ (machine I time),

$3x + 2y \le 4500$ (machine II time),

$x \ge 0$, and $y \ge 0$.

$3x + y = 3000$	F = (0,3000)	B = (1000,0)	
$3x + 2y = 4500$	A = (0,2250)	E = (1500,0)	C = (500,1500)

At A, f = 15(0) + 7(2250) = 15750; at B, f = 15(1000) + 7(0) = 15000; and at C, f = 15(500) + 7(1500) = 18000, so the maximum value of f is 18000 and it occurs at the point C. To ensure the maximum profit of $18000, the company should manufacture 500 of product X and 1500 of product Y.

11.

	hours	cost($)	profit($)	number to ship
X	20	60	40	x
Y	10	10	20	y
totals	≤1200	≤2400		

We are to maximize profit, f = 40x + 20y, subject to the constraints

$20x + 10y \leq 1200$ (divide by 10: $2x + y \leq 120$) (hours),

$60x + 10y \leq 2400$ (divide by 10: $6x + y \leq 240$) (cost),

$x \geq 0$, and $y \geq 0$.

2x + y = 120 A = (0,120) E = (60,0)

6x + y = 240 F = (0,240) B = (40,0) C = (30,60)

At A, f = 40(0) + 20(120) = 2400; at B, f = 40(40) + 20(0) = 1600; and at C, f = 40(30) + 20(60) = 2400, so the maximum value of f is 2400 and it occurs at all points on the line segment joining points A and C. To ensure the maximum profit of $2400, the company should ship x refrigerators from the plant at town X and 120 − 2x refrigerators from the plant at town Y, where $0 \leq x \leq 30$.

12.

	cost($)	profit($)	quantity(barrels)
gasoline	6	3.50	x
heating oil	8	4.00	y
totals	≤9600	≤1400	

We are to maximize profit, f = 3.5x + 4y, subject to the constraints

$x + y \leq 1400$ (quantity),

$6x + 8y \le 9600$ (divide by 2: $3x + 4y \le 4800$) (cost),

$x \ge 0$, and $y \ge 0$.

$x + y = 1400$ F = (0,1400) B = (1400,0)

$3x + 4y = 4800$ A = (0,1200) E = (1600,0) C = (800,600)

At A, f = 3.5(0) + 4(1200) = 4800; at B, f = 3.5(1400) + 4(0) = 4900; and at C, f = 3.5(800) + 4(600) = 5200, so the maximum value of f is 5200 and it occurs at the point C. To ensure the maximum profit of $5200, the company should produce 800 barrels of gasoline and 600 barrels of heating oil.

13.

	cotton(sq. yd.)	wool(sq. yd.)	number to make	income($)
suit	2	1	x	90
dress	1	3	y	90
totals	≤80	≤120		

We are to maximize income, f = 90x + 90y, subject to the constraints

$2x + y \le 80$ (sq. yd. cotton),

$x + 3y \le 120$ (sq. yd. wool),

$x \ge 0$, and $y \ge 0$.

$2x + y = 80$ F = (0,80) B = (40,0)

$x + 3y = 120$ A = (0,40) E = (120,0) C = (24,32)

At A, f = 90(0) + 90(40) = 3600; at B, f = 90(40) + 90(0) = 3600; and at C, f = 90(24) + 90(32) = 5040, so the maximum value of f is 5040 and it occurs at the point C. To ensure the maximum income of $5040, the tailor should make 24 suits and 32 dresses.

14.

	purchase cost($)	maintenance cost($)	number
Arrow	12,000	400	x
Gazelle	15,000	300	y
totals	≤1,800,000	≤40000	

We are to maximize $f = 28x + 25y$, subject to the constraints

$12000x + 15000y \leq 1800000$ (divide by 3000: $4x + 5y \leq 600$),

$400x + 300y \leq 40000$ (divide by 100: $4x + 3y \leq 400$),

$x \geq 0$, and $y \geq 0$.

$4x + 5y = 600$	A = (0,120)	E = (150,0)	
$4x + 3y = 400$	F = (0,400/3)	B = (100,0)	C = (25,100)

At A, $f = 28(0) + 25(120) = 3000$; at B, $f = 28(100) + 25(0) = 2800$; and at C, $f = 28(25) + 25(100) = 3200$, so the maximum value of f is 3200 and it occurs at the point C. To ensure the maximum gasoline efficiency number of 3200, the company should purchase 25 Arrows and 100 Gazelles.

15.

	from A	from B	time
cars to C	x	120 − x	2x + 6(120 − x)
cars to D	y	180 − y	4y + 3(180 − y)
totals	≤100	≤200	≤1030

We are to maximize $f = 120 - x$, subject to the constraints

$x + y \leq 100$,

$120 - x + 180 - y \leq 200$ (simplified: $x + y \geq 100$)

(the first two constraints imply $x + y = 0$),

$2x + 6(120 - x) + 4y + 3(180 - y) \leq 1030$ (simplified: $4x - y \geq 230$),

$x \geq 0$, and $y \geq 0$.

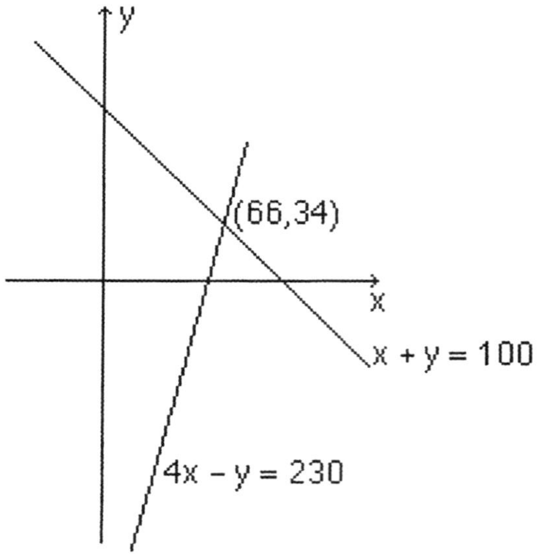

The only point in the feasible region is the point (66,34). Thus from port A, 66 cars should be moved to city C and 34 cars should be moved to city D. From port B, 54 cars should be moved to city C and 146 cars should be moved to city D.

16.

	cost($)	passengers	number to buy
Torro	18000	25	x
Sprite	22000	30	y
totals	≤572000	≤30	

We are to maximize $f = 25x + 30y$, subject to the constraints

$18000x + 22000y \le 572000$ (divide by 2000: $9x + 11y \le 286$),

$x + y \le 30$,

$x \ge 0$, and $y \ge 17$.

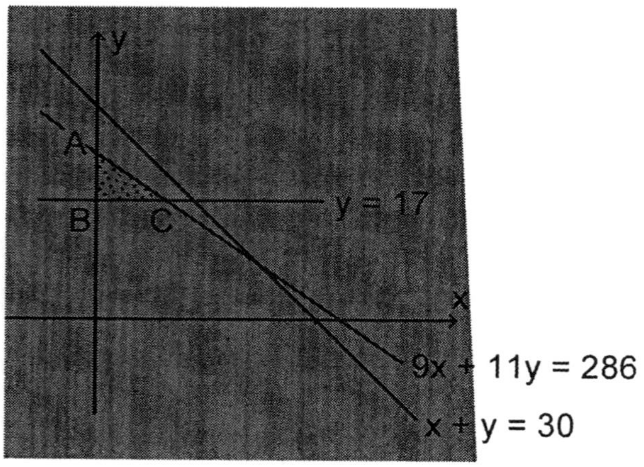

A = (0,26), B = (0,17), and C = (11,17). The values of f at these points are 780, 510, and 785, so to maximize the number of students that can be transported, the school district should purchase 11 Torros and 17 Sprites.

17.

	A (tons)	B (tons)	tons of fertilizer
X	.8	.2	x
Y	.6	.4	y
totals	≤100	≤50	

We are to maximize the amount of fertilizer, f = x + y, subject to the constraints

.8x + .6y ≤ 100 (multiply by 5: 4x + 3y ≤ 500) (amount of A),

.2x + .4y ≤ 50 (multiply by 5: x + 2y ≤ 250) (amount of B),

x ≥ 30, and y ≥ 50.

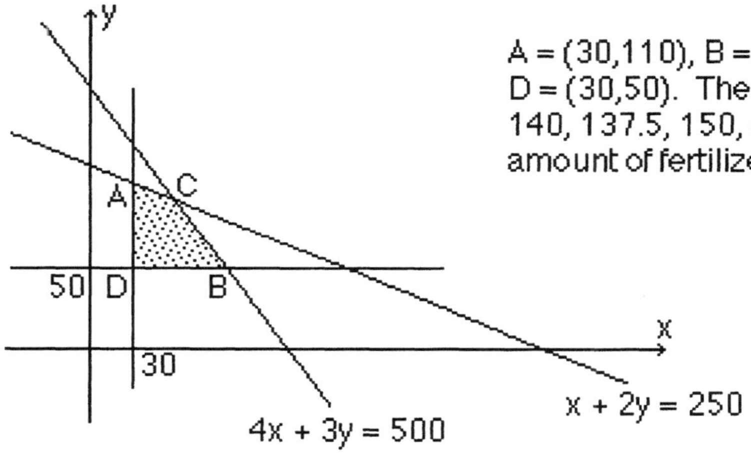

A = (30,110), B = (87.5,50), C = (50,100), and D = (30,50). The values of f at these points are 140, 137.5, 150, and 80. Thus the maximum amount of fertilizer that can be made is 150 tons. To achieve this maximum, the manufacturer should make 50 tons of X and 100 tons of Y.

18.

	units of A per oz.	units of B per oz.	cost(cents per oz.)	oz.
M	1	2	8	x
N	1	1	12	y
totals	≥7	≥10		

We are to minimize the cost, f = 8x + 12y, subject to the constraints

x + y ≥ 7 (units of A),
2x + y ≥ 10 (units of B),

x ≥ 0, and y ≥ 0.

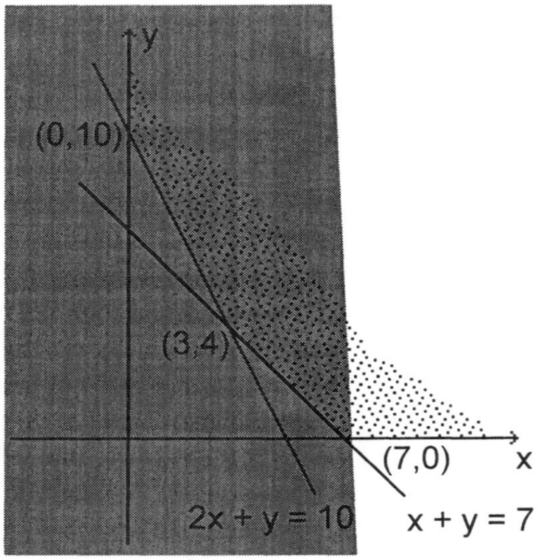

The values of f are 8(0) + 12(10) = 120, 8(3) + 12(4) = 72, and 8(7) + 12(0) = 56. Thus the hospital can achieve the minimum cost of 56 cents by serving 7 oz. of item M and no item N.

19.

	cotton	wool	silk	profit($)	number to make
slacks	1	2	1	3	x
skirts	2	1	1	4	y
totals	≤10	≤10	6		

We are to maximize profit, f = 3x + 4y, subject to the constraints

$x + 2y \le 10$ (sq. yd. of cotton),

$2x + y \le 10$ (sq. yd. of wool),

$x + y \le 6$ (sq. yd. of silk),

$x \ge 0$, and $y \ge 0$.

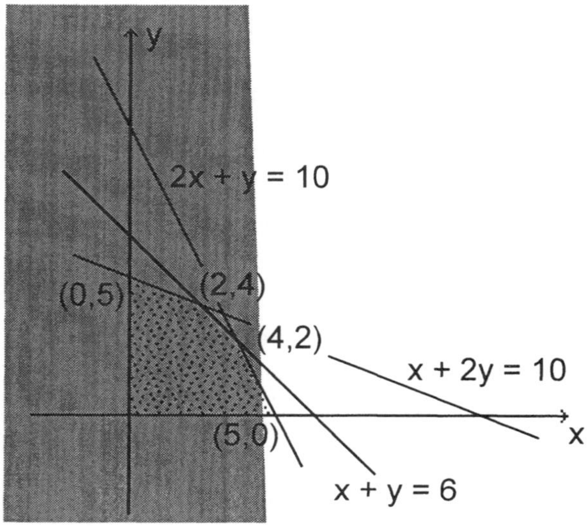

The values of f at (0,5), (2,4), (4,2), and (5,0) are 20, 22, 20, and 15. Thus the maximum profit of $22 can be achieved by making 2 pairs of slacks and 4 skirts.

20. (a)

	pounds	cu. ft.	profit($)	number of packages
Pringle	5	5	.30	x
Williams	6	3	.40	y
totals	≤12000	≤9000		

We are to maximize profit, f = .3x + .4y, subject to the constraints

$5x + 6y \leq 12000$ (weight in pounds),

$5x + 3y \leq 9000$ (volume in cu. ft.),

$x \geq 0$, and $y \geq 0$.

At A, B, and C, f = 800, 540, and 760. Thus the profit is maximized if the shipper carries no packages for Pringle and 2000 packages for Williams.

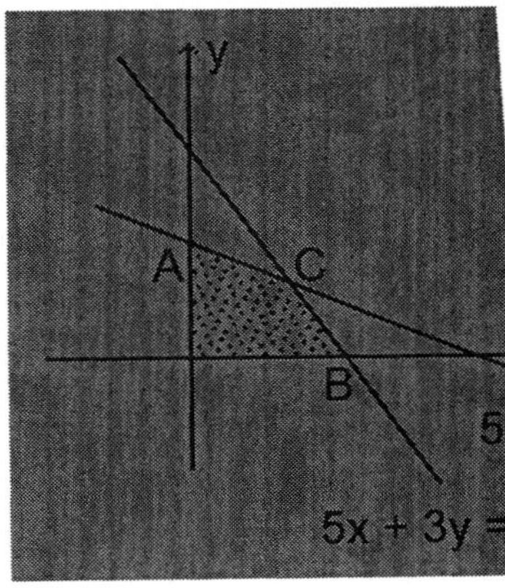

A = (0,2000)
B = (1800,0)
C = (1200,1000)

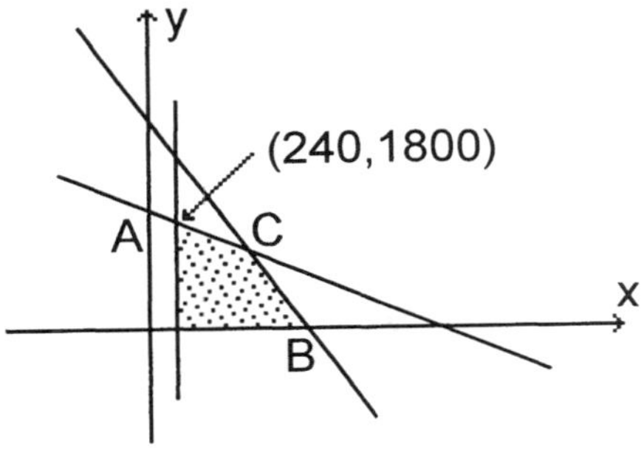

(b) The condition that the shipper must carry at least 240 packages for Pringle is the constraint x ≥ 240. Thus the point A (where the maximum value occurred) is no longer in the feasible region. The maximum value of f now occurs at the point (240, 1800) and is \$792. The clause in the fine print cost the shipper \$8.

Exercise Set 8.2

1. We indicate the row and column of the initial pivot element with arrows. ↑

$$\begin{matrix} & x & y & u & v & f & \\ \rightarrow & \end{matrix}\begin{bmatrix} 4 & 1 & 1 & 0 & 0 & 36 \\ 4 & 3 & 0 & 1 & 0 & 60 \\ -2 & -1 & 0 & 0 & 1 & 0 \end{bmatrix} \approx \begin{bmatrix} 1 & 1/4 & 1/4 & 0 & 0 & 9 \\ 4 & 3 & 0 & 1 & 0 & 60 \\ -2 & -1 & 0 & 0 & 1 & 0 \end{bmatrix} \approx \begin{bmatrix} 1 & 1/4 & 1/4 & 0 & 0 & 9 \\ 0 & 2 & -1 & 1 & 0 & 24 \\ 0 & -1/2 & 1/2 & 0 & 1 & 18 \end{bmatrix}$$

$$\uparrow$$

$$\approx \begin{bmatrix} 1 & 1/4 & 1/4 & 0 & 0 & 9 \\ 0 & 1 & -1/2 & 1/2 & 0 & 12 \\ 0 & -1/2 & 1/2 & 0 & 1 & 18 \end{bmatrix} \approx \begin{bmatrix} 1 & 0 & 3/8 & -1/8 & 0 & 6 \\ 0 & 1 & -1/2 & 1/2 & 0 & 12 \\ 0 & 0 & 1/4 & 1/4 & 1 & 24 \end{bmatrix}.$$ The elements in the last row are all positive, so this is the final tableau. The maximum value of f is 24 when u and v are both zero. With u = v = 0, the first two rows give x = 6 and y = 12.

2.
$$\begin{array}{cc} & \begin{array}{cccccc} x & y & u & v & f & \end{array} \\ \begin{array}{c} \rightarrow \\ \\ \\ \end{array} & \left[\begin{array}{cccccc} 1 & 2 & 1 & 0 & 0 & 4 \\ 1 & 6 & 0 & 1 & 0 & 8 \\ -1 & 4 & 0 & 0 & 1 & 0 \end{array}\right] \\ & \begin{array}{cccccc} \uparrow & & & & & \end{array} \end{array} \approx \begin{bmatrix} 1 & 2 & 1 & 0 & 0 & 4 \\ 0 & 4 & -1 & 1 & 0 & 4 \\ 0 & 6 & 1 & 0 & 1 & 4 \end{bmatrix}.$$ The maximum value of f is 4 when y = u = 0. The first row gives x = 4.

3.
$$\begin{array}{cc} & \begin{array}{cccccc} x & y & u & v & f & \end{array} \\ \begin{array}{c} \rightarrow \\ \\ \\ \end{array} & \left[\begin{array}{cccccc} 1 & 3 & 1 & 0 & 0 & 6 \\ 3 & 1 & 0 & 1 & 0 & 8 \\ -4 & -6 & 0 & 0 & 1 & 0 \end{array}\right] \\ & \begin{array}{cccccc} & \uparrow & & & & \end{array} \end{array} \approx \begin{bmatrix} 1/3 & 1 & 1/3 & 0 & 0 & 2 \\ 3 & 1 & 0 & 1 & 0 & 8 \\ -4 & -6 & 0 & 0 & 1 & 0 \end{bmatrix} \approx \begin{bmatrix} 1/3 & 1 & 1/3 & 0 & 0 & 2 \\ 8/3 & 0 & -1/3 & 1 & 0 & 6 \\ -2 & 0 & 2 & 0 & 1 & 12 \end{bmatrix}$$

$$\approx \begin{bmatrix} 1/3 & 1 & 1/3 & 0 & 0 & 2 \\ 1 & 0 & -1/8 & 3/8 & 0 & 9/4 \\ -2 & 0 & 2 & 0 & 1 & 12 \end{bmatrix} \approx \begin{bmatrix} 0 & 1 & 3/8 & -1/8 & 0 & 5/4 \\ 1 & 0 & -1/8 & 3/8 & 0 & 9/4 \\ 0 & 0 & 7/4 & 3/4 & 1 & 33/2 \end{bmatrix}.$$ The maximum value of f is 33/2 when u = v = 0, y = 5/4, and x = 9/4.

4.
$$\begin{array}{cc} & \begin{array}{cccccc} x & y & u & v & f & \end{array} \\ \begin{array}{c} \\ \rightarrow \\ \\ \end{array} & \left[\begin{array}{cccccc} 1 & 1 & 1 & 0 & 0 & 180 \\ 3 & 2 & 0 & 1 & 0 & 480 \\ -10 & -5 & 0 & 0 & 1 & 0 \end{array}\right] \\ & \begin{array}{cccccc} \uparrow & & & & & \end{array} \end{array} \approx \begin{bmatrix} 1 & 1 & 1 & 0 & 0 & 180 \\ 1 & 2/3 & 0 & 1/3 & 0 & 160 \\ -10 & -5 & 0 & 0 & 1 & 0 \end{bmatrix}$$

$$\approx \begin{bmatrix} 0 & 1/3 & 1 & -1/3 & 0 & 20 \\ 1 & 2/3 & 0 & 1/3 & 0 & 160 \\ 0 & 5/3 & 0 & 10/3 & 1 & 1600 \end{bmatrix}$$. The maximum value of f is 1600 when y = v = 0. From the second row x = 160.

5.
$$\begin{array}{c} \\ \rightarrow \\ \\ \\ \\ \end{array}\begin{array}{c|cccccc|c|} & x & y & z & u & v & f & \\ \hline & 3 & 1 & 1 & 1 & 0 & 0 & 3 \\ & 1 & -10 & -4 & 0 & 1 & 0 & 20 \\ & -1 & -2 & -1 & 0 & 0 & 1 & 0 \\ & & \uparrow & & & & & \end{array} \approx \begin{bmatrix} 3 & 1 & 1 & 1 & 0 & 0 & 3 \\ 31 & 0 & 6 & 10 & 1 & 0 & 50 \\ 5 & 0 & 1 & 2 & 0 & 1 & 6 \end{bmatrix}$$. The maximum value of f is 6 when x = z = u = 0. The first row then gives y = 3.

6.
$$\begin{array}{c} \\ \\ \\ \rightarrow \\ \\ \\ \end{array}\begin{array}{c|ccccccc|c|} & x & y & z & u & v & w & f & \\ \hline & 5 & 5 & 10 & 1 & 0 & 0 & 0 & 1000 \\ & 10 & 8 & 5 & 0 & 1 & 0 & 0 & 2000 \\ & 10 & 5 & 0 & 0 & 0 & 1 & 0 & 500 \\ & -100 & -200 & -50 & 0 & 0 & 0 & 1 & 0 \\ & & \uparrow & & & & & & \end{array} \approx \begin{bmatrix} 5 & 5 & 10 & 1 & 0 & 0 & 0 & 1000 \\ 10 & 8 & 5 & 0 & 1 & 0 & 0 & 2000 \\ 2 & 1 & 0 & 0 & 0 & 1/5 & 0 & 100 \\ -100 & -200 & -50 & 0 & 0 & 0 & 1 & 0 \end{bmatrix}$$

$$\approx \begin{bmatrix} -5 & 0 & 10 & 1 & 0 & -1 & 0 & 500 \\ -6 & 0 & 5 & 0 & 1 & -8/5 & 0 & 1200 \\ 2 & 1 & 0 & 0 & 0 & 1/5 & 0 & 100 \\ 300 & 0 & -50 & 0 & 0 & 40 & 1 & 20000 \end{bmatrix}$$

$$\approx \begin{bmatrix} -1/2 & 0 & 1 & 1/10 & 0 & -1/10 & 0 & 50 \\ -6 & 0 & 5 & 0 & 1 & -8/5 & 0 & 1200 \\ 2 & 1 & 0 & 0 & 0 & 1/5 & 0 & 100 \\ 300 & 0 & -50 & 0 & 0 & 40 & 1 & 20000 \end{bmatrix}$$

$$\approx \begin{bmatrix} -1/2 & 0 & 1 & 1/10 & 0 & -1/10 & 0 & 50 \\ -7/2 & 0 & 0 & -1/2 & 1 & -11/10 & 0 & 950 \\ 2 & 1 & 0 & 0 & 0 & 1/5 & 0 & 100 \\ 275 & 0 & 0 & 5 & 0 & 35 & 1 & 22500 \end{bmatrix}$$. The maximum value of f is 22500 when

x = u = w = 0. Row 3 gives y = 100 and row 1 gives z = 50.

7.

$$\begin{array}{c} \\ \\ \rightarrow \\ \\ \\ \\ \end{array}\begin{array}{cccccccc} x & y & z & u & v & w & f & \\ -1 & 2 & 3 & 1 & 0 & 0 & 0 & 6 \\ -1 & 4 & 5 & 0 & 1 & 0 & 0 & 5 \\ -1 & 5 & 7 & 0 & 0 & 1 & 0 & 7 \\ -2 & -4 & -1 & 0 & 0 & 0 & 1 & 0 \\ & \uparrow & & & & & & \end{array} \approx \begin{bmatrix} -1 & 2 & 3 & 1 & 0 & 0 & 0 & 6 \\ -1/4 & 1 & 5/4 & 0 & 1/4 & 0 & 0 & 5/4 \\ -1 & 5 & 7 & 0 & 0 & 1 & 0 & 7 \\ -2 & -4 & -1 & 0 & 0 & 0 & 1 & 0 \end{bmatrix}$$

$$\approx \begin{bmatrix} -1/2 & 0 & 1/2 & 1 & -1/2 & 0 & 0 & 7/2 \\ -1/4 & 1 & 5/4 & 0 & 1/4 & 0 & 0 & 5/4 \\ 1/4 & 0 & 3/4 & 0 & -5/4 & 1 & 0 & 3/4 \\ -3 & 0 & 4 & 0 & 1 & 0 & 1 & 5 \end{bmatrix} \approx \begin{bmatrix} -1/2 & 0 & 1/2 & 1 & -1/2 & 0 & 0 & 7/2 \\ -1/4 & 1 & 5/4 & 0 & 1/4 & 0 & 0 & 5/4 \\ 1 & 0 & 3 & 0 & -5 & 4 & 0 & 3 \\ -3 & 0 & 4 & 0 & 1 & 0 & 1 & 5 \end{bmatrix}$$

$$\approx \begin{bmatrix} 0 & 0 & 2 & 1 & -3 & 2 & 0 & 5 \\ 0 & 1 & 2 & 0 & -1 & 1 & 0 & 2 \\ 1 & 0 & 3 & 0 & -5 & 4 & 0 & 3 \\ 0 & 0 & 13 & 0 & -14 & 12 & 1 & 14 \end{bmatrix}$$. We can do no more.

There are no positive terms in the pivot column. This means that the feasible region is unbounded and the objective function is unbounded. That is, there is no maximum.

8.

$$\begin{array}{c} \\ \rightarrow \\ \\ \\ \\ \end{array}\begin{array}{cccccccc} x & y & z & w & u & v & f & \\ 5 & 0 & 4 & 6 & 1 & 0 & 0 & 20 \\ 4 & 2 & 2 & 8 & 0 & 1 & 0 & 40 \\ -1 & -2 & -4 & 1 & 0 & 0 & 1 & 0 \\ & & \uparrow & & & & & \end{array} \approx \begin{bmatrix} 5/4 & 0 & 1 & 3/2 & 1/4 & 0 & 0 & 5 \\ 4 & 2 & 2 & 8 & 0 & 1 & 0 & 40 \\ -1 & -2 & -4 & 1 & 0 & 0 & 1 & 0 \end{bmatrix}$$

$$\approx \begin{bmatrix} 5/4 & 0 & 1 & 3/2 & 1/4 & 0 & 0 & 5 \\ 3/2 & 2 & 0 & 5 & -1/2 & 1 & 0 & 30 \\ 4 & -2 & 0 & 7 & 1 & 0 & 1 & 20 \end{bmatrix} \approx \begin{bmatrix} 5/4 & 0 & 1 & 3/2 & 1/4 & 0 & 0 & 5 \\ 3/4 & 1 & 0 & 5/2 & -1/4 & 1/2 & 0 & 15 \\ 4 & -2 & 0 & 7 & 1 & 0 & 1 & 20 \end{bmatrix}$$

$$\approx \begin{bmatrix} 5/4 & 0 & 1 & 3/2 & 1/4 & 0 & 0 & 5 \\ 3/4 & 1 & 0 & 5/2 & -1/4 & 1/2 & 0 & 15 \\ 11/2 & 0 & 0 & 12 & 1/2 & 1 & 1 & 50 \end{bmatrix}$$. The maximum value of f is 50 when

x = w = u = v = 0. The first row then yields z = 5 and the second row gives y = 15.

9.

$$\begin{array}{c} \\ \\ \rightarrow \\ \\ \end{array} \begin{array}{|cccccccc|} x & y & z & w & u & v & f & \\ 2 & 4 & 5 & 6 & 1 & 0 & 0 & 24 \\ 4 & 4 & 2 & 2 & 0 & 1 & 0 & 4 \\ -1 & -2 & 1 & -3 & 0 & 0 & 1 & 0 \\ & & & & & \uparrow & & \end{array} \approx \begin{bmatrix} 2 & 4 & 5 & 6 & 1 & 0 & 0 & 24 \\ 2 & 2 & 1 & 1 & 0 & 1/2 & 0 & 2 \\ -1 & -2 & 1 & -3 & 0 & 0 & 1 & 0 \end{bmatrix}$$

$$\approx \begin{bmatrix} -10 & -8 & -1 & 0 & 1 & -3 & 0 & 12 \\ 2 & 2 & 1 & 1 & 0 & 1/2 & 0 & 2 \\ 5 & 4 & 4 & 0 & 0 & 3/2 & 1 & 6 \end{bmatrix}$$. The maximum value of f is 6 when

x = y = z = v = 0. Row 2 gives w = 2.

10.

	I	II	III	profit($)	number to make
X	2	4		10	x
Y	3		6	8	y
Z	1	2	3	12	z
totals	≤360	≤360	≤360		

We are to maximize profit, f = 10x + 8y + 12z, subject to the constraints

2x + 3y + z ≤ 360, 4x + 2z ≤ 360, 6y +3z ≤ 360, x ≥ 0, y ≥ 0, z ≥ 0.

$$\begin{array}{c} \\ \\ \\ \rightarrow \\ \\ \end{array} \begin{array}{|cccccccc|} x & y & z & u & v & w & f & \\ 2 & 3 & 1 & 1 & 0 & 0 & 0 & 360 \\ 4 & 0 & 2 & 0 & 1 & 0 & 0 & 360 \\ 0 & 6 & 3 & 0 & 0 & 1 & 0 & 360 \\ -10 & -8 & -12 & 0 & 0 & 0 & 1 & 0 \\ & & \uparrow & & & & & \end{array} \approx \begin{bmatrix} 2 & 3 & 1 & 1 & 0 & 0 & 0 & 360 \\ 4 & 0 & 2 & 0 & 1 & 0 & 0 & 360 \\ 0 & 2 & 1 & 0 & 0 & 1/3 & 0 & 120 \\ -10 & -8 & -12 & 0 & 0 & 0 & 1 & 0 \end{bmatrix}$$

$$\approx \begin{bmatrix} 2 & 1 & 0 & 1 & 0 & -1/3 & 0 & 240 \\ 4 & -4 & 0 & 0 & 1 & -2/3 & 0 & 120 \\ 0 & 2 & 1 & 0 & 0 & 1/3 & 0 & 120 \\ -10 & 16 & 0 & 0 & 0 & 4 & 1 & 1440 \end{bmatrix} \approx \begin{bmatrix} 2 & 1 & 0 & 1 & 0 & -1/3 & 0 & 240 \\ 1 & -1 & 0 & 0 & 1/4 & -1/6 & 0 & 30 \\ 0 & 2 & 1 & 0 & 0 & 1/3 & 0 & 120 \\ -10 & 16 & 0 & 0 & 0 & 4 & 1 & 1440 \end{bmatrix}$$

$$\approx \begin{bmatrix} 0 & 3 & 0 & 1 & -1/2 & 0 & 0 & 180 \\ 1 & -1 & 0 & 0 & 1/4 & -1/6 & 0 & 30 \\ 0 & 2 & 1 & 0 & 0 & 1/3 & 0 & 120 \\ 0 & 6 & 0 & 0 & 5/2 & 7/3 & 1 & 1740 \end{bmatrix}$$

. The maximum value of f is 1740 when

y = v = w = 0. Row 2 gives x = 30 and row 3 gives z = 120. So the company should produce 30 of item X, no item Y, and 120 of item Z to make maximum profit of $1740 per day.

11. Let x be the number of desks, y be the number of cabinets, and z be the number of chairs. We are to maximize profit, f = 16x + 12y + 6z, subject to the constraints

$3x + 6y + z \le 800$, $4x + y + 2z \le 400$, $2x + y + 2z \le 100$, $x \ge 0$, $y \ge 0$, $z \ge 0$.

$$\begin{array}{c} \\ \\ \rightarrow \\ \\ \end{array}\begin{array}{cccccccc} x & y & z & u & v & w & f & \\ \hline 3 & 6 & 1 & 1 & 0 & 0 & 0 & 800 \\ 4 & 1 & 2 & 0 & 1 & 0 & 0 & 400 \\ 2 & 1 & 2 & 0 & 0 & 1 & 0 & 100 \\ -16 & -12 & -6 & 0 & 0 & 0 & 1 & 0 \\ \uparrow & & & & & & & \end{array} \approx \begin{bmatrix} 3 & 6 & 1 & 1 & 0 & 0 & 0 & 800 \\ 4 & 1 & 2 & 0 & 1 & 0 & 0 & 400 \\ 1 & 1/2 & 1 & 0 & 0 & 1/2 & 0 & 50 \\ -16 & -12 & -6 & 0 & 0 & 0 & 1 & 0 \end{bmatrix}$$

$$\approx \begin{bmatrix} 0 & 9/2 & -2 & 1 & 0 & -3/2 & 0 & 650 \\ 0 & -1 & -2 & 0 & 1 & -2 & 0 & 200 \\ 1 & 1/2 & 1 & 0 & 0 & 1/2 & 0 & 50 \\ 0 & -4 & 10 & 0 & 0 & 8 & 1 & 800 \end{bmatrix} \approx \begin{bmatrix} 0 & 9/2 & -2 & 1 & 0 & -3/2 & 0 & 650 \\ 0 & -1 & -2 & 0 & 1 & -2 & 0 & 200 \\ 2 & 1 & 2 & 0 & 0 & 1 & 0 & 100 \\ 0 & -4 & 10 & 0 & 0 & 8 & 1 & 800 \end{bmatrix}$$

$$\approx \begin{bmatrix} -9 & 0 & -11 & 1 & 0 & -6 & 0 & 200 \\ 2 & 0 & 0 & 0 & 1 & -1 & 0 & 300 \\ 2 & 1 & 2 & 0 & 0 & 1 & 0 & 100 \\ 8 & 0 & 18 & 0 & 0 & 12 & 1 & 1200 \end{bmatrix}$$

. The maximum value of f is 1200 when

x = z = w = 0. Row 3 then gives y = 100. Thus the company should manufacture 100 cabinets and no desks or chairs to make the maximum profit of $1200.

12.

	cost($)	time(hrs)	profit($)	number to transport
A	10	6	12	x
B	20	4	20	y
C	40	2	16	z
totals	≤6000	≤4000		

We are to maximize profit, f = 12x + 20y − 16z, subject to the constraints

$10x + 20y + 40z \le 6000,\ \ 6x + 4y + 2z \le 4000,\ \ x \ge 0,\ y \ge 0,\ z \ge 0.$

$$\begin{matrix} & x & y & z & u & v & f & \\ \rightarrow & 10 & 20 & 40 & 1 & 0 & 0 & 6000 \\ & 6 & 4 & 2 & 0 & 1 & 0 & 4000 \\ & -12 & -20 & -16 & 0 & 0 & 1 & 0 \\ & & \uparrow & & & & & \end{matrix} \approx \begin{bmatrix} 1/2 & 1 & 2 & 1/20 & 0 & 0 & 300 \\ 6 & 4 & 2 & 0 & 1 & 0 & 4000 \\ -12 & -20 & -16 & 0 & 0 & 1 & 0 \end{bmatrix}$$

$$\approx \begin{bmatrix} 1/2 & 1 & 2 & 1/20 & 0 & 0 & 300 \\ 4 & 0 & -6 & -1/5 & 1 & 0 & 2800 \\ -2 & 0 & 24 & 1 & 0 & 1 & 6000 \end{bmatrix} \approx \begin{bmatrix} 1 & 2 & 4 & 1/10 & 0 & 0 & 600 \\ 4 & 0 & -6 & -1/5 & 1 & 0 & 2800 \\ -2 & 0 & 24 & 1 & 0 & 1 & 6000 \end{bmatrix}$$

$$\approx \begin{bmatrix} 1 & 2 & 4 & 1/10 & 0 & 0 & 600 \\ 0 & -8 & -22 & -3/5 & 1 & 0 & 400 \\ 0 & 4 & 32 & 6/5 & 0 & 1 & 7200 \end{bmatrix}$$

. The maximum value of f is 7200 when $y = z = u = 0$. Row 1 then gives $x = 600$. The maximum profit of $7200 is attained if 600 washing machines are transported from A to P and none is transported from B or C to P.

13.

	material (sq.yd.)	cost($)	profit	number to make
Aspen	60	32	12	x
Alpine	30	20	8	y
Cub	15	12	4	z
totals	≤7800	≤8320		

We are to maximize profit, $f = 12x + 8y + 4z$, subject to the constraints

$60x + 30y + 15z \le 7800,\ \ 32x + 20y + 12z \le 8320,\ \ x \ge 0,\ y \ge 0,\ z \ge 0.$

$$\begin{matrix} & x & y & z & u & v & f & \\ \rightarrow & 60 & 30 & 15 & 1 & 0 & 0 & 7800 \\ & 32 & 20 & 12 & 0 & 1 & 0 & 8320 \\ & -12 & -8 & -4 & 0 & 0 & 1 & 0 \\ & \uparrow & & & & & & \end{matrix} \approx \begin{bmatrix} 1 & 1/2 & 1/4 & 1/60 & 0 & 0 & 130 \\ 32 & 20 & 12 & 0 & 1 & 0 & 8320 \\ -12 & -8 & -4 & 0 & 0 & 1 & 0 \end{bmatrix}$$

$$\approx \begin{matrix} \rightarrow \\ \\ \\ \end{matrix}\begin{bmatrix} 1 & 1/2 & 1/4 & 1/60 & 0 & 0 & 130 \\ 0 & 4 & 4 & -8/15 & 1 & 0 & 4160 \\ 0 & -2 & -1 & 1/5 & 0 & 1 & 1560 \end{bmatrix} \approx \begin{bmatrix} 2 & 1 & 1/2 & 1/30 & 0 & 0 & 260 \\ 0 & 4 & 4 & -8/15 & 1 & 0 & 4160 \\ 0 & -2 & -1 & 1/5 & 0 & 1 & 1560 \end{bmatrix}$$
$$\uparrow$$

$$\approx \begin{bmatrix} 2 & 1 & 1/2 & 1/30 & 0 & 0 & 260 \\ -8 & 0 & 2 & -2/3 & 1 & 0 & 3120 \\ 4 & 0 & 0 & 4/15 & 0 & 1 & 2080 \end{bmatrix}$$. The maximum value of f is 2080 when

x = u = 0. From row 1, y + z/2 = 260. Thus the maximum profit of $2080 will be achieved if no Aspens are manufactured and the number of Alpines manufactured plus one-half the number of Cubs manufactured is 260.

14. To minimize f = −x + 2y, maximize −f = x − 2y.

$$\begin{matrix} x & y & u & v & f & \\ \end{matrix}$$
$$\begin{bmatrix} 1 & 2 & 1 & 0 & 0 & 4 \\ 1 & 4 & 0 & 1 & 0 & 6 \\ -1 & 2 & 0 & 0 & 1 & 0 \end{bmatrix} \approx \begin{bmatrix} 1 & 2 & 1 & 0 & 0 & 4 \\ 0 & 2 & -1 & 1 & 0 & 2 \\ 0 & 4 & 1 & 0 & 1 & 4 \end{bmatrix}$$. The maximum value of −f is 4 when

y = u = 0. Row 1 gives x = 4. Thus the minimum value of f is −4 at x = 4, y = 0.

15. To minimize f = −2x + y, maximize −f = 2x − y.

$$\begin{matrix} x & y & u & v & f & \\ \end{matrix}$$
$$\begin{bmatrix} 2 & 2 & 1 & 0 & 0 & 8 \\ 1 & -1 & 0 & 1 & 0 & 2 \\ -2 & 1 & 0 & 0 & 1 & 0 \end{bmatrix} \approx \begin{bmatrix} 0 & 4 & 1 & -2 & 0 & 4 \\ 1 & -1 & 0 & 1 & 0 & 2 \\ 0 & -1 & 0 & 2 & 1 & 4 \end{bmatrix} \approx \begin{bmatrix} 0 & 1 & 1/4 & -1/2 & 0 & 1 \\ 1 & -1 & 0 & 1 & 0 & 2 \\ 0 & -1 & 0 & 2 & 1 & 4 \end{bmatrix}$$

$$\approx \begin{bmatrix} 0 & 1 & 1/4 & -1/2 & 0 & 1 \\ 1 & 0 & 1/4 & 1/2 & 0 & 3 \\ 0 & 0 & 1/4 & 3/2 & 1 & 5 \end{bmatrix}$$. The maximum value of −f is 5 when u = v = 0.

Row 2 gives x = 3 and row 1 gives y = 1. Thus the minimum value of f is −5 at x = 3, y = 1.

16. To minimize f = 2x + y − z, maximize −f = −2x − y + z.

x y z u v w f

$$\begin{bmatrix} 1 & 2 & -2 & 1 & 0 & 0 & 0 & 20 \\ 2 & 1 & 0 & 0 & 1 & 0 & 0 & 10 \\ 1 & 3 & 4 & 0 & 0 & 1 & 0 & 15 \\ 2 & 1 & -1 & 0 & 0 & 0 & 1 & 0 \end{bmatrix} \approx \begin{bmatrix} 1 & 2 & -2 & 1 & 0 & 0 & 0 & 20 \\ 2 & 1 & 0 & 0 & 1 & 0 & 0 & 10 \\ 1/4 & 3/4 & 1 & 0 & 0 & 1/4 & 0 & 15/4 \\ 2 & 1 & -1 & 0 & 0 & 0 & 1 & 0 \end{bmatrix}$$

$$\approx \begin{bmatrix} 3/2 & 7/2 & 0 & 1 & 0 & 1/2 & 0 & 55/2 \\ 2 & 1 & 0 & 0 & 1 & 0 & 0 & 10 \\ 1/4 & 3/4 & 1 & 0 & 0 & 1/4 & 0 & 15/4 \\ 9/4 & 7/4 & 0 & 0 & 0 & 1/4 & 1 & 15/4 \end{bmatrix}$$. The maximum value of $-f$ is 15/4 when

$x = y = w = 0$. Row 3 gives $z = 15/4$. Thus the minimum value of f is $-15/4$ at $x = 0$, $y = 0$, $z = 15/4$.

Exercise Set 8.3

1.

$$\begin{array}{c} \\ \rightarrow \\ \\ \\ \end{array}\begin{array}{cccccc} x & y & u & v & f & \\ \hline 4 & 1 & 1 & 0 & 0 & 36 \\ 4 & 3 & 0 & 1 & 0 & 60 \\ -2 & -1 & 0 & 0 & 1 & 0 \\ \uparrow & & & & & \end{array} \approx \begin{bmatrix} 1 & 1/4 & 1/4 & 0 & 0 & 9 \\ 4 & 3 & 0 & 1 & 0 & 60 \\ -2 & -1 & 0 & 0 & 1 & 0 \end{bmatrix} \approx \begin{bmatrix} 1 & 1/4 & 1/4 & 0 & 0 & 9 \\ 0 & 2 & -1 & 1 & 0 & 24 \\ 0 & -1/2 & 1/2 & 0 & 1 & 18 \end{bmatrix}$$

basic:	u,v	x,v
nonbasic:	x,y	y,u
entering:	x	y
departing:	u	v

$$\approx \begin{bmatrix} 1 & 1/4 & 1/4 & 0 & 0 & 9 \\ 0 & 1 & -1/2 & 1/2 & 0 & 12 \\ 0 & -1/2 & 1/2 & 0 & 1 & 18 \end{bmatrix} \approx \begin{bmatrix} 1 & 0 & 3/8 & -1/8 & 0 & 6 \\ 0 & 1 & -1/2 & 1/2 & 0 & 12 \\ 0 & 0 & 1/4 & 1/4 & 1 & 24 \end{bmatrix}.$$

basic:	x,y
nonbasic:	u,v

The maximum value of f is 24 when u and v are both zero. With $u = v = 0$, the first two rows give $x = 6$ and $y = 12$. Thus the optimal solution is $f = 24$ at $x = 6$, $y = 12$.

2.

$$\begin{array}{c} \\ \rightarrow \\ \\ \\ \end{array}\begin{array}{cccccc} x & y & u & v & f & \\ \hline 1 & 2 & 1 & 0 & 0 & 4 \\ 1 & 6 & 0 & 1 & 0 & 8 \\ -1 & 4 & 0 & 0 & 1 & 0 \\ \uparrow & & & & & \end{array} \approx \begin{bmatrix} 1 & 2 & 1 & 0 & 0 & 4 \\ 0 & 4 & -1 & 1 & 0 & 4 \\ 0 & 6 & 1 & 0 & 1 & 4 \end{bmatrix}.$$

basic:	u,v	x,v

nonbasic: x,y   y,u
entering: x
departing: u

The maximum value of f is 4 when y = u = 0. The first two rows give x = 4 and v = 4. Thus the optimal solution is f = 4 at x = 4, y = 0. (y is a nonbasic variable.)

3.
$$\begin{array}{c} \\ \rightarrow \\ \\ \\ \end{array}\begin{array}{c} \begin{array}{cccccc} x & y & z & u & v & f \end{array} \\ \left[\begin{array}{cccccc|c} 3 & 1 & 1 & 1 & 0 & 0 & 3 \\ 1 & -10 & -4 & 0 & 1 & 0 & 20 \\ -1 & -2 & -1 & 0 & 0 & 1 & 0 \end{array}\right] \\ \begin{array}{cccccc} & \uparrow & & & & \end{array} \end{array} \approx \left[\begin{array}{ccccccc} 3 & 1 & 1 & 1 & 0 & 0 & 3 \\ 31 & 0 & 6 & 10 & 1 & 0 & 50 \\ 5 & 0 & 1 & 2 & 0 & 1 & 6 \end{array}\right].$$

basic: u,v   y,v
nonbasic: x,y,z   x,z,u
entering: y
departing: u

The maximum value of f is 6 when x = z = u = 0. The first two rows give y = 3 and v = 50. Thus the optimal solution is f = 6 at x = 0, y = 3, z = 0. (x and z are nonbasic variables.)

4.
$$\begin{array}{c} \\ \\ \\ \rightarrow \\ \\ \end{array}\begin{array}{c} \begin{array}{ccccccc} x & y & z & u & v & w & f \end{array} \\ \left[\begin{array}{ccccccc|c} 5 & 5 & 10 & 1 & 0 & 0 & 0 & 1000 \\ 10 & 8 & 5 & 0 & 1 & 0 & 0 & 2000 \\ 10 & 5 & 0 & 0 & 0 & 1 & 0 & 500 \\ -100 & -200 & -50 & 0 & 0 & 0 & 1 & 0 \end{array}\right] \\ \begin{array}{ccccccc} & \uparrow & & & & & \end{array} \end{array} \approx \left[\begin{array}{cccccccc} 5 & 5 & 10 & 1 & 0 & 0 & 0 & 1000 \\ 10 & 8 & 5 & 0 & 1 & 0 & 0 & 2000 \\ 2 & 1 & 0 & 0 & 0 & 1/5 & 0 & 100 \\ -100 & -200 & -50 & 0 & 0 & 0 & 1 & 0 \end{array}\right]$$

basic: u,v,w
nonbasic: x,y,z
entering: y
departing: w

$$\approx \left[\begin{array}{cccccccc} -5 & 0 & 10 & 1 & 0 & -1 & 0 & 500 \\ -6 & 0 & 5 & 0 & 1 & -8/5 & 0 & 1200 \\ 2 & 1 & 0 & 0 & 0 & 1/5 & 0 & 100 \\ 300 & 0 & -50 & 0 & 0 & 40 & 1 & 20000 \end{array}\right]$$

basic: u,v,y
nonbasic: x,z,w
entering: z
departing: u

$$\approx \begin{bmatrix} -1/2 & 0 & 1 & 1/10 & 0 & -1/10 & 0 & 50 \\ -6 & 0 & 5 & 0 & 1 & -8/5 & 0 & 1200 \\ 2 & 1 & 0 & 0 & 0 & 1/5 & 0 & 100 \\ 300 & 0 & -50 & 0 & 0 & 40 & 1 & 20000 \end{bmatrix}$$

$$\approx \begin{bmatrix} -1/2 & 0 & 1 & 1/10 & 0 & -1/10 & 0 & 50 \\ -7/2 & 0 & 0 & -1/2 & 1 & -11/10 & 0 & 950 \\ 2 & 1 & 0 & 0 & 0 & 1/5 & 0 & 100 \\ 275 & 0 & 0 & 5 & 0 & 35 & 1 & 22500 \end{bmatrix}.$$

basic: z,v,y
nonbasic: x,u,w

The maximum value of f is 22500 when x = u = w = 0. Row 1 gives z = 50, row 2 gives v = 950, and row 3 gives y = 100. Thus the optimal solution is f = 22500 at x = 0, y = 100, z = 50. (x is a nonbasic variable.)

5.

$$\begin{array}{c} \\ \\ \rightarrow \\ \\ \\ \end{array}\begin{array}{cccccccc} x & y & z & u & v & w & f & \\ \hline -1 & 2 & 3 & 1 & 0 & 0 & 0 & 6 \\ -1 & 4 & 5 & 0 & 1 & 0 & 0 & 5 \\ -1 & 5 & 7 & 0 & 0 & 1 & 0 & 7 \\ -2 & -4 & -1 & 0 & 0 & 0 & 1 & 0 \\ & \uparrow & & & & & & \end{array} \approx \begin{bmatrix} -1 & 2 & 3 & 1 & 0 & 0 & 0 & 6 \\ -1/4 & 1 & 5/4 & 0 & 1/4 & 0 & 0 & 5/4 \\ -1 & 5 & 7 & 0 & 0 & 1 & 0 & 7 \\ -2 & -4 & -1 & 0 & 0 & 0 & 1 & 0 \end{bmatrix}$$

basic: u,v,w
nonbasic: x,y,z
entering: y
departing: v

$$\approx \begin{bmatrix} -1/2 & 0 & 1/2 & 1 & -1/2 & 0 & 0 & 7/2 \\ -1/4 & 1 & 5/4 & 0 & 1/4 & 0 & 0 & 5/4 \\ 1/4 & 0 & 3/4 & 0 & -5/4 & 1 & 0 & 3/4 \\ -3 & 0 & 4 & 0 & 1 & 0 & 1 & 5 \end{bmatrix} \approx \begin{bmatrix} -1/2 & 0 & 1/2 & 1 & -1/2 & 0 & 0 & 7/2 \\ -1/4 & 1 & 5/4 & 0 & 1/4 & 0 & 0 & 5/4 \\ 1 & 0 & 3 & 0 & -5 & 4 & 0 & 3 \\ -3 & 0 & 4 & 0 & 1 & 0 & 1 & 5 \end{bmatrix}$$

basic: u,y,w
nonbasic: x,z,v
entering: x
departing: v

$$\approx \begin{bmatrix} 0 & 0 & 2 & 1 & -3 & 2 & 0 & 5 \\ 0 & 1 & 2 & 0 & -1 & 1 & 0 & 2 \\ 1 & 0 & 3 & 0 & -5 & 4 & 0 & 3 \\ 0 & 0 & 13 & 0 & -14 & 12 & 1 & 14 \end{bmatrix}.$$

basic: u,y,x
nonbasic: z,v,w

We can do no more. There are no positive terms in the pivot column. This means that the feasible region is unbounded and the objective function is unbounded. That is, there is no optimal solution.

6.

$$\begin{array}{c} \\ \rightarrow \\ \\ \\ \\ \end{array}\begin{array}{ccccccc|c} x & y & z & w & u & v & f & \\ \hline 5 & 0 & 4 & 6 & 1 & 0 & 0 & 20 \\ 4 & 2 & 2 & 8 & 0 & 1 & 0 & 40 \\ -1 & -2 & -4 & 1 & 0 & 0 & 1 & 0 \\ & & \uparrow & & & & & \end{array} \approx \begin{bmatrix} 5/4 & 0 & 1 & 3/2 & 1/4 & 0 & 0 & 5 \\ 4 & 2 & 2 & 8 & 0 & 1 & 0 & 40 \\ -1 & -2 & -4 & 1 & 0 & 0 & 1 & 0 \end{bmatrix}$$

basic: u,v
nonbasic: x,y,z,w
entering: z
departing: u

$$\approx \begin{bmatrix} 5/4 & 0 & 1 & 3/2 & 1/4 & 0 & 0 & 5 \\ 3/2 & 2 & 0 & 5 & -1/2 & 1 & 0 & 30 \\ 4 & -2 & 0 & 7 & 1 & 0 & 1 & 20 \end{bmatrix} \approx \begin{bmatrix} 5/4 & 0 & 1 & 3/2 & 1/4 & 0 & 0 & 5 \\ 3/4 & 1 & 0 & 5/2 & -1/4 & 1/2 & 0 & 15 \\ 4 & -2 & 0 & 7 & 1 & 0 & 1 & 20 \end{bmatrix}$$

basic: z,v
nonbasic: x,y,w,v
entering: y
departing: v

$$\approx \begin{bmatrix} 5/4 & 0 & 1 & 3/2 & 1/4 & 0 & 0 & 5 \\ 3/4 & 1 & 0 & 5/2 & -1/4 & 1/2 & 0 & 15 \\ 11/2 & 0 & 0 & 12 & 1/2 & 1 & 1 & 50 \end{bmatrix}.$$

basic: z,y
nonbasic: x,w,u,v

The maximum value of f is 50 when x = w = u = v = 0. The first row then yields z = 5 and the second row gives y = 15. Thus the optimal solution is f = 50 at x = 0, y = 15, z = 5, w = 0. (x and w are nonbasic variables.

Chapter 8 Review Exercises

In Exercises 1 - 4, we refer to the drawing at the right. The points A, B, C, E, and F will be identified, and the objective function will be evaluated at the three nonzero vertices A, B, and C of the feasible region.

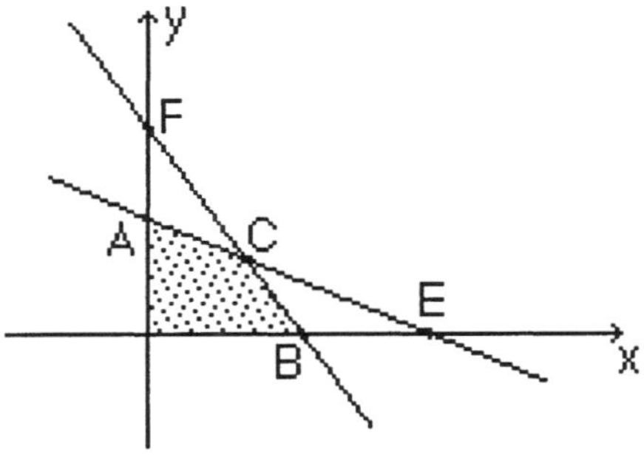

1. $2x + 4y = 16$ A = (0,4) E = (8,0)

   $3x + 2y = 12$ F = (0,6) B = (4,0) C = (2,3)

   At A, f = 2(0) + 3(4) = 12; at B, f = 2(4) + 3(0) = 8; and at C, f = 2(2) + 3(3) = 13, so the maximum value of f is 13 and it occurs at the point C.

2. $x + 2y = 16$ A = (0,8) E = (16,0)

   $3x + 2y = 24$ F = (0,12) B = (8,0) C = (4,6)

   At A, f = 6(0) + 4(8) = 32; at B, f = 6(8) + 4(0) = 48; and at C, f = 6(4) + 4(6) = 48, so the maximum value of f is 48 and it occurs at every point on the line segment joining B and C.

3. $3x + 2y = 21$ F = (0,21/2) B = (7,0)

   $x + 5y = 20$ A = (0,4) E = (20,0) C = (5,3)

   At A, −f = −4(0) − 4 = −4; at B, −f = −4(7) − 0 = −28; and at C, −f = −4(5) − 3 = −23, so the maximum value of −f is zero and it occurs at the origin. Thus the minimum value of f is zero and it occurs at the origin.

4.

	acres	picking time(hrs)	profit($)
strawberries	x	8	700

tomatoes	y	6	600
total	≤40	≤300	

We are to maximize profit, f = 700x + 600y, subject to the constraints

$x + y \le 40$,

$8x + 6y \le 300$ (divide by 2: $4x + 3y \le 150$),

$x \ge 0$, and $y \ge 0$.

$x + y = 40$	A = (0,40)	E = (40,0)	
$4x + 3y = 150$	F = (0,50)	B = (75/2,0)	C = (30,10)

At A, f = 700(0) + 600(40) = 24000; at B, f = 700(75/2) + 600(0) = 26250; and at C, f = 700(30) + 600(10) = 27000, so the maximum value of f is 27000 and it occurs at the point C. To ensure the maximum profit of $27000, the farmer should plant 30 acres of strawberries and 10 acres of tomatoes.

5.

	cu. ft.	sq. ft.	cost($)	number
X	36	6	54	x
Y	44	8	60	y
totals		≤256	≤2100	

We are to maximize volume, f = 36x + 44y, subject to the constraints

$6x + 8y \le 256$ (divide by 2: $3x + 4y \le 128$),

$54x + 60y \le 2100$ (divide by 6: $9x + 10y \le 350$),

$x \ge 0$, and $y \ge 20$.

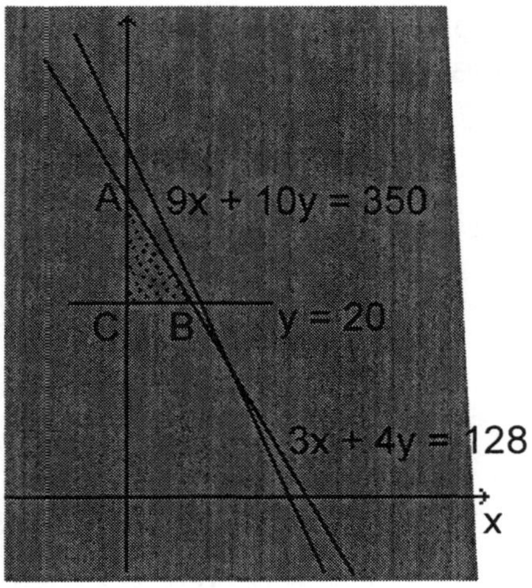

A = (0,32), B = (16,20), and C = (0,20). The values of f at these three points are 1408, 1456, and 880. Thus volume will be maximized if the company purchases 16 lockers of type X and 20 of type Y.

6.

$$\begin{array}{c} \\ \rightarrow \\ \\ \\ \\ \end{array}\begin{array}{cccccccc} x & y & z & u & v & w & f & \\ \left.\begin{array}{cccccccc} 1 & 2 & 4 & 1 & 0 & 0 & 0 & 20 \\ 2 & 4 & 4 & 0 & 1 & 0 & 0 & 60 \\ 3 & 4 & 1 & 0 & 0 & 1 & 0 & 90 \\ -2 & -1 & -1 & 0 & 0 & 0 & 1 & 0 \end{array}\right. \\ \uparrow \end{array} \approx \begin{bmatrix} 1 & 2 & 4 & 1 & 0 & 0 & 0 & 20 \\ 0 & 0 & -4 & -2 & 1 & 0 & 0 & 20 \\ 0 & -2 & -11 & -3 & 0 & 1 & 0 & 30 \\ 0 & 3 & 7 & 2 & 0 & 0 & 1 & 40 \end{bmatrix}.$$

The maximum value of f is 40 when y = z = u = 0. Row 1 gives x = 20.

7. Let x be the number of X tables and y be the number of Y tables.

We are to maximize profit, f = 8x + 4y, subject to the constraints

$10x + 5y \le 300$, $8x + 4y \le 300$, $4x + 8y \le 300$, $x \ge 0$, and $y \ge 0$.

$$\begin{array}{c} \\ \rightarrow \\ \\ \\ \\ \end{array}\begin{array}{ccccccc} x & y & u & v & w & f & \\ \left.\begin{array}{ccccccc} 10 & 5 & 1 & 0 & 0 & 0 & 300 \\ 8 & 4 & 0 & 1 & 0 & 0 & 300 \\ 4 & 8 & 0 & 0 & 1 & 0 & 300 \\ -8 & -4 & 0 & 0 & 0 & 1 & 0 \end{array}\right. \\ \uparrow \end{array} \approx \begin{bmatrix} 1 & 1/2 & 1/10 & 0 & 0 & 0 & 30 \\ 8 & 4 & 0 & 1 & 0 & 0 & 300 \\ 4 & 8 & 0 & 0 & 1 & 0 & 300 \\ -8 & -4 & 0 & 0 & 0 & 1 & 0 \end{bmatrix}$$

$$\approx \begin{bmatrix} 1 & 1/2 & 1/10 & 0 & 0 & 0 & 30 \\ 0 & 0 & -4/5 & 1 & 0 & 0 & 60 \\ 0 & 6 & -2/5 & 0 & 1 & 0 & 180 \\ 0 & 0 & 4/5 & 0 & 0 & 1 & 240 \end{bmatrix}.$$

The maximum value of f is 240 when u = 0.

Row 1 gives x + y/2 = 30. Thus the maximum daily profit of $240 is realized when the number of X tables plus one-half the number of Y tables finished is 30.

8. To minimize f = x − 2y + 4z, maximize −f = −x + 2y − 4z.

$$\begin{array}{c} \\ \\ \\ \rightarrow \\ \\ \end{array}\begin{array}{c} \begin{array}{cccccccc} x & y & z & u & v & w & f & \end{array} \\ \left[\begin{array}{cccccccc} 1 & -1 & 3 & 1 & 0 & 0 & 0 & 4 \\ 2 & 2 & -3 & 0 & 1 & 0 & 0 & 6 \\ -1 & 2 & 3 & 0 & 0 & 1 & 0 & 2 \\ 1 & -2 & 4 & 0 & 0 & 0 & 1 & 0 \end{array}\right] \\ \begin{array}{cccccccc} & \uparrow & & & & & & \end{array} \end{array} \approx \left[\begin{array}{cccccccc} 1 & -1 & 3 & 1 & 0 & 0 & 0 & 4 \\ 2 & 2 & -3 & 0 & 1 & 0 & 0 & 6 \\ -1/2 & 1 & 3/2 & 0 & 0 & 1/2 & 0 & 1 \\ 1 & -2 & 4 & 0 & 0 & 0 & 1 & 0 \end{array}\right]$$

$$\approx \left[\begin{array}{cccccccc} 1/2 & 0 & 9/2 & 1 & 0 & 1/2 & 0 & 5 \\ 3 & 0 & -6 & 0 & 1 & -1 & 0 & 4 \\ -1/2 & 1 & 3/2 & 0 & 0 & 1/2 & 0 & 1 \\ 0 & 0 & 7 & 0 & 0 & 1 & 1 & 2 \end{array}\right].$$ The maximum value of −f is 2 when z = w = 0.

Row 3 then gives −x/2 + y = 1, and rows 1 and 2 give x/2 + u = 5 and 3x + v = 4. Thus the minimum value of f is −2 for all x,y,z where z = 0 and y = 1 + x/2, with 0 ≤ x ≤ 4/3.

Appendix A

We use the symbol x to denote cross product of vectors.

1. (a) $\mathbf{u} \times \mathbf{v} = (2\times4 - 3\times0,\ 3\times{-1} - 1\times4,\ 1\times0 - 2\times{-1}) = (8,-7,2)$.

   (b) $\mathbf{v} \times \mathbf{u} = (0\times3 - 4\times2,\ 4\times1 - (-1)\times3,\ -1\times2 - 0\times1) = (-8,7,-2)$.

   (c) $\mathbf{u} \times \mathbf{w} = (2\times{-1} - 3\times2,\ 3\times1 - 1\times{-1},\ 1\times2 - 2\times1) = (-8,4,0)$.

   (d) $\mathbf{v} \times \mathbf{w} = (0\times{-1} - 4\times2,\ 4\times1 - (-1)\times{-1},\ (-1)\times2 - 0\times1) = (-8,3,-2)$.

   (e) $(\mathbf{u} \times \mathbf{v}) \times \mathbf{w} = (8,-7,2) \times \mathbf{w} = (-7\times{-1} - 2\times2,\ 2\times1 - 8\times{-1},\ 8\times2 - (-7)\times1) = (3,10,23)$.

2. (a) $\mathbf{u} \times \mathbf{v} = \begin{vmatrix} \mathbf{i} & \mathbf{j} & \mathbf{k} \\ -2 & 2 & 4 \\ 3 & 0 & 5 \end{vmatrix} = (10,22,-6)$. (b) $\mathbf{u} \times \mathbf{w} = \begin{vmatrix} \mathbf{i} & \mathbf{j} & \mathbf{k} \\ -2 & 2 & 4 \\ 4 & -2 & 1 \end{vmatrix} = (10,18,-4)$.

   (c) $\mathbf{w} \times \mathbf{v} = \begin{vmatrix} \mathbf{i} & \mathbf{j} & \mathbf{k} \\ 4 & -2 & 1 \\ 3 & 0 & 5 \end{vmatrix} = (-10,-17,6)$. (d) $\mathbf{v} \times \mathbf{w} = \begin{vmatrix} \mathbf{i} & \mathbf{j} & \mathbf{k} \\ 3 & 0 & 5 \\ 4 & -2 & 1 \end{vmatrix} = (10,17,-6)$.

   (e) $(\mathbf{w} \times \mathbf{v}) \times \mathbf{u} = \begin{vmatrix} \mathbf{i} & \mathbf{j} & \mathbf{k} \\ -10 & -17 & 6 \\ -2 & 2 & 4 \end{vmatrix} = (-80,28,-54)$.

3. (a) $\mathbf{u} \times \mathbf{v} = \begin{vmatrix} \mathbf{i} & \mathbf{j} & \mathbf{k} \\ 2 & 3 & 1 \\ -1 & 2 & 4 \end{vmatrix} = 10\mathbf{i} - 9\mathbf{j} + 7\mathbf{k}$.

   (b) $\mathbf{u} \times \mathbf{w} = \begin{vmatrix} \mathbf{i} & \mathbf{j} & \mathbf{k} \\ 2 & 3 & 1 \\ 3 & 0 & -7 \end{vmatrix} = -21\mathbf{i} + 17\mathbf{j} - 9\mathbf{k}$.

   (c) $\mathbf{w} \times \mathbf{u} = \begin{vmatrix} \mathbf{i} & \mathbf{j} & \mathbf{k} \\ 3 & 0 & -7 \\ 2 & 3 & 1 \end{vmatrix} = 21\mathbf{i} - 17\mathbf{j} + 9\mathbf{k}$.

(d) $\mathbf{v} \times \mathbf{w} = \begin{vmatrix} \mathbf{i} & \mathbf{j} & \mathbf{k} \\ -1 & 2 & 4 \\ 3 & 0 & -7 \end{vmatrix} = -14\mathbf{i} + 5\mathbf{j} - 6\mathbf{k}.$

(e) $(\mathbf{w} \times \mathbf{u}) \times \mathbf{v} = \begin{vmatrix} \mathbf{i} & \mathbf{j} & \mathbf{k} \\ 21 & -17 & 9 \\ -1 & 2 & 4 \end{vmatrix} = -86\mathbf{i} - 93\mathbf{j} + 25\mathbf{k}.$

4. (a) $\mathbf{u} \times \mathbf{v} = \begin{vmatrix} \mathbf{i} & \mathbf{j} & \mathbf{k} \\ 3 & 1 & -2 \\ 4 & -1 & 2 \end{vmatrix} = (0,-14,-7).$

(b) $3\mathbf{v} \times 2\mathbf{w} = \begin{vmatrix} \mathbf{i} & \mathbf{j} & \mathbf{k} \\ 12 & -3 & 6 \\ 0 & 6 & -4 \end{vmatrix} = (-24,48,72).$

(c) $(\mathbf{w} \times \mathbf{u})\cdot\mathbf{v} = \begin{vmatrix} \mathbf{i} & \mathbf{j} & \mathbf{k} \\ 0 & 3 & -2 \\ 3 & 1 & -2 \end{vmatrix} \cdot\mathbf{v} = (-4,-6,-9)\cdot(4,-1,2) = -28.$

(d) $(\mathbf{w}+2\mathbf{u}) \times \mathbf{v} = \begin{vmatrix} \mathbf{i} & \mathbf{j} & \mathbf{k} \\ 6 & 5 & -6 \\ 4 & -1 & 2 \end{vmatrix} = (4,-36,-26).$

(e) $\mathbf{u}\cdot(\mathbf{v} \times \mathbf{w}) = \mathbf{u}\cdot\begin{vmatrix} \mathbf{i} & \mathbf{j} & \mathbf{k} \\ 4 & -1 & 2 \\ 0 & 3 & -2 \end{vmatrix} = (3,1,-2)\cdot(-4,8,12) = -28.$

(f) $(\mathbf{v} \times \mathbf{u})\cdot(\mathbf{w} \times \mathbf{v}) = (0,14,7)\cdot(4,-8,-12) = -196.$

5. (a) $\vec{AB} = (-3,4,6)-(1,2,1) = (-4,2,5)$ *and* $\vec{AC} = (1,8,3)-(1,2,1) = (0,6,2).$

$$\vec{AB}\times\vec{AC} = \begin{vmatrix} i & j & k \\ -4 & 2 & 5 \\ 0 & 6 & 2 \end{vmatrix} = (-26,8,-24), \text{ so } area = \frac{1}{2}\,\|(-26,8,-24)\| = \sqrt{329}.$$

(b) $\vec{AB} = (0,2,6)-(3,-1,2) = (-3,3,4)$ *and* $\vec{AC} = (7,1,5)-(3,-1,2) = (4,2,3).$

$$\vec{AB}\times\vec{AC}=\begin{vmatrix} i & j & k \\ -3 & 3 & 4 \\ 4 & 2 & 3 \end{vmatrix}=(1,25,-18),\ so\ area=\frac{1}{2}\|(1,25,-18)\|=\frac{5}{2}\sqrt{38}.$$

(c) $\vec{AB}=(0,5,2)-(1,0,0)=(-1,5,2)$ *and* $\vec{AC}=(3,-4,8)-(1,0,0)=(2,-4,8).$

$$\vec{AB}\times\vec{AC}=\begin{vmatrix} i & j & k \\ -1 & 5 & 2 \\ 2 & -4 & 8 \end{vmatrix}=(48,12,-6),\ so\ area=\frac{1}{2}\|(48,12,-6)\|=3\sqrt{69}.$$

6. (a) **u** = $\vec{AB}=(4,8,1)-(1,2,5)=(3,6,-4),\ v=\vec{AC}=(-3,2,3)-(1,2,5)=(-4,0,-2),$ *and*

w = $\vec{AD}=(0,3,9)-(1,2,5)=(-1,1,4).$

$$\mathbf{w}\cdot(\mathbf{u}\times\mathbf{v})=\begin{vmatrix} 3 & 6 & -4 \\ -4 & 0 & -2 \\ -1 & 1 & 4 \end{vmatrix}=130,\text{ so the volume }=|130|=130.$$

(b) **u** = $\vec{AB}=(-2,3,4)-(3,1,6)=(-5,2,-2),\ v=\vec{AC}=(0,2,-5)-(3,1,6)=(-3,1,-11),$

and **w** = $\vec{AD}=(3,-1,4)-(3,1,6)=(0,-2,-2).$

$$\mathbf{w}\cdot(\mathbf{u}\times\mathbf{v})=\begin{vmatrix} -5 & 2 & -2 \\ -3 & 1 & -11 \\ 0 & -2 & -2 \end{vmatrix}=96,\text{ so the volume }=|96|=96.$$

(c) **u** = $\vec{AB}=(-3,1,4)-(0,1,2)=(-3,0,2),\ v=\vec{AC}=(5,2,3)-(0,1,2)=(5,1,1),$

and **w** = $\vec{AD}=(-3,-2,1)-(0,1,2)=(-3,-3,-1).$

$$\mathbf{w}\cdot(\mathbf{u}\times\mathbf{v})=\begin{vmatrix} -3 & 0 & 2 \\ 5 & 1 & 1 \\ -3 & -3 & -1 \end{vmatrix}=-30,\text{ so the volume }=|-30|=30.$$

7. (a) **i·i** = (1,0,0)·(1,0,0) = 1+0+0 = 1, **j·j** = (0,1,0)·(0,1,0) = 0+1+0 = 1,

$\mathbf{k} \cdot \mathbf{k} = (0,0,1) \cdot (0,0,1) = 0+0+1 = 1.$

(b) $\mathbf{i} \cdot \mathbf{j} = (1,0,0) \cdot (0,1,0) = 0+0+0 = 0$, $\mathbf{i} \cdot \mathbf{k} = (1,0,0) \cdot (0,0,1) = 0+0+0 = 0$,

$\mathbf{j} \cdot \mathbf{k} = (0,1,0) \cdot (0,0,1) = 0+0+0 = 0.$

8. $(\mathbf{i} \times \mathbf{j}) \cdot \mathbf{k} = (0x0 - 0x1, 0x0 - 1x0, 1x1 - 0x0) \cdot (0,0,1) = (0,0,1) \cdot (0,0,1) = 0+0+1 = 1.$

9. $\mathbf{u} = (u_1, u_2, u_3)$ and $\mathbf{0} = (0,0,0)$.

$\mathbf{u} \times \mathbf{0} = (u_2 x0 - u_3 x0, u_3 x0 - u_1 x0, u_1 x0 - u_2 x0) = (0,0,0) = \mathbf{0}.$

$\mathbf{0} \times \mathbf{u} = (0xu_3 - 0xu_2, 0xu_1 - 0xu_3, 0xu_2 - 0xu_1) = (0,0,0) = \mathbf{0}.$

10. $\mathbf{u} = (u_1, u_2, u_3)$, $\mathbf{v} = (v_1, v_2, v_3)$, and $\mathbf{w} = (w_1, w_2, w_3)$.

$$
\begin{aligned}
(\mathbf{u} \times \mathbf{v}) \cdot \mathbf{w} &= (u_2 v_3 - u_3 v_2, u_3 v_1 - u_1 v_3, u_1 v_2 - u_2 v_1) \cdot (w_1, w_2, w_3) \\
&= u_2 v_3 w_1 - u_3 v_2 w_1 + u_3 v_1 w_2 - u_1 v_3 w_2 + u_1 v_2 w_3 - u_2 v_1 w_3 \\
&= u_1 v_2 w_3 - u_1 v_3 w_2 + u_2 v_3 w_1 - u_2 v_1 w_3 + u_3 v_1 w_2 - u_3 v_2 w_1 \\
&= (u_1, u_2, u_3) \cdot (v_2 w_3 - v_3 w_2, v_3 w_1 - v_1 w_3, v_1 w_2 - v_2 w_1) = \mathbf{u} \cdot (\mathbf{v} \times \mathbf{w}).
\end{aligned}
$$

11. $\mathbf{u} = (u_1, u_2, u_3)$, $\mathbf{v} = (v_1, v_2, v_3)$, and $\mathbf{w} = (w_1, w_2, w_3)$.

The coordinates of $\mathbf{u} \times (\mathbf{v} \times \mathbf{w}) =$
$(u_1, u_2, u_3) \times (v_2 w_3 - v_3 w_2, v_3 w_1 - v_1 w_3, v_1 w_2 - v_2 w_1)$
are $u_2(v_1w_2 - v_2w_1) - u_3(v_3w_1 - v_1w_3)$,
$u_3(v_2w_3 - v_3w_2) - u_1(v_1w_2 - v_2w_1)$, and $u_1(v_3w_1 - v_1w_3) - u_2(v_2w_3 - v_3w_2)$.
Rearranging the terms, the coordinates are $(u_2 w_2 + u_3 w_3)v_1 - (u_2 v_2 + u_3 v_3)w_1$,
$(u_1 w_1 + u_3 w_3)v_2 - (u_1 v_1 + u_3 v_3)w_2$, and $(u_1 w_1 + u_2 w_2)v_3 - (u_1 v_1 + u_2 v_2)w_3$.

Now add $0 = u_1 v_1 w_1 - u_1 v_1 w_1$ to the first coordinate, $0 = u_2 v_2 w_2 - u_2 v_2 w_2$ to the second coordinate, and $0 = u_3 v_3 w_3 - u_3 v_3 w_3$ to the third coordinate. The coordinates become $(u_1 w_1 + u_2 w_2 + u_3 w_3)v_1 - (u_1 v_1 + u_2 v_2 + u_3 v_3)w_1$,
$(u_1 w_1 + u_2 w_2 + u_3 w_3)v_2 - (u_1 v_1 + u_2 v_2 + u_3 v_3)w_2$, and

$(u_1 w_1 + u_2 w_2 + u_3 w_3)v_3 - (u_1 v_1 + u_2 v_2 + u_3 v_3)w_3$. These are the coordinates of $(\mathbf{u}\cdot\mathbf{w})\mathbf{v} - (\mathbf{u}\cdot\mathbf{v})\mathbf{w}$, so $\mathbf{u} \times (\mathbf{v} \times \mathbf{w}) = (\mathbf{u}\cdot\mathbf{w})\mathbf{v} - (\mathbf{u}\cdot\mathbf{v})\mathbf{w}$.

12. $c(\mathbf{u} \times \mathbf{v}) = c(u_2 v_3 - u_3 v_2, u_3 v_1 - u_1 v_3, u_1 v_2 - u_2 v_1)$
$= (cu_2 v_3 - cu_3 v_2, cu_3 v_1 - cu_1 v_3, cu_1 v_2 - cu_2 v_1) = c\mathbf{u} \times \mathbf{v}$
$= (u_2 cv_3 - u_3 cv_2, u_3 cv_1 - u_1 cv_3, u_1 cv_2 - u_2 cv_1) = \mathbf{u} \times c\mathbf{v}$.

13. **u** and **v** are parallel if and only if the angle between them is zero. If $0 \le \theta < \pi$ then $\sin\theta = 0$ if and only if $\theta = 0$, so **u** and **v** are parallel if and only if $\sin\theta = 0$, where θ is the angle between the vectors. If neither **u** nor **v** is the zero vector, then $||\mathbf{u}|| \neq 0$ and $||\mathbf{v}|| \neq 0$, so $\sin\theta = \dfrac{||\mathbf{u} \times \mathbf{v}||}{||\mathbf{u}||\,||\mathbf{v}||} = 0$ if and only if $||\mathbf{u} \times \mathbf{v}|| = 0$ if and only if $\mathbf{u} \times \mathbf{v} = \mathbf{0}$.

14. $\mathbf{u} = (u_1, u_2, u_3)$, $\mathbf{v} = (v_1, v_2, v_3)$, and $\mathbf{w} = (w_1, w_2, w_3)$. From Exercise 11, $\mathbf{u} \times (\mathbf{v} \times \mathbf{w}) = (\mathbf{u}\cdot\mathbf{w})\mathbf{v} - (\mathbf{u}\cdot\mathbf{v})\mathbf{w}$. Therefore $\mathbf{v} \times (\mathbf{w} \times \mathbf{u}) = (\mathbf{v}\cdot\mathbf{u})\mathbf{w} - (\mathbf{v}\cdot\mathbf{w})\mathbf{u}$ and $\mathbf{w} \times (\mathbf{u} \times \mathbf{v}) = (\mathbf{w}\cdot\mathbf{v})\mathbf{u} - (\mathbf{w}\cdot\mathbf{u})\mathbf{v}$. Adding gives $\mathbf{u} \times (\mathbf{v} \times \mathbf{w}) + \mathbf{v} \times (\mathbf{w} \times \mathbf{u}) + \mathbf{w} \times (\mathbf{u} \times \mathbf{v})$
$= (\mathbf{u}\cdot\mathbf{w})\mathbf{v} - (\mathbf{u}\cdot\mathbf{v})\mathbf{w} + (\mathbf{v}\cdot\mathbf{u})\mathbf{w} - (\mathbf{v}\cdot\mathbf{w})\mathbf{u} + (\mathbf{w}\cdot\mathbf{v})\mathbf{u} - (\mathbf{w}\cdot\mathbf{u})\mathbf{v} = \mathbf{0}$.

15. $\mathbf{u} = (u_1, u_2, u_3)$, $\mathbf{v} = (v_1, v_2, v_3)$, and $\mathbf{w} = (w_1, w_2, w_3)$.

$$\mathbf{u}\cdot(\mathbf{v} \times \mathbf{w}) = \begin{vmatrix} u_1 & u_2 & u_3 \\ v_1 & v_2 & v_3 \\ w_1 & w_2 & w_3 \end{vmatrix}$$ = 0 if and only if the row vectors **u**, **v**, and **w** are linearly

dependent, i.e., if and only if one is a linear combination of the other two. So $\mathbf{u}\cdot(\mathbf{v} \times \mathbf{w}) = 0$ if and only if one vector lies in the plane of the other two.

16. $(\mathbf{t} \times \mathbf{u})\cdot(\mathbf{v} \times \mathbf{w})$
$= (t_2 u_3 - t_3 u_2, t_3 u_1 - t_1 u_3, t_1 u_2 - t_2 u_1)\cdot(v_2 w_3 - v_3 w_2, v_3 w_1 - v_1 w_3, v_1 w_2 - v_2 w_1)$
$= (t_2 u_3 - t_3 u_2)(v_2 w_3 - v_3 w_2) + (t_3 u_1 - t_1 u_3)(v_3 w_1 - v_1 w_3) +$
$(t_1 u_2 - t_2 u_1)(v_1 w_2 - v_2 w_1)$
$= t_2 u_3 v_2 w_3 + t_3 u_2 v_3 w_2 - t_2 u_3 v_3 w_2 - t_3 u_2 v_2 w_3 + t_3 u_1 v_3 w_1 + t_1 u_3 v_1 w_3$
$- t_3 u_1 v_1 w_3 - t_1 u_3 v_3 w_1 + t_1 u_2 v_1 w_2 + t_2 u_1 v_2 w_1 - t_1 u_2 v_2 w_1 - t_2 u_1 v_1 w_2$.

Now add zero in the form
$t_1 u_1 v_1 w_1 - t_1 u_1 v_1 w_1 + t_2 u_2 v_2 w_2 - t_2 u_2 v_2 w_2 + t_3 u_3 v_3 w_3 - t_3 u_3 v_3 w_3$
an rearrange terms
$(\mathbf{t} \times \mathbf{u})\cdot(\mathbf{v} \times \mathbf{w}) = (t_1 v_1 + t_2 v_2 + t_3 v_3)(u_1 w_1 + u_2 w_2 + u_3 w_3) -$

$$(u_1 v_1 + u_2 v_2 + u_3 v_3)(t_1 w_1 + t_2 w_2 + t_3 w_3) = \begin{vmatrix} \mathbf{t \cdot v} & \mathbf{u \cdot v} \\ \mathbf{t \cdot w} & \mathbf{u \cdot w} \end{vmatrix}$$

.

Appendix B

1. (a) $P_0 = (x_0, y_0, z_0) = (1,-2,4)$ and $(a,b,c) = (1,1,1)$.

point-normal form: $(x - 1) + (y + 2) + (z - 4) = 0$
general form: $x + y + z - 3 = 0$

(b) $P_0 = (x_0, y_0, z_0) = (-3,5,6)$ and $(a,b,c) = (-2,4,5)$.

point-normal form: $-2(x + 3) + 4(y - 5) + 5(z - 6) = 0$
general form: $-2x + 4y + 5z - 56 = 0$

(c) $P_0 = (x_0, y_0, z_0) = (0,0,0)$ and $(a,b,c) = (1,2,3)$.

point-normal form: $x + 2y + 3z = 0$
general form: $x + 2y + 3z = 0$

(d) $P_0 = (x_0, y_0, z_0) = (4,5,-2)$ and $(a,b,c) = (-1,4,3)$.

point-normal form: $-(x - 4) + 4(y - 5) + 3(z + 2) = 0$
general form: $-x + 4y + 3z - 10 = 0$

2. (a) $\overrightarrow{P_1P_2} = (1, 0, 2) - (1, -1, 3) = (0, 2, -1)$ and $\overrightarrow{P_1P_3} = (-1, 4, 6) - (1, -2, 3) = (-2, 6, 3)$.

$$\overrightarrow{P_1P_2} \times \overrightarrow{P_1P_3} = \begin{vmatrix} i & j & k \\ 0 & 2 & -1 \\ -2 & 6 & 3 \end{vmatrix} = (12, 2, 4) \text{ is normal to the plane.}$$

$P_0 = (x_0, y_0, z_0) = (1,-2,3)$ and $(a,b,c) = (12,2,4)$.

point-normal form: $12(x - 1) + 2(y + 2) + 4(z - 3) = 0$
general form: $12x + 2y + 4z - 20 = 0$ or $6x + y + 2z - 10 = 0$

(b) $\overrightarrow{P_1P_2}$ = (1, 2, 4) – (0, 0, 0) = (1, 2, 4) and $\overrightarrow{P_1P_3}$ = (-3, 5, 1) – (0, 0, 0) = (-3, 5, 1).

$$\overrightarrow{P_1P_2} \times \overrightarrow{P_1P_3} = \begin{vmatrix} i & j & k \\ 1 & 2 & 4 \\ -3 & 5 & 1 \end{vmatrix} = (-18, -13, 11) \text{ is normal to the plane.}$$

$P_0 = (x_0, y_0, z_0) = (0,0,0)$ and $(a,b,c) = (-18,-13,11)$.

point-normal form: $-18x - 13y + 11z = 0$
general form: $-18x - 13y + 11z = 0$

(c) $\overrightarrow{P_1P_2}$ = (3, 5, 4) – (-1, -1, 2) = (4, 6, 2) and $\overrightarrow{P_1P_3}$ = (1, 2, 5) – (-1, -1, 2) = (2, 3, 3).

$$\overrightarrow{P_1P_2} \times \overrightarrow{P_1P_3} = \begin{vmatrix} i & j & k \\ 4 & 6 & 2 \\ 2 & 3 & 3 \end{vmatrix} = (12, -8, 0) \text{ is normal to the plane.}$$

$P_0 = (x_0, y_0, z_0) = (-1,-1,2)$ and $(a,b,c) = (12,-8,0)$.

point-normal form: $12(x + 1) - 8(y + 1) = 0$
general form: $12x - 8y + 4 = 0$ or $3x - 2y + 1 = 0$

(d) $\overrightarrow{P_1P_2}$ = (-2, 4, -3) – (7, 1, 3) = (-9, 3, -6) and $\overrightarrow{P_1P_3}$ = (5, 4, 1) – (7, 1, 3) = (-2, 3, -2).

$$\overrightarrow{P_1P_2} \times \overrightarrow{P_1P_3} = \begin{vmatrix} i & j & k \\ -9 & 3 & -6 \\ -2 & 3 & -2 \end{vmatrix} = (12, -6, -21) \text{ is normal to the plane.}$$

$P_0 = (x_0, y_0, z_0) = (7,1,3)$ and $(a,b,c) = (12,-6,-21)$.

point-normal form: $12(x - 7) - 6(y - 1) - 21(z - 3) = 0$
general form: $12x - 6y - 21z - 15 = 0$ or $4x - 2y - 7z - 5 = 0$

3. $(3,-2,4)$ is normal to the plane $3x - 2y + 4z - 3 = 0$ and $(-6,4,-8)$ is normal to the plane $-6x + 4y - 8z + 7 = 0$. $(-6,4,-8) = -2(3,-2,4)$, so the normals are parallel and therefore the planes are parallel.

4. (3,−6,9) is normal to the plane $3x - 6y + 9z - 6 = 0$ and (−1,2,−3) is normal to the plane $-x + 2y - 3z - 2 = 0$. −3(−1,2,−3) = (3,−6,9), so the normals are parallel and therefore the planes are parallel. The planes are distinct so they have no points in common.

5. (2,−3,1) is normal to the plane $2x - 3y + z + 4 = 0$ and therefore to the parallel plane passing through (1,2,−3).

$P_0 = (x_0, y_0, z_0) = (1,2,-3)$ and (a,b,c) = (2,−3,1).

point-normal form: $2(x - 1) - 3(y - 2) + (z + 3) = 0$
general form: $2x - 3y + z + 7 = 0$

6. (a) $P_0 = (x_0, y_0, z_0) = (1,2,3)$ and (a,b,c) = (−1,2,4).

parametric equations: $x = 1 - t,\ y = 2 + 2t,\ z = 3 + 4t,\ -\infty < t < \infty$

symmetric equations: $-x + 1 = \dfrac{y-2}{2} = \dfrac{z-3}{4}$

(b) $P_0 = (x_0, y_0, z_0) = (-3,1,2)$ and (a,b,c) = (1,1,1).

parametric equations: $x = -3 + t,\ y = 1 + t,\ z = 2 + t,\ -\infty < t < \infty$

symmetric equations: $x + 3 = y - 1 = z - 2$

(c) $P_0 = (x_0, y_0, z_0) = (0,0,0)$ and (a,b,c) = (−2,−3,5).

parametric equations: $x = -2t,\ y = -3t,\ z = 5t,\ -\infty < t < \infty$

symmetric equations: $-x/2 = -y/3 = z/5$

(d) $P_0 = (x_0, y_0, z_0) = (-2,-4,1)$ and (a,b,c) = (2,−2,4).

parametric equations: $x = -2 + 2t,\ y = -4 - 2t,\ z = 1 + 4t,\ -\infty < t < \infty$

symmetric equations: $\dfrac{x+2}{2} = -\dfrac{y+4}{2} = \dfrac{z-1}{4}$

7. $P_0 = (x_0, y_0, z_0) = (1,2,-4)$ and (a,b,c) = (2,3,1).

$(x - 1, y - 2, z + 4) = t(2,3,1)$, thus the parametric equations are

$x = 1 + 2t$, $y = 2 + 3t$, $z = -4 + t$, $-\infty < t < \infty$.

8. The direction of the line must be orthogonal to (3,2,5). Let it be (a,b,c). Then $(3,2,5)\cdot(a,b,c) = 3a + 2b + 5c = 0$. There are many solutions; one solution is (1,−4,1). The parametric equations of the line through (2,−3,1) with direction (1,−4,1) are $x = 2 + t$, $y = -3 - 4t$, $z = 1 + t$, $-\infty < t < \infty$.

9. $P_0 = (x_0, y_0, z_0) = (4,-1,3)$ and $(a,b,c) = (2,-1,4)$.

   $(x - 4, y + 1, z - 3) = t(2,-1,4)$, thus the parametric equations are

   $x = 4 + 2t$, $y = -1 - t$, $z = 3 + 4t$, $-\infty < t < \infty$.

10. xy plane: $z = 0$ xz plane: $y = 0$ yz plane: $x = 0$

11. $\overrightarrow{P_1P_2} = (-1, 1, 3) - (3, -5, 5) = (-4, 6, -2)$ and $\overrightarrow{P_1P_3} = (5, -8, 6) - (3, -5, 5) = (2, -3, 1)$.

    $\overrightarrow{P_1P_2} = -2\overrightarrow{P_1P_3}$ so the points P_1, P_2, and P_3 are collinear. There are many planes through the line containing these points.

12. (−1,2,−4) is normal to the plane perpendicular to the given line.

    $P_0 = (x_0, y_0, z_0) = (4,-1,5)$ and $(a,b,c) = (-1,2,-4)$.

    point-normal form: $-(x - 4) + 2(y + 1) - 4(z - 5) = 0$
    general form: $-x + 2y - 4z + 26 = 0$

13. The points on the line are of the form $(1 + t, 14 - t, 2 - t)$. If the line lies in the plane, the points will satisfy the equation of the plane. $2(1 + t) - (14 - t) + 3(2 - t) + 6 = 2 - 14 + 6 + 6 + 2t + t - 3t = 0$, so the line lies in the plane.

14. $3(4 + 2t) + 2(5 + t) - 4(7 + 2t) + 7 = 12 + 10 - 28 + 6t + 2t - 8t = -6$ for all possible values of t, so there is no point on the line that lies in the plane.

15. The direction of the line must be perpendicular to (−2,3,2). Let it be (a,b,c). Then $(-2,3,2)\cdot(a,b,c) = -2a + 3b + 2c = 0$. There are many solutions, all of the form

$\left(\frac{3b+2c}{2}, b, c\right)$; one solution is (1,0,1). The parametric equations of the line through (5,−1,2) with direction (1,0,1) are x = 5 + t, y = −1, z = 2 + t, $-\infty < t < \infty$.

16. The directions of the two lines are (−4,4,8) and (3,−1,2). (−4,4,8)·(3,−1,2) = 0, so the lines are orthogonal. To find the point of intersection, equate the expressions for x, y, and z: −1 − 4t = 4 + 3h, 4 + 4t = 1 − h, 7 + 8t = 5 + 2h. This system of equations has the unique solution t = −1/2, h = −1, so the point of intersection is x = 1, y = 2, z = 3.

17. The directions of the two lines are (4,5,3) and (1,−3,−2). (4,5,3) ≠ c(1,−3,−2), so the lines are not parallel. If the lines intersect there will be a solution to the system of equations 1 + 4t = 2 + h, 2 + 5t = 1 − 3h, 3 + 3t = −1 −2h. However the first equation gives h = 4t − 1 and substituting this into the other two equations gives two different values of t. Thus the system of equations has no solution and the lines do not intersect.

18. (a) Let (e,f,g) and (r,s,t) be two points on the plane and let h be any scalar. Then 2e + 3f − 4g = 0 and 2r + 3s − 4t = 0, so that 2(e+r) + 3(f+s) − 4(g+t) = 0 and 2he + 3hf − 4hg = 0; i.e., (e,f,g) + (r,s,t) = (e+r, f+s, g+t) and h(e,f,g) = (he,hf,hg) are also on the plane. Thus the set of all points on the plane is a subspace of $\mathbf{R}^3$.

    A basis for the space must consist of two linearly independent vectors that are orthogonal to (2,3,−4). Two such vectors are (0,4,3) and (2,0,1).

    (b) The point (0,0,0) is not on the plane so the set of all points on the plane is not a subspace of $\mathbf{R}^3$.

    (c) If d = 0, then the proof of part (a) with 2, 3, and −4 replaced with a, b, and c shows that the plane is a subspace of $\mathbf{R}^3$. If the plane is a subspace of $\mathbf{R}^3$, then (0,0,0) is a point on the plane. Thus ax0 + bx0 + cx0 + d = 0, so that d = 0.

19. Let (a,b,c) be the direction of a line. The line can be written in symmetric form if and only if a, b, and c are all nonzero. But the line is orthogonal to the x axis if and only if a = 0, the line is orthogonal to the y axis if and only if b = 0, and the line is orthogonal to the z axis if and only if c = 0, so the line can be written in symmetric form if and only if it is not orthogonal to any axis.

D2 Solving Systems of Linear Equations (Sections 1.1, 1.2, 1.6)

1. (a) x = 2, y = -1, z = 1; (b) x = 1 - 2r, y = r, z = -2;

   (c) x = 1/2, y = 2/5, z = 3/4 (d) x = 7/10, y = -.2/5, z = 7/20

2. (a) x = 2, y = 5 (b) no solution (c) many solutions (4-2r, r) (d) x = 6/7, y = 1/7

3. (a) x = 4, y = -3, z = -1 (b) x = -1, y = 1, z = 2 (c) x = 2, y = 3, x = -1

4. Start with an appropriate REF and work backward. For example,

$$\begin{bmatrix} 1 & 0 & 0 & 1 \\ 0 & 1 & 0 & 2 \\ 0 & 0 & 1 & 3 \end{bmatrix} \underset{\substack{R1+(2)R3 \\ R2+(-2)R3}}{\approx} \begin{bmatrix} 1 & 0 & 2 & 7 \\ 0 & 1 & -1 & -1 \\ 0 & 0 & 1 & 3 \end{bmatrix} \underset{R1+(-2)R2}{\approx} \begin{bmatrix} 1 & -2 & 4 & 9 \\ 0 & 1 & -1 & -1 \\ 0 & 0 & 1 & 3 \end{bmatrix}$$

$$\underset{(2)R2}{\approx} \begin{bmatrix} 1 & -2 & 4 & 9 \\ 0 & 2 & -2 & -2 \\ 0 & 0 & 1 & 3 \end{bmatrix} \underset{R2 <-> R3}{\approx} \begin{bmatrix} 1 & -2 & 4 & 9 \\ 0 & 0 & 1 & 3 \\ 0 & 2 & -2 & -2 \end{bmatrix} \underset{\substack{R2+(-1)R1 \\ R3+(3)R1}}{\approx} \begin{bmatrix} 1 & -2 & 4 & 9 \\ 1 & -2 & 5 & 12 \\ -1 & 4 & -6 & -11 \end{bmatrix}$$

The following system uses all three types of row operations

$$\begin{aligned} x - 2y + 4z &= 9 \\ x - 2y + 5z &= 12 \\ -x + 4y - 6z &= -11 \end{aligned}$$

6. The matrix corresponds to a system of three equations in two variables. Probability is that such a system written down at random has no solution; the three lines do not cross in a point. Thus REF will be the identity matrix. To get another form of REF, take any three lines that cross in a point. For example, construct a system that crosses at x = 1, y = 2. Let system be

$$\begin{aligned} x + y &= 3 \\ -x + y &= 1 \\ 2x + y &= 4 \end{aligned} . \quad \text{Get} \begin{bmatrix} 1 & 1 & 3 \\ -1 & 1 & 1 \\ 2 & 1 & 4 \end{bmatrix} \approx \begin{bmatrix} 1 & 0 & 1 \\ 0 & 1 & 2 \\ 0 & 0 & 0 \end{bmatrix}$$

7. REF for first matrix is correct, second incorrect due to roundoff error. Crucial error takes place when a small element that should have been 0, in the (3,3) location, is normalized. This can be clearly seen in all steps in Format Long E.

8. See Section 9.1 for discussion of Gaussian elimination and echelon form.

   [GJalg and Galg are useful for class demonstration of difference in algorithms.]

$$\text{Get}\begin{bmatrix}1&1&1&3\\2&3&1&5\\1&-1&-2&-5\end{bmatrix}\underset{\substack{R2+(-2)R1\\R3+(-1)R1}}{\approx}\begin{bmatrix}1&1&1&3\\0&1&-1&-1\\0&-2&-3&-8\end{bmatrix}\underset{R3+(2)R2}{\approx}\begin{bmatrix}1&1&1&3\\0&1&-1&-1\\0&0&-5&-10\end{bmatrix}$$

$$\underset{(1/5)R2}{\approx}\begin{bmatrix}1&1&1&3\\0&1&-1&-1\\0&0&1&2\end{bmatrix}\underset{\substack{R1+(-1)R3\\R2+R3}}{\approx}\begin{bmatrix}1&1&0&1\\0&1&0&1\\0&0&1&2\end{bmatrix}\underset{R1+(-1)R2}{\approx}\begin{bmatrix}1&0&0&0\\0&1&0&1\\0&0&1&2\end{bmatrix}$$

9. (a) 8 mult, 9 add , 0 swap (b) 11 mult, 12 add, 0 swap.
Creation of zeros and leading 1s are not counted as operations - they are substitutions.
Gauss gains one multiplication and one addition in the back substitution. No operations take
place on the (1,3) location when a zero is created in the (1,2) location.

10. Let m denote multiplications, a additions. We get

	Gauss	G/J
n=2	6m,3a	6m,3a
n=3	17m,11a	18m,12a
n=4	36m,26a	40m,30a
n=5	65m,50a	75m,60a

Thus Gauss mult: pts (2,6), (3,17), (4,36), (5,65)
additions: pts (2,3), (3,11), (4,26), (5,50)
G/J multiplications: pts (2,6), (3,18), (4,40), (5,75)
additions: pts (2,3), (3,12), (4,30), (5,60)

Substituting points into the polynomial $a + bn + cn^2 + dn^3$ and solving the systems of linear equations gives:

Gauss mults $\frac{n^3}{3} + n^2 - \frac{n}{3}$, adds $\frac{n^3}{3} + \frac{n^2}{2} - \frac{5n}{6}$ G/J mults $\frac{n^3}{2} + \frac{n^2}{2}$, adds $\frac{n^3}{2} - \frac{n}{2}$

For system with 20x20 matrix of coeffs, Gauss 5910 flops, G/J 8190 flops.

11. (a) $y = 2x^2 + x + 5$ (b) $y = x^3 - 2x^2 + x - 3$.

<u>D3 Dot Product, Norm, Angle, Distance, Projection</u> (Section 1.4)

1. $\mathbf{u}\cdot\mathbf{v} = 2$, $||\mathbf{u}|| = 6.3246$, angle between **u** and **v** = 87.2031°, d(X,Y) = 5.7446.

2. $\mathbf{u}\cdot\mathbf{v} = -26$, $||\mathbf{u}|| = 6.245$, angle between **u** and **v** = 117.0171°, d(X,Y) = 10.3923.

3. $||\mathbf{u}|| = 3.7417$, $||\mathbf{v}|| = 11.0454$, $||\mathbf{u} + \mathbf{v}|| = 11.5758 < 3.7417 + 11.0454$.

4. A is an orthogonal matrix. Rows and columns of A form orthonormal sets. A is invertible with $A^{-1} = A^t$. $|A| = 1$ (in general $|A| = \pm 1$).

D4 Matrix Operations (Sections 2.1 - 2.3)

1. (a) $\begin{bmatrix} 4 & 5 \\ 3 & 1 \end{bmatrix}$ (b) $\begin{bmatrix} 1 & 0 \\ 0 & 1 \end{bmatrix}$ (Explore various powers of matrix B!)

   (c) $\begin{bmatrix} 1 & 0 \\ 2 & -1 \end{bmatrix}$, that is B (d) $\begin{bmatrix} 3 & 5 \\ 5 & 8 \end{bmatrix}$

   (e) $(P + P^t)$ will be symmetric. $(P + P^t)^t = P^t + (P^t)^t = P^t + P = P + P^t$.
   (f) PP^t will be symmetric. $(PP^t)^t = (P^t)^t P^t = PP^t$.

2. (a) $\begin{bmatrix} -9 & 35 \\ 1 & 10 \end{bmatrix}$ (b) $\begin{bmatrix} 265 & 480 \\ 88 & 171 \end{bmatrix}$

4. 29

5. (a)
```
»X = A(3 , :);
»Y = A(: , 4);
» X * A' * Y

ans = 821
```
(b)
```
»B = A(1:3 , 2:4);
»B^7

        1752000  5792960  8249536
ans =   1628992  5367168  7639744
        1768384  5792960  8233152
```
(c)
```
»C= [A(1, :) ; A(3, :) ; A(4, :)];
»C*A
           9    0    25   62
ans =    -23   18     6   22
          13   -6   -11    9
```

D5 Computational Considerations (Section 2.2)

1. Have A 2x2, B 2x3, C 3x1.
   Let m stand for a multiplication. AB is a 2x3 matrix. It has 6 elements. Each element is computed
   by multiplying a row of A by a column of B. Each element requires 2m. Thus 12m for AB.
   (AB)xC is 2x1 matrix, has 2 elements. Each element involves 3m. Therefore 6m to compute (AB)xC. Thus total multiplications for ABxC is 12+6 = 18.
   BC has 2 elements, and requires 3m per element. Thus 6m for BC. Ax(BC) has 2 elements,

and takes 2m per element. Thus 4m for Ax(BC). Total for AxBC is 6+4 = 10.

2. There are mn elements in AB. Each is computed by multiplying a row of A times a column of B, i.e., by doing r multiplications (and r-1 additions). Thus the total number of multiplications is (mn)r or mrn.
The number of multiplications required to compute AB is mrn and the number required to compute BC is rns. (AB)C is the product of an mxn matrix and an nxs matrix, so the number of multiplications required is mns + mrn. A(BC) is the product of an mxr matrix and an rxs matrix, so the number of multiplications is mrs + rns.

3. (a) A(BC) 3,864; (AB)C 7,395 (b) (A(BC))D 4,494; A(B(CD)) 65,058

4. (a) (AB)C 4,320; A(BC) 6,345 (b) (AB)B 8,100; $A(B^2)$ 95,175
(c) ((AB)B)B 12,150; $A(B^3)$ 186,300 (d) ((AB)B)C 8,370; $A((B^2))C$ 97,470

5. Each of the mn elements of AB is computed by multiplying a row of A times a column of B. This requires adding r terms, i.e., doing r-1 additions. Thus the total number of additions required to compute AB is mn(r-1).
To calculate (AB)C requires mn(r-1) additions for AB and ms(n-1) additions for the second product. Thus the total is mn(r-1) + ms(n-1). To calculate A(BC) requires rs(n-1) additions for BC and ms(r-1) for the second product. Thus the total is rs(n-1) + ms(r-1).
If A, B, and C are 2x2, 2x3, and 3x1, respectively, the number of additions is 2x3x1 + 2x1x2 = 10 for (AB)C and 2x1x2 + 2x1x1 = 6 for A(BC).

D6 Inverse of a Matrix (Section 2.4)

2. $\begin{bmatrix} 1 & 0 \\ 2 & -1 \end{bmatrix}^2 = \begin{bmatrix} 1 & 0 \\ 0 & 1 \end{bmatrix}, \begin{bmatrix} 1 & 0 \\ 2 & -1 \end{bmatrix}^3 = \begin{bmatrix} 1 & 0 \\ 2 & -1 \end{bmatrix}, \begin{bmatrix} 1 & 0 \\ 2 & -1 \end{bmatrix}^4 = \begin{bmatrix} 1 & 0 \\ 0 & 1 \end{bmatrix}.$

Further, $\begin{bmatrix} 1 & 0 \\ 2 & -1 \end{bmatrix}^n = \begin{bmatrix} 1 & 0 \\ 2 & -1 \end{bmatrix}$ if n is odd, $\begin{bmatrix} 1 & 0 \\ 2 & -1 \end{bmatrix}^n = \begin{bmatrix} 1 & 0 \\ 0 & 1 \end{bmatrix}$ if n is even.

(a) $\begin{bmatrix} 1 & 0 \\ 2 & -1 \end{bmatrix}^2 = \begin{bmatrix} 1 & 0 \\ 0 & 1 \end{bmatrix}$, thus $B^2 B^2 = I$, implying that $(B^2)^{-1} = B^2$.

(b) $(B^{194})^{-1} = I^{-1} = I$.

3. (a) $A^{-1} = \begin{bmatrix} 1 & 0 \\ -2 & 1 \end{bmatrix}$

(b) $B^{-1} = \begin{bmatrix} 0.15116279069767 & 0.10465116279070 \\ 0.12790697674419 & -0.01162790697674 \end{bmatrix} = \begin{bmatrix} -13/86 & 9/86 \\ 11/86 & -1/86 \end{bmatrix}$

4. (a) $B=\begin{bmatrix} 2 & 0 \\ 0 & 1 \end{bmatrix}$ (b) $B=\begin{bmatrix} 7 & 13 \\ -2 & -3 \end{bmatrix}$ (c) $B=\begin{bmatrix} 6 & 14 & -7 \\ -5 & -11 & 6 \\ -3 & -6 & 5 \end{bmatrix}$

(d) $B=\begin{bmatrix} 42.2000 & -11.6000 & 18.2000 \\ 91.6000 & -24.8000 & 39.6000 \\ -15.4000 & 4.2000 & -5.4000 \end{bmatrix}$

5.
```
function R=sim(P,Q)
%Performs similarity transf inv(B)*A*B
%Calling format: sim(A,B)
R=inv(Q)*P*Q;
```

D7 Solving Systems of Equations Using Matrix Inverse (Section 2.4)

1. x = 10, y = -2
2. x = -1, y = 2.5, z = -1.5
3. x = 9, y = -14
4. x = 7/9, y = 3/9, z = -5/9
5. x=1,y=2,z=3; x=-1,y=2,z=0; x=0,y=1,y=2

D8 Cryptography (Section 2.4)

1. (a) $Y = \begin{bmatrix} -108 & -207 & -120 & -158 & -159 & -112 & -204 \\ 33 & 45 & 17 & 32 & 23 & 28 & 54 \\ 109 & 222 & 140 & 177 & 183 & 115 & 209 \end{bmatrix} = AX$ with $A = \begin{bmatrix} -3 & -3 & 4 \\ 0 & 1 & 1 \\ 4 & 3 & 4 \end{bmatrix}$

$X = inv(A)Y = \begin{bmatrix} 1 & 15 & 20 & 19 & 24 & 3 & 5 \\ 27 & 18 & 8 & 27 & 5 & 9 & 27 \\ 6 & 27 & 9 & 5 & 18 & 19 & 27 \end{bmatrix}$. Original message is;

1	27	6	15	18	27	20	8	9	19	27	5	24	5	18	3	9	19	5	27	27
A	-	F	O	R	-	T	H	I	S	-	E	X	E	R	C	I	S	E	-	-

1. (b) $Y = \begin{bmatrix} -176 & -113 & -150 & -115 & -161 & -92 & -30 & -213 \\ 35 & 9 & 38 & 26 & 29 & 29 & 4 & 54 \\ 195 & 140 & 153 & 119 & 181 & 93 & 35 & 221 \end{bmatrix} = AX$ with $A = \begin{bmatrix} -3 & -3 & 4 \\ 0 & 1 & 1 \\ 4 & 3 & 4 \end{bmatrix}$

$X = inv(A)Y = \begin{bmatrix} 19 & 27 & 3 & 4 & 20 & 1 & 5 & 8 \\ 21 & 4 & 11 & 1 & 15 & 27 & 1 & 27 \\ 14 & 5 & 27 & 25 & 14 & 2 & 3 & 27 \end{bmatrix}$. Original message is;

19	21	14	27	4	5	3	11	27	4	1	25	20	15	14	1	27	2	5	1	3	8	27	27
S	U	N	-	D	E	C	K	-	D	A	Y	T	O	N	A	-	B	E	A	C	H	-	-

D9 Transformations defined by Matrices (Section 2.5, 2.6)

1. »map([cos(2 * pi / 3) -sin(2 * pi / 3); sin(2 * pi / 3) cos(2 * pi / 3)])
   O* = (0, 0), P* = (-0.5000, 0.8600), Q* = (-1.3660, -1.5000)
   R* = (-0.8660, -0.5000)

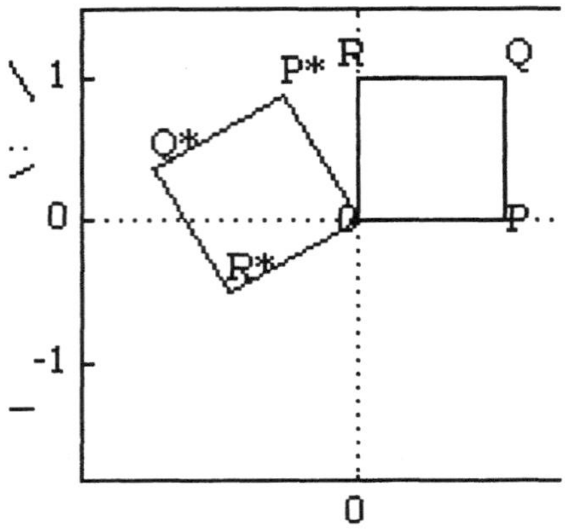

2. O* = (0, 0), P* = (0.8660, -0.5000), Q* = (1.3660, 0.3660),
   R* = (0.5000, 0.8660)

3. A = [3 0;0 3]
   O* = (0, 0), P* = (3, 0), Q* = (3, 3), R* = (0, 3)

4. A = [-1 0;0 1]
   O* = (0, 0), P* = (-1, 0), Q* = (-1, 1), R* = (0, 1)

5. (a) O* = (0, 0), P* = (3, 1), Q* = (3, 5), R* = (0, 4)
   (b) O* = (0, 0), P* = (4, 1), Q* = (3, 6), R* = (-1, 5)
   (c) O* = (0, 0), P* = (0, -3), Q* = (3, -3), R* = (3, 0)

6. (a) A = [cos(pi/2) -sin(pi/2); sin(pi/2) cos(pi/2)], B = [2 0; 0 2]
   O# = (0, 0), P# = (0, 2), Q# = (-2, 2), R# = (-2, 0)

   (b) A = [4 0,0 4], B = [1 0;0 -1]
   O# = (0, 0), P# = (4, 0), Q# = (4, -4), R# = (0, -4)

   (c) A = [cos(pi/3) -sin(pi/3); sin(pi/3) cos(pi/3)], B = [0 1;1 0]
   O# = (0, 0), P# = (0.8660, 0.5000), Q# = (1.3660, -0.3600),
   R# = (0.5000, -0.8660)

7. (a) A = [3 0;0 3], B = [1 2;0 1]
   O# = (0, 0), P# = (3, 0), Q# = (9, 3), R# = (6, 3)

   (b) A = 3 0;0 2], B = 0 1;1 0]

$O^\# = (0, 0)$, $P^\# = (0, 3)$, $Q^\# = (2, 3)$, $R^\# = (2, 0)$

(c) A = [2 0;0 2], B = [1 0, 3 1];
C = [cos(pi/3) -sin(pi/3); sin(pi/3) cos(pi/3)]
compmap can have two arguments only
Let D = B*A the use compmap(D, C) to get
$O^\# = (0, 0)$, $P^\# = (-4.1962, 4.7321)$, $Q^\# = (-5.9282, 5.7321)$,
$R^\# = (-1.7321, 1.0000)$

8. $O^* = (0, 0)$, $P^* = (6, 2)$, $Q^* = (8, 3)$, $R^* = (5, 2)$

9. (a) (2,2) (b) (3, 2)

D10 Fractals (Section 2.6)

1. T_2 gives bottom left leaf. T_1 then gives all the leaves above it. T_3 gives bottom right leaf, T_1 then gives all the leaves above it. T_4 gives bottom of main stem, T_1 then gives the main stem above it.

4. (a) Let $\begin{bmatrix} 0.86 & 0.03 \\ -0.03 & 0.86 \end{bmatrix} = \begin{bmatrix} k & 0 \\ 0 & k \end{bmatrix}\begin{bmatrix} \cos\alpha & -\sin\alpha \\ \sin\alpha & \cos\alpha \end{bmatrix}$. $k\cos\alpha = 0.86$, $k\sin\alpha = -0.03$.
$k^2 = 0.86^2 + 0.03^2$. $k = 0.8605230967$. $\alpha = -1.998^\circ$, to three decimal places.
dilation factor = 0.8605230967, angle of rotation = -1.998°.

(b) $1.5\alpha = -2.997^\circ$.

$$T_1 = \begin{bmatrix} 0.8605230967 & 0 \\ 0 & 0.8605230967 \end{bmatrix}\begin{bmatrix} \cos(-2.997) & -\sin(-2.997) \\ \sin(-2.997) & \cos(-2.997) \end{bmatrix}\begin{bmatrix} x \\ y \end{bmatrix} + \begin{bmatrix} 0 \\ 1.5 \end{bmatrix}$$
$$= \begin{bmatrix} 0.8593461366 & 0.0449913039 \\ 0.0449913039 & 0.8593461366 \end{bmatrix}\begin{bmatrix} x \\ y \end{bmatrix} + \begin{bmatrix} 0 \\ 1.5 \end{bmatrix}$$, T_2, T_3, T_4 remain same.

(c) Write T_1 in the following form for rotation through angle α.

$$T_1 = \begin{bmatrix} 0.8605230967 & 0 \\ 0 & 0.8605230967 \end{bmatrix}\begin{bmatrix} \cos(\alpha) & \sin(\alpha) \\ -\sin(\alpha) & \cos(\alpha) \end{bmatrix}\begin{bmatrix} x \\ y \end{bmatrix} + \begin{bmatrix} 0 \\ 1.5 \end{bmatrix}$$

(d) e.g., let $\alpha>0$ for a rotation through angle to right to the left in formula for (c).
let $\alpha<0$ for a rotation through angle to right to the left.
let $\alpha = 0$ for vertical fern.

5. Tip of fern will be point where $T_1(x,y) = (x,y)$.

$$\begin{bmatrix} 0.86 & 0.03 \\ -0.03 & 0.86 \end{bmatrix}\begin{bmatrix} x \\ y \end{bmatrix} + \begin{bmatrix} 0 \\ 1.5 \end{bmatrix} = \begin{bmatrix} x \\ y \end{bmatrix}. \qquad \begin{aligned} -0.14x + 0.03y &= 0 \\ 0.03x + 0.14y &= 1.5 \end{aligned}$$

$x = 2.1951$, $y = 10.2439$

D11 Leontief I/O Model (Section 2.7)

1. $X = \begin{bmatrix} 165 \\ 480 \\ 250 \end{bmatrix}$ 2. $X_1 = \begin{bmatrix} 60 \\ 40 \end{bmatrix}$, $X_2 = \begin{bmatrix} 22.5 \\ 16.333 \end{bmatrix}$, $X_3 = \begin{bmatrix} 15 \\ 20 \end{bmatrix}$ 3. $D = \begin{bmatrix} 4 \\ .9 \\ .65 \end{bmatrix}$

4. Each column of A adds to 1. A is in fact a stochastic matrix, $0 \le a_i \le 1$, and $\sum_j a_{ij} = 1$. If A is stochastic, $|A - I| = 0$ [Add all columns to 1st column]. Thus $(A - I)^{-1}$ does not exist. In this situation, X = AX. The industrial output is all taken up by industrial demand, and there is nothing left to meet the demand of the open sector.
[Note - prerequisite for this exercise is determinants and the fact that the inverse does not exist if the determinant is zero - this exercise is best left until these topics have been covered. X is in fact an eigenvector of A corresponding to eigenvector 1 here.]

D12 Markov Chains (Sections 2.8, 3.5)

1. (a) $X_{2008} = \begin{bmatrix} 243.59 \\ 53.41 \end{bmatrix} \begin{matrix} City \\ Suburb \end{matrix}$, $X_{2009} = \begin{bmatrix} 242.2223 \\ 54.7777 \end{bmatrix}$, $X_{2010} = \begin{bmatrix} 240.8956 \\ 56.1044 \end{bmatrix}$, $X_{2011} = \begin{bmatrix} 239.6088 \\ 57.3912 \end{bmatrix}$,

$X_{2012} = \begin{bmatrix} 238.3605 \\ 58.6395 \end{bmatrix}$

(b) $\begin{bmatrix} 0.99 & 0.02 \\ 0.01 & 0.98 \end{bmatrix}^5 = \begin{bmatrix} .9529 & .0942 \\ .0471 & .9058 \end{bmatrix}$. $p_{11}^{(5)} = 0.9529$.

2. (a) Using mathematics, solve PX = X. Get $X = \begin{bmatrix} 49 \\ 196 \end{bmatrix}$.

(b) $Q = \begin{bmatrix} 0.2 & 0.2 \\ 0.8 & 0.8 \end{bmatrix}$. $q_{11} = q_{12} = 0.2$, long term probabilities of living in state 1, city, is 0.2. $q_{22} = q_{21} = 0.8$, long term probabilities of living in state 2, suburbs, is 0.8.

3. (a) Use $1.01\begin{bmatrix} 0.96 & 0.01 \\ 0.04 & 0.99 \end{bmatrix} = \begin{bmatrix} .9696 & .0101 \\ .0404 & .9999 \end{bmatrix}$ as transition matrix, to get

$X_{2008} = \begin{bmatrix} 238.0772 \\ 61.8928 \end{bmatrix} \begin{matrix} City \\ Suburb \end{matrix}$, $X_{2009} = \begin{bmatrix} 231.4648 \\ 71.5049 \end{bmatrix}$, $X_{2010} = \begin{bmatrix} 225.1504 \\ 80.8490 \end{bmatrix}$,

$X_{2011} = \begin{bmatrix} 219.1224 \\ 89.9369 \end{bmatrix}$, $X_{2012} = \begin{bmatrix} 213.3695 \\ 98.7805 \end{bmatrix}$.

(b) Use $\begin{bmatrix} 1.012 & 0 \\ 0 & 1.008 \end{bmatrix}\begin{bmatrix} 0.96 & 0.01 \\ 0.04 & 0.99 \end{bmatrix} = \begin{bmatrix} .9715 & .0101 \\ .0403 & .9980 \end{bmatrix}$ as transition matrix, to

get $X_{2008}=\begin{bmatrix} 238.5427 \\ 61.7695 \end{bmatrix}\begin{matrix} City \\ Suburb \end{matrix}$, $X_{2009}=\begin{bmatrix} 232.3681 \\ 71.2592 \end{bmatrix}$, $X_{2010}=\begin{bmatrix} 226.4653 \\ 80.4811 \end{bmatrix}$,

$X_{2011}=\begin{bmatrix} 220.8239 \\ 89.4467 \end{bmatrix}$, $X_{2012}=\begin{bmatrix} 215.4339 \\ 98.1670 \end{bmatrix}$.

4. (a) With P, there is a steady decrease in city population, and steady increase in suburban population. With P', bar graphs show fluctuations. The P model is most realistic. Diagonal elements of P are dominant, non-diagonal elements of P' are dominant. P is more realistic since it reflects the larger probability of a person staying in the same state.

(b) Result: Let $A = \begin{bmatrix} a & b \\ 1-a & 1-b \end{bmatrix}$. If a>b, steady trends. If a<b, fluctuating trends.

5. $X_{n-1}=A^{-1}X_n$. $A^{-1} = \begin{bmatrix} 1.0421 & -0.0105 \\ -0.0421 & 1.0105 \end{bmatrix}$. $X_{2006}=\begin{bmatrix} 83.7368 \\ 161.2632 \end{bmatrix}$, $X_{2005}=\begin{bmatrix} 85.5651 \\ 159.4349 \end{bmatrix}$,

$X_{2004}=\begin{bmatrix} 87.4896 \\ 157.5104 \end{bmatrix}$, $X_{2003}=\begin{bmatrix} 89.5153 \\ 155.4847 \end{bmatrix}$, $X_{2002}=\begin{bmatrix} 91.6477 \\ 153.3523 \end{bmatrix}$

Not stochastic, A^{-1} has negative numbers, but sum of a column is 1.

6. (a) $X_{2008}=\begin{bmatrix} 81.1300 \\ 162.4600 \\ 53.4100 \end{bmatrix}\begin{matrix} City \\ Suburb \\ Nonmetro \end{matrix}$, $X_{2009}=\begin{bmatrix} 80.3105 \\ 161.9117 \\ 54.7777 \end{bmatrix}$, $X_{2010}=\begin{bmatrix} 79.5389 \\ 161.3567 \\ 56.1044 \end{bmatrix}$

$X_{2011}=\begin{bmatrix} 78.8125 \\ 160.7963 \\ 57.3912 \end{bmatrix}$, $X_{2012}=\begin{bmatrix} 78.1288 \\ 160.2317 \\ 58.6395 \end{bmatrix}$

(b) $\begin{bmatrix} 69.3000 \\ 128.7000 \\ 99.0000 \end{bmatrix}\begin{matrix} City \\ Suburb \\ Nonmetro \end{matrix}$ (c) City: 0.2333, Suburb: 0.4333, Nonmetro: 0.3333.

<u>D14 Digraphs</u> (Section 2.9)

1. Distance = 3. Two paths 2 –> 4 –> 5 –> 1 and 2 –> 4 –> 3 –> 1.

2.

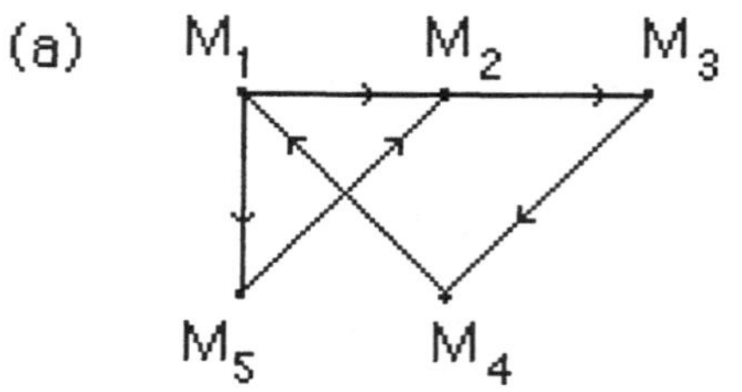

$$D = \begin{bmatrix} 0 & 1 & 2 & 3 & 1 \\ 3 & 0 & 1 & 2 & 4 \\ 2 & 3 & 0 & 1 & 3 \\ 1 & 2 & 3 & 0 & 2 \\ 4 & 1 & 2 & 3 & 0 \end{bmatrix} \begin{matrix} 7 \\ 10 \\ 9 \\ 8 \\ 10 \end{matrix}$$

most to least influential:
M_1, M_4, M_3, with M_2 and M_5 equally uninfluential.

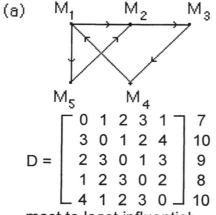

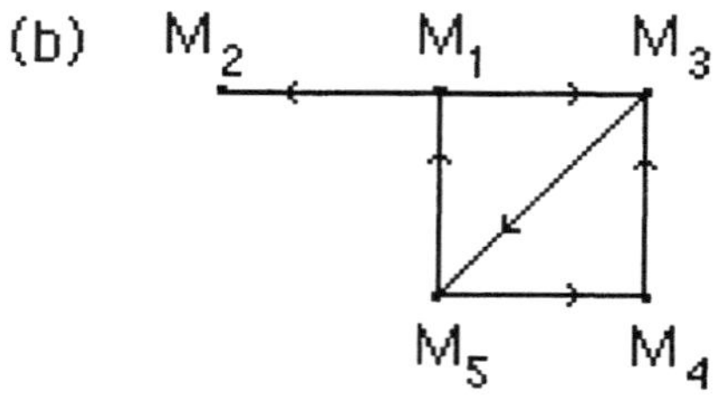

$$D = \begin{bmatrix} 0 & 1 & 1 & 3 & 2 \\ x & 0 & x & x & x \\ 2 & 3 & 0 & 2 & 1 \\ 3 & 4 & 1 & 0 & 2 \\ 1 & 2 & 2 & 1 & 0 \end{bmatrix} \begin{matrix} 7 \\ 4x \\ 8 \\ 9 \\ 6 \end{matrix}$$

most to least influential:
M_5, M_1, M_3, M_4, M_2

M_3 becomes most influential person by influencing M_1 in both groups.

D14 Determinants (Section 3.1, 3.2, 3.3)

1. (a) $|A| = 7$, $M(a_{22}) = -5$, $M(a_{31}) = 5$. (b) $|B| = 12$, $M(b_{21}) = 29$, $M(b_{33}) = 15$.

2. All determinants are zero.
   (a) Row 3 is 3 times row 1. (b) Column 3 is −3/2 times column 1.
   (c) Columns 2 and 3 are equal. (d) Row 3 is −3 times row 1.

3. (a)
$$\begin{vmatrix} 1 & -1 & 0 & 2 \\ -1 & 1 & 0 & 0 \\ 2 & -2 & 0 & 1 \\ 3 & 1 & 5 & -1 \end{vmatrix} \begin{matrix} \approx \\ R2+R1 \\ R3+(-2)R1 \\ R4+(-3)R1 \end{matrix} \begin{vmatrix} 1 & -1 & 0 & 2 \\ 0 & 0 & 0 & 2 \\ 0 & 0 & 0 & -3 \\ 0 & 4 & 5 & -7 \end{vmatrix} \begin{matrix} \approx \\ R2 \leftrightarrow R4 \end{matrix} \begin{vmatrix} 1 & -1 & 0 & 2 \\ 0 & 4 & 5 & -7 \\ 0 & 0 & 0 & -3 \\ 0 & 0 & 0 & 2 \end{vmatrix} = 0.$$

(b)
$$\begin{vmatrix} 2 & 1 & 3 & 1 \\ -2 & 3 & -1 & 2 \\ 2 & 1 & 2 & 3 \\ -4 & -2 & 0 & -1 \end{vmatrix} \begin{matrix} \approx \\ R2+R1 \\ R3+(-1)R1 \\ R4+(2)R1 \end{matrix} \begin{vmatrix} 2 & 1 & 3 & 1 \\ 0 & 4 & 2 & 3 \\ 0 & 0 & -1 & 2 \\ 0 & 0 & 6 & 1 \end{vmatrix} \begin{matrix} \approx \\ R4+(6)R3 \end{matrix} \begin{vmatrix} 2 & 1 & 3 & 1 \\ 0 & 4 & 2 & 3 \\ 0 & 0 & -1 & 2 \\ 0 & 0 & 0 & 13 \end{vmatrix} = -104.$$

D15 Cramer's Rule (Section 3.3)

1. (a) $\frac{-1}{3}\begin{bmatrix} -7 & 9 & 1 \\ 8 & -9 & -2 \\ -4 & 3 & 1 \end{bmatrix}$ (b) $\frac{-1}{3}\begin{bmatrix} 0 & -6 & 3 \\ -3 & -3 & 3 \\ 2 & 3 & -3 \end{bmatrix}$ (c) $\frac{-1}{4}\begin{bmatrix} -6 & 2 & -2 \\ -3 & 1 & 1 \\ -8 & 4 & 0 \end{bmatrix}$

2. (a) 1, 2, -1 (b) -1, 3, 4 (c) 0.5, 0.25, 0.25

D16 Eigenvalues, Eigenvectors, and Applications (Section 3.4, 3.5)

1. $\lambda_1 = 1$, $v_1 = \begin{bmatrix} -.6017 \\ .7453 \\ -.2872 \end{bmatrix}$, $\lambda_2 = 1$, $v_2 = \begin{bmatrix} .4399 \\ .0091 \\ -.8980 \end{bmatrix}$, $\lambda_3 = 10$, $v_3 = \begin{bmatrix} .6667 \\ .6667 \\ .3333 \end{bmatrix}$.

2. $A = \begin{bmatrix} 1 & 2 \\ 1 & 0 \end{bmatrix}$, $\lambda_1 = 2$, $v_1 = \begin{bmatrix} 0.8944 \\ 0.4472 \end{bmatrix}$, $\lambda_2 = -1$, $v_2 = \begin{bmatrix} -.7071 \\ .7071 \end{bmatrix}$.

$A = \begin{bmatrix} 2 & 3 \\ 1 & 0 \end{bmatrix}$, $\lambda_1 = 3$, $v_1 = \begin{bmatrix} 0.9487 \\ 0.3162 \end{bmatrix}$, $\lambda_2 = -1$, $v_2 = \begin{bmatrix} -.7071 \\ .7071 \end{bmatrix}$.

$A = \begin{bmatrix} 3 & 4 \\ 1 & 0 \end{bmatrix}$, $\lambda_1 = 4$, $v_1 = \begin{bmatrix} 0.9701 \\ 0.2425 \end{bmatrix}$, $\lambda_2 = -1$, $v_2 = \begin{bmatrix} -.7071 \\ .7071 \end{bmatrix}$.

Conjecture: $A = \begin{bmatrix} a & a+1 \\ 1 & 0 \end{bmatrix}$, $\lambda_1 = a+1$, $v_1 = \begin{bmatrix} a+1 \\ 1 \end{bmatrix}$, $\lambda_2 = -1$, $v_2 = \begin{bmatrix} -1 \\ 1 \end{bmatrix}$.

Proof: $\begin{vmatrix} a-\lambda & a+1 \\ 1 & -\lambda \end{vmatrix} = 0$ gives $\lambda^2 - \lambda a - (a+1) = 0$. $\lambda = a+1, -1$.

Corresponding eigenvectors are $\begin{bmatrix} a+1 \\ 1 \end{bmatrix}$, $\begin{bmatrix} -1 \\ 1 \end{bmatrix}$.

4. The eigenvectors of $\lambda = 1$ are vectors of the form $r\begin{bmatrix} 2 \\ 1 \end{bmatrix}$. If there is no change in total population $2r + r = 245 + 52 = 297$, so $r = 297/3$. Thus the long-term prediction is that population in metropolitan areas will be 2r=198 million and population in nonmetropolitan areas will be r=99 million.

5. $P = \begin{bmatrix} 0 & 1/3 & 0 & 1/4 \\ 1/2 & 0 & 1/3 & 1/4 \\ 0 & 1/3 & 0 & 1/2 \\ 1/2 & 1/3 & 2/3 & 0 \end{bmatrix}$, P is regular since P^2 is positive. Eigenvectors of P for $\lambda =$
1
are vectors of the form $r\,[2\ \ 3\ \ 3\ \ 4]^t$; distribution of rats in rooms 1, 2, 3, 4 is 2:3:3:4.

Powers of P approach the stochastic matrix $Q = \begin{bmatrix} 2s & 2s & 2s & 2s \\ 3s & 3s & 3s & 3s \\ 3s & 3s & 3s & 3s \\ 4s & 4s & 4s & 4s \end{bmatrix}$, so $2s + 3s + 3s + 4s = 1$, and $s = 1/12$. The long-term probability that a given rat will be in room 4 is $4/12 = 1/3$.

D17 Linear Combinations, Linear Dependence, Basis (Sections 1.3, 4.2 - 4.5)

1. (a) (-3, 3, 7) = 2(1, -1, 2) - (2, 1, 0) + 3(-1, 2, 1) (b) not a combination
   (c) (0, 10,8) = (2-r)(-1, 2, 3) + (2-2r) (1, 3, 1) + r(1, 8, 5)

2. (a) 2(-1, 3, 2) - 3(1, -1, -3) - (-5, 9, 13) = 0 (b) Linearly independent

3. (a) linearly independent, det(A) = 2268 $\neq$ 0. rank(A) = 4
   (b) linearly dependent, det(A) = 0, rank(A) = 3 < 4

4. Rank is 2. R2-R1 gives 5 5 5 5 5, R3-R2 gives 5 5 5 5 5 etc.

   Any matrix of the form $\begin{bmatrix} 1 & : & : & : \\ : & : & : & : \\ : & : & : & n^2 \end{bmatrix}$ is of rank 2.

5. (a) {(1, 0, 1), (0, 1, 1)} (b) {(1, 0, 0.6, -0.2, 1.6), (0, 1, 1.8, 0.4, 0.8)}
   (c) {(1, 0, 0, 0, 0, 0), (0, 1, 0, 0.4, 2.6, 1.4), (0, 0, 1, 0.8, 1.2, 1.8)}

<u>D18 Projection, Gram-Schmidt Orthogonalization</u> (Section 4.6)

1. (0.1905, -0.2381). 2. (0, -0.2857, -0.1429, 0.5714).

3. (a) {(0.5345, 0.8018, 0.2673), (0.3841, -0.5121, 0.7682)}
   (b) {(0.8729, 0.4364, 0.2182, 0), (-0.2264, 0.3843, 0.1369, -0.8845), (-0.4309, 0.6424, 0.4387, 0.4573)}
   (c) {(0.1826, 0.3651, 0.5477, 0.7303), (0.5477, -0.7303, -0.1826, 0.3651)}
   The vectors are linearly dependent. They span a 2D subspace.

<u>D19 Kernel and Range</u> (Section 4.7)

1. (a) kernel $\{(-2, -3, 1)^t\}$, range $\{(1, 0, 2)^t, (0, 1, -1)^t\}$
   (b) kernel $\{(.5, -1.75, 1)^t\}$, range $\{(1, 0, 2)^t, (0, 1, 1)^t\}$
   (c) kernel is zero vector, range $\{(1, 0, 0)^t, (0, 1, 0)^t, (0, 0, 1)^t\}$
   (d) kernel $\{(3, 1, 0)^t, (6, 0, 1)^t\}$, range $\{(1, -2, -1)^t\}$

<u>D20 Inner Product, Non-Euclidean Geometry</u> (Sections 6.1, 6.2)

1. (a) $\| (0, 1) \| = 2$. (b) 90°. The vectors (1,1), and (-4, 1) are at right angles.

(c) dist((1, 0), (0, 1)) = $\sqrt{5}$.

(d) dist((x, y), (0, 0))= 1,2,3; < (x, y), (x, y)>= 1,4,9; [x y]A[x y]^t = 1,4,9; $x^2 + 4y^2$ = 1,4,9.

2. Matrix for dot product is $\begin{bmatrix} 1 & 0 \\ 0 & 1 \end{bmatrix}$.
3. Equations of the circles are [x y]A[x y]^t = r^2; $6x^2 + 4xy + 9y^2$ = 1, 4, 9.

4. (a), (b) All vectors of form a(1, 1) or b(1, -1). i.e., lie on the cone through the origin.

(c) Any two points of the form (x, y) and (x+1, y+1); or (x, y) and (x-1), (y+1).

(d) $-x^2 + y^2$ = 1, 4, 9.

5. (a) y = ± 1, ±$\sqrt{2}$, ±$\sqrt{3}$. (b) x = ± 1, ±$\sqrt{2}$, ±$\sqrt{3}$.
(c) $x^2 + 2xy + y^2$ = 1, 2 or 3; $(x + y)^2$ = 1, 2 or 3; y = -x + 1, $\sqrt{2}$ or $\sqrt{3}$.

6. (a) {(x,y,z): $z^2 - x^2 - y^2 = 0$}. (b) Surfaces $z^2 - x^2 - y^2$ = 1, 4, 9.

D21 Space-Time Travel (Section 6.2)

1. Earth time 120 yrs, spaceship time 79.3725 yrs.

2. Earth time 820.0820 yrs, spaceship time 11.5974 yrs.

D22 Pseudoinverse and Least Squares Curves (Section 6.4)

2. (a) $\begin{bmatrix} 0.3333 & 0 & 0.6667 \\ 0 & 0.1667 & -0.1667 \end{bmatrix}$ (b) $\begin{bmatrix} 0.5325 & 0.3896 & -0.3117 \\ -0.1688 & -0.0260 & 0.2208 \end{bmatrix}$

3. Pseudoinverse of an invertible matrix is that matrix. pinv(A) = $(A^tA)^{-1}A^t = A^{-1}(A^t)^{-1}A^t = A^{-1}$.

4. $(A^tA)^{-1}A^t$ does not exist since rank (A^tA) = rank(A) =2. Thus det(A^tA) = 0 and $(A^tA)^{-1}$ does not exist. Same applies whenever A has more columns than rows. Note that pinv(A) exists. pinv(A) is based on the more general Moore-Penrose definition of pseudoinverse, which agrees with $(A^tA)^{-1}A^t$ when A^tA is invertible. Get

$$\text{pinv}(A) = \begin{bmatrix} -0.9444 & 0.4444 \\ -0.1111 & 0.1111 \\ 0.7222 & -0.2222 \end{bmatrix}$$

5. (a) x = 1.7333, y=0.6 (b) x = 0.4857, y = 1.0857 (c) x = 2.2521, y = -0.8

6. $y = 4x - 3$.

7. $y = 15.25 - 10.05x + 1.75x^2$.

8. $y = 0.0026x+1.7323$. When x=670, y=3.5.

9. $y = 0.000059149x^3-0.0067x^2+0.05234x+14.1469$. In 2005, x=105. y=14.2433, 14.2%.

11. y > 9.4

D23 LU Decomposition (Section 7.2)

1. (a) $L = \begin{bmatrix} 1 & 0 \\ .3333 & 1 \end{bmatrix}$, $U = \begin{bmatrix} 3 & 5 \\ 0 & .3333 \end{bmatrix}$

(b) $L = \begin{bmatrix} 1.0000 & 0 & 0 \\ -0.5000 & 1.0000 & 0 \\ 0.2500 & 0.7308 & 1.0000 \end{bmatrix}$, $U = \begin{bmatrix} 4.0000 & 1.0000 & 3.0000 \\ 0 & 6.5000 & 1.5000 \\ 0 & 0 & -0.8462 \end{bmatrix}$

(c) $L = \begin{bmatrix} 1.0000 & 0 & 0 \\ 0.5000 & 1.0000 & 0 \\ 1.0000 & 1.2500 & 1.0000 \end{bmatrix}$, $U = \begin{bmatrix} 2 & 0 & 2 \\ 0 & 4 & 0 \\ 0 & 0 & 3 \end{bmatrix}$

2. (a) $X = \begin{bmatrix} 1.0000 & 1.0000 \\ 1.0000 & -1.0000 \\ 2.0000 & 1.0000 \end{bmatrix}$ (b) $X = \begin{bmatrix} 1.0000 & -2.0000 \\ 1.0000 & 1.0000 \end{bmatrix}$

D24 Condition Number of a Matrix (Section 7.3)

1. (a) 21 (b) 225 (c) 65 (d) 493.5 2. 4,606

3. Very often, if determinant is very large or small, then the condition number will be large, but not always. For example,
$A = \begin{bmatrix} 2 & 2 \\ 2.0000000001 & 2 \end{bmatrix}$, to make |A| almost zero, columns are almost the same. Have
$|A| =$ -2.000000165490742e-11, c(A) = 7.999999338077087e+11.
If condition number is high, the determinant will be very large or very small. e.g., consider the Hilbert matrix A of Exercise 2. $c(A) = c(A^{-1}) = 4{,}606$. |A| = 2.6455e-06, and
$|A^{-1}| = 1/|A| = 3.7800e{+}05$. However, the converse is not true, since

$$c(kA) = ||kA||\ ||(kA)^{-1}|| = ||kA||\ ||k^{-1}A^{-1}|| = |k|\ |k^{-1}|\ ||A||\ ||A^{-1}|| = c(A)$$

$$|kA| = k|A|$$

Thus condition number is invariant under scalar multiplication of matrix, but determinant is not.

4. $y = 25 - 26\frac{1}{3}x + 9x^2 - \frac{2}{3}x^3$.

$$A = \begin{bmatrix} 1 & 1 & 1 & 1 \\ 1 & 2 & 4 & 16 \\ 1 & 3 & 9 & 27 \\ 1 & 4 & 16 & 64 \end{bmatrix},$$ a Vandermonde matrix. c(A) = 2000, system is ill-conditioned.

<u>D25 Jacobi and Gauss-Seidel Iterative Methods</u> (Section 7.4)

1. (a) Solution x = 1, y = 0.5, z = 0. With tolerance of .0001, Jacobi takes 11 iterations to converge; Gauss-Seidel takes 4 iterations. (b) Does not converge

<u>D26 The Simplex Method in Linear Programming</u> (Section 8.2)

1. f = 24 at x = 6, y = 12. 2. f = 4 at x =4, y = 0. 3. f = 16.5, at x = 1.25, y = 2.25

<u>D27 Cross Product</u> (Appendix A)

1. (a) (36,-27, -26) (b) (-19, 3, 30) 2. (a) 108 (b) 0 (c) 0